DIFFERENTIAL CALCULUS.

DIFFERENTIAL CALCULUS

WITH

APPLICATIONS AND NUMEROUS EXAMPLES:

AN ELEMENTARY TREATISE.

BY

JOSEPH EDWARDS, M.A.,
FORMERLY FELLOW OF SIDNEY SUSSEX COLLEGE, CAMBRIDGE.

London:
MACMILLAN AND CO.,
AND NEW YORK.
1886.

Printed at the University Press
BY ROBERT MACLEHOSE, WEST NILE STREET, GLASGOW.

PREFACE.

THE object of the present volume is to offer to the student a fairly complete account of the elementary portions of the Differential Calculus, unencumbered by such parts of the subject as are not usually read in colleges and schools.

Where a choice of method exists, geometrical proofs and illustrations have been in most cases adopted in preference to purely analytical processes.

It has been the constant endeavour of the author to impress upon the mind of the student the geometrical meaning of differentiation and its aspect as a means of measurement of rates of growth. The purely analytical character of the operator $\frac{d}{dx}$ as a symbol and the laws of combination which it satisfies have also been fully considered.

The applications of the Calculus to the treatment of

plane curves have been introduced at an earlier stage than usual, from the interesting and important nature of the problems to be discussed. At the same time, the chapters on Undetermined Forms and Maxima and Minima, which have been thereby postponed, may be read in their ordinary place if thought desirable.

The direct and inverse hyperbolic functions have been freely used, and the convenient notation $\frac{\partial}{\partial x}$, to denote partial differentiation, has been adopted.

It is hoped that the frequent sets of easy illustrative examples introduced throughout the text will be found useful before attacking the more difficult problems in the copious selections at the ends of the chapters. Many of these examples have been selected from various university and college examination papers, others from papers set in the India and Home Civil Service and Woolwich examinations, and many are new.

I have to thank the Rev. H. P. Gurney, M.A., formerly Senior Fellow of Clare College, Cambridge, for the kind interest he has taken in the preparation of this work, and for many useful suggestions. I have also been much assisted in the revision of proof sheets and in the verification of examples by J. Wilson, Esq., M.A., formerly Fellow of Christ's College, one of H.M. Inspectors of Schools, and also by H. G. Edwards, Esq., B.A., late Scholar of Queen's College. I hope therefore that the book will not be found to contain many serious errors.

JOSEPH EDWARDS.

80 Cambridge Gardens,
North Kensington, W.,
November, 1886.

ERRATA.

Page 82, line 1.—For b^{n+1} read b^n.

„ 258, Ex. 1 (a).—For $-$ read $+$.

CONTENTS.

PRINCIPLES AND PROCESSES OF THE DIFFERENTIAL CALCULUS.

CHAPTER I.

Definitions. Limits.

CHAPTER II.

Fundamental Propositions.

CHAPTER III.

Standard Forms.

CHAPTER IV.

Successive Differentiation.

CHAPTER V.

Expansions.

CHAPTER VI.

Partial Differentiation.

APPLICATIONS TO PLANE CURVES.

CHAPTER VII.

Tangents and Normals.

CHAPTER VIII.

Asymptotes.

CHAPTER IX.

Singular Points.

CHAPTER X.

Curvature.

CHAPTER XI.

Envelopes.

CHAPTER XII.

Curve Tracing.

APPLICATION TO THE EVALUATION OF SINGULAR FORMS AND MAXIMA AND MINIMA VALUES.

CHAPTER XIII.

Undetermined Forms.

CHAPTER XIV.

Maxima and Minima—One Independent Variable.

APPENDIX.

PRINCIPLES AND PROCESSES OF THE DIFFERENTIAL CALCULUS.

CHAPTER I.

DEFINITIONS. LIMITS.

1. **Primary Object of the Differential Calculus.**

In Nature we frequently meet with quantities which, if observed for some period of time, are found to undergo *increase* or *decrease*; for instance, the distance of a moving particle from a known fixed point in its path, the length of a moving ordinate of a given curve, the force exerted upon a piece of soft iron which is gradually made to approach one of the poles of a magnet. When such quantities are made the subject of mathematical investigation, it often becomes necessary to estimate their *rates of growth*. This is the primary object of the Differential Calculus.

2. In the first six chapters we shall be concerned with the description of an instrument for the measurement of such rates, and in framing rules for its formation and use, and the student must make himself as proficient as possible in its manipulation. These chapters contain the whole machinery of the Differential Calculus. The remaining chapters simply consist of various applications of the methods and formulae here established.

3. We commence with an explanation of several technical terms which are of frequent occurrence in this subject, and with the meanings of which the student should be familiar from the outset.

4. **Constants and Variables.**

A CONSTANT *is a quantity which, during any set of mathematical operations, retains the same value.*

A VARIABLE *is a quantity which, during any set of mathematical operations, does not retain the same value, but is capable of assuming different values.*

Ex. The area of any triangle on a given base and between given parallels is a constant quantity; so also the base, the distance between the parallel lines, the sum of the angles of the triangle are constant quantities. But the separate angles, the sides, the position of the vertex are variables.

It has become conventional to make use of the letters $a, b, c, \ldots, \alpha, \beta, \gamma, \ldots$, from the beginning of the alphabet to denote constants; and to retain later letters, such as u, v, w, x, y, z, and the Greek letters ξ, η, ζ, for variables.

5. **Dependent and Independent Variables.**

An INDEPENDENT VARIABLE *is one which may take up any arbitrary value that may be assigned to it.*

A DEPENDENT VARIABLE *is one which assumes its value in consequence of some second variable or system of variables taking up any set of arbitrary values that may be assigned to them.*

6. **Functions.**

When one quantity depends upon another or upon a system of others, so that it assumes a definite value when a system of definite values is given to the others, it is called a FUNCTION *of those others.*

The function itself is a *dependent* variable, and the variables to which values are given are *independent* variables.

The usual notation to express that one variable y is a function of another x is

$$y = f(x); \text{ or } y = F(x), \text{ or } y = \phi(x);$$

the letters $f(\ )$, $F(\ )$, $\phi(\ )$, $\chi(\ )$, . . . being generally retained to represent functions of arbitrary or unknown form. If u be an arbitrary or unknown function of several variables x, y, z, we may express the fact by the equation

$$u = f(x, y, z).$$

Ex. In any triangle, two of whose sides are x and y and the included angle θ, we have $\Delta = \frac{1}{2}xy \sin\theta$ to express the area. Here Δ is the *dependent* variable, and is a *function* of known form—of x, y, and θ, which are the *independent* variables.

7. It will be seen that we could write the same equation in other forms,

e.g., $$\sin\theta = \frac{2\Delta}{xy},$$

which may be regarded as an expression for $\sin\theta$ in terms of the area and two sides; so that now $\sin\theta$ may be regarded as the *dependent* variable, while Δ, x, y, are *independent* variables.

And it is clear that if there be one equation between four variables, as above, it is sufficient to determine one in terms of the other three, so that *any one variable may be regarded as dependent and the others as independent.*

This may be extended. For, if there be one equation between n variables, it will suffice to find one of them in terms of the remaining $(n-1)$, so that *any one variable can be considered dependent and the remaining* $(n-1)$ *independent.*

And, further, if there be r equations connecting n variables (n being greater than r) they will be enough to determine r of the variables in terms of the other $n-r$ variables, so that *any r of the variables can be considered dependent, while the remaining* $(n-r)$ *are independent.*

8. Explicit and Implicit Functions.

A function is said to be EXPLICIT *when expressed directly in terms of the independent variable or variables.*

For example, if $z=x^2$, or $z=r\sin\theta$, or $z=x^2y$,

or $$z=a^ye^x\log x+(a+x)^n:$$

z is expressed *directly* in terms of the independent variables, and is therefore in each of the above cases said to be an *explicit* function of those variables.

But, *if the function be not expressed directly in terms of the independent variable (or variables), the function is said to be* IMPLICIT.

If, for example, $$ax^2+yx-b=0;$$

or $$x^3+y^3=3axy;$$

or $$ax^2+2bxy+cy^2+2dx+2ey+f=0;$$

or $$x^2y^2=(a^2-y^2)(b+y)^2;$$

y in each case is said to be an implicit function of x.

Sometimes, however, we can solve the equation for y: *e.g.*, the first equation we can write as $y=\dfrac{b-ax^2}{x}$, and in this form y is said to be an explicit function of x.

It appears then that if the equation connecting the variables be solved for the dependent variable, that variable is reduced from being an implicit to being an explicit function of the remaining variable or variables. Such solution is not, however, always possible or convenient.

9. Species of Known Functions.

Functions which are made up of powers of variables and constants connected by the signs $+\ -\ \times\ \div$ are classed as *algebraic* functions. If radical signs or fractional indices occur in the function, it is said to be *irrational*; if not, *rational*.

All other functions are classed as *transcendental* functions.

Of transcendental functions, sines, cosines, tangents, etc., are called *trigonometrical* or *circular* functions.

Functions such as $\sin^{-1}x$, $\tan^{-1}x$, etc., are called *inverse trigonometrical* functions.

Functions such as e^x, a^{x^2}, in which the variable occurs in the index, are called *exponential* functions.

While if logarithms are involved, as for instance in $\log_e x$ or $\log_{10}(a+bx)$, etc., the function is called *logarithmic*.

Besides the above we have the *hyperbolic* functions, $\sinh x$, $\cosh x$, etc., of which a short description follows in Art. 25.

10. Limit of a function.

DEF. *When a function can be made to approach continually to equality with some fixed value or condition so as to differ from it by less than any assignable quantity, however small, by making the independent variable or variables approach some assigned value or values, that fixed value or condition is called the* LIMIT *of the function for the value or values of the variable or variables referred to.*

11. Illustrations.

Ex. 1. If an equilateral polygon be inscribed in any closed curve, and the sides of the polygon be decreased indefinitely and at the same time increased in number indefinitely, the polygon continually approximates to the form of the curve, *and ultimately differs from it in area by less than any assignable magnitude*, and the curve is said to be the *limit* of the polygon inscribed in it.

Ex. 2. The limit of $\frac{2x+3}{x+1}$ when x is indefinitely diminished is 3. For the difference between $\frac{2x+3}{x+1}$ and 3 is $\frac{x}{x+1}$; and by diminishing x indefinitely $\frac{x}{x+1}$ *can be made less than any assignable quantity however small.* Hence it is said that the limit of $\frac{2x+3}{x+1}$ when x is indefinitely diminished is 3.

The expression can also be written $\frac{2+\frac{3}{x}}{1+\frac{1}{x}}$, which shows that if x be increased indefinitely it can be made to continually approach and *to differ by less than any assignable quantity* from 2, which is therefore its limit in that case.

Ex. 3. *The limits of some quantities are zero, e.g.,*

$$\left.\begin{array}{r} ax^2+bx, \\ \sin x, \\ 1-\cos x, \end{array}\right\} \text{when } x \text{ is zero,}$$

$$\left.\begin{array}{r} 1-\sin x, \\ \cos x, \end{array}\right\} \text{when } x=\frac{\pi}{2}.$$

When the limit of a quantity is zero for any value or values of the independent variable or variables, the quantity is said to be a *vanishing quantity* for those values.

It is useful to adopt the notation $Lt_{x=a}$ to denote the words "*the limit when $x=a$ of.*"

Ex. 4. The sum of a $G.P.$ of which the first term is a, common ratio r, and n the number of terms, is $a\dfrac{r^n-1}{r-1}$.

If $r<1$, the sum to infinity is $\dfrac{a}{1-r}$. For the difference is $\dfrac{ar^n}{r-1}$; and since $Lt_{n=\infty}\dfrac{ar^n}{r-1}=0$ (when $r<1$), this difference is *a vanishing quantity.*

Ex. 5. We say $\cdot\dot{6}=\frac{2}{3}$, by which we mean that by taking enough sixes we can make ·666... *differ by as little as we please from* $\frac{2}{3}$.

Ex. 6. The DEFINITION OF A TANGENT is another example.

DEF. *Let PQ be a chord joining P, Q, two adjacent points on a curve. Let Q travel along the curve towards P and come so close as ultimately to coincide with P. Then the limiting position of PQ, viz. PT, is called the tangent at P.*

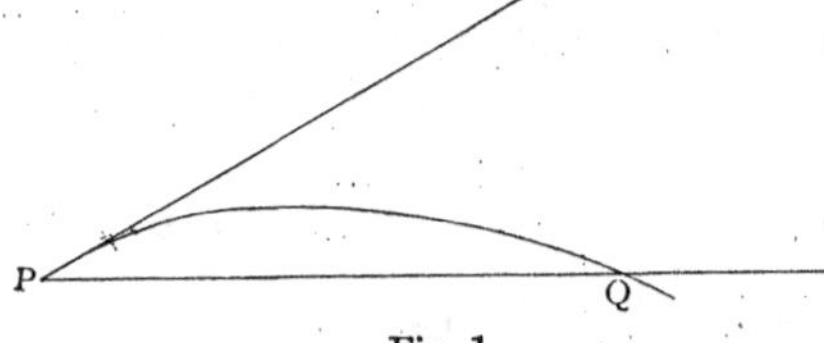

Fig. 1.

The angle QPT is a *vanishing quantity*; for it can be made less than any assignable quantity by making Q move *along the curve* sufficiently close to P.

12. There are several important principles with regard to limits which we shall continually require to use and which we may stay to enunciate here :—

(1) The *limit of the sum* of a finite number of quantities is equal to the *sum of their limits.*

(2) The *limit of the product* of a finite number of quantities is *in general* equal to the *product of their limits.*

(3) The *limit of the ratio* of two quantities (*whose limits are not zero or infinite*) is equal to the *ratio of their limits.*

(4) The *limits of two quantities* (*whose limits are finite*) *are equal when the limit of their difference is zero.*

These statements are almost self-evident. We give however formal proofs of the most important.

(1) Let $u_1, u_2 \dots$ be the variable quantities,

$v_1, v_2 \dots$ their limits.

Let
$$u_1 = v_1 + \alpha_1,$$
$$u_2 = v_2 + \alpha_2,$$
etc.,

where $\alpha_1, \alpha_2, \dots$ become less than any assignable quantities when the variables u_1, u_2, etc., approach their limits.

Then
$$u_1 + u_2 + \dots = (v_1 + \alpha_1) + (v_2 + \alpha_2) + \dots$$
$$= (v_1 + v_2 + \dots) + (\alpha_1 + \alpha_2 + \dots).$$

Now, if α be the greatest of the quantities $\alpha_1, \alpha_2, \dots$, and if n be their number, $\alpha_1 + \alpha_2 + \dots < n\alpha$; and therefore,

$u_1 + u_2 + \dots$ differs from

$v_1 + v_2 + \dots$

by less than $n\alpha$. But by hypothesis $Lt\alpha = 0$; and therefore, if n be finite, $Ltn\alpha = 0$.

Whence
$$Lt(u_1 + u_2 + \dots) = v_1 + v_2 + \dots$$
$$= Ltu_1 + Ltu_2 + Ltu_3 + \dots.$$

(2) Again, with the same notation,
$$u_1 u_2 = (v_1 + \alpha_1)(v_2 + \alpha_2)$$
$$= v_1 v_2 + \alpha_2 v_1 + \alpha_1 v_2 + \alpha_1 \alpha_2.$$

Now $\qquad Lt\alpha_2 v_1 = 0$, unless v_1 be infinite,

$Lt\alpha_1 v_2 = 0$, unless v_2 be infinite,

$Lt\alpha_1\alpha_2 = 0$.

Hence $\qquad Ltu_1u_2 =$ sum of the limits of the above terms

$= v_1v_2 = Ltu_1 \times Ltu_2$.

Similarly $\qquad Lt(u_1u_2u_3\ldots u_n) = Ltu_1 \,.\, Ltu_2 \,.\, Ltu_3 \ldots Ltu_n$,
supposing none of these limits infinite.

(3) Again

$$\frac{u_1}{u_2} = \frac{v_1+\alpha_1}{v_2+\alpha_2} = \frac{v_1}{v_2} + \left\{ \frac{v_1+\alpha_1}{v_2+\alpha_2} - \frac{v_1}{v_2} \right\}$$

$$= \frac{v_1}{v_2} + \frac{\alpha_1 v_2 - \alpha_2 v_1}{v_2(v_2+\alpha_2)},$$

and if v_1, v_2, be finite $Lt(\alpha_1 v_2 - \alpha_2 v_1) = 0$;

and therefore also $\quad Lt\dfrac{\alpha_1 v_2 - \alpha_2 v_1}{v_2(v_2+\alpha_2)} = 0$

provided v_2 does not vanish.

Hence $$Lt\frac{u_1}{u_2} = \frac{v_1}{v_2} = \frac{Ltu_1}{Ltu_2}.$$

The student will find no difficulty in establishing the fourth statement in a similar manner.

In the same way may be proved (with certain exceptions)

(5) $\qquad Ltu^n = (Ltu)^n$ for positive, negative, etc., values of n.

(6) $\qquad Lta^u = a^{Ltu}$.

(7) $\qquad Lt \log u = \log Ltu$.

13. Indeterminate or Illusory Forms.

When a function involves the independent variable (or variables) in such a manner that, for a certain assigned value of that variable, its value *cannot be found by simply substituting* that value of the variable, the function is usually said to take an *indeterminate form* or to assume an *indeterminate value.*

14. The name *indeterminate,* though sanctioned by common use, is open to objection, inasmuch as it will be found that the true values of such forms

can in general be arrived at by means of certain processes which we shall hereafter discuss at length in a special chapter; whereas it would seem to be implied in the name *indeterminate* that it would be *impossible* to obtain the value of a function to which that name was applied. "Undetermined" or "Illusory Forms" appear to be better designations for such cases.

15. One of the commonest forms occurring is when the function takes the form of a fraction whose *Numerator and Denominator both vanish or both become infinite* for the value (or values) of the variable (or variables) assigned.

Several other indeterminate forms are treated fully in Chapter XIII.

16. The *limit of the ratio of two vanishing quantities may be zero, finite, or infinite.*

Ex. (i.) $$Lt_{x=0}\frac{2x^2}{7x}=Lt_{x=0}\frac{2}{7}x=0,$$ zero.

Ex. (ii.) $$Lt_{x=0}\frac{2x}{7x}=Lt_{x=0}\frac{2}{7}=\frac{2}{7},$$ finite,

Ex. (iii.) $$Lt_{x=0}\frac{2x}{7x^2}=Lt_{x=0}\frac{2}{7x}=\infty,$$ infinite.

Ex. (iv.) $$Lt_{x=a}\frac{x^2-a^2}{x-a}=Lt_{x=a}(x+a)=2a,$$ finite.

Ex. (v.) $$Lt_{\theta=0}\frac{\sin\theta}{\theta}=1,$$ finite.

Ex. (vi.) $$Lt_{x=2}\frac{x-2}{x^2-4x+4}=Lt_{x=2}\frac{1}{x-2}=\infty,$$ infinite.

17. Two functions of the same independent variable are said to be *ultimately equal* when, as the independent

variable approaches indefinitely near its assigned value, the *limit of their ratio is unity.*

Thus $$Lt_{\theta=0}\frac{\sin\theta}{\theta}=1;$$

and therefore, when an angle is indefinitely diminished, its sine and its circular measure are ultimately equal.

EXAMPLES.

1. Find the limit when $x=0$ of $\frac{y}{x^2}$,

   (i.) When $y=mx$.

   (ii.) When $y=\frac{x^2}{a}$.

   (iii.) When $y=ax^2+b$.

2. Find $Lt\frac{1+2x}{2+x}$, (i.) when $x=0$; (ii.) when $x=\infty$.

3. Find $Lt_{x=0}\frac{y}{x}$, when $y^2=2ax-x^2$.

4. Find $Lt_{x=0}\frac{y}{x}$, when $\frac{x^2}{a^2}-\frac{y^2}{b^2}=x^3$.

5. Find $Lt_{x=0}\frac{\sqrt{1+x}-1}{x}$.

6. Find $Lt_{x=0}\frac{y^2}{x}$, when $y^2=ax+bx^2+cx^3$.

7. Find $Lt_{x=a}\frac{x^3-a^3}{x-a}$.

8. Find $Lt\frac{ax^2+bx}{bx^2+ax}$, when (i.) $x=0$; (ii.) $x=\infty$.

9 Find $Lt_{x=\infty}\sqrt{x}(\sqrt{x+1}-\sqrt{x})$.

10. Prove that $p-qx$ and $q-px$ tend to equality as x diminishes to zero, but yet that their limits are not equal.

11. The opposite angles of a quadrilateral inscribed in a circle are together equal to two right angles. What does this become when in the limit two angular points coincide?

12. Find the ultimate position of the point of intersection of the diagonals of a rhombus, when one of the angles diminishes indefinitely.

18. We now proceed to consider the limits of four very important indeterminate forms.

19. I. The proofs of the well-known results

$$Lt_{\theta=0}\frac{\sin\theta}{\theta}=1,$$

$$Lt_{\theta=0}\cos\theta=1,$$

$$Lt_{\theta=0}\frac{\tan\theta}{\theta}=1,$$

can be found in any standard book on Plane Trigonometry.

20. II. $$Lt_{x=1}\frac{x^n-1}{x-1}=n.$$

Let $x=1+z$. Then when x approaches the value unity z approaches zero, and we can therefore consider z to be less than 1, and therefore can apply the Binomial Theorem to expand $(1+z)^n$, whatever n may be.

Hence $$Lt_{x=1}\frac{x^n-1}{x-1}=Lt_{z=0}\frac{(1+z)^n-1}{z}$$

$$=Lt_{z=0}\frac{nz+\frac{n(n-1)}{2\,!}z^2+\dots}{z}$$

$$=Lt_{z=0}\left\{n+\frac{n(n-1)}{2\,!}z+\dots\right\}$$

$$=n.$$

21. III. $Lt_{x=\infty}\left(1+\frac{1}{x}\right)^x=e$, where e is the base of the Napierian system of logarithms. This number e is defined as the value of the series $1+1+\frac{1}{2\,!}+\frac{1}{3\,!}+\dots$ to ∞, and it may easily be shown to be 2·7182818....

Since x is to be ultimately infinite, we may throughout

consider $\frac{1}{x}$ to be less than unity, and may therefore apply the Binomial Theorem to the expansion of $\left(1+\frac{1}{x}\right)^x$. We thus obtain

$$\left(1+\frac{1}{x}\right)^x = 1 + x\frac{1}{x} + \frac{x(x-1)}{1\,.\,2}\,\frac{1}{x^2} + \frac{x(x-1)(x-2)}{1\,.\,2\,.\,3}\,\frac{1}{x^3} + \dots$$

$$= 1 + 1 + \frac{1-\frac{1}{x}}{2\,!} + \frac{\left(1-\frac{1}{x}\right)\left(1-\frac{2}{x}\right)}{3\,!} + \dots\dots$$

$$= 1 + 1 + \frac{1}{2\,!} + \frac{1}{3\,!} + \dots$$

in the limit, when x is indefinitely increased.

COR. $$Lt_{x=\infty}\left(1+\frac{a}{x}\right)^x = Lt_{\frac{x}{a}=\infty}\left\{\left(1+\frac{a}{x}\right)^{\frac{x}{a}}\right\}^a = e^a.$$

22. IV. $$Lt_{x=0}\frac{a^x-1}{x} = \log_e a.$$

Assume the expansion for a^x, viz.:

$$a^x = 1 + x\log_e a + \frac{x^2(\log_e a)^2}{2!} + \dots.$$

This is a convergent series, for the test fraction is $\frac{x\log_e a}{n}$, and can be made less than any assignable quantity by making n sufficiently large.

We have therefore

$$\frac{a^x-1}{x} = \log_e a + \frac{x(\log_e a)^2}{2\,!} + \dots$$

and the limit of the right-hand side, when x is indefinitely diminished, is clearly $\log_e a$.

23. The following proof of II. is independent of the Binomial Theorem:

(i.) Let n be a positive integer. Then, by division,

$$\frac{x^n-1}{x-1}=x^{n-1}+x^{n-2}+x^{n-3}+\ldots+x^2+x+1.$$

Putting $x=1$,

$$Lt_{x=1}\frac{x^n-1}{x-1}=1+1+1+\ldots+1+1+1,$$

there being n terms,

$$=n.$$

(ii.) Let n be a positive fraction $=\frac{p}{q}$.

Let $y^q=x$, so that, if $x=1$, $y^q=1$, and $\therefore$ $y=1$.

Then
$$Lt_{x=1}\frac{x^n-1}{x-1}=Lt_{x=1}\frac{x^{\frac{p}{q}}-1}{x-1}$$

$$=Lt_{y=1}\frac{y^p-1}{y^q-1}=Lt_{y=1}\frac{\dfrac{y^p-1}{y-1}}{\dfrac{y^q-1}{y-1}}$$

$$=\frac{Lt_{y=1}\dfrac{y^p-1}{y-1}}{Lt_{y=1}\dfrac{y^q-1}{y-1}}=\frac{p}{q}\text{ by (i.), and }\therefore\ =n.$$

(iii.) Let n be negative and $=-m$.

Then
$$Lt_{x=1}\frac{x^n-1}{x-1}=Lt_{x=1}\frac{x^{-m}-1}{x-1}$$

$$=Lt_{x=1}\left(-\frac{1}{x^m}\right)\times\frac{x^m-1}{x-1}$$

$$=Lt_{x=1}\left(-\frac{1}{x^m}\right)\times Lt_{x=1}\frac{x^m-1}{x-1}=-m\text{ by (i.) and}$$

(ii.), and $\therefore$ $=n$.

(iv.) Finally, if n be incommensurable, two numbers n_1, n_2 can be found, both commensurable, one on each side of n, such that their difference is less than any that can be assigned beforehand, however small.

Hence x^{n_1}, x^n, x^{n_2}, are in order of magnitude; and therefore so also are

$$Lt_{x=1}\frac{x^{n_1}-1}{x-1},\quad Lt_{x=1}\frac{x^n-1}{x-1},\quad Lt_{x=1}\frac{x^{n_2}-1}{x-1},$$

and $$n_1, \quad Lt_{x=1}\frac{x^n-1}{x-1}, \quad n_2;$$

but n_1, n_2, are indefinitely nearly equal, and n lies between them,

$$\therefore \quad Lt_{x=1}\frac{x^n-1}{x-1}=n \text{ in this case also.}$$

24. Also IV. can be deduced from III. thus:

Let $$a^x-1=\frac{1}{y},$$

then $$a^x=1+\frac{1}{y},$$

and therefore when x becomes zero y becomes infinite, and

$$x=\log_a\left(1+\frac{1}{y}\right),$$

$$\therefore \quad Lt_{x=0}\frac{a^x-1}{x}=Lt_{y=\infty}\frac{\frac{1}{y}}{\log_a\left(1+\frac{1}{y}\right)}$$

$$=Lt_{y=\infty}\frac{1}{y\log_a\left(1+\frac{1}{y}\right)}=Lt_{y=\infty}\frac{1}{\log_a\left(1+\frac{1}{y}\right)^y}$$

$$=\frac{1}{\log_a\left[Lt_{y=\infty}\left(1+\frac{1}{y}\right)^y\right]}=\frac{1}{\log_a e} \quad \text{[Arts. 12 (7) and 21.]}$$

$$=\log_e a.$$

EXAMPLES.

1. Prove $$Lt_{x=1}\frac{\log x}{x-1}=1.$$
[Put $x=1+y$.]

2. Prove $$Lt_{x=a}\frac{x^m-a^m}{x^n-a^n}=\frac{m}{n}a^{m-n}.$$

3. Prove $$Lt_{x=0}(1+ax)^{\frac{1}{x}}=e^a.$$

4. Prove $$Lt_{x=0}\frac{\sin mx}{\sin nx}=\frac{m}{n}.$$

5. Prove $$Lt_{x=0}\frac{a^x-1-x\log_e a}{x^2}=\tfrac{1}{2}(\log_e a)^2.$$

25. **Hyperbolic Functions.**

By analogy with the exponential values of the sine, cosine, tangent, etc., the exponential functions

$$\frac{e^\theta - e^{-\theta}}{2}, \frac{e^\theta + e^{-\theta}}{2}, \frac{e^\theta - e^{-\theta}}{e^\theta + e^{-\theta}}, \text{ etc.,}$$

are respectively written

$$\sinh\theta, \cosh\theta, \tanh\theta, \text{ etc.,}$$

and called the *hyperbolic sine, cosine, tangent*, etc., of θ, and as a class are styled *hyperbolic functions.*

Since $\sin\theta = \dfrac{e^{\iota\theta} - e^{-\iota\theta}}{2\iota}$, and $\cos\theta = \dfrac{e^{\iota\theta} + e^{-\iota\theta}}{2}$, where $\iota = \sqrt{-1}$, it will be clear that

$$\sin\iota\theta = \iota\sinh\theta,$$
$$\cos\iota\theta = \cosh\theta,$$

and hence or from the definition

(1) $\tan\iota\theta = \iota\dfrac{\sinh\theta}{\cosh\theta} = \iota\tanh\theta$;

(2) $\cosh^2\theta - \sinh^2\theta = 1$;

(3) $\sin(\theta + \iota\phi) = \sin\theta\cosh\phi + \iota\cos\theta\sinh\phi$;

with many other formulae analogous to, and easily deducible from, the common formulae of Trigonometry.

If $$x = \sinh\theta,$$
we have $$\theta = \sinh^{-1}x,$$
an *inverse hyperbolic* function of x analogous to the inverse trigonometrical function $\sin^{-1}x$.

This species of function however is merely logarithmic;

for, since $$x = \frac{e^\theta - e^{-\theta}}{2},$$
we have $$e^\theta = x + \sqrt{1 + x^2},$$
and $$\theta = \log_e(x + \sqrt{1 + x^2}),$$
while corresponding results hold for $\cosh^{-1}x$, $\tanh^{-1}x$, etc.

EXAMPLES.

1. Prove the following formulae—

(*a*) $\operatorname{cosech}^2\theta = \coth^2\theta - 1$;

(*b*) $\sinh(\theta+\phi) = \sinh\theta\cosh\phi + \cosh\theta\sinh\phi$;

(*c*) $\tanh(\theta+\phi) = \dfrac{\tanh\theta + \tanh\phi}{1+\tanh\theta\tanh\phi}$;

(*d*) $\sinh\theta + \sinh\phi = 2\sinh\dfrac{\theta+\phi}{2}\cosh\dfrac{\theta-\phi}{2}$.

2. Show that the co-ordinates of any point on the rectangular hyperbola $x^2 - y^2 = a^2$ may be denoted by $a\cosh\theta$, $a\sinh\theta$.

3. Prove (*a*) $\sinh^{-1}x = \tanh^{-1}\dfrac{x}{\sqrt{1+x^2}}$;

(*b*) $2\tanh^{-1}x = \log\dfrac{1+x}{1-x}$.

INFINITESIMALS.

26. All measurable quantities are estimated by the ratios which they bear to certain fixed but arbitrary units of their own kind. The whole measure of a quantity thus consists of two factors—the unit itself and an abstract number which represents the ratio of the measured quantity to the unit. The magnitude of the unit should be chosen as something comparable with the quantity to be measured, otherwise the abstract number which measures the ratio of the quantity to the unit will be too large or too small to lie within the limits of comprehension. For instance, the radius of the earth is conveniently estimated in *miles* (roughly 4,000); the moon's distance in *earth's radii* (about 60); the sun's distance in *moon's distances* (about 400); the distance of Sirius in *sun's distances* (at least 200,000). Again, for such relatively small quantities as the wave-length of a particular kind of light, *one ten-millionth of an inch* is found to be a sufficiently large unit: the wave-length for light from the red

end of the spectrum being about 266, that from the violet end 167 such units (Lloyd, "Wave Theory of Light," p. 18).

27. Any comparison of two quantities is equivalent to an estimate of how many times the one is contained in or contains the other; that is, the one quantity is estimated in terms of the other as a unit, and according as the number expressing their ratio is very large compared with unity or a very small fraction, the one is said to be very large or very small in comparison with the other. The *terms great and small are therefore purely relative.*

The standard of smallness is vague and arbitrary. An error of measurement which, centuries ago, would have been reckoned small would now be considered enormous. The accuracy of observation, and therefore the smallness of allowable errors of observation, increases with the continual improvement in the construction of instruments and methods of measurement.

28. **Orders of Smallness.**

If we conceive any magnitude A divided up into any large number of equal parts, say a billion (10^{12}), then each part $\frac{A}{10^{12}}$ is extremely small, and for all practical purposes negligible, in comparison with A. If this part be again subdivided into a billion equal parts, each $= \frac{A}{10^{24}}$, each of these last is extremely small in comparison with $\frac{A}{10^{12}}$, and so on. We thus obtain a series of magnitudes, A, $\frac{A}{10^{12}}, \frac{A}{10^{24}}, \frac{A}{10^{36}}, \ldots$, each of which is excessively small in comparison with the one which precedes it, but very large compared with the one which follows it. This furnishes us with what we may designate *a scale of smallness.*

More generally, if we agree to consider any given fraction f as being small in comparison with unity, then fA will be small in comparison with A, and we may term the expressions $fA, f^2A, f^3A, \ldots$, *small quantities of the first, second, third, etc., orders;* and the numerical quantities $f, f^2, f^3, \ldots$, may be called *small fractions* of the first, second, third, etc., orders.

Thus, supposing A to be any given finite magnitude, any given fraction of A is at our choice to designate a small quantity of the first order in comparison with A. When this is chosen, any quantity which has to this small quantity of the first order a ratio which is a small fraction of the first order, is itself a small quantity of the second order. Similarly, any quantity whose ratio to a small quantity of the second order is a small fraction of the first order is a small quantity of the third order, and so on. So that generally, if a small quantity be such that its ratio to a small quantity of the pth order be a small fraction of the qth order, it is itself termed a small quantity of the $(p+q)^{\text{th}}$ order.

29. **Infinitesimals.**

If these small quantities $Af, Af^2, Af^3, \ldots$, be all quantities whose limits are zero, then, supposing f *made smaller than any assignable quantity* by sufficiently increasing its denominator, these small quantities of the first, second, third, etc., orders are termed *infinitesimals of the first, second, third, etc., orders.*

From the nature of an infinitesimal it is clear that, *if*

any equation contain finite quantities and infinitesimals, the infinitesimals may be rejected.

30. PROP. *In any equation between infinitesimals of different orders, none but those of the lowest order need be retained.*

Suppose, for instance, the equation to be

$$A_1+B_1+C_1+D_2+E_2+F_3+\ldots=0, \quad \ldots\ldots\ldots \text{(i.)}$$

each letter denoting an infinitesimal of the order indicated by the suffix.

Then, dividing by A_1,

$$1+\frac{B_1}{A_1}+\frac{C_1}{A_1}+\frac{D_2}{A_1}+\frac{E_2}{A_1}+\frac{F_3}{A_1}+\ldots=0, \quad \ldots\ldots\ldots\text{(ii.)}$$

the limiting ratios $\frac{B_1}{A_1}$ and $\frac{C_1}{A_1}$ are finite, while $\frac{D_2}{A_1}, \frac{E_2}{A_1}$, are infinitesimals of the first order, $\frac{F_3}{A_1}$ is an infinitesimal of the second order, and so on. Hence, by Art. 29, equation (ii.) may be replaced by

$$1+\frac{B_1}{A_1}+\frac{C_1}{A_1}=0,$$

and therefore equation (i.) by

$$A_1+B_1+C_1=0,$$

which proves the statement.

31. PROP. *In any equation connecting infinitesimals we may substitute for any one of the quantities involved any other which differs from it by a quantity of higher order.*

For if $\qquad A_1+B_1+C_1+D_2+\ldots=0$

be the equation, and if $\qquad A_1=F_1+f_2,$

f_2 denoting an infinitesimal of higher order than F_1, we have $\qquad F_1+B_1+C_1+f_2+D_2+\ldots=0,$

i.e., by the last proposition we may write

$$F_1+B_1+C_1=0,$$

which may therefore, if desirable, replace the equation

$$A_1+B_1+C_1=0.$$

32. **Illustrations.**

Since
$$\sin\theta=\theta-\frac{\theta^3}{3!}+\frac{\theta^5}{5!}-\dots$$

and
$$\cos\theta=1-\frac{\theta^2}{2!}+\frac{\theta^4}{4!}-\dots$$

$\sin\theta$, $1-\cos\theta$, $\theta-\sin\theta$ are respectively of the first, second, and third orders of small quantities, when θ is of the first order; also, 1 may be written instead of $\cos\theta$ if second order quantities are to be rejected, and θ for $\sin\theta$ when cubes and higher powers are rejected.

33. Again, suppose AP the arc of a circle of centre O and radius a. Suppose the angle $AOP(=\theta)$ to be a small quantity of the first order. Let PN be the perpendicular from P upon OA and AQ the tangent at A, meeting OP produced in Q. Join P, A.

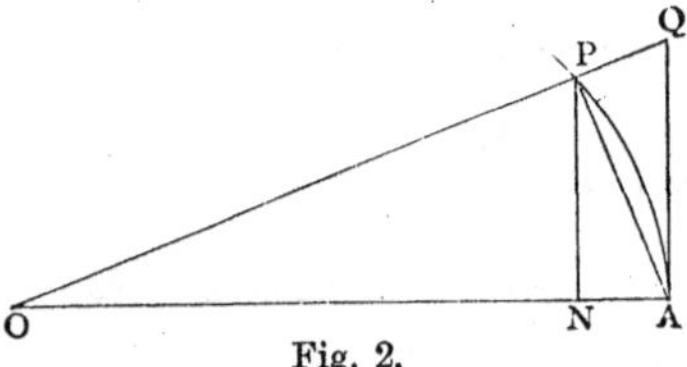

Fig. 2.

Then arc $AP=a\theta$ and is of the first order,

$$NP=a\sin\theta \quad \text{do.} \quad \text{do.},$$
$$AQ=a\tan\theta \quad \text{do.} \quad \text{do.},$$
$$\text{chord } AP=2a\sin\frac{\theta}{2} \quad \text{do.} \quad \text{do.},$$
$$NA=a(1-\cos\theta) \text{ and is of the second order.}$$

So that $OP-ON$ is a small quantity of the second order.

Again, arc $AP-$chord $AP=a\theta-2a\sin\frac{\theta}{2}$

$$=a\theta-2a\left(\frac{\theta}{2}-\frac{\theta^3}{8.3!}+\ldots\right)$$

$$=\frac{a\theta^3}{4.3!}-\text{etc.},$$

and is of the third order.

$$PQ-NA=NA(\sec\theta-1)$$

$$=NA.\frac{2\sin^2\frac{\theta}{2}}{\cos\theta}$$

$$=\text{(second order)(second order)}$$

$$=\text{fourth order of small quantities,}$$

and similarly for others.

34. Such results may also be established *without the use of the series* for $\sin\theta$ and $\cos\theta$.

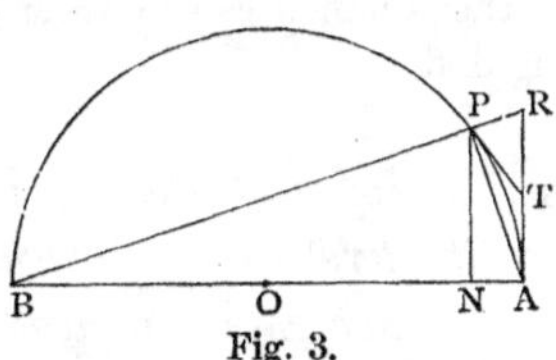

Fig. 3.

For example, let APB be a semicircle, P any point very near to A, so that the arc AP may be considered a small quantity of the first order. Join AP, BP, and let BP produced cut the tangent at A in R, and let the tangent at P cut AR in T, and draw the perpendicular PN upon AB. T will be the middle point of AR, and $AT=TR=TP$.

(1) We may take it as *axiomatic* that the *length of the arc* AP *is intermediate between the chord* AP *and the sum of the tangents* AT, TP; *i.e.*, between chord AP and tangent AR. Hence chord AP, arc AP, tangent AR are in ascending order of magnitude, and

therefore 1, $\dfrac{\text{arc } AP}{\text{chord } AP}$, $\dfrac{\text{tangent } AR}{\text{chord } AP}$ are in ascending order of magnitude.

Now, $$Lt\frac{AR}{\text{chord } AP}=Lt\frac{BA}{BP}=1,$$

whence $$Lt\frac{\text{arc } AP}{\text{chord } AP}=1.$$

and therefore, if arc AP be reckoned a small quantity of the first order, the chord AP and the tangent AR are also of the first order of smallness.

(2) Again, since $\dfrac{AN}{AP}=\dfrac{AP}{AB}$, and since AP is of the first order of smallness, AN is of the second order.

(3) Also $\dfrac{PR}{NA}=\dfrac{BP}{BN}$, which is ultimately a ratio of equality, and therefore PR is also of the second order.

(4) Similarly, since $AR-AP=\dfrac{AR^2-AP^2}{AR+AP}=\dfrac{PR^2}{AR+AP}$, and since PR^2 is a small quantity of the fourth order, and $AR+AP$ is a small quantity of the first order, we see that $AR-AP$ is of the third order of small quantities.

And similarly for other quantities the order of smallness may be geometrically investigated.

35. *The base angles of a triangle being given to be small quantities of the first order, to find the order of the difference between the base and the sum of the sides.*

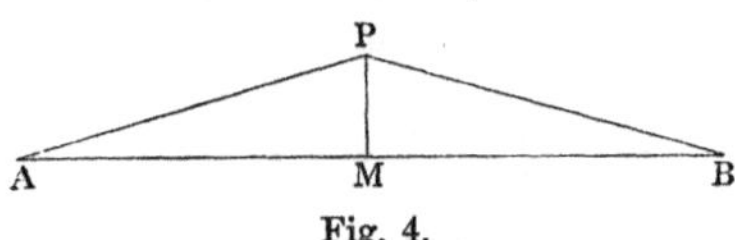

Fig. 4.

By what has gone before (Art. 33), if APB be the triangle and PM the perpendicular on AB, $AP-AM$ and $BP-BM$ are both small quantities of the second order as compared with AB.

Hence $AP+PB-AB$ is of the second order compared with AB.

If AB itself be of the first order of small quantities, then $AP+PB-AB$ is *of the third order*.

36. *Degree of approximation in taking a small chord for a small arc in any curve.*

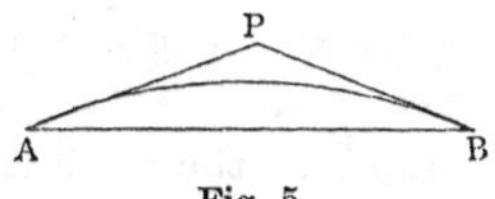

Fig. 5.

Let AB be an arc of a curve supposed continuous between A and B, and so small as to be concave at each point throughout its length to the foot of the perpendicular from that point upon the chord. Let AP, BP be the tangents at A and B. Then, when A and B are taken sufficiently near together, the chord AB and the angles at A and B may each be considered small quantities of at least the first order, and therefore, by what has gone before, $AP+PB-AB$ will be at least of the third order. Now we may take *as an axiom* that the *length of the arc* AB *is intermediate between the length of the chord* AB *and the sum of the tangents* AP, BP. Hence the difference of the arc AB and the chord AB, which is less than that between $AP+PB$ and the chord AB, must be at least *of the third order*.

EXAMPLES.

1. Show that, in the figure of Art. 33, the area of the segment bounded by the chord AP and the arc AP is of the third order of small quantities.

2. In the same figure, if PM be drawn perpendicular to AQ, show that the triangle PMQ is of the fifth order of smallness.

3. A straight line of constant length slides between two straight lines at right angles, viz., CAa, CbB; AB and ab are two positions of the line and P their point of intersection. Show that, in the limit, when the two positions coincide, we have

$$\frac{Aa}{Bb}=\frac{CB}{CA} \text{ and } \frac{PA}{PB}=\frac{CB^2}{CA^2}.$$

4. From a point T in a radius OA of a circle, produced, a tangent TP is drawn to the circle, touching it in P; PN is drawn perpendicular to the radius OA. Show that, in the limit, when P moves up to A, $NA=AT$.

5. If, in the equation $\sin(\omega-\theta)=\sin\omega\cos\alpha$, θ be very small, show that its approximate value is

$$2\tan\omega\sin\frac{\alpha}{2}\left(1-\tan^2\omega\sin^2\frac{\alpha}{2}\right).$$

[I. C. S. EXAM.]

6. Tangents are drawn to a circular arc at its middle point and at its extremities. Show that the area of the triangle formed by the chord of the arc and the two tangents at the extremities is ultimately four times that of the triangle formed by the three tangents. [FROST'S NEWTON.]

7. If G be the centre of gravity of the arc PQ of any uniform curve, and if PT be the tangent at P, prove that, when PQ is indefinitely diminished, the angles GPT and QPT vanish in the ratio of 2 to 3. [I. C. S. EXAM.]

CHAPTER II.

FUNDAMENTAL PROPOSITIONS.

37. Direction of the Tangent of a Curve at a given point.

Let AB be an arc of a curve traced in the plane of the the paper, OX a fixed straight line in the same plane.

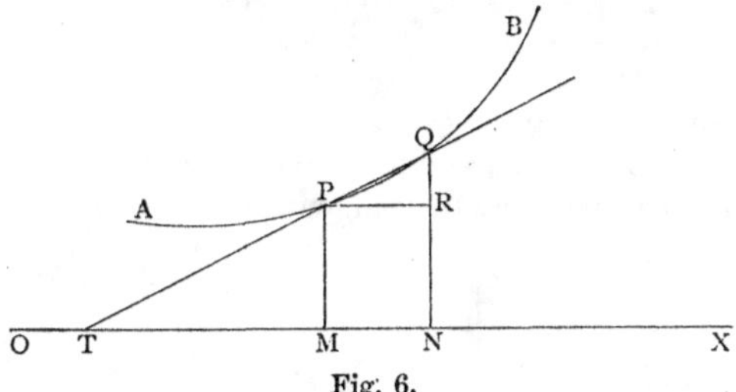

Fig. 6.

Let P, Q, be two points on the curve; PM, QN, perpendiculars on OX, and PR the perpendicular from P on QN. Join P, Q, and let QP be produced to cut OX at T.

When Q, travelling along the curve, approaches indefinitely near to P, the chord QP becomes in the limit the tangent at P. QR and PR both ultimately vanish, but the limit of their ratio *is in general finite*; for

$Lt\dfrac{RQ}{PR} = Lt \tan RPQ = Lt \tan XTP =$ *tangent of the angle which the tangent at* P *to the curve makes with* OX.

Ex. 1. Consider the straight line whose equation is $y=mx+c$.

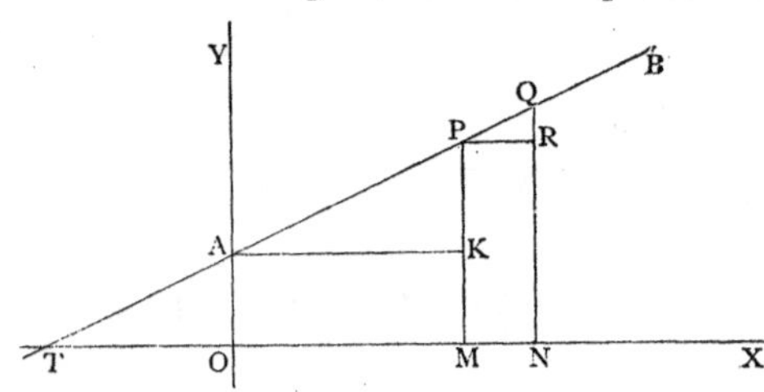

Fig. 7.

Let OX, OY, be the axes, and let the co-ordinates of P be x, y. Then, taking the general construction of the preceding article, the intercept $OA=c$, for $y=c$ when $x=0$.

Draw AK parallel to OX to meet MP in K; then, from similar triangles,

$$\frac{RQ}{PR}=\frac{KP}{AK}=\frac{MP-OA}{OM}$$

$$=\frac{y-c}{x}=\frac{mx}{x}=m.$$

Hence $\tan XTP=\tan RPQ=m$.

Ex. 2. Consider the parabola referred to its usual axes, viz., the axis of the parabola and the tangent at the vertex. With the same

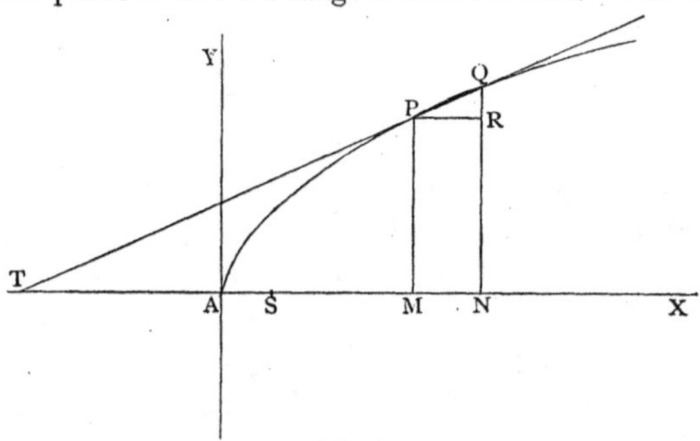

Fig. 8.

construction as before, we have

$$PM^2=4AS.AM,$$

$$QN^2=4AS.AN,$$

$$\therefore \quad QN^2-PM^2=4AS(AN-AM)=4AS.PR.$$

But $QN^2 - PM^2 = (QN - PM)(QN + PM) = RQ \,.\, (QN + PM)$,

$\therefore\; RQ(QN + PM) = 4AS \,.\, PR$,

whence $$Lt\frac{RQ}{PR} = Lt\frac{4AS}{QN + PM} = \frac{4AS}{2PM}$$

when Q comes to coincidence with P,

and therefore in the limit

$$\tan XTP = \frac{2AS}{PM}.$$

Ex. 3. Consider the "curve of sines" whose equation is

$$\frac{y}{b} = \sin\frac{x}{a}.$$

The same construction being made, if P be the point (x, y) on the

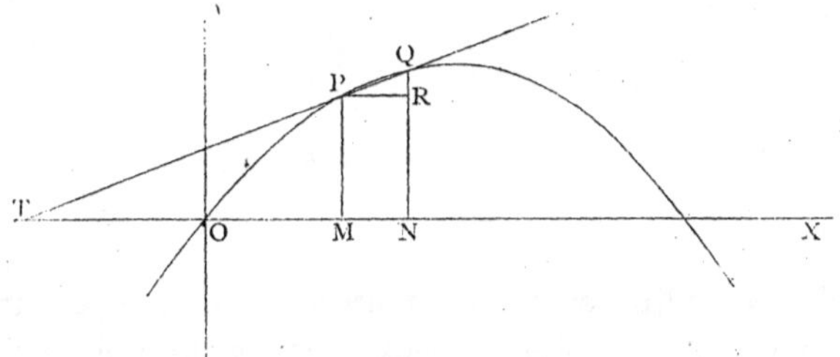

Fig. 9.

curve we have

$$MP = b\sin\frac{x}{a}.$$

Let $MN = h$, then $NQ = b\sin\frac{x+h}{a}$.

Hence $RQ = b\left\{\sin\frac{x+h}{a} - \sin\frac{x}{a}\right\} = 2b\sin\frac{h}{2a}\cos\frac{2x+h}{2a}$,

and therefore

$$Lt\frac{RQ}{PR} = Lt_{h=0}2b\frac{\sin\frac{h}{2a}\cos\frac{2x+h}{2a}}{h} = Lt_{h=0}\frac{b}{a}\left\{\frac{\sin\frac{h}{2a}}{\frac{h}{2a}}\right\}\cos\frac{2x+h}{2a}$$

$$= \frac{b}{a}\cos\frac{x}{a}.$$

Therefore in the limit

$$\tan XTP = \frac{b}{a}\cos\frac{x}{a}.$$

From the above examples it will now be obvious that the direction of the tangent at any point of any curve may be determined in a similar manner.

38. **Equation of Tangent.**

Let us consider the general case in which the equation of the curve is $y=\phi(x)$.

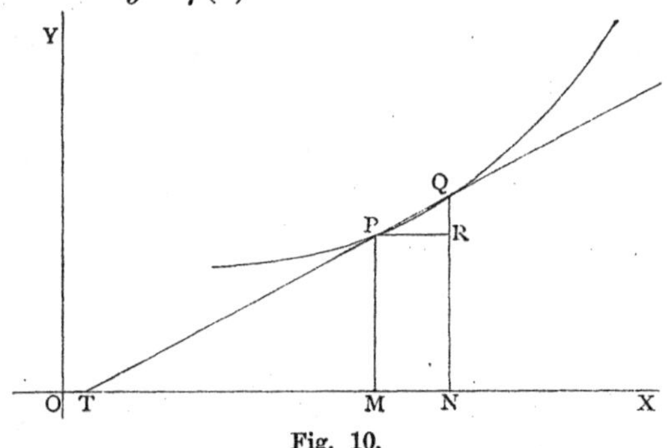

Fig. 10.

Let the co-ordinates of the points P, Q, on the curve be (x, y) $(x+\delta x, y+\delta y)$ respectively, δx and δy being used to denote increments of the variables x and y.

Then, the construction being as before,

$$OM=x, \quad ON=x+\delta x, \text{ therefore } PR=MN=\delta x;$$

also, $MP=y$, $NQ=y+\delta y$, therefore $RQ=\delta y$.

Again, since the point $x+\delta x$, $y+\delta y$, lies on the curve,

$$y+\delta y=\phi(x+\delta x),$$

whence $$RQ=\delta y=\phi(x+\delta x)-\phi(x).$$

Hence we can express $Lt\dfrac{RQ}{PR}$ as $Lt_{\delta x=0}\dfrac{\delta y}{\delta x}$ or

$$Lt_{\delta x=0}\frac{\phi(x+\delta x)-\phi(x)}{\delta x}.$$

Hence, to draw the tangent at any point (x, y) on the curve $y=\phi(x)$, we must draw a line through that point, making with the axis of x an angle whose tangent is $Lt_{\delta x=0}\frac{\phi(x+\delta x)-\phi(x)}{\delta x}$; and if this limit be called m, the equation of the tangent at $P(x, y)$ will be

$$Y-y=m(X-x),$$

X, Y being the current co-ordinates of any point on the tangent; for the line represented by this equation goes through the point (x, y), and makes with the axis of x an angle whose tangent is m.

EXAMPLES.

Find the equation of the tangent at the point (x, y) on each of the following curves :—

1. $x^2+y^2=c^2$.
2. $\frac{x^2}{a^2}+\frac{y^2}{b^2}=1$.
3. $y=e^x$.
4. $y=\log x$.
5. $y=\tan x$.
6. $y=\tan^{-1}x$.

39. DEF.—DIFFERENTIAL COEFFICIENT.

Let $\phi(x)$ denote any function of x, and $\phi(x+h)$ the same function of $x+h$; then $Lt_{h=0}\frac{\phi(x+h)-\phi(x)}{h}$ is called the FIRST DERIVED FUNCTION *or* DIFFERENTIAL CO-EFFICIENT *of $\phi(x)$ with respect to x.*

The operation of finding this limit is called *differentiating $\phi(x)$.*

After reading Chap. V., it will be obvious why the above expression is styled a "coefficient," for it is shown there to be one of a series of coefficients occurring in the expansion of $\phi(x+h)$ in powers of h.

The geometrical meaning of the above limit is indicated in the last article, where it is shown to be *the tangent of the angle ψ which the tangent at any definite point (x, y) on the curve $y=\phi(x)$ makes with the axis of x.*

40. We can now find the differential coefficient of any proposed function by investigating the value of the above limit; but it will be seen later on that, by means of certain rules and a knowledge of the differential coefficients of certain standard forms, we can always avoid the labour of an *a priori* evaluation.

When an *a priori* investigation becomes necessary, it may often be conducted very simply by pure geometry. It is however usual to treat the more complicated functions algebraically. Several examples are appended.

Ex. 1. To find *geometrically* the differential coefficient of $\sin x$.

Let the angle $AOP=x$, $AOQ=x+h$, and let a circle with centre O and radius unity cut the lines OA, OP, OQ, in A, P, Q. Draw perpendiculars PM, QN, to OA, and PR to QN. Join PQ. Then

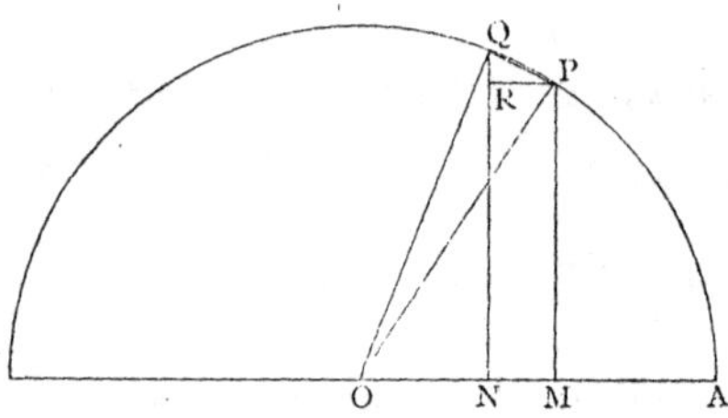

Fig. 11.

$$MP=\sin x,\ NQ=\sin(x+h),$$

$\therefore\ \sin(x+h)-\sin x=RQ.$

Again, $h=\text{angle } POQ=\text{arc } PQ$, the radius being unity.

Hence

$$Lt_{h=0}\frac{\sin(x+h)-\sin x}{h}=Lt\frac{RQ}{\text{arc } PQ}=Lt\frac{RQ}{\text{chord } PQ}$$

(for chord PQ and arc PQ are equal in the limit)

$=Lt \cos RQP=\cos OPR$

(since in the limit QPO is a right angle)

$=\cos AOP=\cos x.$

Ex. 2. To find *geometrically* the differential coefficient of $\sin^{-1}x$.

In Fig. 11 let $$A\hat{O}P=\sin^{-1}x,$$

and $$A\hat{O}Q=\sin^{-1}(x+h).$$

Then, with the same construction as before,

$$MP=x,\quad NQ=x+h,$$

therefore $$RQ=h.$$

Hence

$$Lt_{h=0}\frac{\sin^{-1}(x+h)-\sin^{-1}x}{h}=Lt_{h=0}\frac{A\hat{O}Q-A\hat{O}P}{RQ}$$

$$=Lt\frac{P\hat{O}Q}{RQ}=Lt\frac{\text{chord } PQ}{RQ}=Lt\frac{1}{\cos RQP}=\frac{1}{\cos OPR}$$

$$=\frac{1}{\cos AOP}=\frac{1}{\sqrt{1-\sin^2 AOP}}$$

$$=\frac{1}{\sqrt{1-x^2}}.$$

EXAMPLES.

Find in a similar manner the differential coefficients of

(1) $\tan x$. (3) $\operatorname{cosec} x$.

(2) $\tan^{-1}x$. (4) $\operatorname{cosec}^{-1}x$.

Ex. 3. Find from the definition the differential coefficient of $\frac{x^2}{a}$, where a is a constant.

Here $$\phi(x)=\frac{x^2}{a},$$

$$\phi(x+h)=\frac{(x+h)^2}{a},$$

therefore $$Lt_{h=0}\frac{\phi(x+h)-\phi(x)}{h}=Lt_{h=0}\frac{(x+h)^2-x^2}{ha}$$

$$=Lt_{h=0}\frac{2xh+h^2}{ha}=Lt_{h=0}\frac{(2x+h)}{a}$$

$$=\frac{2x}{a}.$$

The geometrical interpretation of this result is that, if a tangent be drawn to the parabola $ay=x^2$ at the point (x, y), it will be inclined to the axis of x at the angle $\tan^{-1}\frac{2x}{a}$.

Ex. 4. Find from the definition the differential coefficient of $\log\sin\frac{x}{a}$, where a is a constant.

Here $$\phi(x)=\log\sin\frac{x}{a},$$

and $$Lt_{h=0}\frac{\phi(x+h)-\phi(x)}{h}=Lt_{h=0}\frac{\log\sin\frac{x+h}{a}-\log\sin\frac{x}{a}}{h}$$

$$=Lt_{h=0}\frac{1}{h}\log\frac{\sin\frac{x}{a}\cos\frac{h}{a}+\cos\frac{x}{a}\sin\frac{h}{a}}{\sin\frac{x}{a}}$$

$$=Lt_{h=0}\frac{1}{h}\log\left(1+\frac{h}{a}\cot\frac{x}{a}-\text{higher powers of } h\right)$$

$\left[\text{by substituting for } \sin\frac{h}{a} \text{ and } \cos\frac{h}{a} \text{ their expansions in powers of } \frac{h}{a}\right]$

$$=Lt_{h=0}\frac{\frac{h}{a}\cot\frac{x}{a}-\text{higher powers of } h}{h}$$

[by expanding the logarithm]

$$=\frac{1}{a}\cot\frac{x}{a}.$$

Hence the tangent at any point on the curve $\frac{y}{a}=\log\sin\frac{x}{a}$ is inclined to the axis of x at an angle whose tangent is $\cot\frac{x}{a}$; that is at an angle $\frac{\pi}{2}-\frac{x}{a}$.

41. **Notation.**

The *result of the operation* expressed by $Lt_{h=0}\frac{\phi(x+h)-\phi(x)}{h}$ or by $Lt_{\delta x=0}\frac{\delta y}{\delta x}$ is generally denoted by $\frac{d}{dx}y$ or $\frac{dy}{dx}$.

It will be well to note distinctly once for all that in the notation thus introduced, dx and dy, as here used, are *not separate small quantities* as δx and δy are, but that

$\frac{d}{dx}$ is a *symbol of operation* which, when applied to y, denotes the result of *taking the limit of the ratio of the small quantities* δy, δx.

Sometimes $d_x y$ is used to denote the same thing; or, if $y=\phi(x)$, we often meet with the forms $\frac{d\phi(x)}{dx}$, $\frac{d\phi}{dx}$, $\phi'(x)$, ϕ_x, ϕ', or $\dot{\phi}$. Again, as the letters u, v, w, etc., are frequently used to denote functions of x, we shall consequently have the differential coefficient variously expressed, as $\frac{du}{dx}$, u', u_x, or $\dot{u}$, with a similar notation for those of v, w, etc.

42. **Aspect of the Differential Coefficient as a Rate-Measurer.** When a particle is in motion in a given manner the space described is a function of the time of describing it. We may consider the time as an independent variable, and the space described in that time as the dependent variable.

The rate of change of position of the particle is called its velocity.

If *uniform* the velocity is measured by the space described in one second; if *variable*, the velocity at any instant is measured by the space which would be described in one second if, for that second, the velocity remained unchanged.

Suppose a space s to have been described in time t with varying velocity, and an additional space δs to be described in the additional time δt. Let v_1 and v_2 be the greatest and least values of the velocity during the interval δt; then the spaces which would have been described with uniform velocities v_1, v_2, in time δt are $v_1\delta t$

and $v_2\delta t$, and are respectively greater and less than the actual space δs.

Hence $v_1, \frac{\delta s}{\delta t}$, and v_2 are in descending order of magnitude.

If then δt be diminished indefinitely, we have in the limit $v_1 = v_2 =$ the velocity at the instant considered, which is therefore represented by $Lt\frac{\delta s}{\delta t}$, *i.e.*, by $\frac{ds}{dt}$.

43. It appears therefore that we may give another interpretation to a differential coefficient, viz., that $\frac{ds}{dt}$ means the *rate of increase of s in point of time.* Similarly $\frac{dx}{dt}, \frac{dy}{dt}$, mean the *rates of change* of x and y respectively in point of time and *measure the velocities,* resolved parallel to the axes, of a moving particle whose co-ordinates at the instant under consideration are x, y. If x and y be given functions of t, and therefore the path of the particle defined, and if $\delta x, \delta y, \delta t$, be simultaneous infinitesimal increments of x, y, t, then

$$\frac{dy}{dx} = Lt\frac{\delta y}{\delta x} = Lt\frac{\frac{\delta y}{\delta t}}{\frac{\delta x}{\delta t}} = \frac{\frac{dy}{dt}}{\frac{dx}{dt}}$$

and therefore represents *the ratio of the rate of change of y to that of x.* The rate of change of x is arbitrary, and if we choose it to be unit velocity, then $\frac{dy}{dx} = \frac{dy}{dt} =$ absolute rate of change of y.

44. **Meaning of Sign of Differential Coefficient.**

If x be increasing with t, the x-velocity is positive, whilst, if x be decreasing while t increases, that velocity is negative. Similarly for y.

Moreover, since $\dfrac{dy}{dx}=\dfrac{\frac{dy}{dt}}{\frac{dx}{dt}}$, $\dfrac{dy}{dx}$ is *positive when x and y increase or decrease together, but negative when one increases as the other decreases.*

This is obvious also from the geometrical interpretation of $\dfrac{dy}{dx}$. For, if x and y are *increasing together*, $\dfrac{dy}{dx}$ is the *tangent of an acute angle and therefore positive*, while, if as x *increases* y *decreases*, $\dfrac{dy}{dx}$ represents the *tangent of an obtuse angle and is negative.*

EXAMPLES.

Find from the definition the differential coefficient of y with respect to x in each of the following cases :

1. $y=x^3$.
2. $y=2\sqrt{ax}$.
3. $y=\sqrt{a^2+x^2}$.
4. $y=e^x$.
5. $y=e^{\sqrt{x}}$.
6. $y=a^{\sin x}$.
7. $y=a^{\log x}$.
8. $y=\tan^{-1}x^3$.
9. $y=\log\cos x$.
10. $y=\log\tan x$.
11. $y=x^x$.
12. $y=x^{\sin x}$.
13. $y=(\sin x)^x$.
14. $y=(\sin x)^{\sqrt{x}}$.

15. In the curve $y=ce^{\frac{x}{c}}$, if ψ be the angle which the tangent at any point makes with the axis of x, prove $y=c\tan\psi$.

16. In the curve $y=c\cosh\dfrac{x}{c}$, prove $y=c\sec\psi$.

17. In the curve $b^2y=\dfrac{x^3}{3}-ax^2$ find the points at which the tangent is parallel to the axis of x.

[N.B.—This requires that $\tan\psi=0$.]

18. Find at what points of the ellipse $\frac{x^2}{a^2}+\frac{y^2}{b^2}=1$ the tangent cuts off equal intercepts from the axes.

[N.B.—This requires that $\tan\psi=\pm 1$.]

19. Prove that if a particle move so that the space described is proportional to the square of the time of description, the velocity will be proportional to the time, and the rate of increase of the velocity will be constant.

20. Show that if $s \propto \sin \mu t$, where μ is a constant, the rate of increase of the velocity is proportional to the distance of the particle measured along its path from a fixed position.

45. It will often be convenient in proving standard results to denote by a small letter the function of x considered, and by the corresponding capital the same function of $x+h$, *e.g.*, if $u=\phi(x)$, then $U=\phi(x+h)$, or if $u=a^x$, then $U=a^{x+h}$.

Accordingly we shall have

$$\frac{du}{dx}=Lt_{h=0}\frac{U-u}{h},$$

$$\frac{dv}{dx}=Lt_{h=0}\frac{V-v}{h},$$

etc.

46. We now proceed to the consideration of several important propositions.

47. Prop. I. **The Differential Coefficient of any Constant is zero.**

This proposition will be obvious when we refer to the definition of a constant quantity. A constant is essentially a quantity of which there is no variation, so that if $y=c$, $\delta y=$ absolute zero, whatever may be the value of δx. Hence $\frac{\delta y}{\delta x}=0$ and $\frac{dy}{dx}=0$ when the limit is taken.

Or, geometrically; $y=c$ is the equation of a straight line parallel to the axis of x. This makes an angle zero with that axis, and therefore $\tan\psi$ or $\frac{dy}{dx}=0$.

48. Prop. II. Product of Constant and Function.

The differential coefficient of a product of a constant and a function of x is equal to the product of the constant and the differential coefficient of the function, or, stated algebraically,

$$\frac{d}{dx}(cu)=c\frac{du}{dx}.$$

For, with the notation of Art. 45,

$$\frac{d}{dx}(cu)=Lt_{h=0}\frac{cU-cu}{h}=cLt_{h=0}\frac{U-u}{h}$$

$$=c\frac{du}{dx}.$$

49. Prop. III. Differential Coefficient of a Sum.

The differential coefficient of the sum of a set of functions of x is the sum of the differential coefficients of the several functions.

Let $u, v, w, \ldots,$ be the functions of x, and y their sum.

Let $U, V, W, \ldots, Y$ be what these expressions become when x is changed to $x+h$.

Then
$$y=u+v+w+\ldots$$
$$Y=U+V+W+\ldots,$$

and therefore

$$Y-y=(U-u)+(V-v)+(W-w)+\ldots;$$

dividing by h

$$\frac{Y-y}{h}=\frac{U-u}{h}+\frac{V-v}{h}+\frac{W-w}{h}+\ldots$$

and taking the limit

$$\frac{dy}{dx}=\frac{du}{dx}+\frac{dv}{dx}+\frac{dw}{dx}+\dots$$

If some of the connecting signs had been $-$ instead of $+$ a corresponding result would immediately follow, *e.g.*, if

$$y=u+v-w+\dots$$

then

$$\frac{dy}{dx}=\frac{du}{dx}+\frac{dv}{dx}-\frac{dw}{dx}+\dots$$

50. Prop. IV. **The Differential Coefficient of the product of two functions is**

(First Function) × *(Diff. Coeff. of Second)*
+ *(Second Function)* × *(Diff. Coeff. of First)*,

or, stated algebraically,

$$\frac{d(uv)}{dx}=u\frac{dv}{dx}+v\frac{du}{dx}.$$

With the same notation as before, let

$$y=uv, \text{ and therefore } Y=UV;$$

whence

$$\begin{aligned} Y-y&=UV-uv\\ &=u(V-v)+V(U-u); \end{aligned}$$

therefore

$$\frac{Y-y}{h}=u\frac{V-v}{h}+V\frac{U-u}{h},$$

and taking the limit

$$\frac{dy}{dx}=u\frac{dv}{dx}+v\frac{du}{dx}.$$

51. On division by uv the above result may be written

$$\frac{1}{y}\frac{dy}{dx}=\frac{1}{u}\frac{du}{dx}+\frac{1}{v}\frac{dv}{dx}.$$

Hence it is clear that the rule may be extended to products of more functions than two.

For example, if $y=uvw$; let $vw=z$, then $y=uz$.

Whence $$\frac{1}{y}\frac{dy}{dx}=\frac{1}{u}\frac{du}{dx}+\frac{1}{z}\frac{dz}{dx},$$

but $$\frac{1}{z}\frac{dz}{dx}=\frac{1}{v}\frac{dv}{dx}+\frac{1}{w}\frac{dw}{dx},$$

whence by substitution

$$\frac{1}{y}\frac{dy}{dx}=\frac{1}{u}\frac{du}{dx}+\frac{1}{v}\frac{dv}{dx}+\frac{1}{w}\frac{dw}{dx}.$$

Generally, if $$y=uvwt\ldots$$

$$\frac{1}{y}\frac{dy}{dx}=\frac{1}{u}\frac{du}{dx}+\frac{1}{v}\frac{dv}{dx}+\frac{1}{w}\frac{dw}{dx}+\frac{1}{t}\frac{dt}{dx}+\ldots,$$

and if we multiply by $uvwt\ldots$ we obtain

$$\frac{dy}{dx}=(vwt\ldots)\frac{du}{dx}+(uwt\ldots)\frac{dv}{dx}+(uvt\ldots)\frac{dw}{dx}+\ldots,$$

i.e., *multiply the differential coefficient of each separate function by the product of all the remaining functions and add up all the results;* the sum will be the differential coefficient of the product of all the functions.

52. PROP. V. **The Differential Coefficient of a quotient of two functions is**

$$\frac{(\textit{Diff. Coeff. of Num}^{r}.)(\textit{Den}^{r}.)-(\textit{Diff. Coeff. of Den}^{r}.)(\textit{Num}^{r}.)}{\textit{Square of Denominator.}}$$

or, stated algebraically,

$$\frac{d}{dx}\left(\frac{u}{v}\right)=\frac{\frac{du}{dx}v-\frac{dv}{dx}u}{v^2}.$$

With the same notation as before, let

$$y=\frac{u}{v}, \text{ and therefore } Y=\frac{U}{V},$$

whence $$\begin{aligned}Y-y&=\frac{U}{V}-\frac{u}{v}\\&=\frac{Uv-Vu}{Vv};\end{aligned}$$

therefore
$$\frac{Y-y}{h}=\frac{\frac{U-u}{h}v-\frac{V-v}{h}u}{Vv},$$
and taking the limit
$$\frac{dy}{dx}=\frac{\frac{du}{dx}v-\frac{dv}{dx}u}{v^2}.$$

53. This proposition may also be deduced immediately from Prop. IV., thus:

Let $$y=\frac{u}{v};$$

i.e., $$u=vy;$$

whence $$\frac{du}{dx}=v\frac{dy}{dx}+y\frac{dv}{dx}$$
$$=v\frac{dy}{dx}+\frac{u}{v}\frac{dv}{dx};$$

and therefore $$\frac{dy}{dx}=\frac{\frac{du}{dx}-\frac{u}{v}\frac{dv}{dx}}{v}$$
$$=\frac{\frac{du}{dx}.v-\frac{dv}{dx}.u}{v^2}.$$

54. We may also remark that Prop. II. is deducible from Propositions I. and IV. For by Prop. IV.
$$\frac{d}{dx}(cu)=c\frac{du}{dx}+u\frac{dc}{dx},$$
and by Prop. I. $$\frac{dc}{dx}=0.$$

Whence $$\frac{d}{dx}(cu)=c\frac{du}{dx}.$$

The differential coefficient of $\frac{c}{u}$ is also of importance; and it follows immediately from Prop. V. that
$$\frac{d}{dx}\left(\frac{c}{u}\right)=-\frac{c}{u^2}\frac{du}{dx}.$$

55. Prop. VI. **To find the Differential Coefficient of a Function of a Function.**

Let $u=f(v)$ .... (1)
and $v=F(x)$ .... (2)
Then, by elimination of v, we have a result which may be expressed as $u=\phi(x)$ .... (3)

Suppose the independent variable x to change to X in (2) and let *a* value of v *deduced from* (2) *be* V. Let this be substituted for v in (1), and let *a* value of u deduced from (1) be U. Then we have the following equations. $U=f(V)$.......................... (4)
and $V=F(X)$.......................... (5)
and *by the same process by which* (3) *was deduced from* (1) *and* (2) we obtain from (4) and (5)

$$U=\phi(X) \text{ .... (6)}$$

This result proves that if x be changed to X *in equation* (3), then *one of the values thence deduced for* u *will be* U, and therefore $Lt\frac{U-u}{X-x}$ when $X-x$ is diminished indefinitely is *a* value of the differential coefficient of u with respect to x, reckoned as a direct function of x as expressed in equation (3).

Now
$$\frac{U-u}{X-x}=\frac{U-u}{V-v}\cdot\frac{V-v}{X-x}$$

and $Lt_{V-v=0}\frac{U-u}{V-v}$ is *a* value of the differential coefficient of u with respect to v *derived from equation* (1) and denoted by $\frac{du}{dv}$; also, $Lt_{X-x=0}\frac{V-v}{X-x}$ is *a* value of the differential coefficient of v with respect to x *derived from equation* (2) and denoted by $\frac{dv}{dx}$. We therefore have, when we proceed to the limit,

$$\frac{du}{dx}=\frac{du}{dv}\cdot\frac{dv}{dx},$$

a formula already established in a different manner and with different letters in Art. 43.

56. It is obvious that the above result may be extended. For, if $u=\phi(v)$, $v=\psi(w)$, $w=f(x)$, we have

$$\frac{du}{dx}=\frac{du}{dv}\cdot\frac{dv}{dx},$$

but $$\frac{dv}{dx}=\frac{dv}{dw}\cdot\frac{dw}{dx};$$

and therefore $$\frac{du}{dx}=\frac{du}{dv}\cdot\frac{dv}{dw}\cdot\frac{dw}{dx},$$

and a similar result holds however many functions there may be.

Ex. Let $u=b\sin\frac{v}{a}$, $v=\sin^{-1}w$, $w=\frac{x^2}{a}$, that is,

$$u=b\sin\left(\frac{1}{a}\sin^{-1}\frac{x^2}{a}\right).$$

Then, by Ex. 3, Art. 37, $\frac{du}{dv}=\frac{b}{a}\cos\frac{v}{a}$.

Ex. 2, Art. 40, $\frac{dv}{dw}=\frac{1}{\sqrt{1-w^2}}$.

Ex. 3, ibid, $\frac{dw}{dx}=\frac{2x}{a}$.

Hence $$\frac{du}{dx}=\frac{du}{dv}\cdot\frac{dv}{dw}\cdot\frac{dw}{dx}=\frac{b}{a}\cos\frac{v}{a}\cdot\frac{1}{\sqrt{1-w^2}}\cdot\frac{2x}{a}$$

$$=\frac{b}{a}\cos\left(\frac{1}{a}\sin^{-1}\frac{x^2}{a}\right)\cdot\frac{1}{\sqrt{1-\frac{x^4}{a^2}}}\cdot\frac{2x}{a}.$$

The rule may be expressed thus :

$$\frac{d(\mathit{1st\ Func.})}{dx}=\frac{d(\mathit{1st\ Func.})}{d(\mathit{2nd\ Func.})}\cdot\frac{d(\mathit{2nd\ Func.})}{d(\mathit{3rd\ Func.})}\cdots\frac{d(\mathit{Last\ Func.})}{dx}.$$

57. There is a difficulty in Prop. VI. arising from the fact that for one value of x in (2) there may be *several values of* v, and for any value of v in (1) there may be *several values of* u. In fact the $f(v)$ and $F(x)$ may one or both be *many-valued functions* (such, for example, as $\sin^{-1}x$, which denotes any one of the series of angles whose sines are equal to x). But it is clear that the *same values of* u *and* x *will satisfy equation* (3) *as would simultaneously satisfy* (1) *and* (2), and that $Lt\dfrac{U-u}{X-x}$ when $X-x$ is indefinitely diminished is *one* value of the differential coefficient of u considered as a function of x; and it is equally obvious that there may be a *series of such values* for $\dfrac{du}{dx}$, as also for $\dfrac{du}{dv}$ and for $\dfrac{dv}{dx}$, so that in the theorem enunciated and proved above, in Art. 55, a *proper selection of those values is assumed to be made.*

58. If in the theorem $\dfrac{du}{dx}=\dfrac{du}{dy}\cdot\dfrac{dy}{dx}$ (where y is written for v in the result of Art. 55) we suppose $u=x$, then

$$\frac{du}{dx}=\frac{dx}{dx}=Lt_{h=0}\frac{(x+h)-x}{h}=1.$$

Hence we have

$$\frac{dy}{dx}\cdot\frac{dx}{dy}=1,$$

or

$$\frac{dy}{dx}=\frac{1}{\dfrac{dx}{dy}}.$$

59. In this application of the general theorem of Prop. VI. y is assumed to be a function of x and consequently x is the *inverse function of* y. So that $\dfrac{dy}{dx}$ is the differential

coefficient of y with respect to x when y is *considered as a function* of x, and $\frac{dx}{dy}$ is the differential coefficient of x with respect to y when x is considered as the *inverse function* of y:

e.g., if $y = \sin x$, then $x = \sin^{-1} y$,

$$\frac{dy}{dx} = \cos x \text{ (Ex. 1, Art. 40)},$$

and
$$\frac{dx}{dy} = \frac{1}{\sqrt{1-y^2}} \text{ (Ex. 2, ibid)},$$

and
$$\frac{dy}{dx} \cdot \frac{dx}{dy} = \cos x \cdot \frac{1}{\sqrt{1-y^2}} = \frac{\cos x}{\sqrt{1-\sin^2 x}} = 1.$$

60. The same difficulty occurs in Arts. 58 and 59 as that discussed in Art. 57.

If
$$y = f(x) \ldots (1),$$
and this equation be supposed solved for x, the result will be of the form $x = F(y) \ldots (2)$.
Now, if x be changed to X in (1) and Y be a value deduced for y, then if Y be substituted for y in (2), X will be *one of the values* thence deduced for x.

Hence $Lt \frac{X-x}{Y-y}$ when $Y-y$ is indefinitely diminished is *a* value of the differential coefficient of x with respect to y, as derived from equation (2), while $Lt \frac{Y-y}{X-x}$ when $X-x$ is indefinitely diminished is *a* value of the differential coefficient of y with respect to x as derived from equation (1). And since
$$\frac{Y-y}{X-x} \cdot \frac{X-x}{Y-y} = 1,$$
we have
$$\frac{dy}{dx} \cdot \frac{dx}{dy} = 1,$$

when the limit is taken, the *proper selection being made* of the values deduced for $\frac{dy}{dx}$ and $\frac{dx}{dy}$.

61. This may be illustrated geometrically.

Let the curve $y=f(x)$ be drawn. Let the tangent to

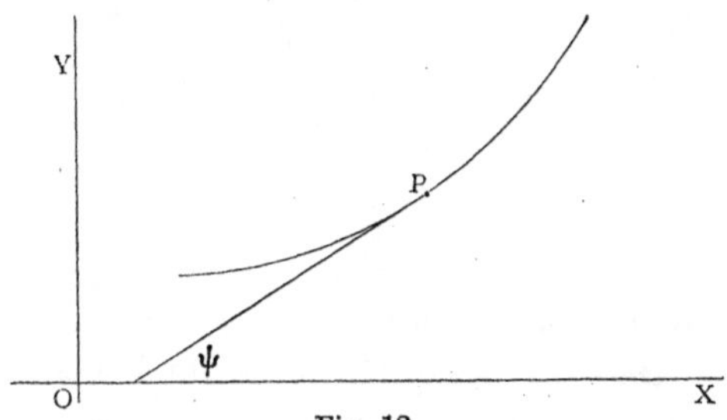

Fig. 12.

the curve at the point P, (x, y), make an angle ψ with the axis of x. Then, by Art. 39, $\frac{dy}{dx}=\tan\psi$; and in the same way it is obvious that $\frac{dx}{dy}=\tan(90-\psi)=\cot\psi$, so that $\frac{dy}{dx}\cdot\frac{dx}{dy}=\tan\psi\,.\cot\psi=1$.

Suppose however that the ordinate through P cuts the curve again at $P_1, P_2, P_3, \ldots$

Then, for a given value of x there are several values of y, and therefore also for a given increase δx in the value of x there may be several values of δy the increment of y. But if it be carefully noted that the δy and δx chosen are to refer to the *same branch of the curve at the same point* when we consider $\frac{dy}{dx}$ as when we consider $\frac{dx}{dy}$, then, under these circumstances, these expressions are respectively the *tangent and cotangent of the same angle,* and

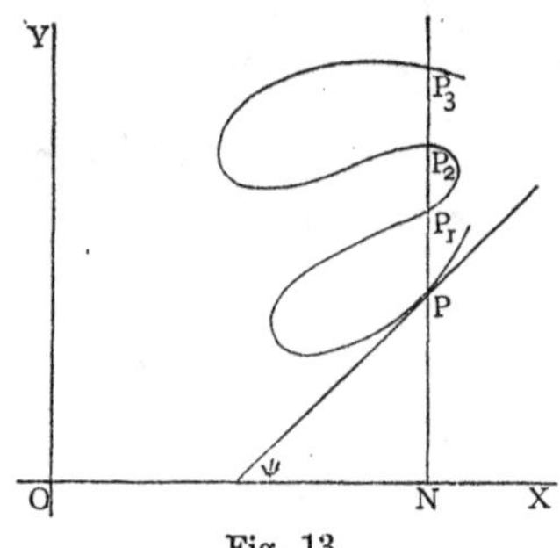

Fig. 13.

therefore their product is unity.

We say *the same branch of the curve,* for it may happen that more than one branch of the curve passes through a

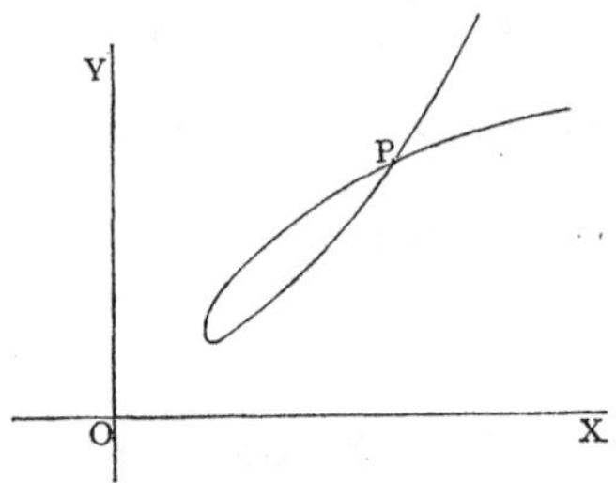

Fig. 14.

given point P, as in Fig. 14, and then there are two or more tangents at P and therefore two or more values of $\frac{dy}{dx}$ and $\frac{dx}{dy}$ at P. But the product of the $\frac{dy}{dx}$ and the $\frac{dx}{dy}$, which belong *to any the same branch through* P, is unity.

62. Differentiation of Inverse Functions.

When the differential coefficient of any function of x is

found, that of the corresponding inverse function is easily deduced by means of the theorem of Art. 58.

For let $x=f(y)$, and therefore $y=f^{-1}(x)$; then

$$\frac{dx}{dy}=f'(y).$$

But
$$\frac{dy}{dx}=\frac{1}{\frac{dx}{dy}};$$

therefore
$$\frac{d}{dx}f^{-1}(x)=\frac{1}{f'(y)}=\frac{1}{f'\{f^{-1}(x)\}}.$$

EXAMPLES.

Differentiate by means of the definition and the foregoing rules:—

1. $y=x\log\sin x.$
2. $y=x\sqrt{a^2-x^2}.$
3. $y=\frac{c^2}{x}e^{\frac{x}{c}}.$
4. $y=\frac{a}{x}\sqrt{a^2-x^2}.$
5. $y=2\sqrt{au}$, where $u=a^{\sin x}$.
6. $y=e^{\sqrt{u}}$, where $u=\log\sin v$, and $v=(\sin w)^w$, and $w=x^2$.

The results of any preceding examples may be assumed.

CHAPTER III.

STANDARD FORMS.

63. It is the object of the present Chapter to investigate and tabulate the results of differentiating the several standard forms referred to in Art. 40.

We shall always consider angles to be measured in circular measure, and all logarithms to be Napierian, unless the contrary is expressly stated.

It will be remembered that if $u=\phi(x)$, then, by the definition of a differential coefficient,

$$\frac{du}{dx}=Lt_{h=0}\frac{\phi(x+h)-\phi(x)}{h}.$$

64. **Differential Coefficient of x^n.**

If $$u=\phi(x)=x^n,$$

then $$\phi(x+h)=(x+h)^n,$$

and $$\frac{du}{dx}=Lt_{h=0}\frac{(x+h)^n-x^n}{h}$$

$$=Lt_{h=0}x^n\frac{\left(1+\frac{h}{x}\right)^n-1}{h}.$$

Now, since h is to be ultimately zero, we may consider $\frac{h}{x}$ to be less than unity, and we can therefore apply the Binomial Theorem to expand $\left(1+\frac{h}{x}\right)^n$, whatever be the value of n; hence

$$\frac{du}{dx}=Lt_{h=0}\frac{x^n}{h}\left\{n\frac{h}{x}+\frac{n(n-1)}{1.2}\frac{h^2}{x^2}+\dots\right\}$$
$$=Lt_{h=0}(nx^{n-1}+\text{powers of } h)$$
$$=nx^{n-1}.$$

65. If it be required to find the differential coefficient of x^n without the use of the Binomial Theorem we quote the result of Art. 23, viz.:

$$Lt_{y=1}\frac{y^n-1}{y-1}=n,$$

and proceed as follows:

$$\frac{dx^n}{dx}=Lt_{h=0}x^n\frac{\left(1+\frac{h}{x}\right)^n-1}{h}\quad\text{[as before]}$$
$$=Lt_{h=0}x^{n-1}\frac{\left(1+\frac{h}{x}\right)^n-1}{\left(1+\frac{h}{x}\right)-1}$$
$$=Lt_{y=1}x^{n-1}\frac{y^n-1}{y-1}\quad\left[\text{where } y=1+\frac{h}{x}\right]$$
$$=nx^{n-1}.$$

66. Differential Coefficient of a^x.

If $$u=\phi(x)=a^x,$$
$$\phi(x+h)=a^{x+h}.$$

and $$\frac{du}{dx}=Lt_{h=0}\frac{a^{x+h}-a^x}{h}$$
$$=a^xLt_{h=0}\frac{a^h-1}{h}$$
$$=a^x\log_e a.\quad\text{[Art. 22.]}$$

Cor. If $u=e^x$, $\frac{du}{dx}=e^x\log_e e=e^x$.

67. Differential Coefficient of $\log_a x$.

If
$$u=\phi(x)=\log_a x,$$
$$\phi(x+h)=\log_a(x+h),$$
and
$$\frac{du}{dx}=Lt_{h=0}\frac{\log_a(x+h)-\log_a x}{h}$$
$$=Lt_{h=0}\frac{1}{h}\log_a\left(1+\frac{h}{x}\right).$$

Let $\frac{x}{h}=z$, so that if $h=0$, $z=\infty$; therefore
$$\frac{du}{dx}=Lt_{z=\infty}\frac{z}{x}\log_a\left(1+\frac{1}{z}\right)$$
$$=\frac{1}{x}Lt_{z=\infty}\log_a\left(1+\frac{1}{z}\right)^z$$
$$=\frac{1}{x}\log_a e. \quad \text{[Arts. 12 (7) and 21.]}$$

Cor. If $u=\log_e x$, $\frac{du}{dx}=\frac{1}{x}\log_e e=\frac{1}{x}$.

68. Differential Coefficient of $\sin x$.

If
$$u=\phi(x)=\sin x,$$
$$\phi(x+h)=\sin(x+h),$$
and
$$\frac{du}{dx}=Lt_{h=0}\frac{\sin(x+h)-\sin x}{h}$$
$$=Lt_{h=0}\frac{2\sin\frac{h}{2}\cos\left(x+\frac{h}{2}\right)}{h}$$
$$=Lt_{h=0}\frac{\sin\frac{h}{2}}{\frac{h}{2}}\cos\left(x+\frac{h}{2}\right)$$
$$=\cos x. \quad \text{[Arts. 12 (2) and 19.]}$$

69. Differential Coefficient of $\cos x$.

If
$$u=\phi(x)=\cos x,$$
$$\phi(x+h)=\cos(x+h),$$
and
$$\frac{du}{dx}=Lt_{h=0}\frac{\cos(x+h)-\cos x}{h}$$
$$=-Lt_{h=0}\frac{2\sin\frac{h}{2}\sin\left(x+\frac{h}{2}\right)}{h}$$
$$=-Lt_{h=0}\frac{\sin\frac{h}{2}}{\frac{h}{2}}\sin\left(x+\frac{h}{2}\right)$$
$$=-\sin x.$$

70. Differential Coefficient of $\tan x$.

If
$$u=\phi(x)=\tan x,$$
$$\phi(x+h)=\tan(x+h),$$
and
$$\frac{du}{dx}=Lt_{h=0}\frac{\tan(x+h)-\tan x}{h}$$
$$=Lt_{h=0}\frac{\sin(x+h)\cos x-\cos(x+h)\sin x}{h\cos x\cos(x+h)}$$
$$=Lt_{h=0}\frac{\sin h}{h}\cdot\frac{1}{\cos x\cos(x+h)}$$
$$=\frac{1}{\cos^2 x}=\sec^2 x.$$

71. Differential Coefficient of $\cot x$.

If
$$u=\phi(x)=\cot x,$$
$$\phi(x+h)=\cot(x+h),$$
and
$$\frac{du}{dx}=Lt_{h=0}\frac{\cot(x+h)-\cot x}{h}$$
$$=Lt_{h=0}\frac{\cos(x+h)\sin x-\cos x\sin(x+h)}{h\sin x\,.\,\sin(x+h)}$$

$$= -Lt_{h=0}\frac{\sin h}{h}\cdot\frac{1}{\sin x \sin (x+h)}$$

$$= -\frac{1}{\sin^2 x} = -\operatorname{cosec}^2 x.$$

72. **Differential Coefficient of sec x.**

If $$u = \phi(x) = \sec x,$$

$$\phi(x+h) = \sec (x+h),$$

and $$\frac{du}{dx} = Lt_{h=0}\frac{\sec (x+h) - \sec x}{h}$$

$$= Lt_{h=0}\frac{\cos x - \cos (x+h)}{h \cos x \cos (x+h)}$$

$$= Lt_{h=0}\frac{\sin \frac{h}{2}}{\frac{h}{2}}\cdot\frac{\sin\left(x+\frac{h}{2}\right)}{\cos x \cos (x+h)}$$

$$= \frac{\sin x}{\cos^2 x}.$$

73. **Differential Coefficient of cosec x.**

If $$u = \phi(x) = \operatorname{cosec} x,$$

$$\phi(x+h) = \operatorname{cosec} (x+h),$$

and $$\frac{du}{dx} = Lt_{h=0}\frac{\operatorname{cosec} (x+h) - \operatorname{cosec} x}{h}$$

$$= Lt_{h=0}\frac{\sin x - \sin (x+h)}{h \sin x \sin (x+h)}$$

$$= -Lt_{h=0}\frac{\sin \frac{h}{2}}{\frac{h}{2}}\cdot\frac{\cos\left(x+\frac{h}{2}\right)}{\sin x \sin (x+h)}$$

$$= -\frac{\cos x}{\sin^2 x}.$$

74. Inverse Trigonometrical Functions.

For the inverse trigonometrical functions it seems useful to recur to the notation of Art. 45, and to denote $\phi(x+h)$ by U.

75. Differential Coefficient of $\sin^{-1}x$.

If $$u=\phi(x)=\sin^{-1}x,$$
$$U=\phi(x+h)=\sin^{-1}(x+h).$$

Hence $x=\sin u$, and $x+h=\sin U$;

therefore $h=\sin U-\sin u$,

and
$$\frac{du}{dx}=Lt_{h=0}\frac{U-u}{h}=Lt_{U=u}\frac{U-u}{\sin U-\sin u}$$
$$=Lt_{U=u}\left\{\frac{\frac{U-u}{2}}{\sin\frac{U-u}{2}}\right\}\frac{1}{\cos\frac{U+u}{2}}$$
$$=\frac{1}{\cos u}=\frac{1}{\sqrt{1-\sin^2 u}}=\frac{1}{\sqrt{1-x^2}}.$$

76. Differential Coefficient of $\cos^{-1}x$.

If $$u=\phi(x)=\cos^{-1}x,$$
$$U=\phi(x+h)=\cos^{-1}(x+h).$$

Hence $x=\cos u$, and $x+h=\cos U$;

therefore $h=\cos U-\cos u$,

and
$$\frac{du}{dx}=Lt_{h=0}\frac{U-u}{h}=Lt_{U=u}\frac{U-u}{\cos U-\cos u}$$
$$=-Lt_{U=u}\left\{\frac{\frac{U-u}{2}}{\sin\frac{U-u}{2}}\right\}\frac{1}{\sin\frac{U+u}{2}}$$
$$=-\frac{1}{\sin u}=-\frac{1}{\sqrt{1-\cos^2 u}}=-\frac{1}{\sqrt{1-x^2}}.$$

77. Differential Coefficient of $\tan^{-1}x$.

If $u=\phi(x)=\tan^{-1}x,$

$U=\phi(x+h)=\tan^{-1}(x+h).$

Hence $x=\tan u$, and $x+h=\tan U$;

therefore $h=\tan U-\tan u,$

and
$$\frac{du}{dx}=Lt_{h=0}\frac{U-u}{h}=Lt_{U=u}\frac{U-u}{\tan U-\tan u}$$
$$=Lt_{U=u}\frac{U-u}{\sin(U-u)}\cos U\cos u$$
$$=\cos^2 u=\frac{1}{\sec^2 u}=\frac{1}{1+\tan^2 u}=\frac{1}{1+x^2}.$$

78. Differential Coefficient of $\cot^{-1}x$.

If $u=\phi(x)=\cot^{-1}x,$

$U=\phi(x+h)=\cot^{-1}(x+h).$

Hence $x=\cot u$, and $x+h=\cot U$;

therefore $h=\cot U-\cot u,$

and
$$\frac{du}{dx}=Lt_{h=0}\frac{U-u}{h}=Lt_{U=u}\frac{U-u}{\cot U-\cot u}$$
$$=-Lt_{U=u}\frac{U-u}{\sin(U-u)}\sin U\sin u$$
$$=-\sin^2 u=-\frac{1}{\operatorname{cosec}^2 u}=-\frac{1}{1+\cot^2 u}=-\frac{1}{1+x^2}.$$

79. Differential Coefficient of $\sec^{-1}x$.

If $u=\phi(x)=\sec^{-1}x,$

$U=\phi(x+h)=\sec^{-1}(x+h).$

Hence $x=\sec u$, and $x+h=\sec U$;

therefore $h=\sec U-\sec u,$

and
$$\frac{du}{dx}=Lt_{h=0}\frac{U-u}{h}=Lt_{U=u}\frac{U-u}{\sec U-\sec u}$$
$$=Lt_{U=u}\frac{U-u}{\cos u-\cos U}\cos u\cos U$$

$$=Lt_{U=u}\left\{\frac{\frac{U-u}{2}}{\sin\frac{U-u}{2}}\right\}\frac{\cos u\cos U}{\sin\frac{U+u}{2}}$$

$$=\frac{\cos^2 u}{\sin u}=\frac{1}{\sec^2 u\sqrt{1-\cos^2 u}}=\frac{1}{x^2\sqrt{1-\frac{1}{x^2}}}$$

$$=\frac{1}{x\sqrt{x^2-1}}.$$

80. **Differential Coefficient of $\operatorname{cosec}^{-1}x$.**

If $\quad u=\phi(x)=\operatorname{cosec}^{-1}x,$

$\quad U=\phi(x+h)=\operatorname{cosec}^{-1}(x+h).$

Hence $\quad x=\operatorname{cosec} u$, and $x+h=\operatorname{cosec} U$;

therefore $\quad h=\operatorname{cosec} U-\operatorname{cosec} u,$

and
$$\frac{du}{dx}=Lt_{h=0}\frac{U-u}{h}=Lt_{U=u}\frac{U-u}{\operatorname{cosec} U-\operatorname{cosec} u}$$

$$=Lt_{U=u}\frac{U-u}{\sin u-\sin U}\sin u\sin U$$

$$=-Lt_{U=u}\left\{\frac{\frac{U-u}{2}}{\sin\frac{U-u}{2}}\right\}\frac{\sin u\sin U}{\cos\frac{U+u}{2}}$$

$$=-\frac{\sin^2 u}{\cos u}=-\frac{1}{\operatorname{cosec}^2 u\sqrt{1-\sin^2 u}}$$

$$=-\frac{1}{x^2\sqrt{1-\frac{1}{x^2}}}=-\frac{1}{x\sqrt{x^2-1}}.$$

81. From the importance of the results it has been thought preferable to deduce the differential coefficients of the inverse functions $\sin^{-1}x$ etc. immediately from the

definition; but by aid of Prop. VI. of the preceding chapter we can simplify the proofs considerably.

Ex. (i.) If $u=\sin^{-1}x$,

we have $x=\sin u$;

whence $\frac{dx}{du}=\cos u$;

and therefore $\frac{du}{dx}=\frac{1}{\frac{dx}{du}}=\frac{1}{\cos u}=\frac{1}{\sqrt{1-\sin^2 u}}=\frac{1}{\sqrt{1-x^2}}$;

and since $\cos^{-1}x=\frac{\pi}{2}-\sin^{-1}x$,

we have $\frac{d\cos^{-1}x}{dx}=-\frac{1}{\sqrt{1-x^2}}$.

Ex. (ii.) If $u=\tan^{-1}x$,

we have $x=\tan u$;

whence $\frac{dx}{du}=\sec^2 u$;

and therefore $\frac{du}{dx}=\frac{1}{\sec^2 u}=\frac{1}{1+\tan^2 u}=\frac{1}{1+x^2}$;

and since $\cot^{-1}x=\frac{\pi}{2}-\tan^{-1}x$,

we have $\frac{d\cot^{-1}x}{dx}=-\frac{1}{1+x^2}$.

Ex. (iii.) If $u=\text{vers}^{-1}x$,

we have $x=\text{vers}\,u=1-\cos u$;

whence $\frac{dx}{du}=\sin u$;

and therefore $\frac{du}{dx}=\frac{1}{\sin u}=\frac{1}{\sqrt{1-\cos^2 u}}=\frac{1}{\sqrt{2x-x^2}}$;

whence also $\dfrac{d\,\text{covers}^{-1}x}{dx} = -\dfrac{1}{\sqrt{2x-x^2}}.$

82. **The Integral Calculus.**

Suppose any expression in terms of x given; *can we find a function of which that expression is the differential coefficient?* The problem here suggested is inverse to that considered in the Differential Calculus. The discovery of such functions is the fundamental aim of the Integral Calculus. The function whose differential coefficient is the given expression is said to be the "integral" of that expression. For example, if $\phi'(x)$ be the differential coefficient of $\phi(x)$, $\phi(x)$ is said to be the integral of $\phi'(x)$. Moreover, since $\phi'(x)$ is also the differential coefficient of $\phi(x)+C$, C being any arbitrary constant disappearing upon differentiation, it is customary to state that the integral of $\phi'(x)$ is $\phi(x)+C$, C being any arbitrary constant.

The notation by which this is expressed is

$$\int\phi'(x)dx = \phi(x) + C,$$

$\int\phi'(x)dx$ being read "integral of $\phi'(x)$ with respect to x."

Thus we have seen

$$\frac{d}{dx}(\sin x) = \cos x,$$

$$\frac{d}{dx}(\tan^{-1}x) = \frac{1}{1+x^2},$$

etc.,

whence it follows immediately that

$$\int\cos x\,dx = \sin x,$$

$$\int\frac{1}{1+x^2}dx = \tan^{-1}x,$$

etc.,

where the arbitrary constant may be added in each case if desired.

83. We do not propose to enter upon any description of the various operations of the Integral Calculus, but it will be found that for integration we shall require to remember the same list of standard forms that is established in the present chapter and tabulated below, and it is advantageous to learn each formula here in its double aspect. We have therefore ventured to tabulate the standard forms for Differentiation and Integration together. Moreover, we shall find it convenient to be able to use the standard forms of integration in several of our subsequent articles.

TABLE OF RESULTS TO BE COMMITTED TO MEMORY.

$u=x^n.$	$\dfrac{du}{dx}=nx^{n-1}.$	$\int x^n dx$	$=\dfrac{x^{n+1}}{n+1}.$
$u=a^x.$	$\dfrac{du}{dx}=a^x\log_e a.$	$\int a^x dx$	$=\dfrac{a^x}{\log_e a}.$
$u=e^x.$	$\dfrac{du}{dx}=e^x.$	$\int e^x dx$	$=e^x.$
$u=\log_a x.$	$\dfrac{du}{dx}=\dfrac{1}{x}\log_a e.$	$\int\dfrac{dx}{x}$	$=\log_e x.$
$u=\log_e x.$	$\dfrac{du}{dx}=\dfrac{1}{x}.$		or $=\dfrac{\log_a x}{\log_a e}.$
$u=\sin x.$	$\dfrac{du}{dx}=\cos x.$	$\int\cos x dx$	$=\sin x.$
$u=\cos x.$	$\dfrac{du}{dx}=-\sin x.$	$\int\sin x dx$	$=-\cos x.$
$u=\tan x.$	$\dfrac{du}{dx}=\sec^2 x.$	$\int\sec^2 x dx$	$=\tan x.$

$u = \cot x.$ $\quad \frac{du}{dx} = -\text{cosec}^2 x.$ $\quad \int \text{cosec}^2 x\,dx = -\cot x.$

$u = \sec x.$ $\quad \frac{du}{dx} = \frac{\sin x}{\cos^2 x}.$ $\quad \int \frac{\sin x}{\cos^2 x}dx = \sec x.$

$u = \text{cosec}\, x.$ $\quad \frac{du}{dx} = -\frac{\cos x}{\sin^2 x}.$ $\quad \int \frac{\cos x}{\sin^2 x}dx = -\text{cosec}\, x.$

$u = \sin^{-1} x.$ $\quad \frac{du}{dx} = \frac{1}{\sqrt{1-x^2}}.$
$u = \cos^{-1} x.$ $\quad \frac{du}{dx} = -\frac{1}{\sqrt{1-x^2}}.$ $\quad \int \frac{dx}{\sqrt{1-x^2}} = \sin^{-1} x,$ or $-\cos^{-1} x.$

$u = \tan^{-1} x.$ $\quad \frac{du}{dx} = \frac{1}{1+x^2}.$
$u = \cot^{-1} x.$ $\quad \frac{du}{dx} = -\frac{1}{1+x^2}.$ $\quad \int \frac{dx}{1+x^2} = \tan^{-1} x,$ or $-\cot^{-1} x.$

$u = \sec^{-1} x.$ $\quad \frac{du}{dx} = \frac{1}{x\sqrt{x^2-1}}.$
$u = \text{cosec}^{-1} x.$ $\quad \frac{du}{dx} = -\frac{1}{x\sqrt{x^2-1}}.$ $\quad \int \frac{dx}{x\sqrt{x^2-1}} = \sec^{-1} x,$ or $-\text{cosec}^{-1} x.$

$u = \text{vers}^{-1} x.$ $\quad \frac{du}{dx} = \frac{1}{\sqrt{2x-x^2}}.$
$u = \text{covers}^{-1} x.$ $\quad \frac{du}{dx} = -\frac{1}{\sqrt{2x-x^2}}.$ $\quad \int \frac{dx}{\sqrt{2x-x^2}} = \text{vers}^{-1} x,$ or $-\text{covers}^{-1} x.$

84. The Form u^v.

In functions of the form u^v, where both u and v are functions of x, it is generally advisable to *take logarithms* before proceeding to differentiate.

Let $\quad y = u^v,$

then $\quad \log_e y = v \log_e u;$

therefore $\quad \frac{1}{y}\frac{dy}{dx} = \frac{dv}{dx}.\log_e u + v.\frac{1}{u}\frac{du}{dx},$ Arts. 50, 55, 67, or $$\frac{dy}{dx}=u^v\left(\log_e u\,.\frac{dv}{dx}+\frac{v}{u}\frac{du}{dx}\right).$$

Three cases of this proposition present themselves.

I. If v *be a constant* and u a function of x, $\frac{dv}{dx}=0$ and the above reduces to

$$\frac{dy}{dx}=v\,.\,u^{v-1}\frac{du}{dx},$$

as might be expected from Arts. 55, 64.

II. If u *be a constant* and v a function of x, $\frac{du}{dx}=0$ and the general form proved above reduces to

$$\frac{dy}{dx}=u^v\log_e u\,.\,\frac{dv}{dx},$$

as might be expected from Arts. 55, 66.

III. If u and v be *both functions of* x, it appears that the general formula

$$\frac{dy}{dx}=u^v\log_e u\frac{du}{dx}+vu^{v-1}\frac{du}{dx}$$

is the sum of the two special forms in I. and II., and therefore we may, instead of taking logarithms in any particular example, *consider first u constant and then v constant and add the results obtained on these suppositions.*

85. **Hyperbolic Functions.**

The differential coefficients of the direct and inverse hyperbolic functions are now appended as additional formulae. Their verification is very simple and is left as an exercise. They will be found useful by the more advanced student by reason of their close analogy of form with the results tabulated above for the direct and inverse trigonometrical functions.

Results for Hyperbolic Functions.

$u=\sinh x=\dfrac{e^x-e^{-x}}{2}.$	$\dfrac{du}{dx}=\cosh x.$	$\int\cosh x\,dx=\sinh x.$
$u=\cosh x=\dfrac{e^x+e^{-x}}{2}.$	$\dfrac{du}{dx}=\sinh x.$	$\int\sinh x\,dx=\cosh x.$
$u=\tanh x=\dfrac{\sinh x}{\cosh x}.$	$\dfrac{du}{dx}=\operatorname{sech}^2 x.$	$\int\operatorname{sech}^2 x\,dx=\tanh x.$
$u=\coth x=\dfrac{\cosh x}{\sinh x}.$	$\dfrac{du}{dx}=-\operatorname{cosech}^2 x.$	$\int\operatorname{cosech}^2 x\,dx=-\coth x.$
$u=\operatorname{sech} x=\dfrac{1}{\cosh x}.$	$\dfrac{du}{dx}=-\dfrac{\sinh x}{\cosh^2 x}.$	$\int\dfrac{\sinh x}{\cosh^2 x}dx=-\operatorname{sech} x.$
$u=\operatorname{cosech} x=\dfrac{1}{\sinh x}.$	$\dfrac{du}{dx}=-\dfrac{\cosh x}{\sinh^2 x}.$	$\int\dfrac{\cosh x}{\sinh^2 x}dx=-\operatorname{cosech} x.$
$u=\sinh^{-1}x=\log(x+\sqrt{1+x^2}).$	$\dfrac{du}{dx}=\dfrac{1}{\sqrt{1+x^2}}.$	$\int\dfrac{dx}{\sqrt{1+x^2}}=\sinh^{-1}x.$
$u=\cosh^{-1}x=\log(x+\sqrt{x^2-1}).$	$\dfrac{du}{dx}=\dfrac{1}{\sqrt{x^2-1}}.$	$\int\dfrac{dx}{\sqrt{x^2-1}}=\cosh^{-1}x.$
$u=\tanh^{-1}x=\frac{1}{2}\log\dfrac{1+x}{1-x}.$	$\dfrac{du}{dx}=\dfrac{1}{1-x^2}(x<1).$	$\int\dfrac{dx}{1-x^2}=\tanh^{-1}x_{(x<1)}$
$u=\coth^{-1}x=\frac{1}{2}\log\dfrac{x+1}{x-1}.$	$\dfrac{du}{dx}=-\dfrac{1}{x^2-1}(x>1).$	$\int\dfrac{dx}{x^2-1}=-\coth^{-1}x_{(x>1)}$
$u=\operatorname{sech}^{-1}x=\cosh^{-1}\dfrac{1}{x}.$	$\dfrac{du}{dx}=-\dfrac{1}{x\sqrt{1-x^2}}.$	$\int\dfrac{dx}{x\sqrt{1-x^2}}=-\operatorname{sech}^{-1}x.$
$u=\operatorname{cosech}^{-1}x=\sinh^{-1}\dfrac{1}{x}.$	$\dfrac{du}{dx}=-\dfrac{1}{x\sqrt{x^2+1}}.$	$\int\dfrac{dx}{x\sqrt{x^2+1}}=-\operatorname{cosech}^{-1}x.$

86. Transformations.

Algebraic or trigonometrical transformations are frequently useful to *shorten the work* of differentiation. For instance, suppose

$$y=\tan^{-1}\frac{2x}{1-x^2}.$$

We observe that $\quad y=2\tan^{-1}x;$

whence $\quad \dfrac{dy}{dx}=\dfrac{2}{1+x^2}.$

Again, suppose $$y=\tan^{-1}\frac{1+x}{1-x}.$$

Here $$y=\tan^{-1}x+\tan^{-1}1,$$

and therefore $$\frac{dy}{dx}=\frac{1}{1+x^2}.$$

87. Examples of Differentiation.

Ex. 1. Let $y=\sqrt{z}$, where z is a known function of x.

Here $$y=z^{\frac{1}{2}},$$

and $$\frac{dy}{dz}=\tfrac{1}{2}z^{-\frac{1}{2}}=\frac{1}{2\sqrt{z}},$$

whence $$\frac{dy}{dx}=\frac{dy}{dz}\cdot\frac{dz}{dx}, \quad \text{(Art. 55.)}$$

$$=\frac{1}{2\sqrt{z}}\cdot\frac{dz}{dx}.$$

This form *occurs so often that it will be found convenient to commit it to memory.*

Ex. 2. Let $$y=e^{\sqrt{\cot x}}.$$

Let $$\sqrt{\cot x}=z \text{ and } \cot x=p,$$

so that $$y=e^z, \text{ where } z=\sqrt{p}.$$

Now $$\frac{dy}{dz}=e^z. \quad \text{(Art. 66.)}$$

$$\frac{dz}{dp}=\frac{1}{2\sqrt{p}}. \quad \text{(Ex. 1 above.)}$$

$$\frac{dp}{dx}=-\operatorname{cosec}^2 x,$$

and (Art. 56) $$\frac{dy}{dx}=\frac{dy}{dz}\cdot\frac{dz}{dp}\cdot\frac{dp}{dx}=-\operatorname{cosec}^2 x\cdot\frac{1}{2\sqrt{\cot x}}\cdot e^{\sqrt{\cot x}}.$$

With a little practice these actual substitutions can be avoided and the following is what passes in the mind :—

$$\frac{d(e^{\sqrt{\cot x}})}{dx}=\frac{d(e^{\sqrt{\cot x}})}{d(\sqrt{\cot x})}\cdot\frac{d(\sqrt{\cot x})}{d(\cot x)}\cdot\frac{d(\cot x)}{dx}$$

$$=e^{\sqrt{\cot x}}\cdot\frac{1}{2\sqrt{\cot x}}\cdot(-\operatorname{cosec}^2 x).$$

Ex. 3. Let $y=(\sin x)^{\log x}\cot\{e^x(a+bx)\}$.

Taking logarithms

$$\log y=\log x\,.\,\log\sin x+\log\cot\{e^x(a+bx)\}.$$

The differential coefficient of $\log y$ is $\frac{1}{y}\frac{dy}{dx}$.

Again, $\log x\,.\log\sin x$ is a product, and when differentiated becomes (Art. 50) $\frac{1}{x}\log\sin x+\log x\,.\,\frac{1}{\sin x}\,.\,\cos x$.

Also, $\log\cot\{e^x(a+bx)\}$ becomes when differentiated

$$\frac{1}{\cot\{e^x(a+bx)\}}\,.\,[-\operatorname{cosec}^2\{e^x(a+bx)\}]\,.\,\{e^x(a+bx)+be^x\};$$

$$\therefore\quad \frac{dy}{dx}=(\sin x)^{\log x}\,.\,\cot\{e^x(a+bx)\}\Big[\frac{1}{x}\log\sin x+\cot x\,.\,\log x$$
$$-2e^x(a+b+bx)\operatorname{cosec}2(e^x\overline{a+bx})\Big].$$

When logarithms are taken before differentiating, the compound process is called *Logarithmic Differentiation.* It is useful to adopt this method when variables occur in the index, or when the function to be differentiated consists of a product of several involved factors.

Ex. 4. Let $y=\sqrt{a^2-b^2\cos^2(\log x)}$.

$$\frac{dy}{dx}=\frac{d\sqrt{a^2-b^2\cos^2(\log x)}}{d\{a^2-b^2\cos^2(\log x)\}}\times\frac{d\{a^2-b^2\cos^2(\log x)\}}{d\{\cos(\log x)\}}$$
$$\times\frac{d\{\cos(\log x)\}}{d(\log x)}\times\frac{d(\log x)}{dx}$$
$$=\tfrac{1}{2}\{a^2-b^2\cos^2(\log x)\}^{-\frac{1}{2}}\times\{-2b^2\cos(\log x)\}$$
$$\times\{-\sin(\log x)\}\times\frac{1}{x}$$
$$=\frac{b^2\sin 2(\log x)}{2x\sqrt{a^2-b^2\cos^2(\log x)}}.$$

Ex. 5. Differentiate x^5 with regard to x^2.

Let $x^2=z$.

Then
$$\frac{dx^5}{dz}=\frac{dx^5}{dx}\cdot\frac{dx}{dz}=\frac{\frac{dx^5}{dx}}{\frac{dz}{dx}}=\frac{5x^4}{2x}$$
$$=\frac{5}{2}x^3.$$

Ex. 6. Given that $x^3+y^3=3axy$, find the value of $\frac{dy}{dx}$.

Here $$3x^2+3y^2\frac{dy}{dx}=3a\left(y+x\frac{dy}{dx}\right),$$

giving $$\frac{dy}{dx}=-\frac{x^2-ay}{y^2-ax}.$$

EXAMPLES.

Find $\frac{dy}{dx}$ in the following cases :

1. $y=\sqrt{x}$.
2. $y=\frac{1}{\sqrt{x}}$.
3. $y=\frac{a+bx}{c}$.
4. $y=x+\frac{1}{x}$.
5. $y=1+x+\frac{x^2}{2!}+\frac{x^3}{3!}+\ldots$
6. $y=x-\frac{x^3}{3!}+\frac{x^5}{5!}-\ldots$
7. $y=1-\frac{x^2}{2!}+\frac{x^4}{4!}-\ldots$
8. $y=\sin(a+bx)$.
9. $y=\sin(a+bx^n)$.
10. $y=\sin\sqrt{x}$.
11. $y=\sqrt{\sin x}$.
12. $y=\sqrt{\sin\sqrt{x}}$.
13. $y=\sin^p x^q$.
14. $y=\sin^{-1}x^2$.
15. $y=(\sin^{-1}x)^2-(\cos^{-1}x)^2$.
16. $y=\tan^{-1}(\log x)$.
17. $y=\sin x^\circ$.
18. $y=x\log x$.
19. $y=e^x\log x$
20. $y=\sin(e^x)\log x$.
21. $y=\tan^{-1}(e^x)\log\cot x$.
22. $y=(x+a)^m(x+b)^n$.
23. $y=\frac{2+x^2}{1+x}$.
24. $y=\sqrt[n]{a+x}$.
25. $y=\sqrt[n]{a^2+x^2}$.
26. $y=\sqrt{\cosh x}$.
27. $y=\log\cosh x$.
28. $y=\tan^{-1}(\tanh x)$.
29. $y=\text{vers}^{-1}x^2$.
30. $y=\text{vers}^{-1}\log(\cot x)$.
31. $y=\cot^{-1}(\text{cosec}\, x)$.
32. $y=\sin^{-1}\frac{1}{\sqrt{1+x^2}}$.
33. $y=\tan^{-1}\frac{1}{\sqrt{x^2-1}}$.
34. $y=\tan^{-1}\frac{\sqrt{x}-x}{1+x^{\frac{3}{2}}}$.
35. $y=\sin^m x\cos^n x$.

36. $y=(\sin^{-1}x)^m(\cos^{-1}x)^n$.

37. $y=\sin(e^x\log x).\sqrt{1-(\log x)^2}$.

38. $y=\sqrt{\dfrac{1-x}{1+x}}$.

39. $y=\dfrac{1-x^2}{\sqrt{1+x^2}}$.

40. $y=\dfrac{x\sqrt{x^2-4a^2}}{\sqrt{x^2-a^2}}$.

41. $y=\sqrt{\dfrac{1-x}{1+x+x^2}}$.

42. $y=\log\dfrac{x^2+x+1}{x^2-x+1}$.

43. $y=\log\dfrac{x}{a^x}$.

44. $y=\cos^{-1}(1-2x^2)$.

45. $y=\left(\dfrac{x}{n}\right)^{nx}\left(1+\log\dfrac{x}{n}\right)$.

46. $y=b\tan^{-1}\left(\dfrac{x}{a}\tan^{-1}\dfrac{x}{a}\right)$.

47. $y=\dfrac{x\cos^{-1}x}{\sqrt{1-x^2}}$.

48. $y=\cos\left(a\sin^{-1}\dfrac{1}{x}\right)$.

49. $y=\sin^{-1}\dfrac{a+b\cos x}{b+a\cos x}$.

50. $y=e^{\tan^{-1}x}\log(\sec^2 x^3)$.

51. $y=e^{ax}\cos(b\tan^{-1}x)$.

52. $y=\tan^{-1}(a^{cx}.x^2)$.

53. $y=\sec(\log_a\sqrt{a^2+x^2})$.

54. $y=\tan^{-1}x+\tanh^{-1}x$.

55. $y=\tanh^{-1}\dfrac{3x+x^3}{1+3x^2}+\tan^{-1}\dfrac{3x-x^3}{1-3x^2}$.

56. $y=\log(\log x)$.

57. $y=\log^n(x)$, where $\log^n$ means $\log\log\log\ldots$ (repeated n times).

58. $y=\dfrac{1}{\sqrt{b^2-a^2}}\log\dfrac{\sqrt{b+a}+\sqrt{b-a}\tan\dfrac{x}{2}}{\sqrt{b+a}-\sqrt{b-a}\tan\dfrac{x}{2}}$.

59. $y=\sin^{-1}(x\sqrt{1-x}-\sqrt{x}\sqrt{1-x^2})$.

60. $y=\tan^{-1}\dfrac{4\sqrt{x}}{1-4x}$.

61. $y=\log\left\{e^x\left(\dfrac{x-2}{x+2}\right)^{\frac{3}{4}}\right\}$.

62. $y=x^x+x^{\frac{1}{x}}$.

63. $y=x^{x^x}$.

64. $y=x^{e^x}$.

65. $y=e^{x^x}$.

66. $y=e^{e^x}$.

67. $y=10^{10^x}$.

68. $y=(\sin x)^{\cos x}+(\cos x)^{\sin x}$.

69. $y=(\cot x)^{\cot x}+(\coth x)^{\coth x}$.

70. $y=\tan^{-1}(a^{cx}x^{\sin x})\dfrac{\sqrt{x}}{1+x^{\frac{3}{2}}}$.

71. $y=\sin^{-1}\left(e^{\tan^{-1}x}\right)$.

72. $y=\sqrt{\left(1+\cos\frac{m}{x}\right)\left(1-\sin\frac{m}{x}\right)}$.

73. $y=\tan^{-1}\sqrt{\sqrt{x}+\cos^{-1}x}$.

74. $y=\left(\frac{1+\sqrt{x}}{1+2\sqrt{x}}\right)^{\sin e^{x^2}}$.

75. $y=(\cos x)^{\cot^2 x}$.

76. $y=(\cot^{-1}x)^{\frac{1}{x}}$.

77. $y=\left(1+\frac{1}{x}\right)^x+x^{1+\frac{1}{x}}$.

78. $y=b\tan^{-1}\left(\frac{x}{a}+\tan^{-1}\frac{y}{x}\right)$.

79. $\tan y=e^{\cos^2 x}\sin x$.

80. $ax^2+2hxy+by^2=1$.

81. $e^y=\frac{(a+bx^n)^{\frac{1}{2}}-a^{\frac{1}{2}}}{(bx^n)^{\frac{1}{2}}}$

82. $(\cos x)^y=(\sin y)^x$.

83. $x=e^{\tan^{-1}\frac{y-x^2}{x^2}}$.

84. $x=y\log xy$.

85. $y=x^y$.

86. $y=x^{y^x}$.

87. $y=x\log\frac{y}{a+bx}$.

88. $ax^2+2hxy+by^2+2gx+2fy+c=0$.

89. $x^m y^n=(x+y)^{m+n}$.

90. $y=e^{\tan^{-1}y}\log\sec^2 x^3$.

91. Differentiate $\log_{10}x$ with regard to x^2.

92. Differentiate $(x^2+ax+a^2)^n\log\cot\frac{x}{2}$ with regard to $\tan^{-1}(a\cos bx)$.

93. Differentiate $\log_e\left\{\frac{a+b\tan\frac{x}{2}}{a-b\tan\frac{x}{2}}\right\}$ with regard to

$$\frac{1}{a^2\cos^2\frac{x}{2}-b^2\sin^2\frac{x}{2}}.$$

94. Differentiate $x^{\sin^{-1}x}$ with regard to $\sin^{-1}x$.

95. Differentiate $\tan^{-1}\frac{\sqrt{1+x^2}-1}{x}$ with regard to $\tan^{-1}x$.

96. Differentiate $\dfrac{\sqrt{1+x^2}+\sqrt{1-x^2}}{\sqrt{1+x^2}-\sqrt{1-x^2}}$ with regard to $\sqrt{1-x^4}$.

97. Differentiate $\sec^{-1}\dfrac{1}{2x^2-1}$ with regard to $\sqrt{1-x^2}$.

98. Differentiate $\tan^{-1}\dfrac{x}{\sqrt{1-x^2}}$ with regard to $\sec^{-1}\dfrac{1}{2x^2-1}$.

99. Differentiate $\tan^{-1}\dfrac{2x}{1-x^2}$ with regard to $\sin^{-1}\dfrac{2x}{1+x^2}$.

100. Differentiate $x^n \log\tan^{-1}x$ with regard to $\dfrac{\sin\sqrt{x}}{x^{\frac{3}{2}}}$.

101. If $y=x^{x^{x^{\cdot\cdot\text{ to }\infty}}}$ prove $x\dfrac{dy}{dx}=\dfrac{y^2}{1-y\log x}$.

102. If $y=\dfrac{x}{1+}\,\dfrac{x}{1+}\,\dfrac{x}{1+}\ldots$ to ∞, prove $\dfrac{dy}{dx}=\dfrac{1}{1+}\,\dfrac{2x}{1+}\,\dfrac{x}{1+}\,\dfrac{x}{1+}\,\dfrac{x}{1+}\ldots$

103. If $y=x+\dfrac{1}{x+}\,\dfrac{1}{x+}\,\dfrac{1}{x+}\ldots$ to ∞, prove $\dfrac{dy}{dx}=\dfrac{1}{2-}\,\dfrac{x}{x+}\,\dfrac{1}{x+}\,\dfrac{1}{x+}\ldots$

104. If $y=\dfrac{\sin x}{1+}\,\dfrac{\cos x}{1+}\,\dfrac{\sin x}{1+}\,\dfrac{\cos x}{1+}\ldots$ to ∞.

prove $$\frac{dy}{dx}=\frac{(1+y)\cos x+y\sin x}{1+2y+\cos x-\sin x}.$$

105. If $y=\sqrt{\sin x+\sqrt{\sin x+\sqrt{\sin x+\sqrt{\text{etc. to }\infty}}}}$,

prove $$\frac{dy}{dx}=\frac{\cos x}{2y-1}.$$

106. If S_n = the sum of a G. P. to n terms of which r is the common ratio, prove that

$$(r-1)\frac{dS_n}{dr}=(n-1)S_n-nS_{n-1}.$$

107. If $\dfrac{P}{Q}=a+\cfrac{1}{a_1+\cfrac{1}{a_2+\cfrac{1}{a_3+\ldots+\cfrac{1}{x},}}}$ prove $\dfrac{d}{dx}\left(\dfrac{P}{Q}\right)=\pm\dfrac{1}{Q^2}$.

108. Given $C=1+r\cos\theta+\dfrac{r^2\cos 2\theta}{2\,!}+\dfrac{r^3\cos 3\theta}{3\,!}+\ldots$

and $$S=r\sin\theta+\frac{r^2\sin 2\theta}{2\,!}+\frac{r^3\sin 3\theta}{3\,!}+\ldots$$

show that

$$C\frac{dC}{dr}+S\frac{dS}{dr}=(C^2+S^2)\cos\theta;$$

$$C\frac{dS}{dr}-S\frac{dC}{dr}=(C^2+S^2)\sin\theta.$$

109. If $y=\sec 4x$, prove that

$$\frac{dy}{dt}=\frac{16t(1-t^4)}{(1-6t^2+t^4)^2}, \text{ where } t=\tan x.$$

110. If $y=e^{-xz}\sec^{-1}(x\sqrt{z})$ and $z^4+x^2z=x^5$, find $\dfrac{dy}{dx}$ in terms of x and z.

111. Prove that if x be less than unity

$$\frac{1}{1+x}+\frac{2x}{1+x^2}+\frac{4x^3}{1+x^4}+\frac{8x^7}{1+x^8}+\ldots \text{ ad inf.}=\frac{1}{1-x}.$$

112. Prove that if x be less than unity

$$\frac{1-2x}{1-x+x^2}+\frac{2x-4x^3}{1-x^2+x^4}+\frac{4x^3-8x^7}{1-x^4+x^8}+\ldots \text{ ad inf.}=\frac{1+2x}{1+x+x^2}.$$

113. Given Euler's Theorem that

$$Lt_{n=\infty}\cos\frac{x}{2}\cos\frac{x}{2^2}\cos\frac{x}{2^3}\ldots\cos\frac{x}{2^n}=\frac{\sin x}{x},$$

prove $\dfrac{1}{2}\tan\dfrac{x}{2}+\dfrac{1}{2^2}\tan\dfrac{x}{2^2}+\dfrac{1}{2^3}\tan\dfrac{x}{2^3}+\ldots$ ad inf. $=\dfrac{1}{x}-\cot x$,

and $\dfrac{1}{2^2}\sec^2\dfrac{x}{2}+\dfrac{1}{2^4}\sec^2\dfrac{x}{2^2}+\dfrac{1}{2^6}\sec^2\dfrac{x}{2^3}+\ldots$ ad inf. $=\operatorname{cosec}^2 x-\dfrac{1}{x^2}$.

114. Given the identity

$$(2\cos 2\theta - 1)(2\cos 2^2\theta - 1)\ldots(2\cos 2^n\theta - 1) = \frac{2\cos 2^{n+1}\theta + 1}{2\cos 2\theta + 1},$$

prove that

$$\sum_{r=1}^{r=n} \frac{2^r \sin 2^r\theta}{2\cos 2^r\theta - 1} = \frac{2^{n+1}\sin 2^{n+1}\theta}{2\cos 2^{n+1}\theta + 1} - \frac{2\sin 2\theta}{2\cos 2\theta + 1}.$$

115. Given

$$\sin\phi \sin(2\alpha + \phi) \sin(4\alpha + \phi)\ldots\sin\{2(n-1)\alpha + \phi\} = \frac{\sin n\phi}{2^{n-1}}$$

where $$2n\alpha = \pi,$$

prove that

$$\cot\phi + \cot(2\alpha + \phi) + \cot(4\alpha + \phi) + \ldots + \cot\{2(n-1)\alpha + \phi\} = n\cot n\phi,$$

and that

$$\operatorname{cosec}^2\phi + \operatorname{cosec}^2(2\alpha + \phi) + \operatorname{cosec}^2(4\alpha + \phi) + \ldots + \operatorname{cosec}^2\{2(n-1)\alpha + \phi\} = n^2\operatorname{cosec}^2 n\phi.$$

116. Given

$$\sin\theta = \theta\left(1 - \frac{\theta^2}{\pi^2}\right)\left(1 - \frac{\theta^2}{2^2\pi^2}\right)\left(1 - \frac{\theta^2}{3^2\pi^2}\right)\ldots,$$

prove $$\theta\cot\theta = 1 + 2\theta^2\sum_{n=1}^{n=\infty}\frac{1}{\theta^2 - n^2\pi^2},$$

and hence that

$$\pi\coth\pi = 1 + \frac{2}{1+1^2} + \frac{2}{1+2^2} + \frac{2}{1+3^2} + \ldots \text{ ad inf.},$$

and that $$\frac{\pi}{2}\coth\frac{\pi}{2} = 1 + \frac{2}{1+2^2} + \frac{2}{1+4^2} + \frac{2}{1+6^2} + \ldots \text{ ad inf.}$$

117. Given

$$\cos\theta = \left(1 - \frac{4\theta^2}{\pi^2}\right)\left(1 - \frac{4\theta^2}{3^2\pi^2}\right)\left(1 - \frac{4\theta^2}{5^2\pi^2}\right)\ldots,$$

prove $$\frac{\tan\theta}{8\theta} = \sum_{n=1}^{n=\infty}\frac{1}{(2n-1)^2\pi^2 - 4\theta^2},$$

and deduce

$$\frac{\pi}{8}\tanh\pi = \frac{1}{2^2+1^2} + \frac{1}{2^2+3^2} + \frac{1}{2^2+5^2} + \ldots \text{ ad inf.},$$

and $$\frac{\pi}{4}\tanh\frac{\pi}{2} = \frac{1}{1+1^2} + \frac{1}{1+3^2} + \frac{1}{1+5^2} + \ldots \text{ ad inf.}$$

118. Prove $$\frac{x}{2}\coth x = \tfrac{1}{2} + x^2\sum_{n=1}^{n=\infty}\frac{1}{x^2+n^2\pi^2}.$$

119. Prove that

$$\frac{nx^{n-1}}{x^n-a^n} = \frac{1}{x-a} + \frac{1}{x+a} + 2\sum_{r=1}^{r=\frac{n-2}{2}}\frac{x-a\cos\frac{2r\pi}{n}}{x^2-2ax\cos\frac{2r\pi}{n}+a^2},$$

if n be even,

but $$= \frac{1}{x-a} + 2\sum_{r=1}^{r=\frac{n-1}{2}}\frac{x-a\cos\frac{2r\pi}{n}}{x^2-2ax\cos\frac{2r\pi}{n}+a^2},$$

if n be odd.

120. Prove that

$$\frac{nx^{n-1}(x^n-a^n\cos\theta)}{x^{2n}-2x^na^n\cos\theta+a^{2n}} = \sum_{r=0}^{r=n-1}\frac{x-a\cos\frac{2r\pi+\theta}{n}}{x^2-2ax\cos\frac{2r\pi+\theta}{n}+a^2}.$$

CHAPTER IV.

SUCCESSIVE DIFFERENTIATION.

88. **Repeated Operations.**

The operation denoted by $\frac{d}{dx}$ is defined in Art. 39 without any reference to the form of the function operated upon, the only assumption made being that the function is a function of the same independent variable as that referred to in the operative symbol, viz. x. It is moreover clear that the result of the operation is also a function of x, and as such is itself capable of being operated upon by the same symbol. That is to say, if y be a function of x, $\frac{dy}{dx}$ is also a function of x, and therefore we can have $\frac{d}{dx}\left(\frac{dy}{dx}\right)$ as a true mathematical quantity. And further, it will be thus seen that the operation $\frac{d}{dx}$ may be performed upon any given function of x any number of times.

89. **Notation.**

The expression $\frac{d}{dx}\left(\frac{dy}{dx}\right)$ is generally abbreviated into

$\left(\frac{d}{dx}\right)^2 y$ or $\frac{d^2y}{dx^2}$, and is called the "*second derived function*" or "*second differential coefficient*" of y with respect to x. And, generally, if the operator $\frac{d}{dx}$ be applied n times, the result is denoted by $\left(\frac{d}{dx}\right)^n y$ or $\frac{d^ny}{dx^n}$, and is called the nth *derived function* or n^{th} *differential coefficient* of y with respect to x.

It will be convenient to denote the operative symbol $\frac{d}{dx}$ by D, which, in addition to being simpler to write, makes no assumption that the independent variable is *denoted by* x; and in many problems the independent variable is more conveniently denoted by some other letter. For example, in dynamical problems the time which has elapsed since a given epoch is frequently taken as the independent variable and is denoted by t, while the letters x, y, z, are reserved to denote the co-ordinates at that time of the point whose motion is considered.

It appears then that if we use indices to denote the number of times an operation has been performed, we may write

$$Dy = \frac{dy}{dx},$$

$$D \,.\, Dy = D^2y = \frac{d^2y}{dx^2},$$

$$D \,.\, D^2y = D^3y = \frac{d^3y}{dx^3},$$

$$\cdots\cdots\cdots$$

$$D \,.\, D^{n-1}y = D^ny = \frac{d^ny}{dx^n}.$$

90. Analogy between the operator $\frac{d}{dx}$ and symbols of quantity.

The index notation employed above to denote the number of times an operation is repeated is exactly analogous to the index notation used in algebra to denote powers of symbols of quantity.

If a be an algebraic quantity, the algebraical notation for $a.a$ is a^2, and for $a.a.a$ is a^3, and so on; the index here denoting the number of factors each equal to a which are multiplied together. But, as defined above, there is no idea of multiplication in $D.D$ or D^2, but a simple *repetition of an operation.* In the same way D^n has no *quantitative* meaning in itself, but represents an *operation* consisting of employing the process of differentiation n times. For example, the difference between such quantities as D^2y, $(Dy)^2$, and D^2y^2 should be carefully noted. The index in the first case has reference only to the *symbol of operation* "D," which is therefore to be applied twice to y.

In $(Dy)^2$ the index is a purely *quantitative one* used in the algebraical sense to denote the product $Dy \times Dy$.

While in D^2y^2 we are to understand that the *square of y is to be differentiated twice.*

That the ultimate results are different may be easily seen by taking any simple case,

e.g., if $\qquad y=x^2,$

then $\qquad Dy=2x,$

and $\qquad D^2y=2. \qquad\qquad (1)$

Again, $\qquad (Dy)^2=4x^2. \qquad\qquad (2)$

whilst $\qquad y^2=x^4;$

and $\qquad Dy^2=4x^3,$

giving $$D^2y^2 = 12x^2. \quad \text{......................} \quad (3)$$

A comparison of the results (1), (2), (3), will at once satisfy the student of the truth of the above remarks.

91. The operator D satisfies the elementary rules of Algebra.

We will next consider how far the analogy goes between symbols of quantity and the symbol of operation which we have denoted by D.

The fundamental rules of algebra are three in number and are known as

(1) The "*Distributive Law,*"
(2) The "*Commutative Law,*" and
(3) The "*Index Law.*"

These three laws form the basis of all subsequent algebraical formulae and investigations.

(1) The *Distributive Law* is that denoted by

$$m(a+b+c+\ldots) = ma+mb+mc+\ldots$$

Now, in Chap. II., Prop. III., it is proved that

$$D(u+v+w+\ldots) = Du+Dv+Dw+\ldots,$$

so that the symbol D is *distributive* in its operation.

(2) The *Commutative Law* in algebra is that expressed by

$$ab = ba.$$

Now, in Chap. II., Prop. II., it is proved that

$$Dcy = cDy,$$

so that the symbol D is *commutative with regard to constants.*

But it is clear that the positions of the D and the y cannot be interchanged; such an error would be similar to writing $\theta \sin$ instead of $\sin \theta$. So that, while D is commutative with regard to constants, *it is not so with regard to variables.*

(3) The *Index Law* in algebra is denoted by

$$a^m \cdot a^n = a^{m+n},$$

m and n being supposed to be positive integers.

Now, to differentiate a result m times which has already been operated upon n times is clearly the same as differentiating $m+n$ times,

i.e., $$D^m \cdot D^n y = D^{m+n} y.$$

So the operator $D^m \cdot D^n$ is equivalent to the operator D^{m+n} where m and n are positive integers.

Hence the symbol D *obeys the Index Law* for a positive integral exponent.

To sum up then, the operative symbol D *satisfies all the elementary rules of combination of algebraical quantities, with the exception that it is not commutative with regard to variables.*

92. It follows from the above remarks that any rational algebraical identity has a corresponding symbolical operative analogue.

For example,

$$(m+a)(m+b) = m^2 + (a+b)m + ab,$$

so also the operation $(D+a)(D+b)$ is exactly equivalent to the operation $D^2 + (a+b)D + ab$.

Similarly, to the identity

$$(m+a)^2 = m^2 + 2am + a^2$$

corresponds the equivalence of the operations $(D+a)^2$ and $D^2 + 2aD + a^2$.

93. It is clear that in cases like the above an *a priori* proof may be given of the identity of the operations represented. For instance, suppose it be required to show that $(D+a)(D+b)y = [D^2 + (a+b)D + ab]y$,

we have $$(D+b)y=Dy+by,$$
and
$$\begin{aligned}(D+a)(D+b)y&=(D+a)(Dy+by)\\&=D(Dy+by)+a(Dy+by)\\&=D^2y+bDy+aDy+aby\\&=D^2y+(a+b)Dy+aby\\&=[D^2+(a+b)D+ab]y,\end{aligned}$$
the result to be proved: and the process of proof is exactly the same as that employed in proving that
$$(m+a)(m+b)=m^2+(a+b)m+ab.$$

However, such proofs are unnecessary after the remarks of Art. 91, for they simply repeat *in form* the proof of the corresponding algebraical theorem.

It will now be obvious, for instance, without further proof, that since
$$(m+a)^n=m^n+nam^{n-1}+\frac{n(n-1)}{1.2}a^2m^{n-2}+\ldots+a^n,$$
we shall also have
$$\begin{aligned}(D+a)^ny&=\left(D^n+naD^{n-1}+\frac{n(n-1)}{1.2}a^2D^{n-2}+\ldots+a^n\right)y\\&=D^ny+naD^{n-1}y+\frac{n(n-1)}{1.2}a^2D^{n-2}y+\ldots+a^ny.\end{aligned}$$

94. **Notation.**

The first derived function of y with respect to the independent variable is often denoted by y_1, y', or $\dot{y}$. This notation can be conveniently extended, and we shall often find it convenient to denote

	Dy,	D^2y,	D^3y,	$\ldots$ D^ny
by	y_1,	y_2,	y_3,	$\ldots$ y_n,
or by	$y^{(1)}$,	$y^{(2)}$,	$y^{(3)}$	$\ldots$ $y^{(n)}$,
or by	y',	y'',	y''',	etc.,
or by	$\dot{y}$,	$\ddot{y}$,	$\dddot{y}$,	etc.

It is clear however that the notation of dashes or dots as used in the last two systems is inconvenient for higher differential coefficients than the fourth or fifth by reason of the number of dashes or dots which it would be necessary to use. The bracketed index notation is a somewhat dangerous one, from the liability of confusion with an algebraical index. The suffix notation appears to be free from objection in cases when there can be no misunderstanding as to which is the independent variable.

95. **Standard Results and Processes.**

The nth differential coefficients of some functions are easy to find.

Ex. 1. If $y=e^{ax}$; $y_1=ae^{ax}$; $y_2=a^2e^{ax}$; ... $y_n=a^ne^{ax}$.

Cor. (i.) If $a=1$

$$y=e^x,\ y_1=e^x,\ \dots\ y_n=e^x.$$

Cor. (ii.) $y=a^x=e^{x\log_e a}$;

$$y_1=(\log_e a)e^{x\log_e a}=(\log_e a)a^x;$$
$$y_2=(\log_e a)^2e^{x\log_e a}=(\log_e a)^2a^x;$$
$$\text{etc.}=\text{etc.},$$
$$y_n=(\log_e a)^ne^{x\log_e a}=(\log_e a)^na^x.$$

Ex. 2. If $y=\log_e(x+a)$;

$$y_1=\frac{1}{x+a};\ y_2=-\frac{1}{(x+a)^2};\ y_3=\frac{(-1)(-2)}{(x+a)^3};$$
$$y_n=\frac{(-1)(-2)(-3)\dots(-n+1)}{(x+a)^n}$$
$$=\frac{(-1)^{n-1}(n-1)!}{(x+a)^n}.$$

Cor. If $y=\dfrac{1}{x+a}$, $y_n=\dfrac{(-1)^nn!}{(x+a)^{n+1}}$.

Ex. 3. If $y = \sin(ax+b)$;

$$y_1 = a\cos(ax+b) = a\sin\left(ax+b+\frac{\pi}{2}\right);$$

$$y_2 = a^2\sin\left(ax+b+\frac{2\pi}{2}\right);$$

$$y_3 = a^3\sin\left(ax+b+\frac{3\pi}{2}\right);$$

. . . .

$$y_n = a^n\sin\left(ax+b+\frac{n\pi}{2}\right).$$

Similarly, if $y = \cos(ax+b)$,

$$y_n = a^n\cos\left(ax+b+\frac{n\pi}{2}\right).$$

Cor. If $a = 1$ and $b = 0$;

then, when $y = \sin x$, $y_n = \sin\left(x+\frac{n\pi}{2}\right)$;

and, when $y = \cos x$, $y_n = \cos\left(x+\frac{n\pi}{2}\right)$.

Ex. 4. If $y = e^{ax}\sin(bx+c)$;

$$y_1 = ae^{ax}\sin(bx+c) + be^{ax}\cos(bx+c).$$

Let $a = r\cos\phi$ and $b = r\sin\phi$,

so that $r^2 = a^2+b^2$ and $\tan\phi = \frac{b}{a}$;

and therefore $y_1 = re^{ax}\sin(bx+c+\phi)$.

Similarly $y_2 = r^2e^{ax}\sin(bx+c+2\phi)$,

. . . .

and finally $y_n = r^ne^{ax}\sin(bx+c+n\phi)$

$$= (a^2+b^2)^{\frac{n}{2}}e^{ax}\sin\left(bx+c+n\tan^{-1}\frac{b}{a}\right).$$

Similarly, if $y = e^{ax}\cos(bx+c)$,

$$y_n = (a^2+b^2)^{\frac{n}{2}}e^{ax}\cos\left(bx+c+n\tan^{-1}\frac{b}{a}\right).$$

As the above results are frequently wanted, it will be well for the student to be able to obtain them immediately.

96. Fractional expressions of the form $\dfrac{f(x)}{\phi(x)}$ (both functions being algebraic and rational) can be differentiated n times by first putting them into *partial fractions.* (See p. 85.)

Ex. 1.

$$y=\frac{x^2}{(x-a)(x-b)(x-c)}=\frac{a^2}{(a-b)(a-c)}\,\frac{1}{x-a}$$

$$+\frac{b^2}{(b-c)(b-a)}\,\frac{1}{x-b}+\frac{c^2}{(c-a)(c-b)}\,\frac{1}{x-c},$$

(see note on partial fractions);

$$\therefore\quad y_n=\frac{a^2}{(a-b)(a-c)}\,\frac{(-1)^n n!}{(x-a)^{n+1}}+\frac{b^2}{(b-c)(b-a)}\,\frac{(-1)^n n!}{(x-b)^{n+1}}$$

$$+\frac{c^2}{(c-a)(c-b)}\,\frac{(-1)^n n!}{(x-c)^{n+1}}.$$

Ex. 2. $$y=\frac{x^2}{(x-1)^2(x+2)}.$$

To put this into Partial Fractions let $x=1+z$;

then $$y=\frac{1}{z^2}\cdot\frac{1+2z+z^2}{3+z}$$

$$=\frac{1}{z^2}\left(\frac{1}{3}+\frac{5z}{9}+\frac{4}{9}\,\frac{z^2}{3+z}\right)\text{ by division}$$

$$=\frac{1}{3z^2}+\frac{5}{9z}+\frac{4}{9}\,\frac{1}{3+z}$$

$$=\frac{1}{3(x-1)^2}+\frac{5}{9(x-1)}+\frac{4}{9(x+2)}$$

whence $$y_n = \frac{(n+1)!(-1)^n}{3(x-1)^{n+2}} + \frac{5n!(-1)^n}{9(x-1)^{n+1}} + \frac{4n!(-1)^n}{9(x+2)^{n+1}}.$$

97. When quadratic factors (which are not resolvable into real linear factors) occur in the denominator, it is often convenient to make use of Demoivre's Theorem.

Ex. Let
$$y = \frac{1}{(x+a)^2+b^2} = \frac{1}{\{(x+a)+\iota b\}\{(x+a)-\iota b\}}$$
$$= \frac{1}{2\iota b}\left\{\frac{1}{x+a-\iota b} - \frac{1}{x+a+\iota b}\right\};$$

then $$y_n = \frac{1}{2\iota b}(-1)^n n!\left\{\frac{1}{(x+a-\iota b)^{n+1}} - \frac{1}{(x+a+\iota b)^{n+1}}\right\}$$
$$= \frac{(-1)^n n!}{2\iota b}\left\{\frac{(x+a+\iota b)^{n+1}-(x+a-\iota b)^{n+1}}{[(x+a)^2+b^2]^{n+1}}\right\}.$$

Let $x+a = r\cos\theta$,
and $b = r\sin\theta$;
whence $r^2 = (x+a)^2+b^2$,
and $\tan\theta = \dfrac{b}{x+a}$.

Hence $$y_n = \frac{(-1)^n n!}{2\iota b}\,\frac{r^{n+1}}{[(x+a)^2+b^2]^{n+1}}$$
$$\times\{(\cos\theta+\iota\sin\theta)^{n+1}-(\cos\theta-\iota\sin\theta)^{n+1}\}$$
$$= \frac{(-1)^n n!}{2\iota b}\,\frac{2\iota\sin(n+1)\theta}{[(x+a)^2+b^2]^{\frac{n+1}{2}}}$$
$$= \frac{(-1)^n n!}{b^{n+2}}\sin(n+1)\theta\,\sin^{n+1}\theta,$$

where $\theta = \tan^{-1}\dfrac{b}{x+a}$.

Cor. If $y = \tan^{-1}\dfrac{x+a}{b}$, $y_1 = \dfrac{b}{(x+a)^2+b^2}$, and therefore $y_n = \dfrac{(-1)^{n-1}(n-1)!}{b^{n+1}} \sin n\theta \sin^n\theta,$

$$\text{where } \tan\theta = \frac{b}{x+a},$$

$$\textit{i.e.,} \qquad \theta = \frac{\pi}{2} - y.$$

EXAMPLES.

Find the nth differential coefficients of y with respect to x in the following cases.

1. $y = \dfrac{x}{(x-a)(x-b)}.$
2. $y = \dfrac{1}{(3x-2)(x-3)}.$
3. $y = \sin^3 x.$
4. $y = e^{2x}\cos^2 x.$
5. $y = e^{ax}\sin^3 bx.$
6. $y = \dfrac{x^2}{(x-1)^3(x-2)}.$
7. $y = \dfrac{1}{x^2-a^2}.$
8. $y = \dfrac{1}{x^2+a^2}.$
9. $y = \tan^{-1}\dfrac{x}{a}.$
10. $y = \dfrac{1}{(x^2+a^2)(x^2+b^2)}.$
11. $y = \dfrac{1}{x^4-a^4}.$
12. $y = \dfrac{x}{x^4+a^2x^2+a^4}.$

98. LEIBNITZ'S THEOREM.

To find the nth differential coefficient of a product of two functions of x in terms of the differential coefficients of the separate functions.

It was proved in Chap. II., Prop. IV., that

$$\frac{d}{dx}(uv) = v\frac{du}{dx} + u\frac{dv}{dx}.$$

It appears from this formula that the operative symbol $\dfrac{d}{dx}$ or D may be considered as the sum of two operative symbols D_1 and D_2, such that D_1 only operates on u and differential coefficients of u, while D_2 operates solely

upon v and differential coefficients of v. For with such symbols
$$D_1(uv) = v\frac{du}{dx},$$
and
$$D_2(uv) = u\frac{dv}{dx},$$
whence
$$D(uv) = v\frac{du}{dx} + u\frac{dv}{dx} = D_1(uv) + D_2(uv)$$
$$= (D_1 + D_2)uv.$$
We may therefore write for D the compound symbol
$$D_1 + D_2.$$
Now, since D_1 and D_2 are symbols which indicate differentiations, they each, like the original symbol D, obey the *distributive and index laws and are commutative with regard to constants and each other.* It therefore follows by formal analogy with the Binomial Theorem that the operations
$$(D_1 + D_2)^n$$
and
$$D_1^n + nD_1^{n-1}D_2 + \frac{n(n-1)}{1.2}D_1^{n-2}D_2^2 + \ldots + D_2^n$$
are identical.

Now
$$D_1^n(uv) = v\frac{d^n u}{dx^n},$$
$$D_1^{n-1}D_2(uv) = \frac{dv}{dx}\cdot\frac{d^{n-1}u}{dx^{n-1}},$$
etc.

Hence
$$\frac{d^n(uv)}{dx^n} = D^n(uv) = (D_1 + D_2)^n(uv)$$
$$= \left(D_1^n + nD_1^{n-1}D_2 + \frac{n(n-1)}{1.2}D_1^{n-2}D_2^2 + \ldots + D_2^n\right)uv$$
$$= v\frac{d^n u}{dx^n} + n\frac{dv}{dx}\cdot\frac{d^{n-1}u}{dx^{n-1}} + \frac{n(n-1)}{1.2}\frac{d^2v}{dx^2}\cdot\frac{d^{n-2}u}{dx^{n-2}}$$
$$+ \ldots + u\frac{d^n v}{dx^n}.$$

It appears therefore from this formula that if all the differential coefficients of u and v be known up to the nth, inclusive, the nth differential coefficient of the product may at once be written down.

99. Another Proof.

From the importance of the above result it is considered useful to add here an inductive proof of the same theorem.

[Lemma. If ${}_nC_r$ denote the number of combinations of n things r at a time, then will

$$ {}_nC_r + {}_nC_{r+1} = {}_{n+1}C_{r+1}. $$

This will form an easy exercise for the student.]

Let $y = uv$, and let suffixes denote differentiations with regard to x.

Then $y_1 = u_1v + uv_1$,

$y_2 = u_2v + 2u_1v_1 + uv_2$, by differentiation.

Assume generally that

$$ y_n = u_nv + {}_nC_1u_{n-1}v_1 + {}_nC_2u_{n-2}v_2 + \ldots + {}_nC_ru_{n-r}v_r + {}_nC_{r+1}u_{n-r-1}v_{r+1} + \ldots + uv_n. \quad \text{(α)} $$

Therefore, differentiating,

$$ y_{n+1} = u_{n+1}v + u_nv_1\left\{\begin{matrix} {}_nC_1 \\ +1 \end{matrix}\right\} + u_{n-1}v_2\left\{\begin{matrix} {}_nC_2 \\ +{}_nC_1 \end{matrix}\right\} + \ldots $$

$$ + u_{n-r}v_{r+1}\left\{\begin{matrix} {}_nC_{r+1} \\ +{}_nC_r \end{matrix}\right\} + \ldots + uv_{n+1} $$

$$ = u_{n+1}v + {}_{n+1}C_1u_nv_1 + {}_{n+1}C_2u_{n-1}v_2 + {}_{n+1}C_3u_{n-2}v_3 + \ldots $$

$$ + {}_{n+1}C_{r+1}u_{n-r}v_{r+1} + \ldots + uv_{n+1}, \text{ by the Lemma;} $$

therefore if the law (α) hold for n differentiations it holds for $n+1$.

But it was proved to hold for two differentiations, and therefore it holds for three; therefore for four; and so on; and therefore it is generally true, *i.e.*,

$$ (uv)_n = u_nv + {}_nC_1u_{n-1}v_1 + {}_nC_2u_{n-2}v_2 + \ldots + {}_nC_ru_{n-r}v_r + \ldots + uv_n. $$

100. Applications.

Ex. 1. Let $y = e^{ax}X$, where $X =$ any function of x.

Since $$ \frac{d^r}{dx^r}(e^{ax}) = a^re^{ax}, $$

$$y_n = e^{ax}\left(a^n X + {}_nC_1 a^{n-1}\frac{d}{dx}X + {}_nC_2 a^{n-2}\frac{d^2}{dx^2}X + \ldots + \frac{d^n}{dx^n}X\right),$$

which may be written by analogy with the Binomial Theorem $\qquad e^{ax}\left(a+\dfrac{d}{dx}\right)^n X$ (Art. 93).

Ex. 2. $y = x^3 \sin ax$.

$$y_n = x^3 a^n \sin\left(ax + \frac{n\pi}{2}\right) + n3x^2 a^{n-1}\sin\left(ax + \frac{n-1}{2}\pi\right)$$
$$+\frac{n(n-1)}{2!}3\,.\,2xa^{n-2}\sin\left(ax+\frac{n-2}{2}\pi\right)$$
$$+\frac{n(n-1)(n-2)}{3!}3\,.\,2\,.\,1a^{n-3}\sin\left(ax+\frac{n-3}{2}\pi\right).$$

Ex. 3. Differentiate n times the equation

$$x^2\frac{d^2y}{dx^2} + x\frac{dy}{dx} + y = 0.$$

$$\frac{d^n}{dx^n}(x^2y_2) = x^2y_{n+2} + n\,.\,2x\,.\,y_{n+1} + \frac{n(n-1)}{2!}2y_n,$$
$$\frac{d^n}{dx^n}(xy_1) = \qquad xy_{n+1} + \qquad ny_n,$$
$$\frac{d^ny}{dx^n} = \qquad y_n\,;$$

therefore by addition

$$x^2y_{n+2} + (2n+1)xy_{n+1} + (n^2+1)y_n = 0,$$

or
$$x^2\frac{d^{n+2}y}{dx^{n+2}} + (2n+1)x\frac{d^{n+1}y}{dx^{n+1}} + (n^2+1)\frac{d^ny}{dx^n} = 0.$$

101. Note on Partial Fractions.

Since a number of examples on successive differentiation and on integration depend on the ability of the student to put certain fractional forms into partial fractions, we give the methods to be pursued in a short note.

Let $\dfrac{f(x)}{\phi(x)}$ be the fraction which is to be resolved into its partial fractions.

1. If $f(x)$ be not already of lower degree than the denominator, *we can divide out until the numerator of the remaining fraction is of lower degree;*

e.g.,
$$\frac{x^2}{(x-1)(x-2)}=1+\frac{3x-2}{(x-1)(x-2)}.$$

Hence we shall consider only the case in which $f(x)$ is of lower degree than $\phi(x)$.

2. If $\phi(x)$ contain a single factor $(x-a)$, not repeated, we proceed thus: suppose
$$\phi(x)=(x-a)\psi(x),$$
and let
$$\frac{f(x)}{(x-a)\psi(x)}\equiv\frac{A}{x-a}+\frac{\chi(x)}{\psi(x)},$$
A being independent of x.

Hence
$$\frac{f(x)}{\psi(x)}\equiv A+(x-a)\frac{\chi(x)}{\psi(x)}.$$

This is an identity and therefore true for all values of the variable x; put $x=a$. Then, since $\psi(x)$ does not vanish when $x=a$ (for by hypothesis $\psi(x)$ does not contain $x-a$ as a factor), we have
$$A=\frac{f(a)}{\psi(a)}.$$

Hence the rule to find A is, "Put $x=a$ in every portion of the fraction except in the factor $x-a$ itself."

Ex. (i.) $$\frac{x-c}{(x-a)(x-b)}=\frac{a-c}{a-b}\cdot\frac{1}{x-a}+\frac{b-c}{b-a}\cdot\frac{1}{x-b}.$$

Ex. (ii.) $$\frac{x^2+px+q}{(x-a)(x-b)(x-c)}=\frac{a^2+pa+q}{(a-b)(a-c)}\,\frac{1}{x-a}+\frac{b^2+pb+q}{(b-c)(b-a)}\,\frac{1}{x-b}+\frac{c^2+pc+q}{(c-a)(c-b)}\,\frac{1}{x-c}.$$

Ex. (iii.) $$\frac{x}{(x-1)(x-2)(x-3)}=\frac{1}{2(x-1)}-\frac{2}{x-2}+\frac{3}{2(x-3)}.$$

Ex. (iv.) $$\frac{x^2}{(x-a)(x-b)}.$$

Here the numerator not being of *lower degree than the denominator*, we divide the numerator by the denominator. The result will then be expressible in the form $1+\dfrac{A}{x-a}+\dfrac{B}{x-b}$, where A and B are found as before and are respectively $\dfrac{a^2}{a-b}$ and $\dfrac{b^2}{b-a}$.

3. Suppose the factor $(x-a)$ in the denominator to be repeated r times so that $\phi(x)=(x-a)^r\psi(x)$.

Put $\qquad x-a=y.$

Then $\qquad \dfrac{f(x)}{\phi(x)}=\dfrac{f(a+y)}{y^r\psi(a+y)},$

or expanding each function by any means in ascending powers of y,

$$=\frac{A_0+A_1y+A_2y^2+\dots}{y^r(B_0+B_1y+B_2y^2+\dots)}.$$

Divide out thus:

$$B_0+B_1y+\dots)\,A_0+A_1y+\dots\,(C_0+C_1y+C_2y^2+\dots,$$
etc.,

and let the division be continued until y^r is a factor of the remainder.

Let the remainder be $y^r\chi(y)$.

$$\text{Hence the fraction}=\frac{C_0}{y^r}+\frac{C_1}{y^{r-1}}+\frac{C_2}{y^{r-2}}+\dots+\frac{C_{r-1}}{y}+\frac{\chi(y)}{\psi(a+y)}$$
$$=\frac{C_0}{(x-a)^r}+\frac{C_1}{(x-a)^{r-1}}+\frac{C_2}{(x-a)^{r-2}}+\dots$$
$$+\frac{C_{r-1}}{x-a}+\frac{\chi(x-a)}{\psi(x)}.$$

Hence the partial fractions corresponding to the factor $(x-a)^r$ are determined by a long division sum.

Ex. Take $$\frac{x^2}{(x-1)^3(x+1)}.$$

Put $\qquad x-1=y.$

$$\text{Hence the fraction}=\frac{(1+y)^2}{y^3(2+y)}.$$

$$\begin{array}{l}
2+y\,\big)\,1+2y+y^2\,\big(\tfrac12+\tfrac34y+\tfrac18y^2-\tfrac18\dfrac{y^3}{2+y}.\\
\underline{1+\tfrac12y}\\
\tfrac32y+y^2\\
\underline{\tfrac32y+\tfrac34y^2}\\
\tfrac14y^2\\
\underline{\tfrac14y^2+\tfrac18y^3}\\
-\tfrac18y^3
\end{array}$$

Therefore the fraction

$$=\frac{1}{2y^3}+\frac{3}{4y^2}+\frac{1}{8y}-\frac{1}{8(2+y)}$$

$$=\frac{1}{2(x-1)^3}+\frac{3}{4(x-1)^2}+\frac{1}{8(x-1)}-\frac{1}{8(x+1)}.$$

4. If a factor, such as x^2+ax+b, which is not resolvable into real linear factors occur in the denominator, the form of the corresponding partial fraction is $\frac{Ax+B}{x^2+ax+b}$. For instance, if the expression be

$$\frac{1}{(x-a)(x-b)^2(x^2+a^2)(x^2+b^2)^2}$$

the proper assumption for the form in partial fractions would be

$$\frac{A}{x-a}+\frac{B}{x-b}+\frac{C}{(x-b)^2}+\frac{Dx+E}{x^2+a^2}+\frac{Fx+G}{x^2+b^2}+\frac{Hx+K}{(x^2+b^2)^2},$$

where A, B, and C can be found according to the preceding methods, and on reduction to a common denominator we can, by equating coefficients of like powers in the two numerators, find the remaining letters D, E, F, G, H, K. Variations upon these methods will suggest themselves to the student.

EXAMPLES.

1. If $y=\tan^{-1}x^2$, find $\frac{d^2y}{dx^2}$.

2. If $y=x^2\log x$, find $\frac{d^3y}{dx^3}$.

3. If $y=\sin mx\cos nx$, find $\frac{d^2y}{dx^2}$.

4. If $y=xe^{ax}$, find $\frac{d^3y}{dx^3}$.

5. If $y=x^n$, find $\frac{d^ry}{dx^r}$, distinguishing the cases in which $r <, =,$ or $> n$.

6. If $y=x\log\frac{x}{a+bx}$, prove that $x^3\frac{d^2y}{dx^2}=\left(y-x\frac{dy}{dx}\right)^2$.

7. If $y = ax \sin x$, prove that $x^2\frac{d^2y}{dx^2} - 2x\frac{dy}{dx} + (x^2 + 2)y = 0.$

8. If $y = a \cos(\log x)$, prove that $x^2\frac{d^2y}{dx^2} + x\frac{dy}{dx} + y = 0.$

9. If $y = ax^{n+1} + bx^{-n}$, prove that $x^2\frac{d^2y}{dx^2} = n(n+1)y.$

10. If $y = \sin^{-1}x$, prove that $(1 - x^2)\frac{d^2y}{dx^2} - x\frac{dy}{dx} = 0.$

11. If $y = (\sin^{-1}x)^2$, prove that $(1 - x^2)\frac{d^2y}{dx^2} = x\frac{dy}{dx} + 2.$

12. If $y = e^{a \sin^{-1}x}$, prove that $(1 - x^2)\frac{d^2y}{dx^2} - x\frac{dy}{dx} = a^2y.$

13. If $y = A \sin mx + B \cos mx$, prove that $\frac{d^2y}{dx^2} + m^2y = 0.$

14. If $y = Ae^{mx} + Be^{-mx}$, prove that $\frac{d^2y}{dx^2} - m^2y = 0.$

15. If $y = A(x + \sqrt{x^2 + a^2})^n$, prove that

$$(x^2 + a^2)\frac{d^2y}{dx^2} + x\frac{dy}{dx} - n^2y = 0.$$

16. If $y = \tan^{-1}x$, prove that $(1 + x^2)\frac{d^2y}{dx^2} + 2x\frac{dy}{dx} = 0.$

17. If $y = e^{-x}\cos x$, prove that $\frac{d^4y}{dx^4} + 4y = 0.$

18. If $y = x^2e^{ax}$, find $\frac{d^ny}{dx^n}.$

19. If $y = x^2 \sin ax$, find $\frac{d^ny}{dx^n}.$

20. If $y = \frac{x^2}{(x-a)(x-b)}$, find $\frac{d^ny}{dx^n}.$

21. If $y = \frac{1}{(x-1)^3(x-2)}$, find $\frac{d^ny}{dx^n}.$

22. If $y = x \tan^{-1}x$, find $\frac{d^ny}{dx^n}.$

23. If $y = \log x^x + x^n \log x$, find $\frac{d^ny}{dx^n}.$

24. If $y=\dfrac{1}{1+x+x^2+x^3}$, show that $\dfrac{d^ny}{dx^n}$ is

$\frac{1}{2}(-1)^n n!\sin^{n+1}\theta\{\sin(n+1)\theta-\cos(n+1)\theta+(\sin\theta+\cos\theta)^{-n-1}\}$
where $\theta=\cot^{-1}x$. [MATH. TRIPOS.

25. If $(1-x^2)\dfrac{d^2y}{dx^2}-x\dfrac{dy}{dx}=0$, prove that

$$(1-x^2)\frac{d^{n+2}y}{dx^{n+2}}-(2n+1)x\frac{d^{n+1}y}{dx^{n+1}}-n^2\frac{d^ny}{dx^n}=0.$$

26. If $(1-x^2)\dfrac{d^2y}{dx^2}-x\dfrac{dy}{dx}=a^2y$, prove that

$$(1-x^2)\frac{d^{n+2}y}{dx^{n+2}}-(2n+1)x\frac{d^{n+1}y}{dx^{n+1}}-(n^2+a^2)\frac{d^ny}{dx^n}=0.$$

27. If $y=\sin(m\sin^{-1}x)$, show that $(1-x^2)\dfrac{d^2y}{dx^2}=x\dfrac{dy}{dx}-m^2y$,

and $(1-x^2)\dfrac{d^{n+2}y}{dx^{n+2}}-(2n+1)x\dfrac{d^{n+1}y}{dx^{n+1}}-(n^2-m^2)\dfrac{d^ny}{dx^n}=0.$

28. If $y=A(x+\sqrt{x^2+a^2})^n+B(x+\sqrt{x^2+a^2})^{-n}$,

then will $$(x^2+a^2)\frac{d^2y}{dx^2}+x\frac{dy}{dx}-n^2y=0,$$

and $(x^2+a^2)\dfrac{d^{m+2}y}{dx^{m+2}}+(2m+1)x\dfrac{d^{m+1}y}{dx^{m+1}}+(m^2-n^2)\dfrac{d^my}{dx^m}=0.$

29. If $y=e^{\tan^{-1}x}=a_0+a_1x+a_2x^2+\ldots$, show that

(i.) $(1+x^2)\dfrac{d^2y}{dx^2}+(2x-1)\dfrac{dy}{dx}=0$;

(ii.) $(1+x^2)\dfrac{d^{n+2}y}{dx^{n+2}}+\{2(n+1)x-1\}\dfrac{d^{n+1}y}{dx^{n+1}}+n(n+1)\dfrac{d^ny}{dx^n}=0$;

(iii.) $(n+2)a_{n+2}+na_n=a_{n+1}$.

The last equation is to be found by substituting the series for y in equation (i.) and equating the coefficients of x^n to zero.

30. If $\sin(m\sin^{-1}x)=a_0+a_1x+a_2x^2+\ldots$, prove

$$(n+1)(n+2)a_{n+2}=(n^2-m^2)a_n.$$

31. If $e^{a\sin^{-1}x}=a_0+a_1x+a_2x^2+\ldots$, prove

$$(n+1)(n+2)a_{n+2}=(n^2+a^2)a_n.$$

32. If $(\sin^{-1}x)^2 = a_0 + a_1x + a_2x^2 + a_3x^3 + \ldots$, show that

$$(n+1)(n+2)a_{n+2} = n^2a_n.$$

33. If $f(z)$ can be expanded in a series of positive integral powers of z, show that

$$f\left(\frac{d}{dx}\right)e^{ax} = f(a)e^{ax}.$$

34. Show that

$$f\left(\frac{d}{dx}\right)e^{ax}X = e^{ax}f\left(a + \frac{d}{dx}\right)X,$$

where X represents any function of x.

35. Show that

$$f\left(\frac{d^2}{dx^2}\right)\frac{\sin}{\cos}mx = f(-m^2)\frac{\sin}{\cos}mx.$$

36. Find the nth differential coefficient of

$$e^{ax}\{a^2x^2 - 2nax + n(n+1)\}.$$

[I. C. S. EXAM.]

37. If $u = \sin nx + \cos nx$, show that

$$\frac{d^ru}{dx^r} = n^r\{1 + (-1)^r \sin 2nx\}^{\frac{1}{2}}.$$

[I. C. S. EXAM.]

38. If $\dfrac{1}{e^x - 1}$ be differentiated i times, the denominator of the result will be $(e^x - 1)^{i+1}$, and the sum of the coefficients of the several powers of e^x in the numerator will be $(-1)^i 1 \,.\, 2 \,.\, 3 \ldots i$.

[CAIUS COLL.]

39. Prove that

$$v\frac{d^nu}{dx^n} = \frac{d^nuv}{dx^n} - n\frac{d^{n-1}}{dx^{n-1}}\left(u\frac{dv}{dx}\right) + \frac{n(n-1)}{1 \,.\, 2}\frac{d^{n-2}}{dx^{n-2}}\left(u\frac{d^2v}{dx^2}\right)$$
$$- \ldots + (-1)^n u\frac{d^nv}{dx^n}.$$

CHAPTER V.

EXPANSIONS.

102. The student will have already met with several expansions of given explicit functions in ascending integral powers of the independent variable; for example, those for $(x+a)^n$, e^x, $\log(1+x)$, $\tan^{-1}x$, $\sin x$, $\cos x$, which occur in ordinary Algebra and Trigonometry.

The principal methods of development in common use may be briefly classified as follows:

I. By purely Algebraical or Trigonometrical processes.

II. By Taylor's or Maclaurin's Theorems.

III. By Differentiation or Integration of a known series, or equivalent process.

IV. By the use of a differential equation.

These methods we proceed to explain and exemplify.

103. Method I. **Algebraic and Trigonometrical Methods.**

Ex. 1. Find the first three terms of the expansion of $\log\sec x$ in ascending powers of x.

By Trigonometry

$$\cos x = 1 - \frac{x^2}{2!} + \frac{x^4}{4!} - \frac{x^6}{6!} + \dots$$

Hence $\quad \log\sec x = -\log\cos x = -\log(1-z),$

where $\quad z = \frac{x^2}{2!} - \frac{x^4}{4!} + \frac{x^6}{6!} - \dots;$

and expanding $\log(1-z)$ by the logarithmic theorem we obtain

$$\log\sec x = z + \frac{z^2}{2} + \frac{z^3}{3} + \ldots$$

$$= \left[\frac{x^2}{2!} - \frac{x^4}{4!} + \frac{x^6}{6!} - \ldots\right] + \frac{1}{2}\left[\frac{x^2}{2!} - \frac{x^4}{4!} + \ldots\right]^2$$

$$+ \frac{1}{3}\left[\frac{x^2}{2!} - \ldots\right]^3 \ldots\ldots$$

$$= \frac{x^2}{2} - \frac{x^4}{24} + \frac{x^6}{720} - \ldots$$

$$+ \frac{x^4}{8} - \frac{x^6}{48} + \ldots$$

$$+ \frac{x^6}{24} - \ldots;$$

hence $\log\sec x = \dfrac{x^2}{2} + \dfrac{x^4}{12} + \dfrac{x^6}{45} \ldots\ldots$

Ex. 2. Expand $\cos^3 x$ in powers of x.

Since $4\cos^3 x = \cos 3x + 3\cos x$

$$= 1 - \frac{3^2x^2}{2!} + \frac{3^4x^4}{4!} - \ldots + (-1)^n \frac{3^{2n}x^{2n}}{(2n)!} + \ldots$$

$$+ 3\left[1 - \frac{x^2}{2!} + \frac{x^4}{4!} - \ldots + (-1)^n \frac{x^{2n}}{(2n)!} + \ldots\right],$$

we obtain $\cos^3 x = \frac{1}{4}\left\{(1+3) - (3^2+3)\dfrac{x^2}{2!} + (3^4+3)\dfrac{x^4}{4!} - \ldots\right.$

$$\left. + (-1)^n(3^{2n}+3)\frac{x^{2n}}{(2n)!} + \ldots\right\}.$$

Similarly $\sin^3 x = \frac{1}{4}\left\{(3^3-3)\dfrac{x^3}{3!} - (3^5-3)\dfrac{x^5}{5!} + (3^7-3)\dfrac{x^7}{7!} - \ldots\right.$

$$\left. + (-1)^n \frac{3^{2n-1}-3}{(2n-1)!} x^{2n-1} - \ldots\right\}.$$

Ex. 3. Expand $\tan x$ in powers of x as far as the term involving x^5.

Since
$$\tan x = \frac{x - \dfrac{x^3}{3!} + \dfrac{x^5}{5!} - \ldots}{1 - \dfrac{x^2}{2!} + \dfrac{x^4}{4!} - \ldots}$$

we may by actual division show that

$$\tan x = x + \frac{x^3}{3} + \frac{2}{15}x^5 + \ldots$$

Ex. 4. Expand $\frac{1}{2}\{\log(1+x)\}^2$ in powers of x.

Since $$(1+x)^y \equiv e^{y\log(1+x)},$$

we have, by expanding each side of this identity,

$$1+yx+\frac{y(y-1)}{2!}x^2+\frac{y(y-1)(y-2)}{3!}x^3+\frac{y(y-1)(y-2)(y-3)}{4!}x^4+\dots$$

$$\equiv 1+y\log(1+x)+\frac{y^2}{2!}\{\log(1+x)\}^2+\dots$$

Hence, equating coefficients of y^2,

$$\tfrac{1}{2}\{\log(1+x)\}^2=\frac{x^2}{2!}-\frac{1+2}{3!}x^3+\frac{1.2+2.3+3.1}{4!}x^4-\text{etc.},$$

a series which may be written in the form

$$\frac{x^2}{2}-(1+\tfrac{1}{2})\frac{x^3}{3}+(1+\tfrac{1}{2}+\tfrac{1}{3})\frac{x^4}{4}-(1+\tfrac{1}{2}+\tfrac{1}{3}+\tfrac{1}{4})\frac{x^5}{5}+\dots$$

Examples.

1. Prove $e^{x\sin x}=1+x^2+\frac{1}{3}x^4+\frac{1}{120}x^6\dots$

2. Prove $\cosh^n x=1+\frac{nx^2}{2!}+n(3n-2)\frac{x^4}{4!}\dots$

3. Prove that

$$\frac{[\log(1+x)]^r}{r!}=\frac{x^r}{r!}-{}_rP_1\frac{x^{r+1}}{(r+1)!}+{}_{r+1}P_2\frac{x^{r+2}}{(r+2)!}-{}_{r+2}P_3\frac{x^{r+3}}{(r+3)!}+\dots,$$

where ${}_rP_k$ denotes the sum of all products k at a time of the first r natural numbers.

104. Method II. Taylor's and Maclaurin's Theorems.

It has been discovered that the Binomial, Exponential, and other well-known expansions are all particular cases of one general theorem known as Taylor's Theorem, which has for its object the *expansion of $f(x+h)$ in ascending integral positive powers of h, $f(x)$ being a function of x of any form whatever.* It will be found that such an expansion is not always possible, but we reserve for later articles [120 to 128] a rigorous discussion of the limitations of the theorem.

105. The theorem referred to is that *under certain circumstances*

$$f(x+h)=f(x)+hf'(x)+\frac{h^2}{2!}f''(x)+\frac{h^3}{3!}f'''(x)+\dots$$
$$+\frac{h^n}{n!}f^n(x)+\dots \text{ to infinity,}$$

an expansion of $f(x+h)$ in powers of h.

This result was first published by Taylor in 1715, in his "Methodus Incrementorum Directa et Inversa." In 1717 Stirling pointed out another form of Taylor's Theorem, viz.,

$$f(x)=f(0)+xf'(0)+\frac{x^2}{2!}f''(0)+\frac{x^3}{3!}f'''(0)+\dots$$
$$+\frac{x^n}{n!}f^n(0)+\dots \text{ to infinity,}$$

which is *easily deducible from Taylor's Series* by writing 0 for x and x for h; the meaning of $f^r(0)$ being that $f(x)$ is to be differentiated r times with respect to x, and then x is to be put equal to zero in the result.

The latter series gives a method of expanding any function of x in positive integral powers of x. Being a form of Taylor's Theorem it is subject to the same limitations. It is generally known as *Maclaurin's Theorem,* though its publication by Maclaurin was not made until twenty-five years after its first discovery by Stirling.

106. *Taylor's Theorem also deducible from Maclaurin's.*

It has been shown that Maclaurin's series is deducible from Taylor's form. Taylor's series is also deducible from Maclaurin's.

For, let $\qquad f(x)=F(x+y)$,

then $\qquad f'(x)=F'(x+y)$, etc.,

so that $\quad f(0)=F(y),\ f'(0)=F'(y),\ f''(0)=F''(y)$, etc.

Hence Maclaurin's Theorem

$$f(x)=f(0)+xf'(0)+\frac{x^2}{2!}f''(0)+\dots$$

becomes $F(y+x)=F(y)+xF'(y)+\frac{x^2}{2!}F''(y)+\dots,$

which is Taylor's form.

Taylor's Theorem.

107. Prop. To prove that, *if $f(x+h)$ can be expanded in a convergent series of positive integral powers of h,* that expansion is

$$f(x+h)=f(x)+hf'(x)+\frac{h^2}{2!}f''(x)+\dots \text{ to } \infty.$$

Put $x+h=X$; then since x *and* h *are independent*

$$\frac{dX}{dh}=1.$$

Hence $$\frac{df(X)}{dh}=\frac{df(X)}{dX}\cdot\frac{dX}{dh}=f'(X).$$

Similarly $$\frac{d^2f(X)}{dh^2}=f''(X), \text{ etc.}$$

Now, *assuming the possibility* of such an expansion,

let $$f(x+h)=A_0+A_1h+A_2\frac{h^2}{2!}+A_3\frac{h^3}{3!}+\dots, \quad\dots\dots\dots(1)$$

where A_0, A_1, A_2, ... are functions *of x alone, not containing h,* and are to be determined.

Differentiating with regard to h we have, by the preceding work,

$$f'(x+h)=\frac{df(x+h)}{dh}=A_1+A_2h+A_3\frac{h^2}{2!}+A_4\frac{h^3}{3!}+\dots \quad (2)$$

Differentiating again

$$f''(x+h)=\frac{df'(x+h)}{dh}=A_2+A_3h+A_4\frac{h^2}{2!}+A_5\frac{h^3}{3!}+\dots, \quad (3)$$

etc.

Put $h=0$, and we have at once from (1), (2), (3), ...

$$A_0=f(x),\ A_1=f'(x),\ A_2=f''(x), \text{ etc.}, \dots$$

Substituting these values in (1)

$$f(x+h)=f(x)+hf'(x)+\frac{h^2}{2!}f''(x)+\dots+\frac{h^r}{r!}f^r(x)+\dots$$

Maclaurin's Theorem.

108. Prop. To prove that *if $f(x)$ can be expanded in a convergent series of positive integral powers of x*, that expansion is

$$f(x)=f(0)+xf'(0)+\frac{x^2}{2!}f''(0)+\frac{x^3}{3!}f'''(0)+\dots \text{ to } \infty.$$

Assuming the possibility of such an expansion, let

$$f(x)=A_0+A_1x+A_2\frac{x^2}{2!}+A_3\frac{x^3}{3!}+\dots, \quad \dots\dots\dots\dots(1)$$

where A_0, A_1, A_2, ..., are constants to be determined, *not containing x*.

Then differentiating we have

$$f'(x)=A_1+A_2x+A_3\frac{x^2}{2!}+A_4\frac{x^3}{3!}+\dots\dots\dots\dots(2)$$

$$f''(x)=A_2+A_3x+A_4\frac{x^2}{2!}+A_5\frac{x^3}{3!}+\dots\dots\dots\dots(3)$$

etc.

Hence putting $x=0$ in (1), (2), (3), ..., we have

$$A_0=f(0),\ A_1=f'(0),\ A_2=f''(0), \text{ etc.}, \dots;$$

and substituting these values in (1)

$$f(x)=f(0)+xf'(0)+\frac{x^2}{2!}f''(0)+\frac{x^3}{3!}f'''(0)+\dots+\frac{x^r}{r!}f^r(0)+\dots$$

109. It will be noticed that in the above proofs there is nothing to indicate in what cases the expansions assumed in the equations numbered (1) in each of the last two

articles are illegitimate, and we shall have to refer the student to Arts. 120 to 128 for a fuller and more rigorous discussion.

110. It is important, before proceeding farther, that the student should satisfy himself that the well known expansions of such functions as $(x+h)^n$, e^x, $\sin x$, etc., are really all included in the general results of Arts. 107, 108.

For example, if $f(x)=x^n$, $f(x+h)=(x+h)^n$, $f'(x)=nx^{n-1}$, $f''(x)=n(n-1)x^{n-2}$, etc. Hence Taylor's Theorem,

$$f(x+h)=f(x)+hf'(x)+\frac{h^2}{2!}f''(x)+\dots,$$

gives the binomial expansion

$$(x+h)^n=x^n+nhx^{n-1}+\frac{n(n-1)}{2!}h^2x^{n-2}+\dots$$

Again, suppose $f(x)=e^x$, then $f'(x)=e^x$, $f''(x)=e^x$, etc., therefore $\quad f(0)=1, f'(0)=1, f''(0)=1$, etc.

Hence Maclaurin's Theorem,

$$f(x)=f(0)+xf'(0)+\frac{x^2}{2!}f''(0)+\dots,$$

gives $\quad e^x=1+x+\frac{x^2}{2!}+\frac{x^3}{3!}+\dots,$

the result known as the Exponential Theorem.

111. We append a few examples which admit of expansion, and to which therefore the results of Arts. 107, 108 apply.

EXAMPLES.

Prove the following results :—

1. $\sin x=x-\frac{x^3}{3!}+\frac{x^5}{5!}-\dots.$

2. $\log(1+x)=x-\frac{x^2}{2}+\frac{x^3}{3}-\ldots.$

3. $\tan^{-1}x=x-\frac{x^3}{3}+\frac{x^5}{5}-\ldots.$

4. $e^x\cos x=1+2^{\frac{1}{2}}\cos\frac{\pi}{4}.x+2^{\frac{2}{2}}\cos\frac{2\pi}{4}\frac{x^2}{2!}+2^{\frac{3}{2}}\cos\frac{3\pi}{4}\frac{x^3}{3!}+\ldots$

$$+2^{\frac{n}{2}}\cos\frac{n\pi}{4}\frac{x^n}{n!}+\ldots.$$

5. $\log(1+e^x)=\log 2+\frac{1}{2}x+\frac{1}{8}x^2-\frac{x^4}{192}\ldots.$

6. $e^{\sin x}=1+x+\frac{1}{2}x^2-\frac{1}{8}x^4-\ldots.$

7. $\sin(x+h)=\sin x+h\cos x-\frac{h^2}{2!}\sin x-\frac{h^3}{3!}\cos x+\ldots.$

8. $\sin^{-1}(x+h)=\sin^{-1}x+\frac{h}{\sqrt{1-x^2}}+\frac{x}{(1-x^2)^{\frac{3}{2}}}\frac{h^2}{2!}+\frac{1+2x^2}{(1-x^2)^{\frac{5}{2}}}\frac{h^3}{3!}+\ldots.$

9. $\log\sin(x+h)=\log\sin x+h\cot x-\frac{h^2}{2}\operatorname{cosec}^2x+\frac{h^3}{3}\frac{\cos x}{\sin^3x}+\ldots.$

10. $\sec^{-1}(x+h)=\sec^{-1}x+\frac{h}{x\sqrt{x^2-1}}-\frac{2x^2-1}{x^2(x^2-1)^{\frac{3}{2}}}\frac{h^2}{2!}+\ldots.$

Method III.

112. *Expansion by Differentiation or Integration of a known series or equivalent process.*

The method of treatment is indicated in the following examples:

Ex. 1. *To expand $\tan^{-1}x$ in powers of x.* Gregory's Series.

Suppose $\qquad f(x)=\tan^{-1}x=a_0+a_1x+a_2x^2+a_3x^3+\ldots,$

then $\qquad f'(x)=\frac{1}{1+x^2}=a_1+2a_2x+3a_3x^2+4a_4x^3+\ldots;$

also $\qquad \frac{1}{1+x^2}=1-x^2+x^4-x^6+\ldots.$

Hence, comparing these expansions, we have

$$a_2=a_4=a_6=a_8=\ldots=0,$$

and $\qquad a_1=1,\ 3a_3=-1,\ 5a_5=1,$ etc.

Also, $\qquad a_0=\tan^{-1}0=n\pi;$

therefore $\qquad \tan^{-1}x=n\pi+x-\frac{x^3}{3}+\frac{x^5}{5}-\frac{x^7}{7}+\ldots.$

This result may be obtained immediately by integration of the series for $\frac{1}{1+x^2}$, viz.,

$$1-x^2+x^4-x^6+\ldots,$$

the constant a_0 being determined as before.

Ex. 2. *To expand* $\sin^{-1} x$.

Suppose $f(x)=\sin^{-1}x=a_0+a_1x+a_2x^2+a_3x^3+\ldots;$

therefore $f'(x)=\frac{1}{\sqrt{1-x^2}}=a_1+2a_2x+3a_3x^2+4a_4x^3+\ldots.$

But $\frac{1}{\sqrt{1-x^2}}=1+\frac{1}{2}x^2+\frac{1.3}{2.4}x^4+\ldots.$

Hence, comparing these series, we have

$$a_2=a_4=a_6=\ldots=0,$$

and $a_1=1,\ 3a_3=\frac{1}{2},\ 5a_5=\frac{1.3}{2.4}\ldots.$

Also $a_0=\sin^{-1}0=n\pi.$

Hence $\sin^{-1}x=n\pi+x+\frac{1}{2}.\frac{x^3}{3}+\frac{1.3}{2.4}\frac{x^5}{5}+\frac{1.3.5}{2.4.6}.\frac{x^7}{7}+\ldots;$

and, as before, this might have been obtained immediately by integration of the expansion of $\frac{1}{\sqrt{1-x^2}}$.

Ex. 3. Again, if a known series be given, we can obtain others from it *by differentiation.*

For example, borrowing the series for $(\sin^{-1}x)^2$ established in Ex. 2 of the next Art., viz.—

$$\tfrac{1}{2}(\sin^{-1}x)^2=\frac{x^2}{2}+\frac{2}{3}\frac{x^4}{4}+\frac{2.4}{3.5}\frac{x^6}{6}+\frac{2.4.6}{3.5.7}.\frac{x^8}{8}+\ldots,$$

we obtain at once by differentiation

$$\frac{\sin^{-1}x}{\sqrt{1-x^2}}=x+\frac{2}{3}x^3+\frac{2.4}{3.5}x^5+\frac{2.4.6}{3.5.7}x^7+\ldots.$$

Examples.

1. Prove $\log(x+\sqrt{1+x^2})=\sinh^{-1}x=x-\frac{1}{2}\frac{x^3}{3}+\frac{1.3}{2.4}.\frac{x^5}{5}-\ldots.$

2. Prove $\tanh^{-1}x=x+\frac{x^3}{3}+\frac{x^5}{5}+\ldots.$

3. Deduce from Ex. 3, Art. 112,

$$(1-x^2)^{\frac{1}{2}}\sin^{-1}x = x - \frac{x^3}{3} - \frac{2}{3}\frac{x^5}{5} - \frac{2.4}{3.5}\cdot\frac{x^7}{7} - \dots.$$

And hence by putting $x=\sin\theta$, prove

$$\theta\cot\theta = 1 - \frac{\sin^2\theta}{3} - \frac{2}{3}\cdot\frac{\sin^4\theta}{5} - \frac{2.4}{3.5}\frac{\sin^6\theta}{7} - \dots.$$

[QUARTERLY JOURNAL, vol. vi.]

IV.—A NEWTONIAN METHOD.

113. It remains to exemplify a fourth method of proceeding which may often be employed with advantage, and moreover is of historical interest, as having been employed by Newton.

Assume a series for the expansion

$$(\text{say } a_0 + a_1x + a_2x^2 + \dots\dots).$$

Then form a differential equation in the way indicated in several of the examples in the preceding chapter. Substitute the series in the differential equation and equate the coefficients of like powers of x on each side of the equation. We shall thus get equations enough to find all the coefficients except one or two of the first which may be easily obtained from the values of $f(0)$ and $f'(0)$.

Ex. 1. *Expand a^x in this manner.*

Let $$a^x = a_0 + a_1x + a_2x^2 + a_3x^3 + \dots \qquad \dots\dots(1)$$

If $$y = a^x,$$

$$\frac{dy}{dx} = a^x\log_e a = y\log_e a. \qquad \dots\dots(2)$$

But $$\frac{dy}{dx} = a_1 + 2a_2x + 3a_3x^2 + \dots; \qquad \dots\dots(3)$$

therefore, substituting from (1) and (3) in the differential equation (2), $a_1 + 2a_2x + 3a_3x^2 + \dots \equiv \log_e a(a_0 + a_1x + a_2v^2 + \dots)$.

Hence, comparing coefficients,

$$a_1 = a_0\log_e a,$$
$$2a_2 = a_1\log_e a,$$
$$3a_3 = a_2\log_e a, \text{ etc.}$$

Now $$a_0=f(0)=a^0=1;$$
therefore $$a_1=\log_e a,\ a_2=\frac{(\log_e a)^2}{2!},\ a_3=\frac{(\log_e a)^3}{3!},\ \ldots$$
and the series is $$a^x=1+x\log_e a+\frac{x^2}{2!}(\log_e a)^2+\ldots.$$

Ex. 2. Let $y=f(x)=(\sin^{-1}x)^2$.

$$y_1=2\sin^{-1}x\frac{1}{\sqrt{1-x^2}},$$
$$y_2=\frac{2}{1-x^2}+\frac{2x}{(1-x^2)^{\frac{3}{2}}}\sin^{-1}x$$
$$=\frac{2}{1-x^2}+\frac{xy_1}{1-x^2};$$
$$\therefore\quad (1-x^2)y_2=xy_1+2. \quad\ldots\ldots(1)$$

Now, let $y=a_0+a_1x+a_2x^2+\ldots+a_nx^n+a_{n+1}x^{n+1}+a_{n+2}x^{n+2}+\ldots,$

therefore $y_1=a_1+2a_2x+\ldots+na_nx^{n-1}+(n+1)a_{n+1}x^n$
$+(n+2)a_{n+2}x^{n+1}+\ldots,$

and $y_2=2a_2+\ldots+n(n-1)a_nx^{n-2}+(n+1)na_{n+1}x^{n-1}$
$+(n+2)(n+1)a_{n+2}x^n+\ldots.$

Picking out the coefficient of x^n in the equation (*which may be done without actual substitution*) we have

$$(n+2)(n+1)a_{n+2}-n(n-1)a_n=na_n;$$
therefore $$a_{n+2}=\frac{n^2}{(n+1)(n+2)}a_n. \quad\ldots\ldots(2)$$

Now, $$a_0=f(0)=(\sin^{-1}0)^2,$$
and if we consider $\sin^{-1}x$ to be the *smallest positive angle* whose sine is x, $$\sin^{-1}0=0.$$

Hence $$a_0=0.$$

Again, $$a_1=f'(0)=2\sin^{-1}0\,.\,\frac{1}{\sqrt{1-0}}=0,$$
and $$a_2=\tfrac{1}{2}f''(0)=\tfrac{1}{2}\left(\frac{2}{1-0}+0\right)=1.$$

Hence, from equation (2), a_3, a_5, a_7, ..., are each $=0$.

and $$a_4=\frac{2^2}{3.4}\cdot a_2=\frac{2^2}{3.4}=\frac{2^2}{4!}2,$$
$$a_6=\frac{4^2}{5.6}\cdot a_4=\frac{2^2.4^2}{3.4.5.6}=\frac{2^2.4^2}{6!}.2,$$
etc. = etc.;

therefore $$(\sin^{-1}x)^2=\frac{2x^2}{2!}+\frac{2^2}{4!}2x^4+\frac{2^2.4^2}{6!}2x^6+\frac{2^2.4^2.6^2}{8!}2x^8+\ldots.$$

A slightly different method of proceeding is indicated in the following example.

Ex. 3. Let $y=\sin(m\sin^{-1}x)=a_0+a_1x+a_2\frac{x^2}{2!}+a_3\frac{x^3}{3!}+\ldots$ ....(1)

Then $$y_1=\cos(m\sin^{-1}x)\frac{m}{\sqrt{1-x^2}},$$

whence $(1-x^2)y_1^2=m^2(1-y^2)$.

Differentiating again, and dividing by $2y_1$, we have

$$(1-x^2)y_2-xy_1+m^2y=0. \qquad\text{.......................(2)}$$

Differentiating this n times by Leibnitz's Theorem

$$(1-x^2)y_{n+2}-(2n+1)xy_{n+1}+(m^2-n^2)y_n=0. \qquad\text{...........(3)}$$

Now, $$a_0=(y)_{x=0}=\sin(m\sin^{-1}0)=0,$$

(assuming that $\sin^{-1}x$ is the smallest positive angle whose sine is x)

$$a_1=(y_1)_{x=0}=m,$$
$$a_2=(y_2)_{x=0}=0,$$
etc.
$$a_n=(y_n)_{x=0}.$$

Hence, putting $x=0$ in equation (3),

$$a_{n+2}=-(m^2-n^2)a_n.$$

Hence $a_4, a_6, a_8, \ldots$, each $=0$,

and $$a_3=-(m^2-1^2)a_1=-m(m^2-1^2),$$
$$a_5=-(m^2-3^2)a_3=m(m^2-1^2)(m^2-3^2),$$
$$a_7=-(m^2-5^2)a_5=-m(m^2-1^2)(m^2-3^2)(m^2-5^2),$$
etc.

Whence

$$\sin(m\sin^{-1}x)=mx-\frac{m(m^2-1^2)}{3!}x^3+\frac{m(m^2-1^2)(m^2-3^2)}{5!}x^5$$
$$-\frac{m(m^2-1^2)(m^2-3^2)(m^2-5^2)}{7!}x^7+\ldots.$$

The corresponding series for $\cos(m\sin^{-1}x)$ is

$$\cos(m\sin^{-1}x)=1-\frac{m^2x^2}{2!}+\frac{m^2(m^2-2^2)}{4!}x^4-\frac{m^2(m^2-2^2)(m^2-4^2)}{6!}x^6+\ldots.$$

If we write $x=\sin\theta$ these series become

$$\sin m\theta=m\sin\theta-\frac{m(m^2-1^2)}{3!}\sin^3\theta+\frac{m(m^2-1^2)(m^2-3^2)}{5!}\sin^5\theta-\text{etc.},$$

$$\cos m\theta=1-\frac{m^2}{2!}\sin^2\theta+\frac{m^2(m^2-2^2)}{4!}\sin^4\theta$$
$$-\frac{m^2(m^2-2^2)(m^2-4^2)}{6!}\sin^6\theta+\text{etc.}$$

EXAMPLES.

1. If $y=(1+x)^n=a_0+a_1x+a_2x^2+a_3x^3+\dots,$

prove that $$(1+x)\frac{dy}{dx}=ny,$$

and hence that $$a_{r+1}=\frac{n-r}{r+1}a_r.$$

In this manner find all the coefficients of the Binomial Theorem.

2. If $y=\sin^{-1}x=a_0+a_1x+a_2x^2+a_3x^3+\dots,$

prove that $$a_{n+2}=\frac{n^2}{(n+1)(n+2)}a_n,$$

and in this manner deduce the expansion given in Ex. 2, Art. 112.

3. If $y=(\tan^{-1}x)^2=a_0+a_1x+a_2x^2+a_3x^3+\dots,$

prove that $(n+2)(n+1)a_{n+2}+2n^2a_n+(n-2)(n-1)a_{n-2}=0.$

CONTINUITY.

114. DEF. A function is said to be *continuous* between any two values of the independent variable involved if, as that variable is made to assume successively all intermediate values from the one assigned value to the other,

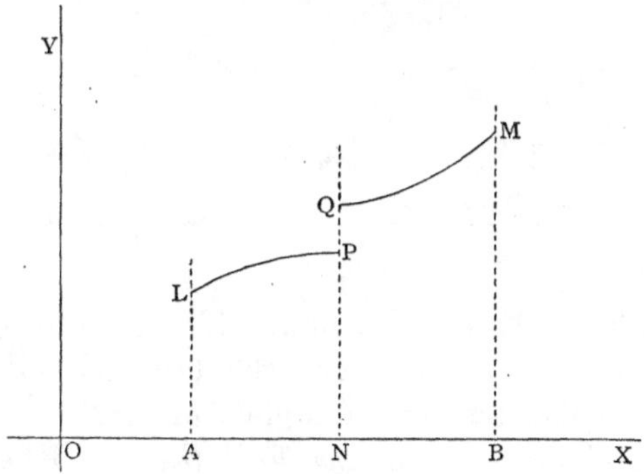

Fig. 15.

the function *does not suddenly change its value*, but changes so that for any indefinitely small change in the variable there is never a change of finite magnitude in the value of the function.

115. Suppose the function to be $\phi(x)$. Trace the curve $y=\phi(x)$ between the ordinates $AL(x=a)$ and $BM(x=b)$. Then if we find that as x increases through some value, as ON (Fig. 15), the ordinate $\phi(x)$ *suddenly changes* from NP to NQ without going through the intermediate values, the function is said to be discontinuous for the value $x=ON$ of the independent variable.

116. Similarly, we may represent geometrically the discontinuity of a differential coefficient. For $\dfrac{dy}{dx}$ represents the tangent of the angle which the tangent line to

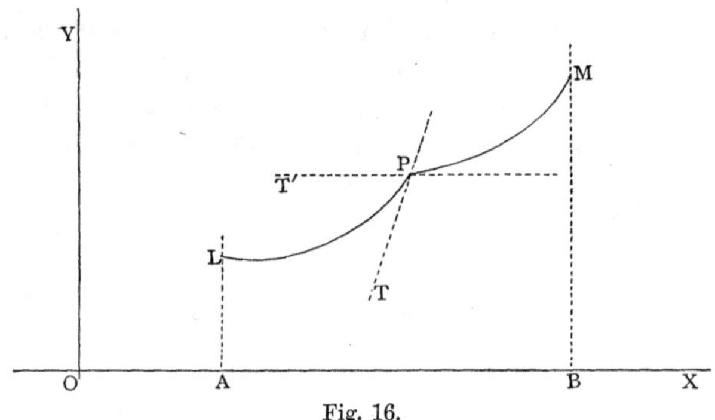

Fig. 16.

the curve makes with the axis of x. If, therefore, as the point P travels along the curve the tangent *suddenly changes its position* (as, for example, from PT to PT' in the figure), *without going through the intermediate positions*, there is a discontinuity in the value of $\dfrac{dy}{dx}$.

117. Prop. *If any function of x, say $\phi(x)$, vanish when $x=a$ and when $x=b$ and is finite and continuous, as*

also its first differential coefficient $\phi'(x)$ *between those values, then will* $\phi'(x)$ *vanish for at least one intermediate value.*

For if $\phi'(x)$ were always positive or always negative between $x=a$ and $x=b$, $\phi(x)$ would be continually increasing or continually decreasing between those values (Art. 44), and therefore could not vanish for both $x=a$ and $x=b$, which would be contrary to the hypothesis. Hence $\phi'(x)$ must change sign and therefore vanish for some value of x intermediate between $x=a$ and $x=b$.

118. The same thing is obvious at once from a figure. For, suppose the curve $y=\phi(x)$ cuts the axis at A

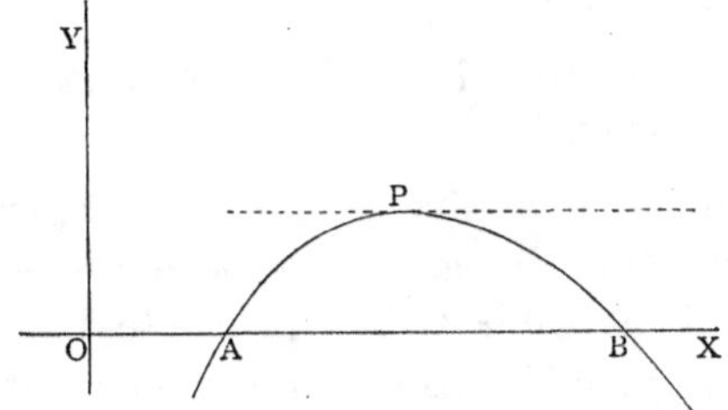

Fig. 17.

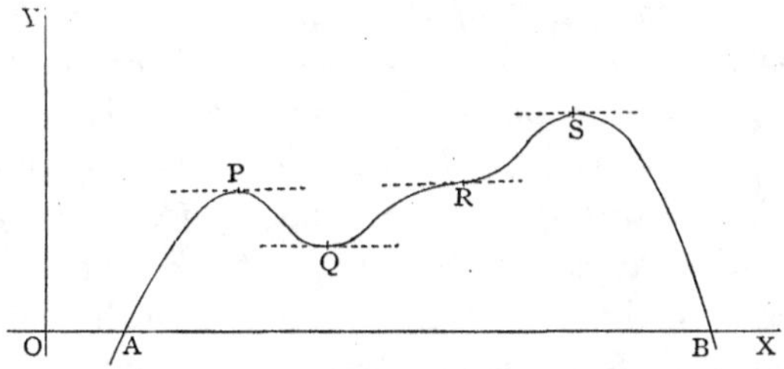

Fig. 18.

$(x=a,\ y=0)$ and B $(x=b,\ y=0)$, then it is obvious that if the curve $y=\phi(x)$ and the inclination of its tangent be continuous between A and B, the tangent line must be *parallel to the axis of x* at some intermediate point P.

It is also clear that the tangent may be parallel to the axis of x at other points between A and B besides P as in Fig. 18, so that it does not follow that $\phi'(x)$ vanishes *only once* between two contiguous roots of $\phi(x)=0$.

119. The same proposition is thus enunciated in books on Theory of Equations: "*A real root of the equation $\phi'(x)=0$ lies between every adjacent two of the real roots of the equation $\phi(x)=0$*"; and is known as Rolle's Theorem.

120. Remainder after the first n terms have been taken from Taylor's Series.

There is much difficulty in giving a rigorous direct proof of Taylor's Series, as might be expected from the highly general character of the result to be established. It is therefore found easier to consider what is left after n terms of Taylor's Series have been taken from $\phi(x+h)$. *If the form of this remainder be such that it can be made smaller than any assignable quantity when sufficient terms of the series are taken, the difference between $\phi(x+h)$ and Taylor's Series for $\phi(x+h)$ will be indefinitely small, and under these circumstances we shall be able to assert the truth of the theorem.* The following investigations of an expression for the remainder are taken, with few changes, from Bertrand's "Traité de Calcul Différentiel et Intégral."

Let R denote the remainder after n terms of Taylor's Series have been taken from $\phi(x+h)$; so that

$$\phi(x+h)=\phi(x)+h\phi'(x)+\frac{h^2}{2!}\phi''(x)+\ldots+\frac{h^{n-1}}{(n-1)!}\phi^{n-1}(x)+R\ \ldots(1)$$

Let $x+h=X$, hence

$$\phi(X)-\phi(x)-\frac{X-x}{1}\phi'(x)-\frac{(X-x)^2}{2!}\phi''(x)-\ldots$$
$$-\frac{(X-x)^{n-1}}{(n-1)!}\phi^{n-1}(x)-R=0.\ \ldots\ldots\ldots\ldots(2)$$

Put $R=\frac{(X-x)^n}{n!}P$, a form suggested by the remaining terms of Taylor's Series. Consider the function formed by writing z instead of x throughout the left-hand member of equation (2) *except in P, which is therefore independent of z.* Call the function thus obtained $F(z)$. Hence

$$F(z)=\phi(X)-\phi(z)-\frac{X-z}{1}\phi'(z)-\frac{(X-z)^2}{2!}\phi''(z)-\ldots$$
$$-\frac{(X-z)^{n-1}}{(n-1)!}\phi^{n-1}(z)-\frac{(X-z)^n}{n!}P.\ \ \ldots\ldots(3)$$

We shall assume that $\phi(z)$ and all its differential coefficients up to the nth inclusive are *finite and continuous* between the values x and X of the variable z.

It is clear from equation (2) *that* $F(x)=0$, also by putting $z=X$ in (3) we have $F(X)=0$; also $F(z)$ and $F'(z)$ are finite and continuous between these values of the variable z. *Hence $F'(z)$ vanishes for some value of z intermediate between* x and X, say for $z=x+\theta(X-x)$, where θ is a *proper fraction.* Differentiating equation (3) with respect to z, the terms alternately destroy each other except at the end of the series, and we have left

$$F'(z)=-\frac{(X-z)^{n-1}}{(n-1)!}\phi^n(z)+\frac{(X-z)^{n-1}}{(n-1)!}P,\ \ldots\ldots\ldots\ldots(4)$$

whence $\quad P=\phi^n\{x+\theta(X-x)\}$,

that value of z being taken which makes $F'(z)$ vanish. Hence, remembering that $X-x=h$, the true value of R sought is

$$\frac{h^n}{n!}\phi^n(x+\theta h). \quad\text{.......................(5)}$$

The theorem may therefore be written

$$\phi(x+h)=\phi(x)+h\phi'(x)+\frac{h^2}{2!}\phi''(x)+\ldots$$
$$+\frac{h^{n-1}}{(n-1)!}\phi^{n-1}(x)+\frac{h^n}{n!}\phi^n(x+\theta h), \quad\text{....(6)}$$

where θ is a *proper fraction.*

If then the form of the function $\phi(x)$ be such that by making n sufficiently great the expression $\frac{h^n}{n!}\phi^n(x+\theta h)$ can be made less than any assignable quantity however small, we can make the true series for $\phi(x+h)$ *differ by as little as we please from Taylor's form*

$$\phi(x)+h\phi'(x)+\frac{h^2}{2!}\phi''(x)+\ldots\text{to }\infty.$$

The above form of the remainder is due to Lagrange, and the investigation is spoken of as *Lagrange's Theorem on the Limits of Taylor's Theorem.*

121. A different form of the remainder is due to Cauchy.

In equation (2) put $R=(X-x)P$ and proceed as before, then, instead of equation (4), we shall have

$$F'(z)=-\frac{(X-z)^{n-1}}{(n-1)!}\phi^n(z)+P,$$

which vanishes as before for some value of z between $z=x$ and $z=X=x+h$, say for $z=x+\theta h$; whence

$$P=\frac{(1-\theta)^{n-1}h^{n-1}}{(n-1)!}\phi^n(x+\theta h),$$

and therefore $R=\dfrac{(1-\theta)^{n-1}h^n}{(n-1)!}\phi^n(x+\theta h).$

122. Another form is obtained by Schlömilch and Roche by assuming a slightly different form for R, viz.,

$$=\frac{(X-x)^{p+1}}{p+1}P.$$

This gives, instead of equation (4),

$$F'(z)=-\frac{(X-z)^{n-1}}{(n-1)!}\phi^n(z)+(X-z)^pP,$$

whence $$P=\frac{(1-\theta)^{n-p-1}h^{n-p-1}}{(n-1)!}\phi^n(x+\theta h),$$

and $$R=\frac{(1-\theta)^{n-p-1}h^n}{(n-1)!\,(p+1)}\phi^n(x+\theta h).$$

The last form includes the two former as particular cases; for putting $p+1=n$ it reduces to Lagrange's result, and putting $p=0$ it reduces to Cauchy's.

123. The corresponding forms of remainder for Maclaurin's Theorem are obtained by writing 0 for x and x for h, when the three expressions investigated above become respectively

$$\frac{x^n}{n!}\phi^n(\theta x),\quad \frac{(1-\theta)^{n-1}x^n}{(n-1)!}\phi^n(\theta x),\text{ and }\frac{(1-\theta)^{n-p-1}x^n}{(n-1)!\,(p+1)}\phi^n(\theta x).$$

124. The student should notice the special cases of equation (6), Art. 120, when $n=1$, 2, 3, etc., viz.,

$$\phi(x+h)=\phi(x)+h\phi'(x+\theta_1 h),$$

$$\phi(x+h)=\phi(x)+h\phi'(x)+\frac{h^2}{2!}\phi''(x+\theta_2 h),$$

etc.

All that is known with respect to the θ in each case being that it is a *proper fraction.*

125. **Geometrical Illustration.**

It is easy to give a geometrical illustration of the equation $$\phi(x+h)=\phi(x)+h\phi'(x+\theta h).$$

For let x, $\phi(x)$, be the co-ordinates of a point P on the curve $y=\phi(x)$, and let $x+h$, $\phi(x+h)$ be the co-ordinates of another point Q, also on the curve. And suppose the curve and the inclination of the tangent to the curve to the axis of x to be continuous and finite between P and Q; draw PM, QN perpendicular to OX and PL perpendicular to QN, then

$$\frac{\phi(x+h)-\phi(x)}{h}=\frac{NQ-MP}{MN}=\frac{LQ}{PL}=\tan LPQ.$$

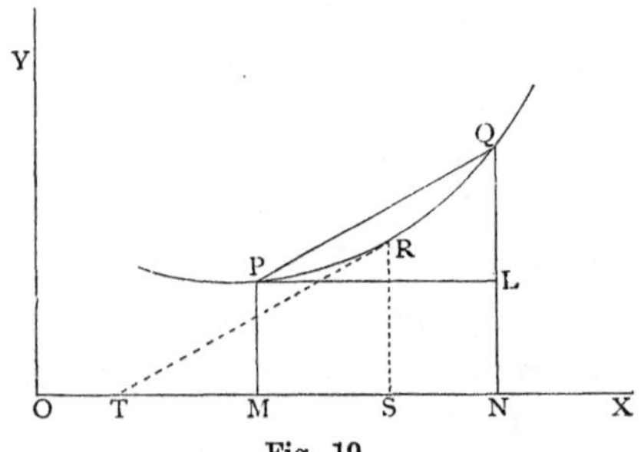

Fig. 19.

Also, $x+\theta h$ is the abscissa of some point R on the curve between P and Q, and $\phi'(x+\theta h)$ is the tangent of the angle which the tangent line to the curve at R makes with the axis of x. Hence the assertion that

$$\frac{\phi(x+h)-\phi(x)}{h}=\phi'(x+\theta h)$$

is equivalent to the obvious geometrical fact that *there must be a point R somewhere between P and Q at which the tangent to the curve is parallel to the chord PQ.*

126. The cases in which Taylor's Theorem is said to fail are those in which it happens

(1) That $\phi(x)$, or one of its differential coefficients,

becomes infinite between the values of the variable considered;

(2) Or that $\phi(x)$, or one of its differential coefficients, becomes *discontinuous* between the same values;

(3) Or that the remainder, $\frac{h^n}{n!}\phi^n(x+\theta h)$, *cannot be made to vanish in the limit* when n is taken sufficiently large, so that the series does not approach a finite limit.

Ex. If
$$\phi(x)=\sqrt{x}$$
$$\phi(x+h)=\sqrt{x+h},\ \phi'(x)=\frac{1}{2\sqrt{x}},\ \text{etc.}$$
Hence Taylor's Theorem gives
$$\phi(x+h)=\sqrt{x+h}=\sqrt{x}+\frac{1}{2\sqrt{x}}h+\dots$$
If, however, we put $x=0$, $\frac{1}{2\sqrt{x}}$ becomes infinite, while $\sqrt{x+h}$ becomes $\sqrt{h}$.

Thus, as we might expect, we fail at the second term to expand $\sqrt{h}$ in a series of integral powers of h.

127. In Art. 107 the proof of Taylor's Theorem is not general, the assumption being made that a convergent expansion in ascending positive integral powers of x is possible. The above article points out clearly when this assumption is legitimate.

For any continuous function in which the $(p+1)^{\text{th}}$ differential coefficient is the first to become infinite or discontinuous for the value x of the variable, the theorem
$$\phi(x+h)=\phi(x)+h\phi'(x)+\dots+\frac{h^p}{p!}\phi^p(x+\theta h),$$
which involves no differential coefficients of higher order than the pth, is rigorously true, although Taylor's Theorem,

$$\phi(x+h)=\phi(x)+h\phi'(x)+\ldots+\frac{h^p}{p!}\phi^p(x)+\frac{h^{p+1}}{(p+1)!}\phi^{p+1}(x)+\ldots$$

fails to furnish us with an intelligible result.

Ex. If $\phi(x)=(x-a)^{\frac{5}{2}}$,

we have $\phi'(x)=\frac{5}{2}(x-a)^{\frac{3}{2}}$,

$$\phi''(x)=\frac{15}{4}(x-a)^{\frac{1}{2}},$$

$$\phi'''(x)=\frac{15}{8}\left(\frac{1}{x-a}\right)^{\frac{1}{2}}, \text{ etc.,}$$

and Taylor's Theorem gives

$$(x+h-a)^{\frac{5}{2}}=(x-a)^{\frac{5}{2}}+\frac{5}{2}(x-a)^{\frac{3}{2}}h+\frac{15}{4}(x-a)^{\frac{1}{2}}\frac{h^2}{2!}+\frac{15}{8}\,\frac{1}{(x-a)^{\frac{1}{2}}}\,\frac{h^3}{3!}+\ldots,$$

which fails at the fourth term when $x=a$.

But Equation 6 of Art. 120 gives the result

$$(x+h-a)^{\frac{5}{2}}=(x-a)^{\frac{5}{2}}+\frac{5}{2}(x-a)^{\frac{3}{2}}h+\frac{15}{4}\,\frac{h^2}{2!}(x+\theta h-a)^{\frac{1}{2}},$$

which, in the case when $x=a$, reduces to

$$h^{\frac{5}{2}}=\frac{15}{8}\theta^{\frac{1}{2}}h^{\frac{5}{2}},$$

or $$\theta=\frac{64}{225},$$

and this obeys the only limitation necessary, viz., that θ should be a proper fraction.

128. The remarks made with respect to the failure of Taylor's Theorem obviously also apply to the particular form of it, Maclaurin's Theorem, so that Maclaurin's Theorem is said to fail when any of the expressions $\phi(0)$, $\phi'(0)$, $\phi''(0)$, ... become *infinite*, or if there be a discontinuity in the function or any of its differential coefficients as x passes through the value zero, or if the remainder $\frac{x^n}{n!}\phi^n(\theta x)$ *does not become infinitely small* when

n becomes infinitely large, for in this case the series is divergent and does not tend to any finite limit.

129. Examples of Expansions by Maclaurin's Theorem, with investigation of Remainder after n terms.

Ex. 1. Let $f(x) = a^x$,

then $f^n(x) = a^x(\log_e a)^n$, and $f^n(0) = (\log_e a)^n$.

Hence the formula

$$f(x) = f(0) + xf'(0) + \frac{x^2}{2!}f''(0) + \dots + \frac{x^{n-1}}{(n-1)!}f^{n-1}(0) + \frac{x^n}{n!}f^n(\theta x)$$

gives

$$a^x = 1 + x\log_e a + \frac{x^2}{2!}(\log_e a)^2 + \dots + \frac{x^{n-1}}{(n-1)!}(\log_e a)^{n-1} + \frac{x^n}{n!}a^{\theta x}(\log_e a)^n.$$

Now $\dfrac{x^n a^{\theta x}(\log_e a)^n}{n!}$ can be made *smaller than any assignable quantity* by sufficiently increasing n; hence the remainder, after n terms of Maclaurin's Theorem have been taken, ultimately vanishes when n is taken very large, and therefore Maclaurin's Theorem is applicable and gives

$$a^x = 1 + x\log_e a + \frac{x^2}{2!}(\log_e a)^2 + \frac{x^3}{3!}(\log_e a)^3 + \dots \text{ to } \infty.$$

Ex. 2. Let $f(x) = \log(1+x)$,

$$f'(x) = \frac{1}{1+x}, \; f''(x) = (-1)\frac{1}{(1+x)^2}, \dots f^n(x) = (-1)^{n-1}\frac{(n-1)!}{(1+x)^n}.$$

Hence $f(0) = 0$, $f'(0) = 1$, $f''(0) = -1$, $f'''(0) = 2 \dots$,

$$f^n(0) = (-1)^{n-1}(n-1)!.$$

And the Lagrange-formula for the remainder, after n terms of Maclaurin's Series have been subtracted from $f(x)$, viz. $\dfrac{x^n f^n(\theta x)}{n!}$, becomes

$$\frac{(-1)^{n-1}}{n} \cdot \left(\frac{x}{1+\theta x}\right)^n;$$

and if x be not greater than 1, and positive, $\dfrac{x}{1+\theta x}$ is a proper fraction, and therefore by making n sufficiently large the above remainder ultimately vanishes, and therefore Maclaurin's Theorem is applicable and gives

$$\log(1+x) = x - \frac{x^2}{2} + \frac{x^3}{3} - \frac{x^4}{4} + \dots \text{ to } \infty,$$

where x lies between 0 and 1 inclusive.

It appears that if we consider $f(x)=\log(1-x)$ the remainder is

$$-\frac{1}{n}\left(\frac{x}{1-\theta x}\right)^n.$$

In this form it is not clear that the limit of the remainder is zero. But if we choose for this example Cauchy's form of remainder, Art. 123, it reduces to

$$-\frac{1}{1-\theta}\left(\frac{x-\theta x}{1-\theta x}\right)^n;$$

and if x be positive and less than unity, $\frac{x-\theta x}{1-\theta x}$ is also less than unity, and therefore $\frac{1}{1-\theta}\left(\frac{x-\theta x}{1-\theta x}\right)^n$ can be made as small as we like by sufficiently increasing n. Hence Maclaurin's series is applicable and gives $\quad \log(1-x)=-x-\frac{x^2}{2}-\frac{x^3}{3}-\frac{x^4}{4}-\ldots$ to ∞.

Bernoulli's Numbers.

130. *To expand* $u=f(x)=\frac{x}{2}\frac{e^x+1}{e^x-1}$ *in powers of* x.

Let

$$u=f(x) \text{ and } u'=f(0),$$
$$u_1=f'(x) \text{ and } u'_1=f'(0),$$
$$u_2=f''(x) \text{ and } u'_2=f''(0),$$

with a similar notation for higher differential coefficients. Then Maclaurin's Theorem gives

$$u=\frac{x}{2}\frac{e^x+1}{e^x-1}=u'+xu'_1+\frac{x^2}{2!}u'_2+\ldots$$

Changing the sign of x we see that the left hand member of this equation remains unaltered; hence we have

$$u=u'-xu'_1+\frac{x^2}{2!}u'_2-\ldots,$$

and by subtraction

$$0=2xu'_1+2\frac{x^3}{3!}u'_3+2\frac{x^5}{5!}u'_5+\ldots,$$

whence, by equating to zero the coefficients of the several powers of x, we infer that $\quad u'_1=u'_3=u'_5=\ldots=0,$
so that the expansion contains no odd powers of x.

Again, since $$e^x u = u + \frac{x}{2} + x\frac{e^x}{2},$$
we have, by differentiating,
$$e^x(u_1+u) = u_1 + \tfrac{1}{2} + (x+1)\frac{e^x}{2}$$
$$e^x(u_2+2u_1+u) = u_2 + (x+2)\frac{e^x}{2}$$
$$e^x(u_3+3u_2+3u_1+u) = u_3 + (x+3)\frac{e^x}{2},$$
etc.,

and putting $x=0$ in these equations we obtain from the first, third, fifth, etc.,
$$u' = \tfrac{1}{2} + \tfrac{1}{2},$$
$$3u'_2 + u' = \tfrac{3}{2},$$
$$5u'_4 + 10u'_2 + u' = \tfrac{5}{2},$$
$$7u'_6 + 35u'_4 + 21u'_2 + u' = \tfrac{7}{2},$$
etc.,

giving $u'=1$, $u'_2 = \frac{1}{6}$, $u'_4 = -\frac{1}{30}$, $u'_6 = \frac{1}{42}$, $u'_8 = -\frac{1}{30}$, etc.

Hence $$\frac{x}{2}\,\frac{e^x+1}{e^x-1} = 1 + \frac{1}{6}\,\frac{x^2}{2!} - \frac{1}{30}\,\frac{x^4}{4!} + \frac{1}{42}\,\frac{x^6}{6!} - \frac{1}{30}\,\frac{x^8}{8!} + \dots$$

This series introduces a set of coefficients which are found of great importance in the higher branches of analysis. The series is frequently written in the form
$$\frac{x}{2}\,\frac{e^x+1}{e^x-1} \text{ or } \left(\frac{x}{e^x-1} + \frac{x}{2}\right) = 1 + B_1\frac{x^2}{2!} - B_3\frac{x^4}{4!} + B_5\frac{x^6}{6!} - B_7\frac{x^8}{8!} + \dots$$
and the numbers B_1, B_3, B_5, ..., which are calculated above are called Bernoulli's numbers, having been first discovered and used by James Bernoulli.

The coefficients of this expansion have been investigated as far as the term containing x^{32} by Rothe, and published in *Crelle's Journal.*

131. Many important expansions can be deduced from that of $\frac{x}{2}\,\frac{e^x+1}{e^x-1}$.

For example, $$x\coth x = x\frac{e^x+e^{-x}}{e^x-e^{-x}} = x\frac{e^{2x}+1}{e^{2x}-1}$$
$$= 1 + B_1\frac{2^2x^2}{2!} - B_3\frac{2^4x^4}{4!} + \dots$$

Writing ιx for x, $\iota x \coth \iota x$ becomes $x \cot x$, and we have

$$x \cot x = 1 - B_1\frac{2^2x^2}{2!} - B_3\frac{2^4x^4}{4!} - \dots$$

Again,

$$\begin{aligned}\tan x &= \cot x - 2\cot 2x \\ &= \frac{1}{x} - B_1\frac{2^2x}{2!} - B_3\frac{2^4x^3}{4!} - \dots \\ &\quad - 2\left[\frac{1}{2x} - B_1\frac{2^3x}{2!} - B_3\frac{2^7x^3}{4!} - \dots\right] \\ &= B_1\frac{2^2(2^2-1)}{2!}x + B_3\frac{2^4(2^4-1)}{4!}x^3 + \dots\end{aligned}$$

EXAMPLES.

1. Prove $\sin ax = ax - \dfrac{a^3x^3}{3!} + \dfrac{a^5x^5}{5!} - \dots + \dfrac{a^nx^n}{n!}\sin\dfrac{n\pi}{2} + \dots$, and that the remainder after r terms may be expressed as

$$\frac{a^rx^r}{r!}\sin\left(a\theta x + \frac{r\pi}{2}\right).$$

2. Prove $\cos ax = 1 - \dfrac{a^2x^2}{2!} + \dfrac{a^4x^4}{4!} - \dots + \dfrac{a^nx^n}{n!}\cos\dfrac{n\pi}{2} + \dots$, and that the remainder after r terms may be expressed as

$$\frac{a^rx^r}{r!}\cos\left(a\theta x + \frac{r\pi}{2}\right).$$

3. Prove $(1-x)^{-n} = 1 + nx + \dfrac{n(n+1)}{2!}x^2 + \dots$

$$+ \frac{n(n+1)\dots(n+r-2)}{(r-1)!}x^{r-1}$$
$$+ \frac{n(n+1)\dots(n+r-1)}{r!}\,\frac{x^r}{(1-\theta x)^{n+r}}.$$

4. Expand and find the general term of the expansion of $e^{ax}\cos bx$.

RESULTS. $1 + ax + \dfrac{a^2-b^2}{2!}x^2 + \dfrac{a(a^2-3b^2)}{3!}x^3 + \dots$

$$\text{General term} = \frac{(a^2+b^2)^{\frac{n}{2}}}{n!}\cos\left(n\tan^{-1}\frac{b}{a}\right)x^n.$$

5. Expand $\sinh^3 x$ and $\cosh^3 x$, giving the general term in each case.

6. Find the first three terms of the expansion in powers of x of $\log(1+\tan x)$. RESULT. $x - \frac{1}{2}x^2 + \frac{2}{3}x^3 + \ldots$.

7. Expand $\log(1 + x^3 e^x)$ as far as the term containing x^5.

RESULT. $x^3 + x^4 + \dfrac{x^5}{2} \ldots$.

8. Expand as far as the term containing x^4 (1) $\log(1+\cos x)$ and (2) $\log(1 + x\sin x)$.

RESULTS. $\begin{cases} (1)\ \log 2 - \dfrac{x^2}{4} - \dfrac{x^4}{96} \ldots \\ (2)\ x^2 - \frac{2}{3}x^4 + \ldots \end{cases}$

9. Prove $\log\cos x = -\dfrac{x^2}{2!} - 2\dfrac{x^4}{4!} - 16\dfrac{x^6}{6!} - 272\dfrac{x^8}{8!} \ldots$

10. Prove $\log\dfrac{\sin x}{x} = -\dfrac{x^2}{6} - \dfrac{x^4}{180} \ldots$

11. Prove $\log x\cot x = -\dfrac{x^2}{3} - \dfrac{7}{90}x^4 \ldots$

12. Prove $\log\dfrac{\sinh x}{x} = \dfrac{x^2}{6} - \dfrac{x^4}{180} \ldots$

13. Prove $\log\dfrac{\tan^{-1}x}{x} = -\dfrac{x^3}{3} + \dfrac{13}{90}x^4 - \dfrac{251}{5 \,.\, 7 \,.\, 9^2}x^6 \ldots$

14. Prove $e^{x\cos x} = 1 + x + \dfrac{x^2}{2} - \dfrac{x^3}{3} - \dfrac{11x^4}{24} - \dfrac{x^5}{5} \ldots$

15. Prove $e^{x\sec x} = 1 + x + \dfrac{1}{2}x^2 + \dfrac{2x^3}{3} \ldots$

16. Prove $\log\dfrac{xe^x}{e^x - 1} = \dfrac{x}{2} - \dfrac{x^2}{24} + \dfrac{x^4}{2880} \ldots$

17. Prove $\log\left\{\log(1+x)^{\frac{1}{x}}\right\} = -\dfrac{x}{2} + \dfrac{5x^2}{24} - \dfrac{x^3}{8} + \dfrac{251x^4}{2880} \ldots$

18. Prove

$$\log(1 - x + x^2) = -x + \frac{x^2}{2} + \frac{2x^3}{3} + \frac{x^4}{4} - \frac{x^5}{5} - \frac{x^6}{3} - \frac{x^7}{7} + \frac{x^8}{8} \ldots$$

19. Prove

$$\log(1+x+x^2+x^3+x^4)=x+\frac{x^2}{2}+\frac{x^3}{3}+\frac{x^4}{4}-\frac{4x^5}{5}+\frac{x^6}{6}+\dots$$

20. Prove $(1+x)^x=1+x^2-\frac{1}{2}x^3+\frac{5}{6}x^4-\frac{3}{4}x^5\dots$

Expand Examples 21 to 30 in ascending integral powers of x.

21. $\tan^{-1}x+\tanh^{-1}x$.

22. $\tan^{-1}\dfrac{2x}{1-x^2}+\sinh^{-1}\dfrac{2x}{1-x^2}$.

23. $\tan^{-1}\dfrac{3x-x^3}{1-3x^2}+\tanh^{-1}\dfrac{3x+x^3}{1+3x^2}$.

24. $\tan^{-1}\dfrac{p-qx}{q+px}$.

25. $\tan^{-1}\dfrac{\sqrt{1+x^2}-1}{x}$.

26. $\tan^{-1}\dfrac{x}{\sqrt{1-x^2}}$.

27. $\sec^{-1}\dfrac{1}{1-2x^2}$.

28. $\sin^{-1}\dfrac{2x}{1+x^2}$.

29. $\cos^{-1}\dfrac{x-x^{-1}}{x+x^{-1}}$.

30. $\sinh^{-1}(3x+4x^3)$.

31. If $y=e^{a\sin^{-1}x}=a_0+a_1x+a_2x^2+a_3x^3+\dots$, prove

(1) $(1-x^2)\dfrac{d^2y}{dx^2}=x\dfrac{dy}{dx}+a^2y$;

(2) $a_{n+2}=\dfrac{n^2+a^2}{(n+1)(n+2)}a_n$;

(3) $$e^{a\sin^{-1}x}=1+ax+\frac{a^2x^2}{2!}+\frac{a(a^2+1)}{3!}x^3+\frac{a^2(a^2+2^2)}{4!}x^4+\frac{a(a^2+1)(a^2+3^2)}{5!}x^5+\dots$$

(4) Deduce from (3), by expanding the left side according

to the exponential theorem and equating the coefficients of a, a^2, ... the series for $\sin^{-1}x$, $(\sin^{-1}x)^2$, ..., and show that if in the development of $\frac{(\sin^{-1}x)^1}{1}$, viz.,

$$\frac{x^1}{1}+\frac{1}{2}\cdot\frac{x^3}{3}+\frac{1\,.\,3}{2\,.\,4}\cdot\frac{x^5}{5}+\ldots,$$

every number which occurs be increased by unity, the result, viz.,
$$\frac{x^2}{2}+\frac{2}{3}\frac{x^4}{4}+\frac{2\,.\,4}{3\,.\,5}\cdot\frac{x^6}{6}+\ldots$$
is equal to $\frac{(\sin^{-1}x)^2}{2}$. [Professor Cayley.]

32. Prove that if $\log_e y = \tan^{-1}x$
$$(1+x^2)\frac{d^ny}{dx^n}=\{1-2(n-1)x\}\frac{d^{n-1}y}{dx^{n-1}}-(n-1)(n-2)\frac{d^{n-2}y}{dx^{n-2}},$$
and hence find the coefficient of x^5 in the expansion of y by Maclaurin's Theorem. [I. C. S. Exam.]

33. If y satisfy the equation $\frac{d^2y}{dx^2}-m^2y=0$, and if the first and second terms of its expansion be respectively $A+B$ and $(Am-Bm)x$, show that the general term is $\{A+(-1)^kB\}\frac{m^kx^k}{k!}$. Hence show that $y=Ae^{mx}+Be^{-mx}$.

34. If y satisfy the differential equation
$$\frac{d^2y}{dx^2}+2k\frac{dy}{dx}+(k^2+b^2)y=0,$$
and the first terms of the expansion of y are
$$1-kx+\frac{k^2-b^2}{2}x^2+\ldots$$
continue the expansion.

35. If a_n be the coefficient of x^n in the expansion of $e^x\sin x$ show that
$$a_n-\frac{a_{n-1}}{1!}+\frac{a_{n-2}}{2!}-\frac{a_{n-3}}{3!}+\ldots=\frac{\sin\frac{n\pi}{2}}{n!}.$$
[I. C. S. Exam.]

36. From $y=(x+\sqrt{1+x^2})^n$ obtain a linear differential equation with rational algebraic coefficients, and by means of it find the expansion of y in ascending powers of x.

37. From the relation $y=\dfrac{(1+x)^{\frac{1}{2}}}{1-x}$ obtain a linear differential equation with rational algebraic coefficients, and by means of it find the expansion of y in ascending powers of x.

38. If $\tan y = 1+ax+bx^2$, expand y in powers of x as far as x^3. [I. C. S. EXAM.]

39. If A_0, A_1, etc., be the successive coefficients in the expansion of $y=e^{\cos mx+\sin mx}$ prove

$$A_{n+1}=\frac{m}{n+1}\left\{A_n+\Sigma_1^n\frac{m^r}{r!}A_{n-r}\left(\cos\frac{r\pi}{2}-\sin\frac{r\pi}{2}\right)\right\}.$$

[I. C. S. EXAM.]

40. If $a_nx^n+a_{n+1}x^{n+1}+a_{n+2}x^{n+2}$ be three consecutive terms of the expansion of $(1-x^2)^{\frac{1}{2}}\sin^{-1}x$ in powers of x, prove that

$$a_{n+2}=\frac{n-1}{n+2}a_n;$$

also that all even terms vanish, and that the expansion is

$$x-\frac{1}{3}x^3-\frac{2}{3\,.\,5}\,.\,x^5-\frac{2\,.\,4}{3\,.\,5\,.\,7}x^7-\dots$$

[QUARTERLY JOURNAL.]

41. Show that if a rational integral function of x vanish for n values between given limits, its first and second differential coefficients will vanish for at least $(n-1)$ and $(n-2)$ values of x respectively between the same limits. Illustrate these results geometrically. [I. C. S. EXAM.]

42. Prove that no more than one root of an equation $f(x)=0$ can lie between any adjacent two of the roots of the equation $f'(x)=0$.

43. Show that the following expressions are positive for all positive values of x:

(i.) $(x-1)e^x+1$;

(ii.) $(x-2)e^x+x+2$;

(iii.) $(x-3)e^x+\frac{x^2}{2}+2x+3$;

(iv.) $x-\log(1+x)$.

[N.B.—By Art. 44, if $\frac{dy}{dx}$ be positive, y is increasing when x is increasing. Hence, if y be positive when $x=0$, and if also $\frac{dy}{dx}$ be positive as x increases from 0 to ∞, it follows that y will be positive for all positive values of x.]

44. Show for what values of x and at what differential coefficient Taylor's Theorem will fail if

$$f(x)=\frac{(x-a)^5(x-b)^{\frac{7}{2}}(x-c)^{\frac{9}{2}}}{(x-d)^{14}}.$$

45. Can $\log x$ or $\tan^{-1}\sqrt{\frac{x}{a}}$ be expanded by Maclaurin's Theorem in a series of ascending positive integral powers of x?

46. If $f(x)=e^{-\frac{1}{x}}$, how does Maclaurin's Theorem fail for an expansion in ascending powers of x? Is $f(x)$ continuous as x passes through zero?

47. If $f(x)=\frac{x}{1+e^{\frac{1}{x}}}$, show that there is a discontinuity in $\frac{df(x)}{dx}$ as x passes through zero.

48. Prove

$$\frac{f(x+h)+f(x-h)}{2}=f(x)+\frac{h^2}{2!}f''(x)+\frac{h^4}{4!}f''''(x)+\dots$$

49. If $$\theta=\log x,$$

prove that
$$u+x\frac{du}{dx}+\frac{x^2}{2!}\frac{d^2u}{dx^2}+\dots$$
$$=u+\log 2\,.\,\frac{du}{d\theta}+\frac{(\log 2)^2}{2!}\frac{d^2u}{d\theta^2}+\dots$$

50. Deduce from Taylor's Theorem, by putting $h = -x$, the series

$$f(x) = f(0) + xf'(x) - \frac{x^2}{2!}f''(x) + \frac{x^3}{3!}f'''(x) - \text{etc.}$$

[JOHN BERNOULLI.]

51. Prove

$$\tan^{-1}(x+h) = \tan^{-1}x + (h\sin\theta)\sin\theta - \frac{(h\sin\theta)^2}{2}\sin 2\theta$$
$$+ \frac{(h\sin\theta)^3}{3}\sin 3\theta - \frac{(h\sin\theta)^4}{4}\sin 4\theta + \text{etc.},$$

where $x = \cot\theta$.

52. Verify the following deductions from Ex. 51 :—

(1) $\frac{\pi}{2} = \theta + \cos\theta \, . \, \sin\theta + \frac{\cos^2\theta}{2}\sin 2\theta + \frac{\cos^3\theta}{3}\sin 3\theta + \frac{\cos^4\theta}{4}\sin 4\theta + \ldots$

by putting $h = -x = -\cot\theta$.

(2) $\frac{\pi}{2} = \frac{\theta}{2} + \sin\theta + \frac{1}{2}\sin 2\theta + \frac{1}{3}\sin 3\theta + \frac{1}{4}\sin 4\theta + \ldots$

by putting $h = -\sqrt{1+x^2} = -\frac{1}{\sin\theta}$.

(3) $\frac{\pi}{2} = \frac{\sin\theta}{\cos\theta} + \frac{1}{2} \cdot \frac{\sin 2\theta}{\cos^2\theta} + \frac{1}{3} \cdot \frac{\sin 3\theta}{\cos^3\theta} + \frac{1}{4} \cdot \frac{\sin 4\theta}{\cos^4\theta} + \ldots$

by putting $h = -x - \frac{1}{x} = -\frac{1}{\sin\theta \, . \, \cos\theta}$. [EULER.]

53. If $\frac{f(x)}{F(x)}$ be a rational fraction in which the denominator has n factors, each equal to $x - a$, and the remaining factors are $x - h$, $x - k$, etc., so that $F(x) = (x-a)^n \phi(x)$ where

$$\phi(x) = (x-h)(x-k)\ldots,$$

prove that

$$\frac{f(x)}{F(x)} = \frac{1}{(x-a)^n}\frac{f(a)}{\phi(a)} + \frac{1}{(x-a)^{n-1}}\frac{d}{da}\left\{\frac{f(a)}{\phi(a)}\right\}$$
$$+ \frac{1}{2!(x-a)^{n-2}}\frac{d^2}{da^2}\left\{\frac{f(a)}{\phi(a)}\right\} + \ldots + \frac{H}{x-h} + \ldots$$
$$= \frac{1}{(n-1)!}\frac{d^{n-1}}{da^{n-1}}\left\{\frac{f(a)}{\phi(a)(x-a)}\right\} + \frac{H}{x-h} + \ldots$$

54. Establish the following approximations to the length of a circular arc:—

Let C be the chord of the whole arc,
H do. half the arc,
Q do. quarter the arc.

(1) $\text{Arc} = \dfrac{8H - C}{3}$ nearly. [HUYGHENS.]

(2) $\text{Arc} = \dfrac{C + 256Q - 40H}{45}$ nearly.

Examine the closeness of the approximation in each case.

55. In the equation

$$f(x+h) = f(x) + hf'(x+\theta h),$$

show that the limiting value of θ as h is indefinitely diminished is $\frac{1}{2}$.

[Expand $f'(x+\theta h)$ in the above in powers of θh, and also $f(x+h)$ in powers of h, and compare the two series, remembering that θ itself, being a function of x and h, may be written $= A_0 + A_1 h + A_2 h^2 + \dots$ where A_0, $A_1 \dots$ are functions of x. The term A_0 will be the limiting value of θ when $h = 0$.]

56. In the equation

$$f(x+h) = f(x) + hf'(x+\theta h),$$

if θ be expanded in powers of h, the first four terms will be

$$\theta = \frac{1}{2} + \frac{1}{24}\frac{f_3}{f_2}h + \frac{1}{48}\frac{f_2 f_4 - f_3^2}{f_2^2}h^2 + \frac{33 f_5 f_2^2 - 90 f_2 f_3 f_4 + 55 f_3^3}{5760 f_2^3}h^3 + \dots$$

suffixes being used to denote differentiations.

57. Find by division the first six of Bernoulli's coefficients.

They are $\dfrac{1}{6}, \dfrac{1}{30}, \dfrac{1}{42}, \dfrac{1}{30}, \dfrac{5}{66}, \dfrac{691}{2730}.$

58. Prove by continuing the differentiations in Art. 130 that

$$\frac{1}{n+1} + \frac{1}{2} + \frac{n}{2!}B_1 - \frac{n(n-1)(n-2)}{4!}B_3 + \dots = 1,$$

a formula from which the values of the coefficients B_1, $B_3 \dots$ can be successively deduced by putting $n = 2, 4, 6,$ etc.

[DE MOIVRE.]

59. Expand $\left(\dfrac{\theta}{\sin\theta}\right)^2$ in powers of θ.

[Differentiate expansion of $\cot\theta$, Art. 131.]

60. Prove

$$\frac{\theta}{\sin\theta} = 1 + 2(2-1)\frac{B_1}{2!}\theta^2 + 2(2^3-1)\frac{B_3}{4!}\theta^4 + \ldots$$

[Use $\operatorname{cosec}\theta = \cot\dfrac{\theta}{2} - \cot\theta$ and Art. 131.]

61. Prove

$$\tanh x = \frac{2^2(2^2-1)}{2!}B_1 x - \frac{2^4(2^4-1)}{4!}B_3 x^3 + \ldots$$

62. By taking the logarithmic differential of the expression for $\sin\theta$ in factors and comparison of the expansion of the result with that of $\theta\cot\theta$ (Art. 131), show that

$$B_{2n-1} = \frac{2(2n)!}{(2\pi)^{2n}}\left\{1 + \frac{1}{2^{2n}} + \frac{1}{3^{2n}} + \ldots\right\}$$

$$= \frac{2(2n)!}{(2\pi)^{2n}}\,\frac{1}{\Pi\left(1 - \dfrac{1}{r^{2n}}\right)},$$

[RAABE.]

where $\Pi\left(1 - \dfrac{1}{r^{2n}}\right)$ denotes the continued product of such factors as $1 - \dfrac{1}{r^{2n}}$ for all integral prime values of r from 2 to ∞.

63. Expand $\sin(m\tan^{-1}x)(1+x^2)^{\frac{m}{2}}$ in powers of x.

CHAPTER VI.

PARTIAL DIFFERENTIATION.

132. Functions of Several Independent Variables.

Our attention has hitherto been confined to methods for the differentiation of functions of a single independent variable. In the present chapter we propose to discuss the case in which several such variables occur. Such functions are common; for instance, the area of a triangle depends upon two variables, viz., the base and the altitude; while the volume of a rectangular box depends upon three, viz., its length, breadth, and depth; and it is plain that each of these variables may vary independently of the others.

133. Partial Differentiation.

If a differentiation of a function of several independent variables be performed with regard to any one of them just as if the others were constants, it is said to be a *partial differentiation.*

The symbols $\frac{\partial}{\partial x}$, $\frac{\partial}{\partial y}$, etc., are used to denote such differentiations, and the expressions $\frac{\partial u}{\partial x}$, $\frac{\partial u}{\partial y}$, etc., are called *partial differential coefficients* with regard to x, y, etc., respectively.

Thus if, for instance,

$$u = e^{xy} \sin z,$$

we have
$$\frac{\partial u}{\partial x} = y e^{xy} \sin z,$$
$$\frac{\partial u}{\partial y} = x e^{xy} \sin z,$$
$$\frac{\partial u}{\partial z} = e^{xy} \cos z.$$

134. **Analytical Meaning.**

The meanings of the differential coefficients thus formed are clear; for if we denote u by $f(x, y, z)$ the operation denoted by $\frac{\partial u}{\partial x}$ may be expressed as

$$Lt_{h=0} \frac{f(x+h, y, z) - f(x, y, z)}{h},$$

and similarly for $\frac{\partial u}{\partial y}$ or $\frac{\partial u}{\partial z}$.

135. **Geometrical Illustration.**

It will throw additional light upon the subject of partial differentiation if we explain the geometrical meaning of the process for the case of two independent variables.

Let $PQRS$ be an elementary portion of the surface $z = f(x, y)$ cut off by the four planes

$$\left.\begin{array}{l} Y = y, \quad Y = y + \delta y \\ X = x, \quad X = x + \delta x \end{array}\right\} \text{[Capital letters representing current co-ordinates]},$$

so that the co-ordinates of the corners P, Q, R, S are

for P $\qquad x, y, f(x, y),$

for Q $\quad x+\delta x,\ y,\ f(x+\delta x,\ y)$,
for S $\quad x,\ y+\delta y,\ f(x,\ y+\delta y)$,
and for R $\quad x+\delta x,\ y+\delta y,\ f(x+\delta x,\ y+\delta y)$.

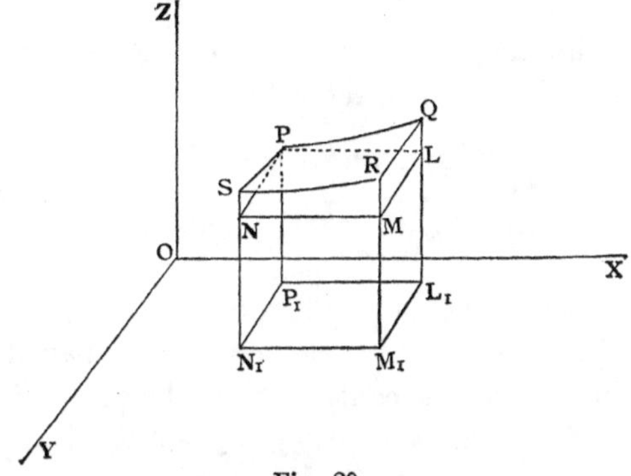

Fig. 20.

If $PLMN$ be a plane through P, parallel to the plane of xy, and cutting the ordinates of P, Q, R, S in P, L, M, N respectively, we have

$$\left.\begin{aligned} LQ&=f(x+\delta x,\ y)-f(x,\ y),\\ NS&=f(x,\ y+\delta y)-f(x,\ y),\\ MR&=f(x+\delta x,\ y+\delta y)-f(x,\ y).\end{aligned}\right\}\ldots\ldots\ldots(1)$$

Hence the partial differential coefficient $\frac{\partial z}{\partial x}$ obtained by *considering y a constant* is

$$=Lt_{\delta x=0}\frac{f(x+\delta x,\ y)-f(x,\ y)}{\delta x}=Lt\frac{LQ}{PL}=Lt\tan LPQ.\ (2)$$

=*tangent of the angle which the tangent at P to the curved section PQ (parallel to the plane xz) makes with a line drawn parallel to the axis of x.*

Similarly $\frac{\partial z}{\partial y}$, which is obtained on the supposition that x is constant

$=Lt \tan NPS,$ ....(3)

$=$ *tangent of the angle which the tangent at P to a section parallel to the plane of yz makes with a parallel to the axis of y.*

136. If the tangent plane at P to the surface cut LQ, MR, NS in Q', R', S' respectively,

$$LQ' = PL \tan LPQ' = \frac{\partial z}{\partial x} \cdot \delta x, \quad \text{...........(4)}$$

$$NS' = PN \tan NPS' = \frac{\partial z}{\partial y} \cdot \delta y, \quad \text{...........(5)}$$

Also the section made on the tangent plane by the four bounding planes of the element is a parallelogram, and the height of its centre above the plane $PLMN$ is given by $\frac{1}{2}MR'$ and also by $\frac{1}{2}(LQ'+NS')$, which proves that

$$MR' = LQ' + NS'$$

$$= \frac{\partial z}{\partial x}\delta x + \frac{\partial z}{\partial y}\delta y. \quad \text{....................(6)}$$

The expressions proved in (4), (5), and (6) are *first approximations* to the lengths LQ, NS, and MR respectively, and differ from those lengths by small quantities of higher order than PL and PN, and which are therefore negligible in the limit when δx and δy are taken very small. The investigation of the total values of LQ, NS, MR must be postponed until we have investigated the extension of Taylor's Theorem to functions of several variables. (Art. 156.)

137. **Differentials.**

It is useful at this point to introduce a new notation, which will prove especially convenient from considerations of symmetry.

Let Dx, Dy, Dz be quantities either finite or infinitesimally small whose ratios to one another are the *same as the limiting ratios of* δx, δy, δz, when these latter are ultimately diminished indefinitely. We shall call the quantities thus defined the *differentials* of x, y, z. Also, as we shall be merely concerned with the ratios of these quantities, and any equation into which they may enter will be homogeneous in them, it is unnecessary to define them farther or to obtain absolute values for them. The student is warned again (see Art. 41) that the differential coefficient $\frac{dy}{dx}$ is to be considered as the result of performing the operation represented by $\frac{d}{dx}$ upon y, an operation described in Art. 39. The *dy* and *dx of the symbol* $\frac{dy}{dx}$ cannot therefore be separated, and have separately no meaning, and hence have no connection with the differentials Dx and Dy as defined in the present article; but at the same time we have by definition

$$Dy : Dx = \text{Limit of the ratio } \delta y : \delta x$$

$$= Lt\frac{\delta y}{\delta x} : 1$$

$$= \frac{dy}{dx} : 1,$$

and therefore $$Dy = \frac{dy}{dx} Dx,$$

and $\frac{Dy}{Dx}$ (which is a fraction)

$= \frac{dy}{dx}$ (which is the result of the process of Art. 39).

We have used a capital in the differentials Dx, Dy, Dz for the purpose of explanation, and for the avoidance of any confusion between the notation for differentials and for differential coefficients; but when once understood there is no necessity for the continuance of the capital letter, and it is usual in the higher branches of mathematics to denote the same quantities by dx, dy, dz. Hence we shall in future adopt this notation.

138. Equation 6 of Art. 136 may now be written

$$dz = \frac{\partial z}{\partial x}dx + \frac{\partial z}{\partial y}dy$$

when δx, δy, δz become infinitesimally small. This value of dz is termed the total differential of z with regard to x and y. The *total* differential of z is therefore equal to the *sum of the partial differentials* formed under the supposition that y and x are alternately constant.

Ex. Consider the surface

$$z = xy,$$

then $$\frac{\partial z}{\partial x} = y \text{ and } \frac{\partial z}{\partial y} = x,$$

whence $$dz = xdy + ydx.$$

139. It is easy to pass from a form in which differentials are used to the equivalent form in terms of differential coefficients. For instance, the equation

$$dz = \frac{\partial z}{\partial x}dx + \frac{\partial z}{\partial y}dy$$

may be at once written

$$\frac{dz}{dt} = \frac{\partial z}{\partial x}\frac{dx}{dt} + \frac{\partial z}{\partial y}\frac{dy}{dt},$$

where t is some fourth variable in terms of which

each of the variables x, y, z may be expressed; for $dz=\frac{dz}{dt}\cdot dt$, $dx=\frac{dx}{dt}\cdot dt$, $dy=\frac{dy}{dt}\cdot dt$ (Art. 137).

Similarly the equation

$$ds^2=dx^2+dy^2$$

may, by the same article, be written in the language of differential coefficients as

$$\left(\frac{dx}{ds}\right)^2+\left(\frac{dy}{ds}\right)^2=1,$$

or

$$\left(\frac{ds}{dt}\right)^2=\left(\frac{dx}{dt}\right)^2+\left(\frac{dy}{dt}\right)^2,$$

or

$$\left(\frac{ds}{dx}\right)^2=1+\left(\frac{dy}{dx}\right)^2,$$

or

$$\left(\frac{ds}{dy}\right)^2=1+\left(\frac{dx}{dy}\right)^2.$$

140. **Total Differential (Analytical).** *Two independent variables.* We may investigate the total differential of the function $\phi(x, y)$ analytically as follows:

Let $u=\phi(x, y)$,

and when x becomes $x+h$ and y becomes $y+k$, let u become $u+\delta u$, then

$$u+\delta u=\phi(x+h, y+k)$$

and

$$\delta u=\phi(x+h, y+k)-\phi(x, y)$$

$$=\frac{\phi(x+h, y+k)-\phi(x, y+k)}{h}\cdot h$$

$$+\frac{\phi(x, y+k)-\phi(x, y)}{k}\cdot k,$$

and when we proceed to the limit in which h and k become indefinitely small we have

$$Lt_{\substack{h=0\\k=0}}\frac{\phi(x+h, y+k)-\phi(x, y+k)}{h}=Lt_{k=0}\frac{\partial}{\partial x}\phi(x, y+k)=\frac{\partial u}{\partial x};$$

and $$Lt_{k=0}\frac{\phi(x, y+k)-\phi(x, y)}{k}=\frac{\partial u}{\partial y}.$$

Also $du:dx:dy=$ the ultimate ratios of $\delta u:h:k$, hence

$$du=\frac{\partial u}{\partial x}dx+\frac{\partial u}{\partial y}dy.$$

141. *Several independent variables.*

We may readily extend this result to a function of three or of any number of variables.

Let $$u=\phi(x_1, x_2, x_3),$$

and let the increments of x_1, x_2, x_3, be respectively h_1, h_2, h_3, and let the corresponding increment of u be δu; then

$$\begin{aligned}\delta u&=\phi(x_1+h_1, x_2+h_2, x_3+h_3)-\phi(x_1, x_2, x_3)\\&=\frac{\phi(x_1+h_1, x_2+h_2, x_3+h_3)-\phi(x_1, x_2+h_2, x_3+h_3)}{h_1}h_1\\&\quad+\frac{\phi(x_1, x_2+h_2, x_3+h_3)-\phi(x_1, x_2, x_3+h_3)}{h_2}h_2\\&\quad+\frac{\phi(x_1, x_2, x_3+h_3)-\phi(x_1, x_2, x_3)}{h_3}h_3;\end{aligned}$$

whence, on taking the limit and substituting the ratios $du:dx_1:dx_2:dx_3$ instead of the ultimate ratios of $\delta u:h_1:h_2:h_3$, we have

$$du=\frac{\partial u}{\partial x_1}dx_1+\frac{\partial u}{\partial x_2}dx_2+\frac{\partial u}{\partial x_3}dx_3,$$

i.e., the total differential of u when x_1, x_2, x_3, all vary is the sum of the partial differentials obtained under the supposition that when each one in turn varies the others are constant.

142. And in exactly the same way if

$$u=\phi(x_1, x_2, \ldots x_n),$$

we have $$du=\frac{\partial u}{\partial x_1}dx_1+\frac{\partial u}{\partial x_2}dx_2+\frac{\partial u}{\partial x_3}dx_3+\ldots+\frac{\partial u}{\partial x_n}dx_n.$$

143. Total Differential Coefficient.

If
$$u=\phi(x_1, x_2)$$
where x_1 and x_2 are known functions of a single variable x, we have
$$du=\frac{\partial u}{\partial x_1}dx_1+\frac{\partial u}{\partial x_2}dx_2,$$
and remembering that
$$du=\frac{du}{dx}dx,\ dx_1=\frac{dx_1}{dx}dx,\ dx_2=\frac{dx_2}{dx}dx,$$
we obtain
$$\frac{du}{dx}=\frac{\partial u}{\partial x_1}\cdot\frac{dx_1}{dx}+\frac{\partial u}{\partial x_2}\cdot\frac{dx_2}{dx}.$$

And similarly, if $u=\phi(x_1, x_2, \ldots, x_n)$, where $x_1, x_2, \ldots, x_n$, are known functions of x, we obtain
$$\frac{du}{dx}=\frac{\partial u}{\partial x_1}\cdot\frac{dx_1}{dx}+\frac{\partial u}{\partial x_2}\cdot\frac{dx_2}{dx}+\ldots\frac{\partial u}{\partial x_n}\cdot\frac{dx_n}{dx}.$$

And further, if $x_1, x_2, x_3, \ldots, x_n$ be each known functions of several variables $x, y, z, \ldots$, we shall have in the same way the series of relations
$$\frac{\partial u}{\partial x}=\frac{\partial u}{\partial x_1}\cdot\frac{\partial x_1}{\partial x}+\frac{\partial u}{\partial x_2}\cdot\frac{\partial x_2}{\partial x}+\ldots\frac{\partial u}{\partial x_n}\cdot\frac{\partial x_n}{\partial x},$$
$$\frac{\partial u}{\partial y}=\frac{\partial u}{\partial x_1}\cdot\frac{\partial x_1}{\partial y}+\frac{\partial u}{\partial x_2}\cdot\frac{\partial x_2}{\partial y}+\ldots\frac{\partial u}{\partial x_n}\cdot\frac{\partial x_n}{\partial y},$$
etc.

144. An Important Case.

The case in which $u=\phi(x, y)$, y being a function of x, is from its frequent occurrence worthy of special notice.

Here, by Art. 143, $\dfrac{du}{dx}=\dfrac{\partial\phi}{\partial x}+\dfrac{\partial\phi}{\partial y}\cdot\dfrac{dy}{dx}$

since
$$\frac{dx}{dx}=1.$$

145. Differentiation of an Implicit Function.

If we have
$$\phi(x, y) = 0,$$
then
$$\phi(x+h, y+k) = 0,$$
and the δu of Art. 140 vanishes. Proceeding as in that article we obtain
$$\frac{\partial\phi}{\partial x} + \frac{\partial\phi}{\partial y}\cdot\frac{dy}{dx} = 0,$$
or
$$\frac{dy}{dx} = -\frac{\dfrac{\partial\phi}{\partial x}}{\dfrac{\partial\phi}{\partial y}}.$$

This is a *very useful formula* for the determination of $\dfrac{dy}{dx}$ in cases in which the relation between x and y is *an implicit one*, of which the solution is inconvenient or impossible.

Ex. $\phi(x, y) \equiv x^3 + y^3 - 3axy = 0$; find $\dfrac{dy}{dx}$.

Here
$$\left.\begin{aligned}\frac{\partial\phi}{\partial x} &= 3(x^2 - ay)\\ \frac{\partial\phi}{\partial y} &= 3(y^2 - ax)\end{aligned}\right\} \quad \therefore \quad \frac{dy}{dx} = -\frac{x^2 - ay}{y^2 - ax}.$$
and

146. Order of Partial Differentiations Commutative.

Suppose we have any relation
$$y = \phi(x, a),$$
where a is a constant, and that by differentiation we obtain
$$\frac{dy}{dx} = F(x, a),$$
it is obvious that the result of differentiating $\phi(x, a')$ would be $F(x, a')$; that is, the operation of changing a to a' may be performed either before or after the differentiation, with the same result. We may put this statement into another form, thus: Let E_a be an

operative symbol such that when applied to any function of a it will change a to a', *i.e.*, such that

$$E_a f(a) = f(a').$$

then in operating upon the function $\phi(x, a)$ the operations E_a and $\frac{d}{dx}$ are commutative, that is,

$$E_a \frac{d}{dx}\phi(x, a) = \frac{d}{dx}E_a\phi(x, a) = F(x, a').$$

Next, suppose $\qquad z = \phi(x, y).$

The partial differential operations $\frac{\partial}{\partial x}$ and $\frac{\partial}{\partial y}$ have been defined to be such that when the operation with regard to either variable is performed the other variable is to be considered constant. We propose to show that these operations are *commutative*, *i.e.*, that

$$\frac{\partial}{\partial x}\frac{\partial}{\partial y}z = \frac{\partial}{\partial y}\frac{\partial}{\partial x}z.$$

Let E_y denote the operation of changing y to $y+\delta y$ in any function to which it is applied; then E_y and the partial operation $\frac{\partial}{\partial x}$ are *commutative symbols*. And

$$\frac{\partial}{\partial y}\frac{\partial}{\partial x}\phi(x,y) = Lt_{\delta y=0}\frac{E_y\frac{\partial\phi(x, y)}{\partial x} - \frac{\partial\phi(x, y)}{\partial x}}{\delta y}, \text{ by Def.,}$$

$$= Lt_{\delta y=0}\frac{\frac{\partial}{\partial x}E_y\phi(x, y) - \frac{\partial\phi(x, y)}{\partial x}}{\delta y}$$

$$= Lt_{\delta y=0}\frac{\partial}{\partial x}\frac{E_y\phi(x, y) - \phi(x, y)}{\delta y}$$

$$= \frac{\partial}{\partial x}Lt_{\delta y=0}\frac{E_y\phi(x, y) - \phi(x, y)}{\delta y}$$

$$= \frac{\partial}{\partial x}\frac{\partial}{\partial y}\phi(x, y).$$

147. Another Proof.

The symbols $\frac{\partial}{\partial x}$ and $\frac{\partial}{\partial y}$ may also be shown to be commutative as follows:

By definition

$$\frac{\partial\phi(x, y)}{\partial x} = Lt_{h=0}\frac{\phi(x+h, y)-\phi(x, y)}{h},$$

and $\frac{\partial}{\partial y}\frac{\partial}{\partial x}\phi(x, y)$

$$= Lt_{k=0}\frac{Lt_{h=0}\frac{\phi(x+h, y+k)-\phi(x, y+k)}{h} - Lt_{h=0}\frac{\phi(x+h, y)-\phi(x, y)}{h}}{k}$$

$$= Lt_{\substack{h=0\\k=0}}\frac{\phi(x+h, y+k)-\phi(x, y+k)-\phi(x+h, y)+\phi(x, y)}{h.k}.$$

And, similarly, $\frac{\partial}{\partial x}\frac{\partial}{\partial y}\phi(x, y)$ may be shown equal to the same expression.

148. Extension of Rule.

This rule admits of easy extension by its repeated application. Thus

$$\left(\frac{\partial}{\partial x}\right)^2\left(\frac{\partial}{\partial y}\right)\phi = \left(\frac{\partial}{\partial x}\right)\left(\frac{\partial}{\partial y}\right)\left(\frac{\partial}{\partial x}\right)\phi$$

$$= \frac{\partial}{\partial y}\left(\frac{\partial}{\partial x}\right)^2\phi.$$

Similarly $\left(\frac{\partial}{\partial x}\right)^m\left(\frac{\partial}{\partial y}\right)^n\phi = \left(\frac{\partial}{\partial y}\right)^n\left(\frac{\partial}{\partial x}\right)^m\phi.$

Also if we have more than two independent variables for instance, if $u = \phi(x, y, z)$

$$\left(\frac{\partial}{\partial x}\right)\left(\frac{\partial}{\partial y}\right)\left(\frac{\partial}{\partial z}\right)u = \left(\frac{\partial}{\partial y}\right)\left(\frac{\partial}{\partial x}\right)\left(\frac{\partial}{\partial z}\right)u$$

$$= \left(\frac{\partial}{\partial y}\right)\left(\frac{\partial}{\partial z}\right)\left(\frac{\partial}{\partial x}\right)u = \text{etc.}$$

so that the order in which the differentiations are performed is immaterial in the final result.

149. Notation.

It is usual to adopt for

$$\left(\frac{\partial}{\partial x}\right)^2 u,\ \left(\frac{\partial}{\partial x}\right)\left(\frac{\partial}{\partial y}\right)u,\ \left(\frac{\partial}{\partial x}\right)^p\left(\frac{\partial}{\partial y}\right)^q\left(\frac{\partial}{\partial z}\right)^r u,\ \text{etc.},$$

the more convenient notation

$$\frac{\partial^2 u}{\partial x^2},\ \frac{\partial^2 u}{\partial x \partial y},\ \frac{\partial^{p+q+r} u}{\partial x^p \partial y^q \partial z^r},\ \text{etc.},$$

and the propositions above enunciated will then be written

$$\frac{\partial^2 u}{\partial x \partial y} = \frac{\partial^2 u}{\partial y \partial x},$$

$$\frac{\partial^3 u}{\partial x^2 \partial y} = \frac{\partial^3 u}{\partial y \partial x^2},$$

$$\frac{\partial^{m+n} u}{\partial x^m \partial y^n} = \frac{\partial^{m+n} u}{\partial y^n \partial x^m},$$

etc.

150. The formulae here established may be easily verified in any particular example.

Ex. Let $u = \sin(xy)$,

then $\frac{\partial u}{\partial x} = y\cos(xy)$,

and $\frac{\partial^2 u}{\partial y \partial x} = \cos xy - xy \sin xy$. ....(1)

Again $\frac{\partial u}{\partial y} = x\cos xy$,

and $\frac{\partial^2 u}{\partial x \partial y} = \cos xy - xy \sin xy$, ...(2)

and the agreement of equations (1) and (2) verifies for this example the result of Arts. 146, 147.

151. It is convenient to use the letters p, q, r, s, t, to denote the partial differential coefficients

$$\frac{\partial\phi}{\partial x}, \frac{\partial\phi}{\partial y}, \frac{\partial^2\phi}{\partial x^2}, \frac{\partial^2\phi}{\partial x\partial y}, \frac{\partial^2\phi}{\partial y^2},$$

where ϕ is a given function of the two variables x and y.

Hence we have, if $\quad z=\phi(x, y)$,

$$dz = pdx + qdy, \text{ Art. } 140;$$

and to obtain $\frac{dy}{dx}$ from the implicit relation $\phi(x, y)=0$,

we have
$$\frac{dy}{dx} = -\frac{p}{q}.$$

152. **To obtain the Second Differential Coefficient of an Implicit Function.**

To obtain $\frac{d^2y}{dx^2}$ we have only to differentiate the last result of the preceding article; thus,

$$\frac{d^2y}{dx^2} = -\frac{q\frac{dp}{dx} - p\frac{dq}{dx}}{q^2}.$$

Now
$$\frac{dp}{dx} = \frac{\partial p}{\partial x} + \frac{\partial p}{\partial y}\frac{dy}{dx} = r + s\left(-\frac{p}{q}\right) = \frac{qr-ps}{q},$$

and
$$\frac{dq}{dx} = \frac{\partial q}{\partial x} + \frac{\partial q}{\partial y}\frac{dy}{dx} = s + t\left(-\frac{p}{q}\right) = \frac{qs-pt}{q},$$

giving
$$\frac{d^2y}{dx^2} = -\frac{q\left(\frac{qr-ps}{q}\right) - p\left(\frac{qs-pt}{q}\right)}{q^2}$$

$$= -\frac{q^2r - 2pqs + p^2t}{q^3}.$$

Similarly $\frac{d^3y}{dx^3}$, etc., may be found, but the results are complicated.

EXAMPLES.

1. If $$u=x^m y^n,$$
prove $$\frac{du}{u}=m\frac{dx}{x}+n\frac{dy}{y},$$
and verify the formula $$\frac{\partial^2 u}{\partial x\partial y}=\frac{\partial^2 u}{\partial y\partial x}.$$

2. Verify the formula $\frac{\partial^2 u}{\partial x\partial y}=\frac{\partial^2 u}{\partial y\partial x}$ in each of the following cases:—

(1) $u=\sin^{-1}\frac{y}{x}$.

(2) $u=\frac{x^2y^2}{a^2-z^2}$.

(3) $u=\log\frac{x^2+y^2}{xy}$.

(4) $u=x^y$.

3. If $$u=\frac{xy}{2x+z},$$
show that $$\frac{\partial^3 u}{\partial y\partial z^2}=\frac{\partial^3 u}{\partial z^2\partial y}.$$

4. If $$x=r\cos\theta \text{ and } y=r\sin\theta,$$
prove $$dx=\cos\theta dr-r\sin\theta d\theta,$$
and $$dy=\sin\theta dr+r\cos\theta d\theta;$$
and hence that $$dx^2+dy^2=dr^2+r^2d\theta^2,$$
and that $$xdy-ydx=r^2d\theta.$$

5. If $$u=\log(x^2+y^2+z^2),$$
prove $$x\frac{\partial^2 u}{\partial y\partial z}=y\frac{\partial^2 u}{\partial z\partial x}=z\frac{\partial^2 u}{\partial x\partial y}.$$

6. Prove that if
$$\frac{x^2}{a^2}+\frac{y^2}{b^2}=1,$$
$$\frac{dy}{dx}=-\frac{b^2x}{a^2y} \text{ and } \frac{d^2y}{dx^2}=-\frac{b^4}{a^2y^3}.$$

7. Show that if
$$x^m+y^m=a^m,$$
$$\frac{d^2y}{dx^2}=-(m-1)a^m\frac{x^{m-2}}{y^{2m-1}}.$$

153. *To find* $\frac{dy}{dx}$ *and* $\frac{dz}{dx}$ *from the equations*

$$F_1(x, y, z)=0,$$
$$F_2(x, y, z)=0.$$

Here, as in Art. 145,

$$\frac{\partial F_1}{\partial x}+\frac{\partial F_1}{\partial y}\cdot\frac{dy}{dx}+\frac{\partial F_1}{\partial z}\cdot\frac{dz}{dx}=0,$$

$$\frac{\partial F_2}{\partial x}+\frac{\partial F_2}{\partial y}\cdot\frac{dy}{dx}+\frac{\partial F_2}{\partial z}\cdot\frac{dz}{dx}=0.$$

Solving these equations we obtain

$$\frac{\frac{dy}{dx}}{\frac{\partial F_1}{\partial z}\cdot\frac{\partial F_2}{\partial x}-\frac{\partial F_2}{\partial z}\cdot\frac{\partial F_1}{\partial x}}=\frac{\frac{dz}{dx}}{\frac{\partial F_1}{\partial x}\cdot\frac{\partial F_2}{\partial y}-\frac{\partial F_2}{\partial x}\cdot\frac{\partial F_1}{\partial y}}$$

$$=\frac{1}{\frac{\partial F_1}{\partial y}\cdot\frac{\partial F_2}{\partial z}-\frac{\partial F_2}{\partial y}\cdot\frac{\partial F_1}{\partial z}},$$

which give the values of $\frac{dy}{dx}$ and $\frac{dz}{dx}$.

Ex. Given $y=F_1(x, z)$,
and $z=F_2(x, y)$,
prove

$$\frac{dy}{dx}=\frac{\frac{\partial F_1}{\partial x}+\frac{\partial F_1}{\partial z}\cdot\frac{\partial F_2}{\partial x}}{1-\frac{\partial F_1}{\partial z}\cdot\frac{\partial F_2}{\partial y}}.$$

154. *Given that*

$$V=\phi(x+\xi t, y+\eta t, z+\zeta t, \ldots)$$

where $x, y, z \ldots, \quad \xi, \eta, \zeta \ldots,$ and t
form a system of independent variables, to show that

$$\frac{\partial V}{\partial t}=\xi\frac{\partial V}{\partial x}+\eta\frac{\partial V}{\partial y}+\zeta\frac{\partial V}{\partial z}+\ldots$$

Let $$x_1 = x + \xi t,$$
$$y_1 = y + \eta t,$$
etc.,

so that $$\frac{\partial x_1}{\partial x} = 1, \quad \frac{\partial y_1}{\partial y} = 1, \text{ etc.,}$$
$$\frac{\partial x_1}{\partial t} = \xi, \quad \frac{\partial y_1}{\partial t} = \eta, \text{ etc.,}$$
$$\frac{\partial x_1}{\partial y} = 0, \quad \frac{\partial y_1}{\partial x} = 0, \text{ etc.;}$$

then $$V = \phi(x_1, y_1, z_1, \ldots),$$

and $$\frac{\partial V}{\partial x} = \frac{\partial V}{\partial x_1} \cdot \frac{\partial x_1}{\partial x} + \frac{\partial V}{\partial y_1} \cdot \frac{\partial y_1}{\partial x} + \ldots \text{ (Art. 143)}$$
$$= \frac{\partial V}{\partial x_1}.$$

Similarly $$\frac{\partial V}{\partial y} = \frac{\partial V}{\partial y_1}, \text{ etc.,}$$

and $$\frac{\partial V}{\partial t} = \frac{\partial V}{\partial x_1} \cdot \frac{\partial x_1}{\partial t} + \frac{\partial V}{\partial y_1} \cdot \frac{\partial y_1}{\partial t} + \ldots$$
$$= \xi \frac{\partial V}{\partial x} + \eta \frac{\partial V}{\partial y} + \zeta \frac{\partial V}{\partial z} + \ldots.$$

155. Hence we have the following identity of operators, viz. :—

$$\frac{\partial}{\partial t} \equiv \xi \frac{\partial}{\partial x} + \eta \frac{\partial}{\partial y} + \zeta \frac{\partial}{\partial z} + \ldots,$$

and as the *variables are all independent and the operators partial*,

$$\left(\frac{\partial}{\partial t}\right)^n \equiv \left(\xi \frac{\partial}{\partial x} + \eta \frac{\partial}{\partial y} + \zeta \frac{\partial}{\partial z} + \ldots\right)^n,$$

the development being made in *formal analogy with the Multinomial Theorem.*

For example, in the case of

$$V = \phi(x + \xi t, y + \eta t),$$

we shall have $$\frac{\partial V}{\partial t}=\xi\frac{\partial V}{\partial x}+\eta\frac{\partial V}{\partial y},$$
$$\frac{\partial^2 V}{\partial t^2}=\xi^2\frac{\partial^2 V}{\partial x^2}+2\xi\eta\frac{\partial^2 V}{\partial x\partial y}+\eta^2\frac{\partial^2 V}{\partial y^2},$$
etc.

TAYLOR'S THEOREM. EXTENSION.

156. **To expand $\phi(x+h, y+k)$ in powers of h and k.**

By Taylor's Theorem we obtain
$$\phi(x+h, y+k)=\phi(x+h, y)+k\frac{\partial\phi(x+h, y)}{\partial y}+\frac{k^2}{2!}\frac{\partial^2\phi(x+h, y)}{\partial y^2}+\dots$$
and expanding each term we have
$$\begin{aligned}\phi(x+h, y+k)&=\phi(x, y)+h\frac{\partial\phi}{\partial x}+\frac{h^2}{2!}\frac{\partial^2\phi}{\partial x^2}+\dots\\&\qquad+k\frac{\partial\phi}{\partial y}+hk\frac{\partial^2\phi}{\partial x\partial y}+\dots\\&\qquad\qquad+\frac{k^2}{2!}\frac{\partial^2\phi}{\partial y^2}+\dots\\&=\phi(x, y)+\left(h\frac{\partial\phi}{\partial x}+k\frac{\partial\phi}{\partial y}\right)\\&\qquad+\frac{1}{2!}\left(h^2\frac{\partial^2\phi}{\partial x^2}+2hk\frac{\partial^2\phi}{\partial x\partial y}+k^2\frac{\partial^2\phi}{\partial y^2}\right)+\dots\end{aligned}$$
or, as it may be written symbolically,
$$\phi(x+h, y+k)=\phi(x, y)+\left(h\frac{\partial}{\partial x}+k\frac{\partial}{\partial y}\right)\phi+\frac{1}{2!}\left(h\frac{\partial}{\partial x}+k\frac{\partial}{\partial y}\right)^2\phi+\dots$$

157. Since it is immaterial whether we first expand with regard to k and then with regard to h, or in the opposite order, we obtain by comparison of the coefficient of hk in the two results the important theorem
$$\frac{\partial^2\phi}{\partial x\partial y}=\frac{\partial^2\phi}{\partial y\partial x}$$
already established in Arts. 146, 147.

158. **Further Extension. Several Variables.**

The form of the general term in the preceding case and the further extension of Taylor's Theorem to the expansion of a function of *several* variables is more readily investigated as follows:

Let $$\phi(x+\xi t,\ y+\eta t,\ \ldots)$$
be called $F(t)$. Then Maclaurin's Theorem gives

$$F(t)=F(0)+tF'(0)+\frac{t^2}{2!}F''(0)+\ldots+\frac{t^n}{n!}F^n(\theta t),$$

and by Art. 155

$$F^r(t)=\left(\xi\frac{\partial}{\partial x}+\eta\frac{\partial}{\partial y}+\ldots\right)^r\phi(x+\xi t,\ \ldots),$$

and since the variables $x, y, \ldots$, are independent of t, we may put $t=0$ either before or after the operation has been performed.

Hence $$F^r(0)=\left(\xi\frac{\partial}{\partial x}+\eta\frac{\partial}{\partial y}+\ldots\right)^r\phi(x,\ y,\ \ldots).$$

We thus obtain

$$\begin{aligned}\phi(x+\xi t,\ y+\eta t,\ldots)=\phi(x,y,z,\ldots)&+t\left(\xi\frac{\partial}{\partial x}+\eta\frac{\partial}{\partial y}+\ldots\right)\phi(x,y,\ldots)\\&+\frac{t^2}{2!}\left(\xi\frac{\partial}{\partial x}+\ldots\right)^2\phi(x,\ \ldots)+\ldots\\&+\frac{t^n}{n!}\left(\xi\frac{\partial}{\partial x}+\ldots\right)^n\phi(x+\xi\theta t,\ y+\eta\theta t,\ldots).\end{aligned}$$

Now, putting $h=\xi t,\ k=\eta t,\ l=\zeta t,\ \ldots$, we obtain

$$\begin{aligned}\phi(x+h,\ y+k,\ z+l,\ \ldots)&=\phi(x,\ y,\ z,\ \ldots)\\&+\left(h\frac{\partial}{\partial x}+k\frac{\partial}{\partial y}+l\frac{\partial}{\partial z}+\ldots\right)\phi(x,\ \ldots)\\&+\frac{1}{2!}\left(h\frac{\partial}{\partial x}+k\frac{\partial}{\partial y}+\ldots\right)^2\phi(x,\ \ldots)+\ldots\\&+\frac{1}{n!}\left(h\frac{\partial}{\partial x}+k\frac{\partial}{\partial y}+\ldots\right)^n\phi(x+\theta h,y+\theta k,\ldots).\end{aligned}$$

159. Extension of Maclaurin's Theorem.

Moreover, if we put $x=0$ and $y=0$, and then write x for h and y for k, we have an *extension of Maclaurin's Theorem* which, for two independent variables, may be written

$$\phi(x, y)=\phi(0, 0)+x\left(\frac{\partial\phi}{\partial x}\right)_0+y\left(\frac{\partial\phi}{\partial y}\right)_0$$
$$+\frac{1}{2!}\left\{x^2\left(\frac{\partial^2\phi}{\partial x^2}\right)_0+2xy\left(\frac{\partial^2\phi}{\partial x\partial y}\right)_0+y^2\left(\frac{\partial^2\phi}{\partial y^2}\right)_0\right\}$$
$$+\text{etc.}$$

160. If we now recur to Art. 136 we see that the true value of MR is $\quad f(x+\delta x, y+\delta y)-f(x, y)$

$$=\frac{\partial f}{\partial x}\delta x+\frac{\partial f}{\partial y}\delta y+\frac{1}{2!}\left(\frac{\partial^2 f}{\partial x^2}\delta x^2+2\frac{\partial^2 f}{\partial x\partial y}\delta x\delta y+\frac{\partial^2 f}{\partial y^2}\delta y^2\right)+\text{etc.},$$

showing what error was made in that article in taking MR' as an approximation to the correct value.

The student will find no difficulty in writing down the true values of the lengths of LQ or NS.

EULER'S THEOREMS ON HOMOGENEOUS FUNCTIONS.

161. *If* $u=Ax^\alpha y^\beta+Bx^{\alpha'}y^{\beta'}+\ldots=\Sigma Ax^\alpha y^\beta$, *say, where*

$$\alpha+\beta=\alpha'+\beta'=\ldots=n,$$

to show that $$x\frac{\partial u}{\partial x}+y\frac{\partial u}{\partial y}=nu.$$

By differentiation we obtain

$$\frac{\partial u}{\partial x}=\Sigma A\alpha x^{\alpha-1}y^\beta,$$

$$\frac{\partial u}{\partial y}=\Sigma A\beta x^\alpha y^{\beta-1},$$

then $$x\frac{\partial u}{\partial x}+y\frac{\partial u}{\partial y}=\Sigma A\alpha x^\alpha y^\beta+\Sigma A\beta x^\alpha y^\beta$$
$$=\Sigma A(\alpha+\beta)x^\alpha y^\beta$$
$$=n\Sigma Ax^\alpha y^\beta=nu.$$

It is clear that this theorem can be extended to the case of three or of any number of independent variables, and that if, for example,

$$u=Ax^\alpha y^\beta z^\gamma+Bx^{\alpha'}y^{\beta'}z^{\gamma'}+\dots$$

where $\alpha+\beta+\gamma=\alpha'+\beta'+\gamma'=\dots=n$,

then will $$x\frac{\partial u}{\partial x}+y\frac{\partial u}{\partial y}+z\frac{\partial u}{\partial z}=nu.$$

The functions thus described are called *homogeneous functions of the* n^{th} *degree.*

162. We now put the same theorem in a more general form.

DEF. *A homogeneous function of the* n^{th} *degree is one which can be put in the form*

$$x^nF\left(\frac{y}{x}, \frac{z}{x}, \dots\right).$$

Let $$u=x^nF\left(\frac{y}{x}, \frac{z}{x}, \dots\right).$$

Put $$\frac{y}{x}=Y,\ \frac{z}{x}=Z,\ \text{etc.},$$

whence $$\frac{\partial Y}{\partial x}=-\frac{y}{x^2},\ \frac{\partial Z}{\partial x}=-\frac{z}{x^2},\ \dots$$
$$\frac{\partial Y}{\partial y}=\frac{1}{x},\ \frac{\partial Z}{\partial y}=0,\ \text{etc.}$$

Now, since $u=x^nF(Y, Z, \dots)$,

$$\frac{\partial u}{\partial x}=nx^{n-1}F(Y, Z, \dots)+x^n\left\{\frac{\partial F}{\partial Y}\cdot\frac{\partial Y}{\partial x}+\frac{\partial F}{\partial Z}\cdot\frac{\partial Z}{\partial x}+\dots\right\}$$
$$=nx^{n-1}F(Y, Z, \dots)-x^{n-2}\left\{y\frac{\partial F}{\partial Y}+z\frac{\partial F}{\partial Z}+\dots\right\},$$

$$\frac{\partial u}{\partial y} = x^n \frac{\partial F}{\partial Y} \cdot \frac{\partial Y}{\partial y} = x^{n-1} \frac{\partial F}{\partial Y},$$

$$\frac{\partial u}{\partial z} = x^n \frac{\partial F}{\partial Z} \cdot \frac{\partial Z}{\partial y} = x^{n-1} \frac{\partial F}{\partial Z},$$

$$\text{etc.} = \text{etc.}$$

Finally, multiplying by $x, y, z, \ldots$ respectively, and adding

$$x\frac{\partial u}{\partial x} + y\frac{\partial u}{\partial y} + z\frac{\partial u}{\partial z} + \ldots = nx^n F(Y, Z, \ldots) = nu.$$

163. If u be a homogeneous function of x and y of the nth degree, $\frac{\partial u}{\partial x}, \frac{\partial u}{\partial y}$ will be homogeneous functions of the $(n-1)^{\text{th}}$ degree, and applying the result of Art. 162 to these we have $\left(x\frac{\partial}{\partial x} + y\frac{\partial}{\partial y}\right)\frac{\partial u}{\partial x} = (n-1)\frac{\partial u}{\partial x},$

$$\left(x\frac{\partial}{\partial x} + y\frac{\partial}{\partial y}\right)\frac{\partial u}{\partial y} = (n-1)\frac{\partial u}{\partial y}.$$

Multiplying by x and y we have on addition

$$x^2\frac{\partial^2 u}{\partial x^2} + 2xy\frac{\partial^2 u}{\partial x \partial y} + y^2\frac{\partial^2 u}{\partial y^2} = (n-1)\left(x\frac{\partial u}{\partial x} + y\frac{\partial u}{\partial y}\right)$$
$$= n(n-1)u.$$

Similarly we may proceed and finally by induction establish a general theorem of similar character, but of higher order; but it is better to adopt the method hereafter applied in Art. 166.

164. If $V = u_n + u_{n-1} + u_{n-2} + \ldots + u_2 + u_1 + u_0$, where $u_n, u_{n-1}, \ldots$ are homogeneous functions of degrees $n, n-1, \ldots$ respectively. Then

$$x\frac{\partial V}{\partial x} + y\frac{\partial V}{\partial y} + \ldots$$
$$= \left(x\frac{\partial}{\partial x} + \ldots\right)u_n + \left(x\frac{\partial}{\partial x} + \ldots\right)u_{n-1} + \text{ etc.}$$

$$=nu_n+(n-1)u_{n-1}+(n-2)u_{n-2}+\dots+2u_2+u_1$$
$$=nV-\{u_{n-1}+2u_{n-2}+3u_{n-3}+\dots+(n-1)u_1+nu_0\}.$$

Hence if $V=0$

$$x\frac{\partial V}{\partial x}+y\frac{\partial V}{\partial y}+\dots+u_{n-1}+2u_{n-2}+\dots+nu_0=0.$$

165. Let $u=\phi(H_n)$, where H_n is a homogeneous function of the nth degree.

Suppose we obtain from this equation

$$H_n=F(u);$$

then $$x\frac{\partial}{\partial x}F(u)+y\frac{\partial}{\partial y}F(u)+\dots=nH_n,$$

or $$F'(u)\left\{x\frac{\partial u}{\partial x}+y\frac{\partial u}{\partial y}+\dots\right\}=nF(u),$$

or $$x\frac{\partial u}{\partial x}+y\frac{\partial u}{\partial y}+\dots=n\frac{F(u)}{F'(u)}\dots\dots\dots \quad (1)$$

In the *particular case* in which $n=0$ we therefore have

$$x\frac{\partial u}{\partial x}+y\frac{\partial u}{\partial y}+\dots=0.\ \dots\dots\ \dots\dots\dots\dots \quad (2)$$

Examples.

Verify the following results by differentiation.

1. Let $$u=x^3+y^3+3xyz.$$

This is clearly homogeneous and of the 3rd degree, whence

$$x\frac{\partial u}{\partial x}+y\frac{\partial u}{\partial y}+z\frac{\partial u}{\partial z}=3u.$$

2. Let $$u=\frac{x^{\frac{1}{4}}+y^{\frac{1}{4}}}{x^{\frac{1}{5}}+y^{\frac{1}{5}}}=x^{\frac{1}{20}}\frac{1+\left(\frac{y}{x}\right)^{\frac{1}{4}}}{1+\left(\frac{y}{x}\right)^{\frac{1}{5}}}.$$

This is a homogeneous expression of degree $\frac{1}{20}$, whence

$$x\frac{\partial u}{\partial x}+y\frac{\partial u}{\partial y}=\frac{1}{20}u.$$

3. Let $$u=\sin^{-1}\frac{\sqrt{x}-\sqrt{y}}{\sqrt{x}+\sqrt{y}}.$$

Here Art. 162 gives $x\frac{\partial u}{\partial x}+y\frac{\partial u}{\partial y}=0.$

4. Let $$u=\tan^{-1}\frac{x^3+y^3}{x-y}.$$

Here Art. 165 gives $x\frac{\partial u}{\partial x}+y\frac{\partial u}{\partial y}=\sin 2u.$

5. Find which of the following functions are homogeneous, and in cases of homogeneity verify Euler's Theorem of the first degree:

(α) xe^{-y}.

(β) $ye^{-\frac{x}{y}}$.

(γ) $(x-y)(\log x-\log y)$.

(δ) $\sin^{-1}\frac{\sqrt{x^2+y^2}}{x+y}$.

6. Given $z=x^2+y$ and $y=z^2+x$, find the differential coefficients of the first order

(1) when x is the independent variable,
(2) when y is the independent variable,
(3) when z is the independent variable.

7. Given $xyz=a^3$, find all the differential coefficients of the first and second orders, taking x and y for independent variables.

8. If $$u=\sin^{-1}\frac{x+y}{\sqrt{x}+\sqrt{y}},$$

prove that $$x\frac{\partial u}{\partial x}+y\frac{\partial u}{\partial y}=\tfrac{1}{2}\tan u.$$

9. If $$u=ax^2+by^2+cz^2+2fyz+2gzx+2hxy,$$
show that, if it be possible to find values of x, y, z which will simultaneously satisfy

$$\frac{\partial u}{\partial x}=\frac{\partial u}{\partial y}=\frac{\partial u}{\partial z}=0,$$

then will $$\begin{vmatrix} a, & h, & g \\ h, & b, & f \\ g, & f, & c \end{vmatrix}=0.$$

10. If u be a homogeneous function of the nth degree of any number of variables, prove that

$$\left(x\frac{\partial}{\partial x}+y\frac{\partial}{\partial y}+\dots\right)^m u=n^m u.$$

11. If $u=\phi(x, y)$ and $\psi(x, y)=0$, prove that

$$\frac{du}{dx}=\frac{\dfrac{\partial\phi}{\partial x}\dfrac{\partial\psi}{\partial y}-\dfrac{\partial\psi}{\partial x}\dfrac{\partial\phi}{\partial y}}{\dfrac{\partial\psi}{\partial y}}.$$

166. **General Proof of Euler's Theorems.**

We now proceed to give a more complete investigation of Euler's results.

Let $u\equiv\phi(x, y, z \dots)$ be any function expressible in the form

$$x^n F\left(\frac{y}{x}, \frac{z}{x}, \dots\right).$$

It is observable that if $x+xt$, $y+yt$, $z+zt$, ... be written instead of x, y, z, ... in any such function we obtain the result

$$\begin{aligned}\phi(x+xt, y+yt, \dots)&=x^n(1+t)^n F\left(\frac{y+yt}{x+xt}, \dots\right)\\&=x^n(1+t)^n F\left(\frac{y}{x}, \frac{z}{x}, \dots\right)\\&=(1+t)^n u;\end{aligned}$$

so that the effect is simply that of multiplying the original function by $(1+t)^n$.

Now, let V_m denote the symbol of operation obtained by expanding $(xX+yY+zZ+\dots)^m$ by the Multinomial Theorem, and *after* expansion writing $\frac{\partial}{\partial x}, \frac{\partial}{\partial y}, \frac{\partial}{\partial z}, \dots$ in place of X, Y, Z, etc.; then we have, upon expansion of each side of the above equality,

$$u+tV_1u+\frac{t^2}{2!}V_2u+\frac{t^3}{3!}V_3u+\ldots+\frac{t^r}{r!}V_ru+\ldots$$
$$=\left\{1+nt+\frac{n(n-1)}{2!}t^2+\frac{n(n-1)(n-2)}{3!}t^3+\ldots\right.$$
$$\left.+\frac{n(n-1)\ldots(n-r+1)}{r!}t^r+\ldots\right\}u.$$

And on equating coefficients of like powers of t

$$V_1u=nu,$$
$$V_2u=n(n-1)u,$$
$$V_3u=n(n-1)(n-2)u,$$
etc.
$$V_ru=n(n-1)\ldots(n-r+1)u.$$

167. When there are two independent variables, x and y, these become

$$x\frac{\partial u}{\partial x}+y\frac{\partial u}{\partial y}=nu,$$
$$x^2\frac{\partial^2 u}{\partial x^2}+2xy\frac{\partial^2 u}{\partial x\partial y}+y^2\frac{\partial^2 u}{\partial y^2}=n(n-1)u,$$
etc.;

and for the case of three independent variables

$$x\frac{\partial u}{\partial x}+y\frac{\partial u}{\partial y}+z\frac{\partial u}{\partial z}=nu,$$
$$x^2\frac{\partial^2 u}{\partial x^2}+y^2\frac{\partial^2 u}{\partial y^2}+z^2\frac{\partial^2 u}{\partial z^2}+2yz\frac{\partial^2 u}{\partial y\partial z}+2zx\frac{\partial^2 u}{\partial z\partial x}+2xy\frac{\partial^2 u}{\partial x\partial y}$$
$$=n(n-1)u,$$
etc.

168. Care must be taken to distinguish between the expressions
$$x^2\frac{\partial^2 u}{\partial x^2}+2xy\frac{\partial^2 u}{\partial x\partial y}+y^2\frac{\partial^2 u}{\partial y^2}$$
and
$$\left(x\frac{\partial}{\partial x}+y\frac{\partial}{\partial y}\right)^2u,$$

which might at first sight be thought to be identical. However, it is apparent that the latter

$$=\left(x\frac{\partial}{\partial x}+y\frac{\partial}{\partial y}\right)\left(x\frac{\partial u}{\partial x}+y\frac{\partial u}{\partial y}\right)$$

$$=\left(x\frac{\partial}{\partial x}+y\frac{\partial}{\partial y}\right)\left(x\frac{\partial u}{\partial x}\right)+\left(x\frac{\partial}{\partial x}+y\frac{\partial}{\partial y}\right)y\frac{\partial u}{\partial y}$$

$$=\left(x^2\frac{\partial^2 u}{\partial x^2}+x\frac{\partial u}{\partial x}+xy\frac{\partial^2 u}{\partial y\partial x}\right)$$

$$+\left(xy\frac{\partial^2 u}{\partial x\partial y}+y\frac{\partial u}{\partial y}+y^2\frac{\partial^2 u}{\partial y^2}\right)$$

$$=x^2\frac{\partial^2 u}{\partial x^2}+2xy\frac{\partial^2 u}{\partial x\partial y}+y^2\frac{\partial^2 u}{\partial y^2}+x\frac{\partial u}{\partial x}+y\frac{\partial u}{\partial y},$$

and therefore differs from the first expression by the addition of the two terms

$$x\frac{\partial u}{\partial x},\ y\frac{\partial u}{\partial y}.$$

EXAMPLES.

1. Verify the formula $\dfrac{\partial^2 u}{\partial x\partial y}=\dfrac{\partial^2 u}{\partial y\partial x}$ in the following cases:

 (α) $u=\sin\dfrac{y}{x}$.

 (β) $u=\log\{x\tan^{-1}\sqrt{x^2+y^2}\}$.

2. Find $\dfrac{dy}{dx}$ (α) if $ax^2+2hxy+by^2=1$.

 (β) if $x^4+y^4=5a^2xy$.

 (γ) if $(\cos x)^y=(\sin y)^x$.

 (δ) if $y^x+x^y=(x+y)^{x+y}$.

 (ε) if $x^y.\,y^x=x^{\cos y}+y^{\log x}$.

3. If $u = \sin^{-1}\frac{x}{y} + \tan^{-1}\frac{y}{x}$, show that $x\frac{\partial u}{\partial x} + y\frac{\partial u}{\partial y} = 0$.

4. If u, y, z be functions of x such that

$$y = \frac{d}{dx}\left(u\frac{dy}{dx}\right),\ z = \frac{d}{dx}\left(u\frac{dz}{dx}\right),$$

prove that $$\frac{d}{dx}u\left(y\frac{dz}{dx} - z\frac{dy}{dx}\right) = 0.$$

5. If u and v be both functions of the same function of x and y, prove that $\frac{\partial u}{\partial x}\cdot\frac{\partial v}{\partial y} = \frac{\partial u}{\partial y}\cdot\frac{\partial v}{\partial x}$, and that $\frac{\partial}{\partial x}\left(u\frac{\partial v}{\partial y}\right) = \frac{\partial}{\partial y}\left(u\frac{\partial v}{\partial x}\right)$.

6. If $V = f(u, v)$, $u = f_1(x, y)$, $v = f_2(x, y)$, show how to find $\frac{\partial V}{\partial u}$ in terms of $\frac{\partial V}{\partial x}$ and $\frac{\partial V}{\partial y}$.

Ex. Given $u = x^2 + y^2$, $v = 2xy$, show that

$$x\frac{\partial V}{\partial x} - y\frac{\partial V}{\partial y} = 2(u^2 - v^2)^{\frac{1}{2}}\frac{\partial V}{\partial u}.$$

7. Verify Euler's Theorem

$$x\frac{\partial u}{\partial x} + y\frac{\partial u}{\partial y} = nu$$

for the functions (α) $u = \sin\left(\frac{x - y}{x + y}\right)^{\frac{1}{2}}$.

(β) $u = x^3 \log\frac{\sqrt[3]{y} - \sqrt[3]{x}}{\sqrt[3]{y} + \sqrt[3]{x}}$.

8. If $u = \phi(y + ax) + \psi(y - ax)$, prove $\frac{\partial^2 u}{\partial x^2} = a^2\frac{\partial^2 u}{\partial y^2}$.

9. If $u = x\phi\left(\frac{y}{x}\right) + \psi\left(\frac{y}{x}\right)$, prove $x^2\frac{\partial^2 u}{\partial x^2} + 2xy\frac{\partial^2 u}{\partial x \partial y} + y^2\frac{\partial^2 u}{\partial y^2} = 0$.

10. If $u = \frac{(x^2 + y^2)^m}{2m(2m - 1)} + x\phi\left(\frac{y}{x}\right) + \psi\left(\frac{y}{x}\right)$, prove that

$$x^2\frac{\partial^2 u}{\partial x^2} + 2xy\frac{\partial^2 u}{\partial x \partial y} + y^2\frac{\partial^2 u}{\partial y^2} = (x^2 + y^2)^m.$$

11. If $\nabla^2 \equiv \frac{\partial^2}{\partial x^2} + \frac{\partial^2}{\partial y^2} + \frac{\partial^2}{\partial z^2}$ and $r^2 \equiv x^2 + y^2 + z^2$, show that each of the functions $\tan^{-1}\frac{y}{x}$, $\frac{1}{r}$, $\frac{xz}{x^2+y^2}$, $\frac{1}{r}\log\frac{r+z}{r-z}$, satisfy the equation

$$\nabla^2 u = 0.$$

12. Prove $\nabla^2 r^m = m(m+1)r^{m-2}$.

13. If u and u' be functions of x, y, z, both satisfying $\nabla^2 V = 0$, prove that
$$\nabla^2(uu') = 2\left(\frac{\partial u}{\partial x}\cdot\frac{\partial u'}{\partial x} + \frac{\partial u}{\partial y}\cdot\frac{\partial u'}{\partial y} + \frac{\partial u}{\partial z}\cdot\frac{\partial u'}{\partial z}\right).$$

14. If V_n be a homogeneous function of the nth degree, satisfying $\nabla^2 V = 0$, then will $\frac{V_n}{r^{2n+1}}$ also satisfy the same equation.

15. If $f(x, y) = 0$, $\phi(x, z) = 0$, show that
$$\frac{\partial \phi}{\partial x}\cdot\frac{\partial f}{\partial y}\cdot\frac{dy}{dz} = \frac{\partial f}{\partial x}\cdot\frac{\partial \phi}{\partial z}.$$

16. Find $\frac{dy}{dz}$ in terms of y and z from the equations:
$$a\sin x + b\sin y = c.$$
$$a\cos x + b\cos z = c.$$
[I. C. S. Exam.]

17. If $x^4 + y^4 + 4a^2xy = 0$, show that
$$(y^3 + a^2x)^3\frac{d^2y}{dx^2} = 2a^2xy(x^2y^2 + 3a^4).$$

18. If $\left(\frac{x}{a}\right)^n + \left(\frac{y}{b}\right)^n + \left(\frac{z}{c}\right)^n = 1$, find $\frac{\partial z}{\partial x}$ and $\frac{\partial^2 x}{\partial y \partial z}$. Also, find $\frac{dy}{dx}$ when the variables are connected by the two equations
$$\left(\frac{z}{c}\right)^n = \left(\frac{x}{a}\right)^n - \left(\frac{y}{b}\right)^n, \quad \frac{x}{a} + \frac{y}{b} + \frac{z}{c} = 1.$$
[H.C.S. Exam.]

19. If $u = F(x-y, y-z, z-x)$, prove $\frac{\partial u}{\partial x} + \frac{\partial u}{\partial y} + \frac{\partial u}{\partial z} = 0.$

20. If $u = \begin{vmatrix} x^2, & y^2, & z^2 \\ x, & y, & z \\ 1, & 1, & 1 \end{vmatrix}$, prove $\dfrac{\partial u}{\partial x} + \dfrac{\partial u}{\partial y} + \dfrac{\partial u}{\partial z} = 0$.

21. If $u = \operatorname{cosec}^{-1}\sqrt{\dfrac{x^{\frac{1}{2}} + y^{\frac{1}{2}}}{x^{\frac{1}{3}} + y^{\frac{1}{3}}}}$, show that

$$x^2\frac{\partial^2 u}{\partial x^2} + 2xy\frac{\partial^2 u}{\partial x \partial y} + y^2\frac{\partial^2 u}{\partial y^2} = \frac{\tan u}{12}\left(\frac{13}{12} + \frac{\tan^2 u}{12}\right).$$

22. Find the value of the expression $\dfrac{\partial^2 z}{\partial x^2} + \dfrac{\partial^2 z}{\partial y^2}$ when

$$a^2x^2 + b^2y^2 - c^2z^2 = 0.$$ [I. C. S. EXAM.]

23. If $V = Ax^2 + 2Bxy + Cy^2$, prove

$$\left(\frac{\partial V}{\partial x}\right)^2 \cdot \frac{\partial^2 V}{\partial y^2} - 2\frac{\partial V}{\partial x} \cdot \frac{\partial V}{\partial y} \cdot \frac{\partial^2 V}{\partial x \partial y} + \left(\frac{\partial V}{\partial y}\right)^2 \cdot \frac{\partial^2 V}{\partial x^2} = 8V(AC - B^2).$$

24. If $u + \sqrt{-1}v = f(x + \sqrt{-1}y)$, prove $\dfrac{\partial u}{\partial y} = -\dfrac{\partial v}{\partial x}$, $\dfrac{\partial u}{\partial x} = \dfrac{\partial v}{\partial y}$,

$\dfrac{\partial^2 u}{\partial x^2} + \dfrac{\partial^2 u}{\partial y^2} = 0$, and $\dfrac{\partial^2 v}{\partial x^2} + \dfrac{\partial^2 v}{\partial y^2} = 0$.

25. If $u + \sqrt{-1}v$ be a homogeneous function of x, y, z, of degree $p + \sqrt{-1}q$, then $\quad x\dfrac{\partial u}{\partial x} + y\dfrac{\partial u}{\partial y} + z\dfrac{\partial u}{\partial z} = pu - qv$,

and $$x\frac{\partial v}{\partial x} + y\frac{\partial v}{\partial y} + z\frac{\partial v}{\partial z} = pv + qu.$$

26. If $V = (1 - 2xy + y^2)^{-\frac{1}{2}}$, prove that

$$x\frac{\partial V}{\partial x} - y\frac{\partial V}{\partial y} = y^2 V^3.$$

Also that $$\frac{\partial}{\partial x}\left\{(1 - x^2)\frac{\partial V}{\partial x}\right\} + \frac{\partial}{\partial y}\left\{y^2\frac{\partial V}{\partial y}\right\} = 0.$$

27. If $\frac{x^2}{a^2}+\frac{y^2}{b^2}+\frac{z^2}{c^2}=1$, and $lx+my+nz=0$, prove that

$$\frac{dx}{\frac{ny}{b^2}-\frac{mz}{c^2}}=\frac{dy}{\frac{lz}{c^2}-\frac{nx}{a^2}}=\frac{dz}{\frac{mx}{a^2}-\frac{ly}{b^2}}.$$

28. If $\frac{x^2}{a^2}+\frac{y^2}{b^2}+\frac{z^2}{c^2}=1$, and $\frac{x^2}{a^2+\lambda}+\frac{y^2}{b^2+\lambda}+\frac{z^2}{c^2+\lambda}=1$, prove that $\frac{x(b^2-c^2)}{dx}+\frac{y(c^2-a^2)}{dy}+\frac{z(a^2-b^2)}{dz}=0$.

29. If $Pdx+Qdy$ be a perfect differential of some function of x, y, prove that $\frac{\partial P}{\partial y}=\frac{\partial Q}{\partial x}$.

30. If $Pdx+Qdy+Rdz$ can be made a perfect differential of some function of x, y, z by multiplying each term by a common factor, show that

$$P\left(\frac{\partial Q}{\partial z}-\frac{\partial R}{\partial y}\right)+Q\left(\frac{\partial R}{\partial x}-\frac{\partial P}{\partial z}\right)+R\left(\frac{\partial P}{\partial y}-\frac{\partial Q}{\partial x}\right)=0.$$

APPLICATIONS TO PLANE CURVES.

CHAPTER VII.

TANGENTS AND NORMALS.

169. Equation of TANGENT.

It was shown in Art. 38 that the equation of the tangent at the point (x, y) on the curve $y=f(x)$ is

$$Y-y=\frac{dy}{dx}(X-x), \quad \text{.................} \quad (1)$$

X and Y being the current co-ordinates of any point on the tangent.

Suppose the equation of the curve to be given in the form $f(x, y)=0$.

It is shown in Art. 145 that

$$\frac{dy}{dx}=-\frac{\dfrac{\partial f}{\partial x}}{\dfrac{\partial f}{\partial y}}.$$

Substituting this expression for $\frac{dy}{dx}$ in (1) we obtain

$$Y-y=-\frac{\dfrac{\partial f}{\partial x}}{\dfrac{\partial f}{\partial y}}(X-x),$$

or $$(X-x)\frac{\partial f}{\partial x}+(Y-y)\frac{\partial f}{\partial y}=0 \text{ .... } (2)$$

for the equation of the tangent.

170. **Simplification for Algebraic Curves.**

If $f(x, y)$ be an algebraic function of x and y of degree n, suppose it made *homogeneous in x, y, and z by the introduction of a proper power of the linear unit z* wherever necessary. Call the function thus altered $f(x, y, z)$. Then $f(x, y, z)$ is a homogeneous algebraic function of the nth degree; hence we have by Euler's Theorem

(Art. 161) $$x\frac{\partial f}{\partial x}+y\frac{\partial f}{\partial y}+z\frac{\partial f}{\partial z}=nf(x, y, z)=0,$$

by virtue of the equation to the curve.

Adding this to equation (2), the equation of the tangent takes the form

$$X\frac{\partial f}{\partial x}+Y\frac{\partial f}{\partial y}+z\frac{\partial f}{\partial z}=0 \text{ .... } (3)$$

where the z is to be put $=1$ after the differentiations have been performed.

Ex. $$f(x, y)\equiv x^4+a^2xy+b^3y+c^4=0.$$

The equation, when made *homogeneous in x, y, z by the introduction of a proper power of z*, is

$$f(x, y, z)\equiv x^4+a^2xyz^2+b^3yz^3+c^4z^4=0,$$

and $$\frac{\partial f}{\partial x}=4x^3+a^2yz^2,$$

$$\frac{\partial f}{\partial y}=a^2xz^2+b^3z^3,$$

$$\frac{\partial f}{\partial z}=2a^2xyz+3b^3yz^2+4c^4z^3.$$

Substituting these in Equation 3, and putting $z=1$, we have for the equation of the tangent to the curve at the point (x, y)

$$X(4x^3+a^2y)+Y(a^2x+b^3)+2a^2xy+3b^3y+4c^4=0.$$

With very little practice the introduction of the z can

be performed *mentally.* It is generally *more advantageous* to use equation (3) than equation (2), because (3) gives the result *in its simplest form,* whereas if (2) be used it is often necessary to reduce by substitutions from the equation of the curve.

171. Application to General Rational Algebraic Curve.

If the equation of the curve be written in the form

$$f(x, y) \equiv u_n + u_{n-1} + u_{n-2} + \dots + u_2 + u_1 + u_0 = 0$$

(where u_r represents the sum of all the terms of the rth degree), then when made homogeneous by the introduction where necessary of a proper power of z we shall have

$$f(x, y, z) \equiv u_n + u_{n-1}z + u_{n-2}z^2 + \dots + u_2 z^{n-2} + u_1 z^{n-1} + u_0 z^n,$$

and
$$\frac{\partial f}{\partial z} = u_{n-1} + 2u_{n-2}z + 3u_{n-3}z^2 + \dots + (n-2)u_2 z^{n-3} + (n-1)u_1 z^{n-2} + nu_0 z^{n-1},$$

and therefore substituting in (3) and putting $z=1$, the equation of the tangent is

$$X\frac{\partial f}{\partial x} + Y\frac{\partial f}{\partial y} + u_{n-1} + 2u_{n-2} + 3u_{n-3} + \dots + (n-2)u_2 + (n-1)u_1 + nu_0 = 0 \dots\dots \quad (4)$$

172. NORMAL.

DEF.—*The normal at any point of a curve is a straight line through that point and perpendicular to the tangent to the curve at that point.*

Let the axes be assumed rectangular. The equation of the normal may then be at once written down. For if the equation of the curve be

$$y = f(x),$$

the tangent at (x, y) is $Y-y=\frac{dy}{dx}(X-x)$,

and the normal is therefore

$$(X-x)+(Y-y)\frac{dy}{dx}=0.$$

If the equation of the curve be given in the form

$$f(x, y)=0,$$

the equation of the tangent is

$$(X-x)\frac{\partial f}{\partial x}+(Y-y)\frac{\partial f}{\partial y}=0,$$

and therefore that of the normal is

$$\frac{X-x}{\frac{\partial f}{\partial x}}=\frac{Y-y}{\frac{\partial f}{\partial y}}.$$

Ex. 1. Consider the ellipse

$$\frac{x^2}{a^2}+\frac{y^2}{b^2}=1.$$

This requires z^2 in the last term to make a homogeneous equation in x, y, and z. We have then

$$\frac{x^2}{a^2}+\frac{y^2}{b^2}-z^2=0.$$

Hence the equation of the tangent is

$$X.\frac{2x}{a^2}+Y.\frac{2y}{b^2}-z.2z=0,$$

where z is to be put $=1$. Hence we get

$$\frac{Xx}{a^2}+\frac{Yy}{b^2}=1 \text{ for the tangent,}$$

and therefore $$\frac{X-x}{\frac{x}{a^2}}=\frac{Y-y}{\frac{y}{b^2}} \text{ for the normal.}$$

Ex. 2. Take the general equation of a conic

$$ax^2+2hxy+by^2+2gx+2fy+c=0.$$

When made homogeneous this becomes

$$ax^2+2hxy+by^2+2gxz+2fyz+cz^2=0.$$

The equation of the tangent is therefore

$$X(ax+hy+g)+Y(hx+by+f)+gx+fy+c=0,$$

and that of the normal is

$$\frac{X-x}{ax+hy+g}=\frac{Y-y}{hx+by+f}.$$

Ex. 3. Consider the curve

$$\frac{y}{a}=\log\sec\frac{x}{a}.$$

Then $$\frac{dy}{dx}=\tan\frac{x}{a},$$

and the equation of the tangent is

$$Y-y=\tan\frac{x}{a}(X-x),$$

and of the normal

$$(Y-y)\tan\frac{x}{a}+(X-x)=0.$$

EXAMPLES.

1. Find the equations of the tangents and normals at the point (x, y) on each of the following curves :—

(1) $x^2+y^2=c^2$.
(2) $y^2=4ax$.
(3) $xy=k^2$.
(4) $y=c\cosh\frac{x}{c}$.
(5) $x^2y+xy^2=a^3$.
(6) $e^y=\sin x$.
(7) $x^3-3axy+y^3=0$.
(8) $(x^2+y^2)^2=a^2(x^2-y^2)$.

2. Write down the equations of the tangents and normals to the curve $y(x^2+a^2)=ax^2$ at the points where $y=\frac{a}{4}$.

3. Prove that $\frac{x}{a}+\frac{y}{b}=1$ touches the curve $y=be^{-\frac{x}{a}}$ at the point where the curve crosses the axis of y.

4. If $p=x\cos\alpha+y\sin\alpha$ touch the curve

$$\frac{x^m}{a^m}+\frac{y^m}{b^m}=1,$$

prove that $$p^{\frac{m}{m-1}}=(a\cos\alpha)^{\frac{m}{m-1}}+(b\sin\alpha)^{\frac{m}{m-1}}.$$

Hence write down the polar equation of the locus of the foot of the perpendicular from the origin on the tangent to this curve.

Examine the cases of an ellipse and of a rectangular hyperbola.

5. Prove that, if the axes be oblique and inclined at an angle ω, the equation of the normal to $y=f(x)$ at (x, y) is

$$(Y-y)\left(\cos\omega+\frac{dy}{dx}\right)+(X-x)\left(1+\cos\omega\frac{dy}{dx}\right)=0.$$

173. **Tangents at the Origin.**

It will be shown by a general method in a subsequent article (254) that in the case in which a curve, whose equation is given in the rational algebraic form, passes through the origin, the equation of the tangent or tangents at that point can be at once written down; the rule being to *equate to zero the terms of lowest degree* in the equation of the curve.

Ex. In the curve $x^2+y^2+ax+by=0$, $ax+by=0$ is the equation of the tangent at the origin; and in the curve $(x^2+y^2)^2=a^2(x^2-y^2)$, $x^2-y^2=0$ is the equation of a pair of tangents at the origin.

It is easy to deduce this result from the equation of the tangent established in Chapter II. That equation is

$$Y-y=m(X-x) \text{ where } m=\frac{dy}{dx}.$$

At the origin this becomes $\qquad Y=mX,$

where the limiting value or values of m are to be found.

Let the equation of the curve be arranged in homogeneous sets of terms, and suppose the lowest set to be of the rth degree. The equation may be written

$$x^r f_r\left(\frac{y}{x}\right)+x^{r+1}f_{r+1}\left(\frac{y}{x}\right)+\ldots x^n f_n\left(\frac{y}{x}\right)=0.$$

Dividing by x^r, and putting $y=mx$, and then $x=0$ and $y=0$, the above reduces to the form

$$f_r(m)=0,$$

an equation which has r roots giving the directions in which the several branches of the curve pass through the origin. If $m_1, m_2, m_3, \ldots m_r$ be the roots, the equations of the several tangents are

$$y = m_1 x,\ y = m_2 x,\ \ldots\ y = m_r x.$$

These are all contained in the one equation $f_r\left(\frac{y}{x}\right) = 0$; and this is the result obtained by "*equating to zero the terms of lowest degree*" in the equation of the curve, thus proving the rule. In this manner all the trouble of *differentiation is avoided,* and the result written down *by inspection.*

GEOMETRICAL RESULTS.

174. **Cartesians. Intercepts.**

From the equation $Y - y = \frac{dy}{dx}(X - x)$

it is clear that the *intercepts* which the tangent cuts off from the axes of x and y are respectively

$$x - \frac{y}{\frac{dy}{dx}} \text{ and } y - x\frac{dy}{dx},$$

for these are respectively the values of X when $Y = 0$ and of Y when $X = 0$.

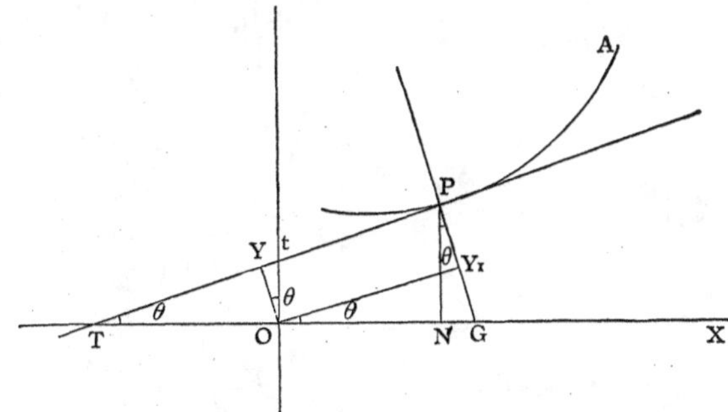

Fig. 21.

Let PN, PT, PG be the ordinate, tangent, and normal to the curve, and let PT make an angle θ with the axis of x; then $\tan\theta = \frac{dy}{dx}$. Let the tangent cut the axis of y in t, and let OY, OY_1 be perpendiculars from O, the origin, on the tangent and normal. Then the above values of the intercepts are also obvious from the figure.

175. Subtangent, etc.

DEF. The line TN is called the *subtangent* and the line NG is called the *subnormal.*

From the figure

$$Subtangent = TN = y\cot\theta = \frac{y}{\frac{dy}{dx}}.$$

$$Subnormal = NG = y\tan\theta = y\frac{dy}{dx}.$$

$$Normal = PG = y\sec\theta = y\sqrt{1+\tan^2\theta} = y\sqrt{1+\left(\frac{dy}{dx}\right)^2}.$$

$$Tangent = TP = y\operatorname{cosec}\theta = y\frac{\sqrt{1+\tan^2\theta}}{\tan\theta} = y\frac{\sqrt{1+\left(\frac{dy}{dx}\right)^2}}{\frac{dy}{dx}}.$$

$$OY = Ot\cos\theta = \frac{y - x\frac{dy}{dx}}{\sqrt{1+\tan^2\theta}} = \frac{y - x\frac{dy}{dx}}{\sqrt{1+\left(\frac{dy}{dx}\right)^2}}.$$

$$OY_1 = OG\cos\theta = \frac{ON+NG}{\sqrt{1+\tan^2\theta}} = \frac{x + y\frac{dy}{dx}}{\sqrt{1+\left(\frac{dy}{dx}\right)^2}}.$$

These results may of course also be obtained analytically from the equation of the tangent.

176. **Values of** $\dfrac{ds}{dx}, \dfrac{dx}{ds}$, **etc.**

Let P, Q be contiguous points on a curve. Let the co-ordinates of P be (x, y) and of Q $(x+\delta x, y+\delta y)$. Then the perpendicular $PR=\delta x$, and $RQ=\delta y$. Let the

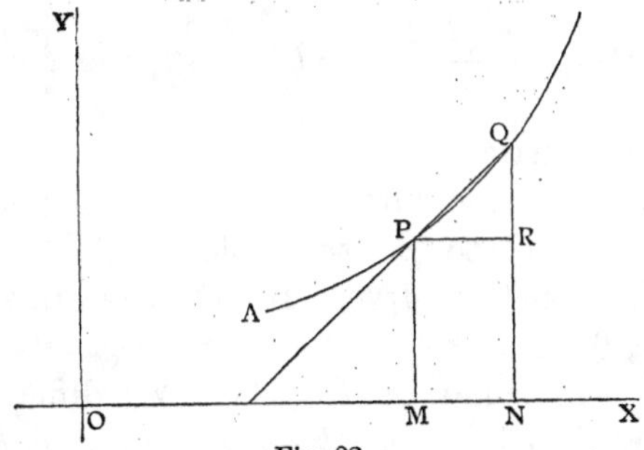

Fig. 22.

arc AP measured from some fixed point A on the curve be called s and the arc $AQ=s+\delta s$. Then arc $PQ=\delta s$. When Q travels along the curve so as to come indefinitely near to P, the arc PQ and the chord PQ ultimately differ by a small quantity of higher order than the arc PQ itself (Art. 36).

Hence, rejecting infinitesimals of order higher than the second, we have

$$\delta s^2 = (\text{chord } PQ)^2 = (\delta x^2 + \delta y^2),$$

or
$$1 = Lt\left(\frac{\delta x^2}{\delta s^2} + \frac{\delta y^2}{\delta s^2}\right) = \left(\frac{dx}{ds}\right)^2 + \left(\frac{dy}{ds}\right)^2.$$

Similarly $Lt\dfrac{\delta s^2}{\delta x^2} = Lt\left(1 + \dfrac{\delta y^2}{\delta x^2}\right)$,

or
$$\left(\frac{ds}{dx}\right)^2 = 1 + \left(\frac{dy}{dx}\right)^2;$$

and in the same manner

$$\left(\frac{ds}{dy}\right)^2 = 1 + \left(\frac{dx}{dy}\right)^2.$$

If ψ be the angle which the tangent makes with the axis of x as in Art. 39,

$$\tan\psi = Lt\frac{RQ}{PR} = Lt\frac{\delta y}{\delta x} = \frac{dy}{dx},$$

and also $$\cos\psi = Lt\frac{PR}{\text{chord } PQ} = Lt\frac{PR}{\text{arc } PQ} = Lt\frac{\delta x}{\delta s} = \frac{dx}{ds},$$

and $$\sin\psi = Lt\frac{RQ}{\text{chord } PQ} = Lt\frac{RQ}{\text{arc } PQ} = Lt\frac{\delta y}{\delta s} = \frac{dy}{ds}.$$

177. **Polar Co-ordinates.**

If the equation of the curve be referred to polar co-ordinates, suppose O to be the pole and P, Q two contiguous points on the curve. Let the co-ordinates of P and Q be (r, θ) and $(r+\delta r, \theta+\delta\theta)$ respectively. Let PN be the perpendicular on OQ, then NQ differs from δr and NP from $r\delta\theta$ *by small quantities of a higher order than* $\delta\theta$ (Art. 33).

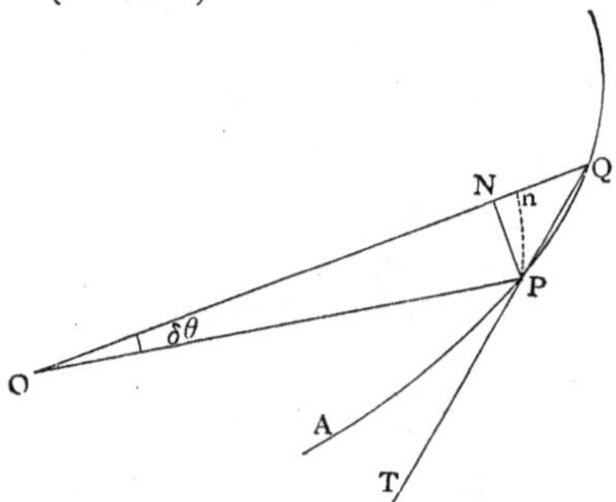

Fig. 23.

Let the arc measured from some fixed point A to P be called s, and from A to Q, $s+\delta s$. Then arc $PQ=\delta s$. Hence, rejecting infinitesimals of order higher than the second, we have

$$\delta s^2 = (\text{chord } PQ)^2 = (NQ^2 + PN^2) = (\delta r^2 + r^2\delta\theta^2),$$

and therefore

$$\left(\frac{dr}{ds}\right)^2 + r^2\left(\frac{d\theta}{ds}\right)^2 = 1, \text{ or } \left(\frac{ds}{dr}\right)^2 = 1 + r^2\left(\frac{d\theta}{dr}\right)^2,$$

or

$$\left(\frac{ds}{d\theta}\right)^2 = r^2 + \left(\frac{dr}{d\theta}\right)^2,$$

according as we divide by δs^2, δr^2, or $\delta\theta^2$ before proceeding to the limit.

178. **Inclination of the Radius Vector to the Tangent.**

Next, let ϕ be the angle which the tangent at any point P makes with the radius vector, then

$$\tan\phi = r\frac{d\theta}{dr}, \quad \cos\phi = \frac{dr}{ds}, \quad \sin\phi = \frac{rd\theta}{ds}.$$

For, with the figure of the preceding article, since, when Q has moved along the curve so near to P that Q and P may be considered as ultimately coincident, QP becomes the tangent at P and the angles OQT and OPT are each of them ultimately equal to ϕ, and

$$\tan\phi = Lt\tan NQP = Lt\frac{NP}{QN} = Lt\frac{r\delta\theta}{\delta r} = r\frac{d\theta}{dr};$$

$$\cos\phi = Lt\cos NQP = Lt\frac{NQ}{\text{chord } QP} = Lt\frac{NQ}{\text{arc } QP} = Lt\frac{\delta r}{\delta s} = \frac{dr}{ds};$$

$$\sin\phi = Lt\sin NQP = Lt\frac{NP}{\text{chord } QP} = Lt\frac{NP}{\text{arc } QP} = Lt\frac{r\delta\theta}{\delta s} = \frac{rd\theta}{ds}.$$

179. **Polar Subtangent, Subnormal, etc.**

DEF. Let OY be the perpendicular from the origin on the tangent at P. Let TOt be drawn through O perpendicular to OP and cutting the tangent in T and the normal in t. Then OT is called the "*Polar Subtangent*" and Ot is called the "*Polar Subnormal.*"

It is clear that $OT = OP \tan\phi = r^2 \dfrac{d\theta}{dr}$, ... (1)

and that $Ot = OP \cot\phi = \dfrac{dr}{d\theta}$ .... (2)

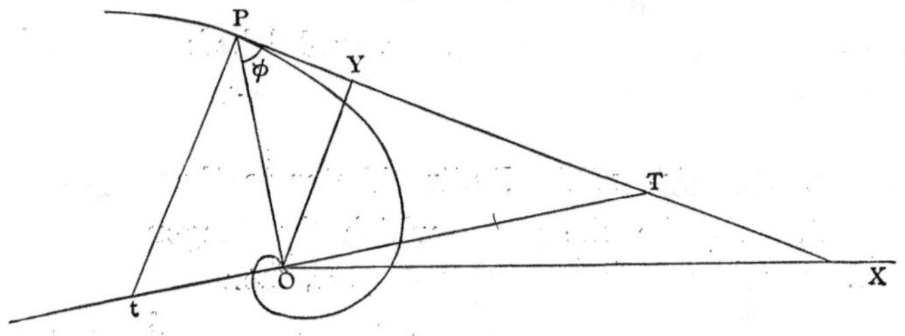

Fig. 24.

180. It is often found convenient when using polar co-ordinates to write $\dfrac{1}{u}$ for r, and therefore $-\dfrac{1}{u^2}\dfrac{du}{d\theta}$ for $\dfrac{dr}{d\theta}$. With this notation

$$\text{Polar Subtangent} = r^2\frac{d\theta}{dr} = -\frac{d\theta}{du}.$$

181. **Perpendicular from Pole on Tangent, etc.**

Let $OY = p$ and $PY = t$.

Then $p = r\sin\phi$,

and therefore

$$\frac{1}{p^2} = \frac{1}{r^2}\operatorname{cosec}^2\phi = \frac{1}{r^2}(1+\cot^2\phi) = \frac{1}{r^2}\left\{1+\frac{1}{r^2}\left(\frac{dr}{d\theta}\right)^2\right\};$$

therefore $\dfrac{1}{p^2} = \dfrac{1}{r^2} + \dfrac{1}{r^4}\left(\dfrac{dr}{d\theta}\right)^2$ ....(1)

$$= u^2 + \left(\frac{du}{d\theta}\right)^2 \text{ ....(2)}$$

Similarly $t = r\cos\phi$;

therefore $$\frac{1}{t^2}=\frac{1}{r^2}\sec^2\phi=\frac{1}{r^2}(1+\tan^2\phi)$$

$$=\frac{1}{r^2}\left\{1+r^2\left(\frac{d\theta}{dr}\right)^2\right\};$$

therefore $$\frac{1}{t^2}=\frac{1}{r^2}+\left(\frac{d\theta}{dr}\right)^2 \ldots\ldots\ldots\ldots\ldots (3)$$

$$=u^2+u^4\left(\frac{d\theta}{du}\right)^2 \ldots\ldots\ldots\ldots (4)$$

182. **Polar Equation of the Tangent.**

Let the polar co-ordinates of the point of contact be $\left(\frac{1}{U}, \alpha\right)$; and let U' be the value of $\frac{du}{d\theta}$ for the curve at that point.

The equation of any straight line may be written in the form $$u=A\cos(\theta-\alpha)+B\sin(\theta-\alpha), \ldots\ldots(1)$$ A and B being the arbitrary constants. Let this straight line represent the required tangent.

By differentiation

$$\frac{du}{d\theta}=-A\sin(\theta-\alpha)+B\cos(\theta-\alpha). \ldots.(2)$$

Now, since the tangent touches the curve, the value of $\frac{du}{d\theta}$ at the point of contact is the same for the curve and for the tangent. Hence, putting $\theta=\alpha$ in equations (1) and (2), we have

$$U=A \text{ and } U'=B,$$

whence the required equation will be

$$u=U\cos(\theta-\alpha)+U'\sin(\theta-\alpha). \ldots\ldots(3)$$

183. **Polar Equation of the Normal.**

The equation of *any* straight line at right angles to the

tangent given by equation (3) of the preceding article may be written in the form

$$Cu = U'\cos(\theta - \alpha) - U\sin(\theta - \alpha),$$

C being an arbitrary constant.

This equation is to be satisfied by $u = U$, $\theta = \alpha$ for the point of contact; therefore substituting we have

$$CU = U',$$

whence the required equation of the normal is

$$\frac{U'}{U}u = U'\cos(\theta - \alpha) - U\sin(\theta - \alpha).$$

184. **Class of a Curve of the nth degree.**

DEF. *The number of tangents which can be drawn from a given point to a rational algebraic curve is called its class.*

Let the equation of the curve be $f(x, y) = 0$. The equation of the tangent at the point (x, y) is

$$X\frac{\partial f}{\partial x} + Y\frac{\partial f}{\partial y} + z\frac{\partial f}{\partial z} = 0,$$

where z is to be put equal to unity after the differentiation is performed. If this pass through the point h, k we have

$$h\frac{\partial f}{\partial x} + k\frac{\partial f}{\partial y} + z\frac{\partial f}{\partial z} = 0.$$

This is an equation of the $(n-1)^{\text{th}}$ degree in x and y and represents a curve of the $(n-1)^{\text{th}}$ degree *passing through the points of contact* of the tangents drawn from the point (h, k) to the curve $f(x, y) = 0$. These two curves have $n(n-1)$ points of intersection, and therefore there are $n(n-1)$ *points of contact* corresponding to $n(n-1)$ *tangents, real or imaginary,* which can be drawn from a given point to a curve of the nth degree.*

It appears then that if the degree of a curve be n, *its*

* Poncelet, "Annales de Gergonne," vol. VIII.

class is $n(n-1)$; for example, the classes of a conic, a cubic, a quartic are the second, sixth, twelfth respectively.

185. **Number of Normals which can be drawn to a Curve to pass through a given point.**

Let h, k be the point through which the normals are to pass.

The equation of the normal to the curve $f(x, y)=0$ at the point (x, y) is
$$\frac{X-x}{\frac{\partial f}{\partial x}}=\frac{Y-y}{\frac{\partial f}{\partial y}}.$$

If this pass through h, k,
$$(h-x)\frac{\partial f}{\partial y}=(k-y)\frac{\partial f}{\partial x}.$$

This equation is of the nth degree in x and y and represents a curve which goes *through the feet of all normals* which can be drawn from the point h, k to the curve. Combining this with $f(x, y)=0$, which is also of the nth degree, it appears that there are n^2 points of intersection, and that therefore there can be n^2 *normals, real or imaginary,* drawn to a given curve to pass through a given point.

For example, if the curve be an ellipse, $n=2$, and the number of normals is 4. Let $\frac{x^2}{a^2}+\frac{y^2}{b^2}=1$ be the equation of the curve, then
$$(h-x)\frac{y}{b^2}=(k-y)\frac{x}{a^2}$$
is the curve which, with the ellipse, determines the feet of the normals drawn from the point (h, k). This is a rectangular hyperbola which passes through the origin and through the point (h, k).

186. The curves
$$(h-x)\frac{\partial f}{\partial x}+(k-y)\frac{\partial f}{\partial y}=0 \quad \text{..............(1)}$$

and $$(h-x)\frac{\partial f}{\partial y}-(k-y)\frac{\partial f}{\partial x}=0, \quad \text{..............(2)}$$

on which lie the points of contact of tangents and the feet of the normals respectively, which can be drawn to the curve $f(x, y)=0$ so as to pass through the point (h, k), are the same for the curve $f(x, y)=a$. And, as equations (1) and (2) do not depend on a, they represent *the loci of the points of contact and of the feet of the normals* respectively for all values of a, that is, for all members of the family of curves obtained by varying a in $f(x, y)=a$ in any arbitrary manner.

187. **Polar Curves.**

The curve $$h\frac{\partial f}{\partial x}+k\frac{\partial f}{\partial y}+z\frac{\partial f}{\partial z}=0$$

is called the "*First Polar Curve*" of the point h, k with regard to the curve $f(x, y)=0$; z being a linear unit introduced as explained previously to make $f(x, y)$ homogeneous in x, y, z, and put equal to unity after the differentiation is performed.

As this is a curve of the $(n-1)^{\text{th}}$ degree it is clear that the first polar of a point with regard to a conic is a *straight line*, the first polar with regard to a cubic is a *conic*, and so on.

The first polar *of the origin* is given by

$$\frac{df}{dz}=0.$$

If the curve be put in the form

$$u_n+u_{n-1}+u_{n-2}+\dots+u_2+u_1+u_0=0$$

the first polar of the origin is

$$u_{n-1}+2u_{n-2}+3u_{n-3}+\dots+(n-1)u_1+nu_0=0.$$

In the particular case of the conic

$$u_2+u_1+u_0=0$$

the polar line of the origin has for its equation

$$u_1+2u_0=0.$$

For the cubic $u_3+u_2+u_1+u_0=0$
the polar conic of the origin is

$$u_2+2u_1+3u_0=0.$$

188. **The p, r or Pedal Equation of a Curve.**

In many curves the relation between the perpendicular on the tangent and the radius vector of the point of contact from some given point is very simple, and when known it frequently forms a very useful equation to the curve; especially indeed in investigating certain Statical and Dynamical properties.

189. **Pedal Equation deduced from Cartesian.**

Suppose the curve to be given by its Cartesian Equation and the origin to be taken at the point with regard to which it is required to find the Pedal Equation of the curve. Let x, y be the co-ordinates of any point on the curve; then, if $F(x, y)=0$ be the equation of the curve, the equation of the tangent is

$$X\frac{\partial F}{\partial x}+Y\frac{\partial F}{\partial y}+z\frac{\partial F}{\partial z}=0,$$

where z is as usual to be put equal unity after the differentiation is performed.

If p be the perpendicular from the origin on the tangent at (x, y) we have

$$p^2=\frac{\left(\frac{\partial F}{\partial z}\right)^2}{\left(\frac{\partial F}{\partial x}\right)^2+\left(\frac{\partial F}{\partial y}\right)^2}\quad\text{.......................(1)}$$

Also $$r^2=x^2+y^2, \qquad \text{.............................(2)}$$
and $$F(x, y)=0. \qquad \text{..................................(3)}$$

If x and y be eliminated between these three equations the required relation between p and r is obtained.

Ex. If $F(x, y)=0$ be $$\frac{x^2}{a^2}+\frac{y^2}{b^2}=1$$
we have $$\frac{x^2}{a^4}+\frac{y^2}{b^4}=\frac{1}{p^2}$$
and $$x^2+y^2=r^2;$$
therefore $$\begin{vmatrix} \frac{1}{a^2}, & \frac{1}{b^2}, & 1 \\ \frac{1}{a^4}, & \frac{1}{b^4}, & \frac{1}{p^2} \\ 1, & 1, & r^2 \end{vmatrix}=0,$$
or $$\frac{a^2b^2}{p^2}+r^2=a^2+b^2.$$

This result might be at once obtained by eliminating CD from the equations $$CP^2+CD^2=a^2+b^2$$
and $$CD.p=ab,$$
CP and CD being conjugate semi-diameters.

190. Pedal Equation deduced from Polar.

Let the curve be given in Polar co-ordinates and the pole be taken at the point with regard to which it is required to find the pedal equation of the curve. Let r, θ be the co-ordinates of any point on the curve, and p the length of the perpendicular from the pole on the tangent at r, θ. If
$$F(r, \theta)=0 \qquad \text{........................(1)}$$
be the equation of the curve, then we have (see Fig. 24)
$$p=r\sin\phi, \qquad \text{..................(2)}$$
and $$\tan\phi=\frac{rd\theta}{dr}. \qquad \text{......................(3)}$$

Eliminate θ and ϕ between the equations (1), (2), (3), and the required equation between p and r will be obtained.

Ex. Given $r^m = a^m \sin m\theta$, required its pedal equation.

Taking logarithms and differentiating,

$$\frac{m}{r}\frac{dr}{d\theta} = m\frac{\cos m\theta}{\sin m\theta};$$

therefore $\cot\phi = \cot m\theta$, or $\phi = m\theta$.

Again, $p = r\sin\phi = r\sin m\theta$

$$= r \cdot \frac{r^m}{a^m};$$

therefore $$p = \frac{r^{m+1}}{a^m}.$$

The following special cases of this example are worthy of notice, and will furnish exercises for the student.

Value of m.	Equation.	Name.	Pedal Equation.
-2	$r^2 \sin 2\theta + a^2 = 0$	Rectangular Hyperbola	$rp = a^2$
-1	$r\sin\theta + a = 0$	Straight line	$p = a$
$-\frac{1}{2}$	$\frac{2a}{r} = 1 - \cos\theta$	Parabola	$p^2 = ar$
$\frac{1}{2}$	$r = \frac{a}{2}(1 - \cos\theta)$	Cardioide	$p^2 a = r^3$
1	$r = a\sin\theta$	Circle	$pa = r^2$
2	$r^2 = a^2 \sin 2\theta$	Lemniscate of Bernoulli	$pa^2 = r^3$

PEDAL CURVES.

191. DEF. *If a perpendicular be drawn from a fixed point on a variable tangent to a curve, the locus of the*

foot of the perpendicular is called the "FIRST POSITIVE PEDAL" *of the original curve with regard to the given point.*

To find the first positive pedal with regard to the origin of any curve whose **Cartesian Equation** *is given.*

Let $$F(x, y)=0 \quad\text{..........................(1)}$$
be the equation of the curve.

Suppose $X\cos\alpha+Y\sin\alpha=p$ touches this curve.

By comparison of this equation with

$$X\frac{\partial F}{\partial x}+Y\frac{\partial F}{\partial y}+z\frac{\partial F}{\partial z}=0$$

we have $$\frac{\frac{\partial F}{\partial x}}{\cos\alpha}=\frac{\frac{\partial F}{\partial y}}{\sin\alpha}=\frac{\frac{\partial F}{\partial z}}{-p}=\lambda,\text{ say } \quad\text{...........(2)}$$

If x, y, λ be eliminated between the four equations (1) and (2) a result will remain which depends on p and α only. And since p, α are the polar co-ordinates of the foot of the perpendicular, if r be written for p and θ for α, the polar equation of the locus required will be obtained.

Ex. Find the first positive pedal of the curve

$$Ax^m+By^m=1.$$

The tangent is $AXx^{m-1}+BYy^{m-1}=1$.

Compare this with $X\cos\alpha+Y\sin\alpha=p$,

$$Ax^{m-1}=\frac{\cos\alpha}{p},\text{ and } By^{m-1}=\frac{\sin\alpha}{p}.$$

Hence $$A\left(\frac{\cos\alpha}{Ap}\right)^{\frac{m}{m-1}}+B\left(\frac{\sin\alpha}{Bp}\right)^{\frac{m}{m-1}}=1.$$

Therefore the polar equation of the locus required is

$$r^{\frac{m}{m-1}}=\frac{\cos^{\frac{m}{m-1}}\theta}{A^{\frac{1}{m-1}}}+\frac{\sin^{\frac{m}{m-1}}\theta}{B^{\frac{1}{m-1}}}.$$

192. *To find the Pedal with regard to the Pole of any curve whose* **Polar Equation** *is given.*

Let $$F(r, \theta)=0 \qquad \text{.................................(1)}$$
be the equation of the curve.

Let r', θ' be the polar co-ordinates of the point Y, which is the foot of the perpendicular OY drawn from the pole

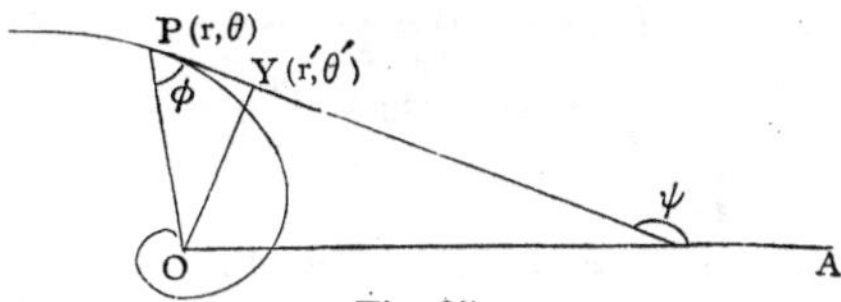

Fig. 25.

on a tangent. Let OA be the initial line. Then

$$\theta = AOP = AOY + YOP$$

$$= \theta' + \frac{\pi}{2} - \phi; \qquad \text{...................(2)}$$

also $$\tan\phi = r\frac{d\theta}{dr}, \qquad \text{............................(3)}$$

and $$r' = r\sin\phi,$$

or $$\left.\frac{1}{r'^2} = \frac{1}{r^2} + \frac{1}{r^4}\left(\frac{dr}{d\theta}\right)^2\right\} \text{(Art. 181). .....(4)}$$

If r, θ, ϕ be eliminated from equations 1, 2, 3, and 4 there will remain an equation in r', θ'. The dashes may then be dropped and the required equation will be obtained.

Ex. To find the equation of the first positive pedal of the curve

$$r^m = a^m \cos m\theta.$$

Taking the logarithmic differential

$$\frac{m}{r}\frac{dr}{d\theta} = -m\tan m\theta;$$

therefore $$\cot\phi = -\tan m\theta;$$

therefore $$\phi = \frac{\pi}{2} + m\theta.$$

But $$\theta = \theta' + \frac{\pi}{2} - \phi,$$

therefore $$\theta = \theta' - m\theta, \text{ or } \theta = \frac{\theta'}{m+1}.$$

Again
$$r' = r\sin\phi = r\cos m\theta$$
$$= a\cos^{\frac{1}{m}} m\theta \,.\, \cos m\theta$$
$$= a\cos^{\frac{m+1}{m}}\frac{m\theta'}{m+1}.$$

Hence the equation of the pedal curve is
$$r^{\frac{m}{m+1}} = a^{\frac{m}{m+1}}\cos\frac{m}{m+1}\theta.$$

193. Def. If there be a series of curves which we may designate as
$$A,\ A_1,\ A_2,\ A_3,\ \ldots\ A_n,\ \ldots$$
such that each is the *first positive pedal* curve of the one which immediately precedes it; then A_2, A_3, etc., are respectively called the *second, third, etc., positive pedals* of A. Also, any one of this series of curves may be regarded as the original curve, *e.g.*, A_3; then A_2 is called the *first negative pedal of* A_3, A_1 *the second negative pedal*, and so on.

Ex. 1. Find the kth positive pedal of
$$r^m = a^m\cos m\theta.$$
It has been shown that the first positive pedal is
$$r^{m_1} = a^{m_1}\cos m_1\theta,$$
where
$$m_1 = \frac{m}{1+m}.$$

Similarly the second positive pedal is
$$r^{m_2} = a^{m_2}\cos m_2\theta,$$
where
$$m_2 = \frac{m_1}{1+m_1} = \frac{m}{1+2m};$$
and generally the kth positive pedal is
$$r^{m_k} = a^{m_k}\cos m_k\theta,$$
where
$$m_k = \frac{m}{1+km}.$$

Ex. 2. Find the kth negative pedal of the curve
$$r^m = a^m\cos m\theta.$$

We have shown above that $r^m = a^m \cos m\theta$ is the kth positive pedal of the curve $r^n = a^n \cos n\theta$, provided $m = \dfrac{n}{1+kn}$.

This gives $$n = \frac{m}{1-km}.$$

Hence the kth negative pedal of $r^m = a^m \cos m\theta$ is
$$r^n = a^n \cos n\theta,$$
where $$n = \frac{m}{1-km}.$$

194. Tangential-Polar, or p, ψ Equation of a Curve.

If ψ be the angle which the tangent to a curve makes with any fixed straight line, the relation between p and ψ often forms a very simple and elegant equation of the curve. This relation has been called by Dr. Ferrers the Tangential-Polar Equation.

The p, ψ equation may be deduced at once from the equation of the first positive pedal.

If $r = f(\theta)$ be the pedal curve, then, since $\psi = \dfrac{\pi}{2} + \theta$ (see Fig. 25, Art. 192), the equation between p and ψ is clearly
$$p = f\left(\psi - \frac{\pi}{2}\right).$$

Ex. 1. The p, ψ equation of $Ax^2 + By^2 = 1$ is
$$p^2 = \frac{\sin^2\psi}{A} + \frac{\cos^2\psi}{B} \text{ (Art. 191).}$$

Ex. 2. The pedal of $\dfrac{2a}{r} = 1 + \cos\theta$ with regard to the origin is $r\cos\theta = a$, and therefore its p, ψ equation is $p\sin\psi = a$.

195. Relations between p, t, ρ, etc.

Let PY, QY' be tangents at the contiguous points P, Q on the curve, and let OY, OY' be perpendiculars from O upon these tangents. Let OZ be drawn at right angles to

$Y'Y$ produced. Let the tangents at P and Q intersect at T, and let them cut the initial line OX in R and S. Let the normals at P and Q intersect in C.

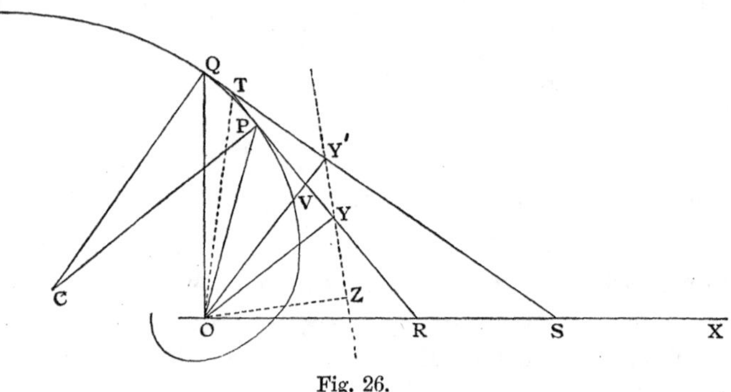

Fig. 26.

Let the co-ordinates of P be (r, θ), and those of Q $(r+\delta r,\ \theta+\delta\theta)$. Let $OY=p$, $OY'=p+\delta p$, $P\hat{R}X=\psi$, $Q\hat{S}X=\psi+\delta\psi$. Then $S\hat{T}R$, $P\hat{C}Q$, $Y\hat{O}Y'$ each $=\delta\psi$. Let $PY=t$, and arc $PQ=\delta s$. Let OY' cut TY in V; then, since $O\hat{Y}V$ is a right angle and $Y\hat{O}V=\delta\psi$ a small angle of the first order, OV differs from OY by a quantity of higher order than the first (Art. 33).

Hence VY' differs from δp by a quantity of higher order than δp, and

$$TY'\tan\delta\psi = VY',$$

therefore $$TY'\frac{\tan\delta\psi}{\delta\psi}=\frac{VY'}{\delta\psi},$$

and proceeding to the limit $t=\dfrac{dp}{d\psi}$. .... (1)

Similarly, if PC be called ρ we have

$$\text{arc } PQ = PC\,.\,\delta\psi,$$

neglecting infinitesimals of higher order than $\delta\psi$, therefore

$$PC = \frac{\text{arc } PQ}{\delta\psi},$$

and proceeding to the limit,

$$\rho = \frac{ds}{d\psi}. \quad \ldots\ldots\ldots\ldots\ldots\ldots\ldots\ldots\ldots\ldots \quad (2)$$

Again $\quad \delta t = Y'Q - YP$

$$= (Y'T + TQ) - (YV + VT - PT)$$

$$= (PT + TQ) + (Y'T - VT) - YV.$$

Now $\quad YV = p \tan \delta\psi,$

and remembering that when $\delta\psi$ is an infinitesimal of the first order, VT and $Y'T$, $PT+TQ$ and δs, $\tan \delta\psi$ and $\delta\psi$, each differ by quantities of order higher than the first, we have, upon dividing by $\delta\psi$ and proceeding to the limit,

$$\frac{dt}{d\psi} = \frac{ds}{d\psi} - p,$$

or

$$\rho = p + \frac{d^2p}{d\psi^2}, \text{ by (1) and (2).} \quad \ldots\ldots\ldots\ldots (3)$$

196. Perpendicular on Tangent to Pedal.

From the same figure it is clear that since $Y\hat{O}Y' = Y\hat{T}Y'$, the points O, Y, Y', T are concyclic, and therefore $O\hat{Y}Z = \pi - O\hat{Y}Y' = O\hat{T}Y'$; and the triangles OYZ and OTY' are similar. Therefore $\dfrac{OZ}{OY} = \dfrac{OY'}{OT}$.

And in the limit when Q comes into coincidence with P, Y' comes into coincidence with Y, and the limiting position of $Y'Y$ is the tangent to the pedal curve. Let the perpendicular on the tangent at Y to the pedal curve be called p_1, then the above ratio becomes

$$\frac{p_1}{p}=\frac{p}{r},$$

or $$p_1 r = p^2.$$

197. Circle on Radius Vector for Diameter touches Pedal.

It is clear also from the figure of Art. 195 that the circle on the radius vector as diameter touches the first positive pedal of the curve. For OT is in the limit a radius vector; and the circle on OT as diameter passing through Y and Y', two contiguous points on the pedal, must in the limit have the same tangent at Y as the pedal curve, and must therefore touch it.

198. Pedal Equation of Pedal Curve.

Let $r=f(p)$ be the pedal equation of a given curve. Then, since $p_1 r=p^2$, we have $p_1=\frac{p^2}{f(p)}$, and therefore, writing r for p and p for p_1, the pedal equation of the first positive pedal curve is $p=\frac{r^2}{f(r)}$.

Ex. The first positive pedal of the rectangular hyperbola $r=\frac{a^2}{p}$ is

$$p=\frac{r^2}{\frac{a^2}{r}}=\frac{r^3}{a^2},$$

which is the p, r equation of Bernoulli's Lemniscate, as is also obvious from Art. 190.

EXAMPLES.

Write down the pedal equations of the first positive pedals of the curves given in the table of Art. 190.

199. We may also prove the results of Art. 195 as follows :—

Let the tangent P_1T make an angle ψ with the initial line. Then the perpendicular makes an angle $a=\psi-\frac{\pi}{2}$ with the same line. Let $OY=p$. Let P_1P_2 be the normal, and P_2 its point of intersection with the normal at the contiguous point Q. Let OY_1 be the perpendicular from O upon the normal. Call this p_1. Let P_2P_3 be drawn at right angles to P_1P_2, and let the length of OY_2, the perpendicular upon it from O, be p_2.

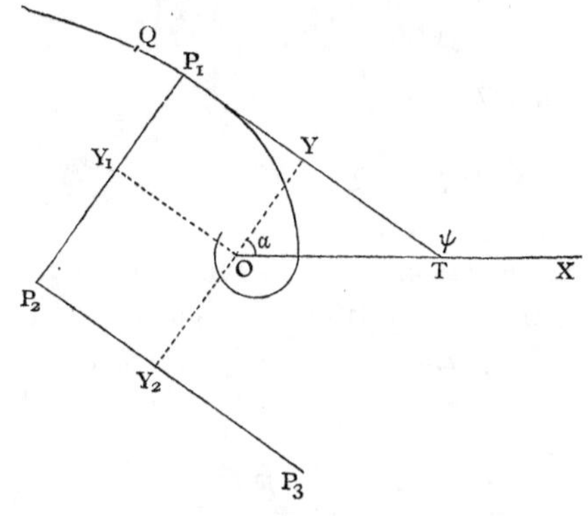

Fig. 27.

The equation of P_1T is clearly

$$p=x\cos a+y\sin a \;\ldots\ldots\ldots\ldots\ldots\ldots(1)$$

The contiguous tangent at Q has for its equation

$$p+\delta p=x\cos(a+\delta a)+y\sin(a+\delta a)\ldots\ldots\ldots(2)$$

Hence subtracting and proceeding to the limit it appears that

$$\frac{dp}{da}=-x\sin a+y\cos a\ldots\ldots\ldots\ldots\ldots\ldots(3)$$

is a straight line passing through the point of intersection of (1) and (2); also being perpendicular to (1) it is the equation of the normal P_1P_2.

Similarly $$\frac{d^2p}{da^2} = -x\cos a - y\sin a \quad \text{.................}(4)$$

represents a straight line through the point of intersection of two contiguous positions of the line P_1P_2 and perpendicular to P_1P_2, viz., the line P_2P_3, and so on for further differentiations.

From this it is obvious that

$$OY_1 = \frac{dp}{da} = \frac{dp}{d\psi}, \text{ since } \frac{d\psi}{da} = 1;$$

$$OY_2 = \frac{d^2p}{da^2} = \frac{d^2p}{d\psi^2},$$

etc.

Hence $$t = P_1Y = \frac{dp}{d\psi},$$

and $$\rho = P_1P_2 = OY + OY_2 = p + \frac{d^2p}{d\psi^2}.$$

200. **Tangential Equation of a Curve.**

DEF. The tangential equation of a curve is the *condition that the line $lx+my+n=0$ may touch* the curve.

Method 1. Let $F(x, y)=0$ be the curve, then the tangent at x, y is

$$X\frac{\partial F}{\partial x} + Y\frac{\partial F}{\partial y} + z\frac{\partial F}{\partial z} = 0.$$

Comparing this with $lX+mY+n=0$,

$$\frac{\frac{\partial F}{\partial x}}{l} = \frac{\frac{\partial F}{\partial y}}{m} = \frac{\frac{\partial F}{\partial z}}{n} = \lambda, \text{ say.}$$

If x, y, λ be eliminated between these equations, and $F(x, y)=0$, or $lx+my+n=0$, a relation between l, m, n will result. This is the equation required.

Method 2. We may also proceed thus. Eliminate y between $F(x, y)=0$ and $lx+my+n=0$; we obtain an equation in x, say $\phi(x)=0$. For tangency this equation must have a pair of equal roots. The condition for this will be found by eliminating x between $\phi(x)=0$ and $\phi'(x)=0$.

In following this method, instead of eliminating y it is often better to make a homogeneous equation between $F(x, y)=0$ and $lx+my+n=0$, and then express that the resulting equation for the ratio $y:x$ has a pair of equal roots.

Ex. Find the tangential equation of the conic

$$ax^2+2hxy+by^2+2gx+2fy+c=0.$$

The first process gives us

$$ax+hy+g=\frac{\lambda}{2}l,$$

$$hx+by+f=\frac{\lambda}{2}m,$$

$$gx+fy+c=\frac{\lambda}{2}n.$$

Also $$lx+my+n=0.$$

The eliminant from these four equations is

$$\begin{vmatrix} a, & h, & g, & l \\ h, & b, & f, & m \\ g, & f, & c, & n \\ l, & m, & n, & 0 \end{vmatrix}=0,$$

which may be written

$$Al^2+Bm^2+Cn^2+2Fmn+2Gnl+2Hlm=0,$$

where A, B, C, ... are the minors of the determinant

$$\begin{vmatrix} a, & h, & g \\ h, & b, & f \\ g, & f, & c \end{vmatrix}.$$

INVERSION.

201. DEF. Let O be the pole, and suppose any point P be given; then if a second point Q be taken on OP, or OP produced, such that $OP \,.\, OQ=\text{constant}$, k^2 say, then Q is said to be the *inverse of the point P with respect to a circle of radius k and centre O.*

If the point P move in any given manner, the *path of Q is said to be inverse to the path of P.* If (r, θ) be the polar co-ordinates of the point P, and (r', θ) those of the inverse point Q, then $rr'=k^2$. Hence, if the locus of P be $f(r, \theta)=0$, that of Q will be $f\left(\frac{k^2}{r}, \theta\right)=0$.

For example, the curves $r^m=a^m \cos m\theta$ and $r^m \cos m\theta=a^m$ are inverse to each other with regard to a circle of radius a.

202. Again, if (x, y) be the Cartesian co-ordinates of P, and (x', y') those of Q, then

$$x=r\cos\theta=\frac{k^2}{r'}\cos\theta=k^2\frac{r'\cos\theta}{r'^2}=k^2\frac{x'}{x'^2+y'^2},$$

and similarly $$y=\frac{k^2y'}{x'^2+y'^2}.$$

Hence, if the locus of P be given in Cartesians as

$$F(x, y)=0,$$

the locus of Q will be

$$F\left(\frac{k^2x}{x^2+y^2}, \frac{k^2y}{x^2+y^2}\right)=0.$$

Ex. The inverse of the straight line $x=a$ with regard to a circle radius k and centre at the origin is

$$\frac{k^2x}{x^2+y^2}=a,$$

or $$x^2+y^2=\frac{k^2}{a}x,$$

a circle which touches the axis of y at the origin.

203. Tangents to Curve and Inverse inclined to Radius Vector at Supplementary Angles.

If P, P' be two contiguous points on a curve, and Q, Q' the inverse points, then, since $OP.OQ=OP'.OQ'$, the points P, P', Q', Q are concyclic; and since the angles

OPT and $OQ'T$ are therefore supplementary, it follows that in the limit when P' ultimately coincides with P

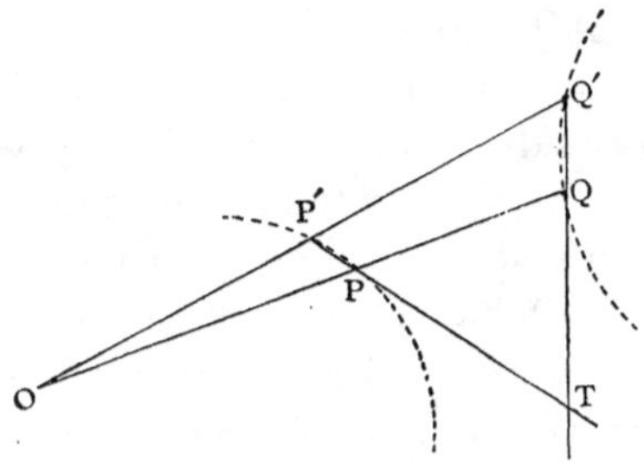

Fig. 28.

and Q' with Q the tangents at P and Q make supplementary angles with OPQ.

The ultimate ratio of corresponding elementary arcs, viz.,

$$\frac{ds}{ds'} = Lt\frac{PP'}{QQ'} = Lt\frac{OP}{OQ'} = \frac{OP}{OQ} = \frac{OP \cdot OQ}{OQ^2} = \frac{k^2}{r'^2} = \frac{r^2}{k^2}.$$

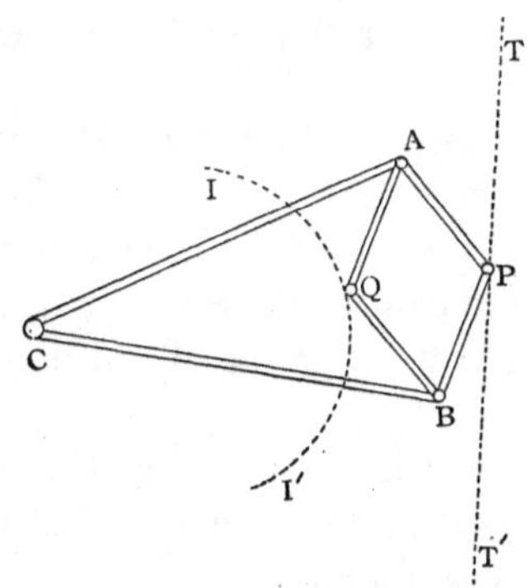

Fig. 29.

204. Mechanical Construction of the Inverse of a Curve. In the accompanying figure AC, CB, BQ, QA, PA, PB

is a system of freely jointed rods, of which $AC=BC$, and

$$AQ=QB=BP=PA.$$

At P and Q sockets are placed to carry tracing pencils. A pin fixes C to the drawing board. The system is then movable about C. It is clear from elementary geometry that C, Q, P are in a straight line, and that

$$CP \,.\, CQ=CA^2-AQ^2,$$

and is therefore constant. Hence whatever curve P is made to trace out, Q *will trace out its inverse*, the point C being the pole of inversion.

In the figure P is represented as tracing a *straight line*, in which case Q will trace an *arc of a circle*, as shown in Art. 202.

Peaucellier has utilized this construction for the conversion of circular into rectilinear motion.

Polar Reciprocals.

205. **Polar Reciprocal of a Curve with regard to a given Circle.**

Def. If OY be the perpendicular from the pole upon the tangent to a given curve, and if a point Z be taken on OY or OY produced *such that* $OY \,.\, OZ$ *is constant* ($=k^2$ say), the locus of Z is called the *polar reciprocal* of the given curve with regard to a circle of radius k and centre at O.

From the definition it is obvious that this curve is the *inverse of the first positive pedal* curve, and therefore its equation can at once be found.

Ex. *Polar reciprocal of an ellipse with regard to its centre.*

For the ellipse $\dfrac{x^2}{a^2}+\dfrac{y^2}{b^2}=1,$

the condition that $p=x\cos\alpha+y\sin\alpha$ touches the curve is

$$p^2=a^2\cos^2\alpha+b^2\sin^2\alpha.$$

Hence the polar equation of the pedal with regard to the origin is

$$r^2 = a^2\cos^2\theta + b^2\sin^2\theta.$$

Again, the inverse of this curve is

$$\frac{k^4}{r^2} = a^2\cos^2\theta + b^2\sin^2\theta,$$

or $$a^2x^2 + b^2y^2 = k^4,$$

which is therefore the equation of the polar reciprocal of the ellipse with regard to a circle with centre at the origin and radius k.

206. The method may therefore be stated thus:—

First *find the condition that* $p = x\cos\alpha + y\sin\alpha$ *will touch the given curve.* Then *write* $\frac{k^2}{r}$ *for* p *and* θ *for* α *in that condition.* The result is the required polar reciprocal with regard to a circle of radius k and centre at the origin.

207. **Polar Reciprocal with regard to a given Conic.**

DEF. If $S=0$ be any curve and $U=0$ a given conic, the *locus of the poles with regard to* U *of tangents to* S is called the *Polar Reciprocal of the curve* S *with regard to the conic* U.

Let the equation of a tangent to S be

$$p = X\cos\alpha + Y\sin\alpha,$$

and the condition of tangency

$$p = f(\alpha).$$

If x, y be the pole of this tangent with regard to $U=0$, the tangent must be coincident with the polar

$$X\frac{\partial U}{\partial x} + Y\frac{\partial U}{\partial y} + z\frac{\partial U}{\partial z} = 0;$$

therefore $$\frac{\cos\alpha}{p} = -\frac{\frac{\partial U}{\partial x}}{\frac{\partial U}{\partial z}}, \quad \frac{\sin\alpha}{p} = -\frac{\frac{\partial U}{\partial y}}{\frac{\partial U}{\partial z}}.$$

Hence $$\frac{1}{p^2}=\frac{\left(\frac{\partial U}{\partial x}\right)^2+\left(\frac{\partial U}{\partial y}\right)^2}{\left(\frac{\partial U}{\partial z}\right)^2} \text{ and } \tan\alpha=\frac{\frac{\partial U}{\partial y}}{\frac{\partial U}{\partial x}}.$$

Hence the equation of the Polar Reciprocal is

$$\left(\frac{\partial U}{\partial x}\right)^2+\left(\frac{\partial U}{\partial y}\right)^2=\frac{\left(\frac{\partial U}{\partial z}\right)^2}{\left[f\left\{\tan^{-1}\left(\frac{\frac{\partial U}{\partial y}}{\frac{\partial U}{\partial x}}\right)\right\}\right]^2}.$$

For further information on the subject of reciprocal polars and the methods of reciprocation the student is referred to Dr. Salmon's *Treatise on Conic Sections*, Chap. XV.

EXAMPLES.

1. Find where the tangent is parallel to the axis of x and where it is perpendicular to that axis for the following curves:—

(a) $ax^2+2hxy+by^2=1$.

(β) $y=\dfrac{x^3-a^3}{ax}$.

(γ) $y^3=x^2(2a-x)$.

2. Find the equations of the tangents at the origin in the following curves:—

(a) $(x^2+y^2)^2=a^2x^2-b^2y^2$.

(β) $(y-a)^2\dfrac{x^2+y^2}{y^2}=b^2$.

(γ) $x^5+y^5=5ax^2y^2$.

3. Find the length of the perpendicular from the origin on the tangent at the point x, y of the curve

$$x^4+y^4=c^4.$$

4. Show that in the curve $y = be^{\frac{x}{a}}$ the subtangent is of constant length.

5. Show that in the curve $by^2 = (x+a)^3$ the square of the subtangent varies as the subnormal.

6. For the parabola $y^2 = 4ax$, prove

$$\frac{ds}{dx} = \sqrt{\frac{a+x}{x}}.$$

7. Prove that for the ellipse $\frac{x^2}{a^2} + \frac{y^2}{b^2} = 1$, if $x = a \sin \phi$,

$$\frac{ds}{d\phi} = a\sqrt{1 - e^2 \sin^2 \phi}.$$

8. For the cycloid

$$\left.\begin{array}{l} x = a \text{ vers } \theta \\ y = a(\theta + \sin\theta) \end{array}\right\},$$

prove

$$\frac{ds}{dx} = \sqrt{\frac{2a}{x}}.$$

9. In the curve

$$y = a \log \sec \frac{x}{a},$$

prove

$$\frac{ds}{dx} = \sec\frac{x}{a}, \quad \frac{ds}{dy} = \operatorname{cosec}\frac{x}{a}, \text{ and } x = a\psi.$$

10. Show that the portion of the tangent to the curve

$$x^{\frac{2}{3}} + y^{\frac{2}{3}} = a^{\frac{2}{3}},$$

which is intercepted between the axes, is of constant length.

Find the area of the portion included between the axes and the tangent.

11. Find for what value of n the length of the subnormal of the curve $xy^n = a^{n+1}$ is constant. Also for what value of n the area of the triangle included between the axes and any tangent is constant.

12. Prove that for the catenary $y = c \cosh \frac{x}{c}$, the length of the normal $= \frac{y^2}{c}$.

Prove also that the length of the perpendicular from the foot of the ordinate on the tangent is of constant length.

13. In the tractory

$$x = \sqrt{c^2 - y^2} + \frac{c}{2} \log \frac{c - \sqrt{c^2 - y^2}}{c + \sqrt{c^2 - y^2}},$$

prove that the portion of the tangent intercepted between the point of contact and the axis of x is of constant length.

14. In the spiral $r = ae^{\theta \cot \alpha}$, prove

$$\frac{dr}{ds} = \cos \alpha \text{ and } p = r \sin \alpha.$$

15. For the involute of a circle, viz.,

$$\theta = \frac{\sqrt{r^2 - a^2}}{a} - \cos^{-1}\frac{a}{r},$$

prove

$$\cos \phi = \frac{a}{r}.$$

16. In the parabola $\frac{2a}{r} = 1 - \cos \theta$, prove the following results :—

(α) $\phi = \pi - \frac{\theta}{2}$.

(β) $p = \frac{a}{\sin \frac{\theta}{2}}$.

(γ) $p^2 = ar$.

(δ) Polar subtangent $= 2a \operatorname{cosec} \theta$.

17. For the cardioide $r = a(1 - \cos \theta)$, prove

(α) $\phi = \frac{\theta}{2}$.

(β) $p = 2a \sin^3 \frac{\theta}{2}$.

(γ) $p^2 = \frac{r^3}{2a}$.

(δ) Polar subtangent $= 2a \frac{\sin^3 \frac{\theta}{2}}{\cos \frac{\theta}{2}}$.

18. If the curves $f(x, y)=0$, $F(x, y)=0$ touch at the point x, y, prove

$$\frac{\partial f}{\partial x}\cdot\frac{\partial F}{\partial y}-\frac{\partial f}{\partial y}\cdot\frac{\partial F}{\partial x}=0$$

at the point of contact.

19. If the curves $f(x, y)=0$, $F(x, y)=0$, cut orthogonally, prove that at the point of intersection

$$\frac{\partial f}{\partial x}\cdot\frac{\partial F}{\partial x}+\frac{\partial f}{\partial y}\cdot\frac{\partial F}{\partial y}=0.$$

20. If the form of a curve be given by the equations

$$x=\phi(t),\quad y=\psi(t),$$

prove that the equation of the tangent at the point determined by the third variable t is

$$X\psi'(t)-Y\phi'(t)=\phi(t)\psi'(t)-\psi(t)\phi'(t),$$

and that the corresponding normal is

$$X\phi'(t)+Y\psi'(t)=\phi(t)\phi'(t)+\psi(t)\psi'(t).$$

21. Apply the preceding example to find the tangent and normal at the point determined by θ on

(α) The ellipse $\left.\begin{array}{l}x=a\cos\theta\\ y=b\sin\theta\end{array}\right\}.$

(β) The cycloid $\left.\begin{array}{l}x=a(\theta+\sin\theta)\\ y=a(1-\cos\theta)\end{array}\right\}.$

(γ) The epicycloid $\left.\begin{array}{l}x=A\cos\theta-B\cos\dfrac{A}{B}\theta\\ y=A\sin\theta-B\sin\dfrac{A}{B}\theta\end{array}\right\}.$

22. In the four-cusped hypocycloid

$$x^{\frac{2}{3}}+y^{\frac{2}{3}}=a^{\frac{2}{3}},$$

show that if $x=a\cos^3\alpha$ then $y=a\sin^3\alpha$,

and that the equation of the tangent at the point determined by α is $x\sin\alpha+y\cos\alpha=a\sin\alpha\cos\alpha.$

Hence show that the locus of intersection of tangents at right angles to one another is

$$r^2=\frac{a^2}{2}\cos^2 2\theta.$$

23. If p_1 and p_2 be the perpendiculars from the origin on the tangent and normal respectively at the point (x, y), and if $\tan\psi = \frac{dy}{dx}$, prove that $p_1 = x \sin\psi - y \cos\psi$,

and $$p_2 = x \cos\psi + y \sin\psi.$$

Hence prove that $$p_2 = \frac{dp_1}{d\psi}.$$

24. Through the point h, k tangents are drawn to the curve
$$Ax^3 + By^3 = 1\,;$$
show that the points of contact lie on a conic.

25. If from any point P normals be drawn to the curve whose equation is $y^m = max^n$, show that the feet of the normals lie on a conic, of which the straight line joining P to the origin is a diameter. Find the position of the axes of this conic.

26. The points of contact of tangents from the point h, k to the curve $x^3 + y^3 = 3axy$ lie on a conic which passes through the origin.

27. Through a given point h, k tangents are drawn to curves where the ordinate varies as the cube of the abscissa. Show that the locus of the points of contact is the rectangular hyperbola
$$2xy + kx - 3hy = 0,$$
and the locus of the remaining point in which each tangent cuts the curve is the rectangular hyperbola
$$xy - 4kx + 3hy = 0.$$

28. Prove that the locus of the extremity of the polar subtangent of the curve $u + f(\theta) = 0$ is
$$u = f'\left(\frac{\pi}{2} + \theta\right).$$

Ex. Find this locus in the case of the conic
$$\frac{l}{r} = 1 + e\cos\theta.$$

29. Prove that the locus of the extremity of the polar sub-

normal of the curve $r=f(\theta)$ is

$$r=f'\left(\theta-\frac{\pi}{2}\right).$$

Hence show that the locus of the extremity of the polar subnormal in the equiangular spiral $r=ae^{m\theta}$ is another equiangular spiral.

30. In the curve

$$r=\frac{1+\tan\frac{\theta}{2}}{m+n\tan\frac{\theta}{2}}$$

the locus of the extremity of the polar subtangent is a cardioide.

[PROFESSOR WOLSTENHOLME.]

31. Show geometrically that the pedal equation of a circle with regard to a point on the circumference is $pd=r^2$, d being the diameter of the circle.

32. Show that the pedal equation of the ellipse

$$\frac{x^2}{a^2}+\frac{y^2}{b^2}=1$$

with regard to a focus is

$$\frac{b^2}{p^2}=\frac{2a}{r}-1.$$

33. Show that the pedal equation of the parabola $y^2=4ax$ with regard to its vertex is

$$a^2(r^2-p^2)^2=p^2(r^2+4a^2)(p^2+4a^2).$$

34. Show that the pedal equation of the curve $r=a^\theta$ is of the form $p=mr$ where m is a constant.

35. Show that the pedal equation of the tetracuspidal hypocycloid $x^{\frac{2}{3}}+y^{\frac{2}{3}}=a^{\frac{2}{3}}$ is $r^2+3p^2=a^2$.

36. Show that for the epicycloid given by

$$\left.\begin{aligned} x&=(a+b)\cos\theta-b\cos\frac{a+b}{b}\theta \\ y&=(a+b)\sin\theta-b\sin\frac{a+b}{b}\theta \end{aligned}\right\},$$

$$p=(a+2b)\sin\frac{a}{2b}\theta;\quad \psi=\frac{a+2b}{2b}\theta;\quad p=(a+2b)\sin\frac{a\psi}{a+2b};$$

and that the pedal equation is

$$r^2=a^2+4\frac{(a+b)b}{(a+2b)^2}p^2.$$

37. Show that the first positive pedal of the parabola $y^2=4ax$ with regard to the vertex is the cissoid

$$x(x^2+y^2)+ay^2=0.$$

38. Show that the first positive pedal of the curve

$$x^3+y^3=a^3$$

is
$$(x^2+y^2)^{\frac{3}{2}}=a^{\frac{3}{2}}(x^{\frac{3}{2}}+y^{\frac{3}{2}}).$$

39. Show that the first positive pedal of the curve

$$x^{\frac{2}{3}}+y^{\frac{2}{3}}=a^{\frac{2}{3}}$$

is
$$r=\pm a\sin\theta\cos\theta.$$

Also that the tangential polar equation of the curve is

$$p=\mp\frac{a}{2}\sin 2\psi.$$

40. Show that the first positive pedal of the curve

$$x^m y^n=a^{m+n}$$

is
$$r^{m+n}=a^{m+n}\frac{(m+n)^{m+n}}{m^m\,.\,n^n}\cos^m\theta\sin^n\theta.$$

41. Show that the fourth negative pedal of the cardioide $r=a(1+\cos\theta)$ is a parabola.

42. Show that the fourth and fifth positive pedals of the curve
$$r^{\frac{2}{9}}\cos\frac{2}{9}\theta=a^{\frac{2}{9}}$$

are respectively a rectangular hyperbola and a Lemniscate.

43. Show that the nth positive pedal of the spiral $r=ae^{\theta\cot a}$

is
$$r=a\sin^n a e^{n\left(\frac{\pi}{2}-a\right)\cot a}e^{\theta\cot a}.$$

44. Show that, if the curves $r=f(\theta)$, $r=F(\theta)$ intersect at (r,θ), the angle between their tangents at the point of intersection is
$$\tan^{-1}\frac{F(\theta)f'(\theta)-F'(\theta)f(\theta)}{F'(\theta)f'(\theta)+F(\theta)f(\theta)}.$$

45. Show that the inverse of the parabola $y^2=4ax$ with regard to a circle whose centre is at the origin and radius the semilatus rectum is the pedal of the parabola $y^2+4ax=0$ with regard to the vertex.

46. Show that the inverse of the conic $u_2+u_1+u_0=0$ with regard to the origin is the bicircular quartic curve

$$k^4u_2+k^2u_1(x^2+y^2)+u_0(x^2+y^2)^2=0.$$

47. Show that the inverse of the general curve of the nth degree, viz.,

$$u_n+u_{n-1}+u_{n-2}+\ldots+u_1+u_0=0,$$

with regard to the origin is

$$k^{2n}u_n+k^{2n-2}u_{n-1}r^2+k^{2n-4}u_{n-2}r^4+\ldots+k^2u_1r^{2n-2}+u_0r^{2n}=0,$$

where $r^2=x^2+y^2$.

48. Show that the inverse of a conic with regard to the focus is a Limaçon (Equation $r=a+b\cos\theta$), which becomes a cardioide if the conic be a parabola.

49. Show that the inverse of a conic with regard to the centre is an oval of Cassini (Equation $r^2=a+b\cos 2\theta$), which becomes a Lemniscate of Bernoulli if the conic be a rectangular hyperbola.

50. If P_1, P_2 be two points whose inverses are Q_1, Q_2 with regard to any origin O, prove that

$$P_1P_2=\frac{OP_1}{OQ_2}\cdot Q_1Q_2.$$

51. The locus of a point X is defined by the equation

$$F(\rho_1, \rho_2, \rho_3, \ldots \rho_n)=a,$$

where $\rho_1, \rho_2, \ldots$ are the distances of X from n fixed points P_1, $P_2, \ldots P_n$. Show that the equation of its inverse with regard to any origin O is

$$F\left(\frac{r_1\rho_1'}{R}, \frac{r_2\rho_2'}{R}, \ldots \frac{r_n\rho_n'}{R}\right)=a,$$

where $\rho_1', \rho_2', \ldots$ are the distances of X', the inverse of X from the n fixed points $Q_1, Q_2, \ldots$ which are the respective inverses of P_1, P_2, ... ; r_1, r_2, r_3, ... are the lengths of OP_1, OP_2, ... ; and $R = OX'$.

52. Show that the inverse with regard to any pole O of the Cartesian oval whose equation is $lr + mr' = n$, where r, r' are the distances of any point on the curve from two fixed points F_1, F_2, is
$$l \, . \, OF_1 \, . \, \rho_1 + m \, . \, OF_2 \, . \, \rho_2 = n\rho_3,$$
where ρ_1, ρ_2 are the distances of any point on the inverse curve from the points which are the inverses of F_1, F_2, and ρ_3 is the distance of the same point from the pole of inversion.

53. Show that the inverse of a Cassini's oval defined by the equation
$$rr' = \text{constant}$$
is of the form
$$\rho_1\rho_2 = A\rho_3^2,$$
the letters ρ_1, ρ_2, ρ_3 denoting the distances of any point on the inverse curve from certain fixed points.

54. Show that the inverses of two curves intersect at the same angle as the original curves; and as particular cases that if two curves touch their inverses also touch, and if two curves cut orthogonally their inverses cut orthogonally.

55. It is an obvious property of two confocal and co-axial parabolas whose concavities are turned in opposite directions that they cut at right angles. By inverting this proposition, the focus being the pole of inversion, show that the curves which cut orthogonally each member of the family of cardioides $r = a(1 + \cos\theta)$ found by giving different values to a, are also cardioides.

56. Show by inverting a conic with regard to its focus that the circle
$$x^2 + y^2 = l(e + \cos\alpha)x + l\sin\alpha \, . \, y$$
touches the Limaçon
$$r = l + le\cos\theta$$
at the point given by
$$\theta = \alpha.$$

57. Show that the polar reciprocal of the curve $r^m = a^m \cos m\theta$ with regard to a circle whose centre is at the pole is of the form
$$r^{\frac{m}{m+1}} \cos\frac{m}{m+1}\theta = b^{\frac{m}{m+1}}.$$

58. Show that the polar reciprocal of the curve $x^m y^n = a^{m+n}$ with regard to a circle whose centre is at the origin is another curve of the same kind.

59. Show that the first positive pedal of the curve $p = \dfrac{r^{m+1}}{a^m}$ is

$$p^{m+1}a^m = r^{2m+1},$$

and that its polar reciprocal with regard to a circle of radius a whose centre is at the origin is $p^{m+1} = a^m r$.

60. Show that the inverse of the curve $p = f(r)$ with regard to a circle whose radius is k and centre at the pole is

$$p = \frac{r^2}{k^2} f\left(\frac{k^2}{r}\right),$$

and that the polar reciprocal is

$$\frac{k^2}{r} = f\left(\frac{k^2}{p}\right).$$

61. Show that the pedal of the inverse of $p = f(r)$ with regard to a circle whose radius is k and centre at the origin is

$$\frac{k^2 p^2}{r^3} = f\left(\frac{k^2 p}{r^2}\right).$$

62. Show that the pedal of the inverse of $p = \dfrac{r^{m+1}}{a^m}$ with regard to a circle whose radius is k and centre at the origin is

$$p = \left(\frac{a}{k^2}\right)^{\frac{m}{m-1}} r^{\frac{2m-1}{m-1}}.$$

63. Show that the polar reciprocal of the curve $r^m = a^m \cos m\theta$ with regard to the hyperbola $r^2 \cos 2\theta = a^2$ is

$$r^{\frac{m}{m+1}} \cos \frac{m}{m+1}\theta = a^{\frac{m}{m+1}}.$$

64. In the semicubical parabola $ay^2 = x^3$ the tangent at any point P cuts the axis of y in M and the curve in Q. O is the origin and N the foot of the ordinate of P. Prove that MN and OQ are equally inclined to the axis of x.

65. At any point of a curve where the ordinate varies as the cube of the abscissa, a tangent is drawn; where it cuts the curve another tangent is drawn; where this cuts the curve a

third is drawn, and so on. Prove that the abscissae of the points of contact form a geometrical progression, and also the ordinates.

66. A straight line AOP of given length always passes through a fixed point O, while A describes a given straight line AT; show that if PT be the tangent at P to the locus of P, the projection of PT on $AOP = AO$.

67. The point P moves so that $OP \,.\, O'P =$ constant, O, O' being fixed points. If OY, $O'Y'$ be the perpendiculars from O and O' on the tangent at P to the locus of P, prove that

$$PY : PY' :: OP^2 : O'P^2.$$

68. O and O' are two fixed points, P any point in a curve defined by the equation $$\frac{1}{r} - \frac{1}{r'} = \frac{1}{c},$$
where $r = OP$, $r' = O'P$, and c is constant. Prove that the distance between P and the consecutive curve obtained by changing c to $c + \delta c$ is ultimately

$$\frac{\delta c}{\sqrt{1 + \frac{3c^2}{rr'} + \frac{a^2c^4}{r^3r'^3}}},$$

where $a = OO'$. [SMITH'S PRIZE.]

69. In a system of curves defined by an equation containing a variable parameter investigate at any point the normal distance between two consecutive curves, and determine the form of the equation for a system of parallel curves.

[PROFESSOR CAYLEY, *Messenger of Mathematics*, Vol. V.]

CHAPTER VIII.

ASYMPTOTES.

208. Def. *If a straight line cut a curve in two points at an infinite distance from the origin and yet is not itself wholly at infinity, it is called an asymptote to the curve.*

209. Equations of the Asymptotes.

Let the equation of any curve of the nth degree be arranged in homogeneous sets of terms and expressed as

$$x^n\phi_n\left(\frac{y}{x}\right)+x^{n-1}\phi_{n-1}\left(\frac{y}{x}\right)+x^{n-2}\phi_{n-2}\left(\frac{y}{x}\right)+\ldots=0. \quad \ldots\ldots\ldots \text{(A)}$$

To find where this curve is cut by any straight line whose equation is

$$y=\mu x+\beta \quad \ldots\ldots\ldots\ldots\ldots\ldots\ldots \text{(B)}$$

subst[illegible]te $\mu+\frac{\beta}{x}$ for $\frac{y}{x}$ in equation (A), and the resulting equati[illegible]

$$[\text{illegible}]\left(\ldots+\frac{\beta}{x}\right)+x^{n-1}\phi_{n-1}\left(\mu+\frac{\beta}{x}\right)+x^{n-2}\phi_{n-2}\left(\mu+\frac{\beta}{x}\right)\ldots=0\ldots \text{(C)}$$

gives the abscissae of the points of intersection.

Applying Taylor's Theorem to expand each of these

functional forms, equation (C) may be written

$$x^n\phi_n(\mu)+x^{n-1}\left|\begin{matrix}\beta\phi'_n(\mu)\\+\phi_{n-1}(\mu)\end{matrix}\right.+x^{n-2}\left|\begin{matrix}\frac{\beta^2}{2!}\phi''_n(\mu)\\+\beta\phi'_{n-1}(\mu)\\+\phi_{n-2}(\mu)\end{matrix}\right.+\dots=0. \quad (D)$$

This is an equation of the nth degree, *proving that a straight line will in general intersect a curve of the nth degree in n points real or imaginary.*

The straight line $y=\mu x+\beta$ is *at our choice*, and therefore the two constants μ and β may be chosen, so as to satisfy any pair of consistent equations. Suppose we choose μ and β, so that

$$\phi_n(\mu)=0 \dots\dots\dots\dots \quad (E)$$

and
$$\beta\phi'_n(\mu)+\phi_{n-1}(\mu)=0 \dots\dots\dots\dots \quad (F)$$

The two highest powers of x now disappear from equation (D), and that equation has therefore *two infinite roots.*

If, then, $\mu_1, \mu_2, \dots, \mu_n$ be the n values of μ deduced from equation (E) (which is of the nth degree in μ), the corresponding values of β will in general be given by

$$\beta_1=-\frac{\phi_{n-1}(\mu_1)}{\phi'_n(\mu_1)},\ \beta_2=-\frac{\phi_{n-1}(\mu_2)}{\phi'_n(\mu_2)},\dots$$

and the n straight lines

$$\left.\begin{matrix}y=\mu_1x+\beta_1\\y=\mu_2x+\beta_2\\\cdot\ \cdot\ \cdot\ \cdot\ \cdot\\y=\mu_nx+\beta_n\end{matrix}\right\}$$ are the *asymptotes of the curve.*

210. Rule.

Hence, in order to find the asymptotes of any given curve, we may either *substitute $\mu x+\beta$ for y* in the equation of the curve, and then *by equating the coefficients of* *the two highest powers of x to zero find μ and β.* Or we may assume the result of the preceding article, which may be enunciated in the following practical way:—*In the highest degree terms put $x=1$ and $y=\mu$ [the result of this is to form $\phi_n(\mu)$] and equate to zero. Hence find μ. Form $\phi_{n-1}(\mu)$ in a similar way from the terms of degree $n-1$, and differentiate $\phi_n(\mu)$, then the values of β are found by substituting the several values of μ in the formula*

$$\beta=-\frac{\phi_{n-1}(\mu)}{\phi'_n(\mu)}.$$

Ex. *Find the asymptotes of the cubic*

$$2x^3-x^2y-2xy^2+y^3+2x^2+xy-y^2+x+y+1=0.$$

Here $\qquad \phi_3(\mu)=\mu^3-2\mu^2-\mu+2=0;$

therefore $\qquad (\mu-1)(\mu+1)(\mu-2)=0;$

giving $\qquad \mu=1,\ -1,\text{ or }2.$

Again, $\qquad \phi_2(\mu)=2+\mu-\mu^2,$

and $\qquad \phi'_3(\mu)=3\mu^2-4\mu-1;$

therefore $\qquad \beta=\dfrac{\mu^2-\mu-2}{3\mu^2-4\mu-1}.$

Hence if $\qquad \mu=1,\quad \beta=1,$

if $\qquad \mu=-1,\ \beta=0,$

and if $\qquad \mu=2,\quad \beta=0.$

Hence the asymptotes of the curve are

$$y=x+1,$$
$$y=-x,$$
$$y=2x.$$

Examples.

1. The asymptotes of

$$y^3-6xy^2+11x^2y-6x^3+x+y=0$$

are $\qquad y=x,\ y=2x,\ y=3x.$

2. The asymptotes of

$$y^3-x^2y+2y^2+4y+x=0$$

are $\qquad y=0,\ y-x+1=0,\ y+x+1=0.$

211. Number of Asymptotes to a Curve of the nth Degree.

It is clear that since $\phi_n(\mu)=0$ is in general of the nth degree in μ, and $\beta\phi'_n(\mu)+\phi_{n-1}(\mu)=0$ is of the first degree in β, that *n values of μ*, and no more, can be found from the first equation, while the *n corresponding values of β* can be found from the second. Hence *n asymptotes, real or imaginary, can be found for a curve of the nth degree.*

212. If the degree of an equation be odd it is proved in Theory of Equations that there must be one real root at least. Hence any curve of an odd degree must have at least one real asymptote, and therefore must extend to infinity. *No curve therefore of an odd degree can be closed.* Neither can a curve of odd degree have an even number of real asymptotes, or a curve of even degree an odd number.

213. If, however, the term y^n be missing from the terms of the nth degree in the equation of the curve, the term μ^n will also be missing from the equation $\phi_n(\mu)=0$, and there will therefore be an *apparent loss of degree* in this equation. It is clear, however, that in this case, since the coefficient of μ^n is zero, one root of the equation $\phi_n(\mu)=0$ is infinite, and therefore the corresponding asymptote is at right angles to the axis of x; *i.e., parallel to that of y.* This leads us to the special consideration of such asymptotes as may be parallel to either of the axes of co-ordinates.

214. Asymptotes Parallel to the Axes.

Let the curve arranged as in equation (A), Art. 209, be

$$\begin{aligned}a_0x^n+a_1x^{n-1}y+a_2x^{n-2}y^2+\ldots+a_{n-1}xy^{n-1}+a_ny^n\\+b_1x^{n-1}+b_2x^{n-2}y+\ldots\ldots\ldots\ldots\ldots\ldots+b_{n-1}y^{n-1}\\+c_2x^{n-2}\quad+\ldots\\+\ldots=0.\ldots\ldots\ldots\ldots\ldots\ldots\quad(\text{A}')\end{aligned}$$

If arranged in descending powers of x this is

$$a_0x^n+(a_1y+b_1)x^{n-1}+\ldots=0.\ \ldots\ldots\ldots\ldots\ldots\quad(\text{B}')$$

Hence, if a_0 vanish, and y be so chosen that

$$a_1y+b_1=0,$$

the coefficients of the two highest powers of x in equation (B′) vanish, and therefore *two of its roots are infinite.* Hence the straight line $a_1y+b_1=0$ is an asymptote.

In the same way, if $a_n=0$, $a_{n-1}x+b_{n-1}=0$ is an asymptote.

Again, if $a_0=0$, $a_1=0$, $b_1=0$, and if y be so chosen that

$$a_2y^2+b_2y+c_2=0,$$

three roots of equation (B′) become infinite, and the lines represented by

$$a_2y^2+b_2y+c_2=0$$

represent a pair of asymptotes, real or imaginary, parallel to the axis of y.

Hence the rule to find those asymptotes which are parallel to the axes is, "*equate to zero the coefficients of the highest powers of x and y.*"

Ex. *Find the asymptotes of the curve*

$$x^2y^2-x^2y-xy^2+x+y+1=0.$$

Here the coefficient of x^2 is y^2-y and the coefficient of y^2 is x^2-x. Hence $x=0$, $x=1$, $y=0$, and $y=1$ are asymptotes. Also, since the curve is one of the fourth degree, we have thus obtained all the asymptotes.

EXAMPLES.

1. The asymptotes of $y^2(x^2-a^2)=x$ are

$$\left.\begin{array}{l} y=0 \\ x=\pm a \end{array}\right\}.$$

2. The co-ordinate axes are the asymptotes of

$$xy^3+x^3y=a^4.$$

3. The asymptotes of the curve $x^2y^2=c^2(x^2+y^2)$ are the sides of a square.

215. **Partial Fractions Method.**

The values of β, viz.,

$$-\frac{\phi_{n-1}(\mu_1)}{\phi'_n(\mu_1)},\quad -\frac{\phi_{n-1}(\mu_2)}{\phi'_n(\mu_2)},\text{ etc.,}$$

are *exactly the constants required in putting*

$$-\frac{\phi_{n-1}(t)}{\phi_n(t)}$$

*into partial fractions.**

This gives a very easy way of obtaining the asymptotes. For if

$$-\frac{\phi_{n-1}(t)}{\phi_n(t)}=\frac{\beta_1}{t-\mu_1}+\frac{\beta_2}{t-\mu_2}+\frac{\beta_3}{t-\mu_3}+\dots$$

the asymptotes will be

$$y=\mu_1x+\beta_1,$$
$$y=\mu_2x+\beta_2,$$
etc.

* Suppose the single factor $t-\mu_1$ to occur in $\phi_n(t)$. Let

$$\phi_n(t)=(t-\mu_1)\chi(t).$$

Hence, differentiating

$$\phi'_n(t)=\chi(t)+(t-\mu_1)\chi'(t),$$

and putting $t=\mu_1$,

$$\phi'_n(\mu_1)=\chi(\mu_1).$$

But if $\dfrac{A}{t-\mu_1}$ be the partial fraction corresponding to the factor $t-\mu_1$,

$$A=-\frac{\phi_{n-1}(\mu_1)}{\chi(\mu_1)}\text{ (Art. 101).}$$

$$=-\frac{\phi_{n-1}(\mu_1)}{\phi'_n(\mu_1)}.$$

Ex. *Find the asymptotes of the curve,*

$$(x^2-y^2)(x+2y)+5(x^2+y^2)+x+y=0.$$

Here $$-\frac{\phi_{n-1}(t)}{\phi_n(t)}=5\frac{t^2+1}{(2t+1)(t-1)(t+1)}=\frac{-\frac{25}{3}}{2t+1}+\frac{\frac{5}{3}}{t-1}+\frac{5}{t+1}.$$

Hence the asymptotes are

$$2y+x=-\frac{25}{3},$$

$$y-x=\frac{5}{3},$$

$$y+x=5.$$

216. **Particular Cases of the General Theorem.**

We return to a closer consideration of the equations

$$\phi_n(\mu)=0, \ldots\ldots\ldots\ldots\ldots\ldots\text{(E)}$$

$$\beta\phi_n'(\mu)+\phi_{n-1}(\mu)=0, \ldots\ldots\ldots\ldots\ldots\ldots\text{(F)}$$

of Art. 209.

It is proved in Theory of Equations that if an equation such as $\phi_n(\mu)=0$ have a pair of roots equal, say μ_1, then $\phi_n'(\mu_1)=0$.

I. Let the roots of $\phi_n(\mu)=0$ be $\mu_1, \mu_2, \ldots, \mu_n$, supposed *all different*, so that $\phi_n'(\mu)$ does not vanish for any of these roots. Also, suppose $\phi_n(\mu)$ *and* $\phi_{n-1}(\mu)$ *to contain a common factor* $\mu-\mu_1$ say, then $\phi_{n-1}(\mu_1)=0$, and therefore $\beta_1=0$.

Hence the corresponding asymptote is $y=\mu_1 x$ and *passes through the origin.*

II. Next, suppose *two of the roots* of the equation $\phi_n(\mu)=0$ *to be equal, e.g.,* $\mu_2=\mu_1$, then $\phi_n'(\mu_1)=0$. In this case, if $\phi_{n-1}(\mu)$ do not contain $\mu-\mu_1$ as one of its factors, the value β determined from equation (F) is infinite. The line $y=\mu_1 x+\beta_1$ then does indeed cut the curve in two points at an infinite distance from the origin, but it *makes an infinite intercept* on the axis of y and therefore this line *lies wholly at infinity.* Such a straight line is not in general called an asymptote, but it *will however count as one of the n theoretical asymptotes discussed in Art.* 211.

III. But if $\phi_n(\mu)=0$ have *a pair of equal roots* each $=\mu_1$, we have $\phi_n'(\mu_1)=0$, and *if* μ_1 *be also a root of* $\phi_{n-1}(\mu)=0$ the value of β cannot be determined from equation (F). We may however choose β so that the coefficient of x^{n-2} in equation (D) of Art. 209 vanishes, that is so that

$$\frac{\beta^2}{2}\phi_n''(\mu)+\beta\phi'_{n-1}(\mu)+\phi_{n-2}(\mu)=0,$$

from which two values of β, real or imaginary, may be deduced. Let the roots of this equation be β_1 and β_1'. We thus obtain the equations of *two parallel straight lines*

$$y=\mu_1x+\beta_1,$$
$$y=\mu_1x+\beta_1',$$

which each cut the curve in three points at an infinite distance from the origin. In this case there is a double point on the curve at infinity (see Art. 249).

It is clear that in this case *any* straight line parallel to $y=\mu_1x$ will cut the curve in *two* points at infinity. But of all this system of parallel straight lines the two whose equations we have just found are *the only ones which cut the curve in three points at infinity,* and therefore the *name asymptote is confined to them.* The one equation which includes both straight lines is obtained at once by substituting $y-\mu_1x$ for β in the equation to obtain β and is

$$(y-\mu_1x)^2\phi_n''(\mu_1)+2(y-\mu_1x)\phi'_{n-1}(\mu_1)+2\phi_{n-2}(\mu_1)=0.$$

Ex. *Find the asymptotes of the cubic curve*

$$x^3+2x^2y+xy^2-x^2-xy+2=0.$$

Equating to zero the coefficient of y^2 we obtain $x=0$, the only asymptote parallel to either axis.

Putting $\mu x+\beta$ for y, $x^3+2x^2(\mu x+\beta)+x(\mu x+\beta)^2-x^2-x(\mu x+\beta)+2=0$, or rearranging $x^3(1+2\mu+\mu^2)+x^2(2\beta+2\mu\beta-1-\mu)+x(\beta^2-\beta)+2=0$.

$1+2\mu+\mu^2=0$ gives two roots $\mu=-1$. $2\beta+2\mu\beta-1-\mu=0$ is an identity if $\mu=-1$, and this fails to find β.

Proceeding to the next coefficient, $\beta^2-\beta=0$ gives $\beta=0$ or 1.

Hence the three asymptotes are $x=0$, and the pair of parallel lines

$$y+x=0,$$
$$y+x=1.$$

217. **Form of the Curve at Infinity. Another Method for Oblique Asymptotes.**

Let P_r, F_r be used to denote rational algebraical expressions which contain terms of the rth and lower, but of no higher degrees.

Suppose the equation of a curve of the nth degree to be thrown into the form

$$(ax+by+c)P_{n-1}+F_{n-1}=0. \quad \text{............(1)}$$

Then *any* straight line parallel to $ax+by=0$ obviously cuts the curve in *one* point at infinity; and to find the particular member of this family of parallel straight lines which cuts the curve in a second point at infinity, let us examine what is the ultimate linear form to which the curve gradually approximates as we travel to infinity in the above direction, thus obtaining the ultimate direction of the curve and forming the equation of the tangent at infinity. To do this we make the x and y of the curve become *large in the ratio given by* $x:y=-b:a$, and we obtain the equation

$$ax+by+c+Lt_{y=-\frac{a}{b}x=\infty}\left(\frac{F_{n-1}}{P_{n-1}}\right)=0.$$

If this limit be finite we have arrived at the equation of a straight line which at infinity represents the limiting form of the curve, and which satisfies the definition of an asymptote.

To obtain the value of the limit it is advantageous to put $x=-\frac{b}{t}$ and $y=\frac{a}{t}$, and then after simplification make $t=0$.

Ex. *Find the asymptote of*

$$x^3+3x^2y+3xy^2+2y^3=x^2+y^2+x.$$

We may write this curve as

$$(x+2y)(x^2+xy+y^2)=x^2+y^2+x,$$

whence the equation of the asymptote is given by

$$x+2y=Lt_{x=-2y=\infty}\frac{x^2+y^2+x}{x^2+xy+y^2},$$

and putting $x=\frac{-2}{t}$, $y=\frac{1}{t}$ we have

$$x+2y=Lt_{t=0}\frac{\frac{4}{t^2}+\frac{1}{t^2}-\frac{2}{t}}{\frac{4}{t^2}-\frac{2}{t^2}+\frac{1}{t^2}}=Lt_{t=0}\frac{5-2t}{3}=\frac{5}{3},$$

i.e., $$x+2y=\frac{5}{3}.$$

Example. Show that $x+y=\frac{a}{2}$ is the only real asymptote of the curve $$(x+y)(x^4+y^4)=a(x^4+a^4).$$

218. Next, suppose the equation of a curve put into the form $$(ax+by+c)F_{n-1}+F_{n-2}=0,$$
then the line $ax+by+c=0$ cuts the curve in two points at infinity, for no terms of the nth or $(n-1)^{\text{th}}$ degrees remain in the equation determining the points of intersection. Hence in general the line $ax+by+c=0$ is an asymptote. We say *in general*, because if F_{n-1} be of the

form $(ax+by+c)P_{n-2}$, itself containing a factor $ax+by+c$, there will, as in Art. 216, III., be a *pair of asymptotes* parallel to $ax+by+c=0$, each cutting the curve in *three* points at infinity. The equation of the curve then becomes

$$(ax+by+c)^2P_{n-2}+F_{n-2}=0,$$

and the equations of the parallel asymptotes are

$$ax+by+c=\pm\sqrt{-Lt\frac{F_{n-2}}{P_{n-2}}},$$

where x and y in the limit on the right-hand side become infinite in the ratio $\frac{x}{y}=-\frac{b}{a}$.

And other particular forms which the equation of the curve may assume may be treated similarly.

Ex. *To find the pair of parallel asymptotes of the curve*

$$(2x-3y+1)^2(x+y)-8x+2y-9=0.$$

Here $$2x-3y+1=\pm\sqrt{Lt\frac{8x-2y+9}{x+y}},$$

where x and y become infinite in the direction of the line $2x=3y$.

Putting $x=\frac{3}{t}$, $y=\frac{2}{t}$, the right side becomes ± 2. Hence the asymptotes required are $2x-3y=1$ and $2x-3y+3=0$.

219. Asymptotes by Inspection.

It is now clear that if the equation $F_n=0$ break up into linear factors so as to represent a system of n straight lines no two of which are parallel, they will be the asymptotes of any curve of the form

$$F_n+F_{n-2}=0.$$

Ex. 1. $$(x-y)(x+y)(x+2y-1)=3x+4y+5$$

is a cubic curve whose asymptotes are obviously

$$x-y=0,$$
$$x+y=0,$$
$$x+2y-1=0.$$

Ex. 2. $(x-y)^2(x+2y-1)=3x+4y+5.$

Here $x+2y-1=0$ is one asymptote. The other two asymptotes are parallel to $y=x$. Their equations are

$$x-y=\pm\sqrt{Lt_{t=0}\frac{3+4+5t}{1+2-t}}=\pm\sqrt{\frac{7}{3}}.$$

220. Case in which all the Asymptotes pass through the Origin.

If then, when the equation of a curve is arranged in homogeneous set of terms, as

$$u_n+u_{n-2}+u_{n-3}+\ldots=0,$$

it be found that there are no terms of degree $n-1$, and if also u_n contain no repeated factor, the n straight lines passing through the origin, and whose equation is $u_n=0$, are the n asymptotes.

221. Intersections of a Curve with its Asymptotes.

If a curve of the nth degree have n asymptotes, no two of which are parallel, we have seen in Art. 219 that the equations of the asymptotes and of the curve may be respectively written $F_n=0,$

and $F_n+F_{n-2}=0.$

The n asymptotes therefore intersect the curve again at points lying upon the curve $F_{n-2}=0$. Now each asymptote cuts its curve in two points at infinity, and therefore in $n-2$ other points. Hence these $n(n-2)$ points lie on a certain curve of degree $n-2$. For example,

1. The asymptotes of a *cubic* will cut the curve again in *three points lying in a straight line;*

2. The asymptotes of a *quartic* curve will cut the curve again in *eight points lying on a conic section;*

and so on with curves of higher degree.

222. **Common Transversal of a Curve and its Asymptotes.** The equation of the asymptotes and that of the curve coincide in the terms of the nth and $(n-1)^{\text{th}}$ degrees. Hence, if we put both equations into polars, the sums of the roots of the two equations for r are equal; also, the origin is arbitrary. Hence, if through any point O a line $OP_1P_2P_3\ldots$ be drawn to cut the curve in the points P_1, P_2, P_3, ... and the asymptotes in p_1, p_2, p_3 then $\Sigma OP = \Sigma Op$, whence, if $\Sigma OP = 0$, it follows that $\Sigma Op = 0$, so that both systems of points have the same centre of mean position. Hence also the algebraical sum of the intercepts between the curve and the asymptote is zero. [NEWTON.]

A well known case of this is that of the hyperbola, where, if O be the middle point of P_1P_2, $OP_1 + OP_2 = 0$, and therefore $Op_1 + Op_2 = 0$, and therefore O is also the middle point of p_1p_2, whence it follows that in that case $P_1p_1 = p_2P_2$.

223. **Other Definitions of "Asymptotes."**

Other definitions have been given of an asymptote, *e.g.*, (a) That an asymptote is the *limiting position of the tangent* to a curve when the point of contact moves away along the curve to an infinite distance from the origin, while the tangent itself does not ultimately lie wholly at infinity; again, (β) That an asymptote is a straight line whose *distance from a point on the curve diminishes indefinitely* as the point moves away along the curve to an infinite distance from the origin.

224. **To prove the Consistency of the Several Definitions.** We propose to show that the results derived from these definitions are the same as those derived from our definition in Art. 208.

Consider definition (α).

Let the curve be $\quad U \equiv u_n + u_{n-1} + u_{n-2} + \ldots + u_0 = 0.$

The equation of the tangent is

$$X\frac{\partial U}{\partial x} + Y\frac{\partial U}{\partial y} + u_{n-1} + 2u_{n-2} + \ldots + nu_0 = 0.$$

We shall now suppose the point of contact x, y to move to ∞ along some branch of the curve. We shall therefore only retain the highest powers of x and y which occur, viz., those of the $(n-1)^{\text{th}}$ degree. Thus we must retain only $\frac{\partial u_n}{\partial x}$ for $\frac{\partial U}{\partial x}$, $\frac{\partial u_n}{\partial y}$ for $\frac{\partial U}{\partial y}$, and u_{n-1} for $u_{n-1} + 2u_{n-2} + \ldots + nu_0$. Hence in the limit we shall have

$$Lt\left\{X\frac{\partial u_n}{\partial x} + Y\frac{\partial u_n}{\partial y} + u_{n-1}\right\} = 0,$$

or

$$Y = X\left\{-Lt\frac{\frac{\partial u_n}{\partial x}}{\frac{\partial u_n}{\partial y}}\right\} - Lt\left\{\frac{u_{n-1}}{\frac{\partial u_n}{\partial y}}\right\},$$

and it is easy to see that this agrees with the equation of an asymptote found in Art. 209.

225. We next consider definition (β); we have already shown that $ax + by + c = 0$ is, according to our definition, in general an asymptote of the curve $\quad (ax + by + c)F_{n-1} + F_{n-2} = 0.$

The perpendicular from any point x, y of this curve upon the line

$$ax + by + c = 0$$

is

$$\frac{ax + by + c}{\sqrt{a^2 + b^2}} = -\frac{1}{\sqrt{a^2 + b^2}}\frac{F_{n-2}}{F_{n-1}},$$

and the limit of this expression is clearly zero when x and y become infinite in the ratio $-b : a$, provided that the terms of degree $n-1$ in F_{n-1} do not contain $ax + by$ as a factor, for the degree of the denominator is higher than that of the numerator. Hence the *distance between the curve and the asymptote is ultimately a vanishing quantity*, and the line $ax + by + c = 0$ is such as to satisfy definition β.

226. The Curve in General lies on Opposite Sides of the Asymptote at Opposite Extremities.

Let the straight line $ax + by + c = 0$ be an asymptote of

the curve, and suppose there is no other asymptote of the curve parallel to this. The equation of the curve is of the form $(ax+by+c)F_{n-1}+F_{n-2}=0$; and, as in the last article, the perpendicular from any point x, y of the curve on this asymptote is given by

$$P=-\frac{1}{\sqrt{a^2+b^2}}\frac{F_{n-2}}{F_{n-1}}.$$

When x and y become very large in the ratio given by

$$\frac{y}{x}=-\frac{a}{b},$$

this may ultimately be written as

$$P=\frac{k}{x}f\left(\frac{y}{x}\right),$$

where k is a constant, and it is therefore obvious that P changes sign with x.

Hence in general the curve at the opposite extremities of this asymptote lies on opposite sides of it.

227. **Exceptions.**

If, however, $ax+by$ be a factor of the terms of highest degree in F_{n-2}, we may write the equation of the curve

$$(ax+by+c)F_{n-1}+F_{n-3}=0,$$

so that the perpendicular on the asymptote is now given by

$$P=\frac{ax+by+c}{\sqrt{a^2+b^2}}=-\frac{1}{\sqrt{a^2+b^2}}\frac{F_{n-3}}{F_{n-1}};$$

and when x and y become very large in the ratio given by

$$\frac{y}{x}=-\frac{a}{b},$$

this can be ultimately written

$$\frac{k}{x^2}f\left(\frac{y}{x}\right).$$

This, however, though ultimately vanishing, does not change sign with x, so that in this case the curve at *opposite extremities* of the asymptote *lies on the same side* of it.

228. Again, if the equation of the curve be expressible in the form $(ax+by+c)^2 P_{n-2}+F_{n-2}=0$,
the expression for the length of the perpendicular is in the limit of the form $f\left(\frac{y}{x}\right)$. This does not in general ultimately vanish, and therefore in general $ax+by+c=0$ is not an asymptote, but is parallel to a pair of asymptotes. This case has been discussed in Art. 218.

229. If, however, the curve assumes the form

$$(ax+by+c)^2 F_{n-2}+F_{n-3}=0,$$

the length of the perpendicular is given by

$$(\text{Perpendicular})^2 = -\frac{1}{a^2+b^2}\frac{F_{n-3}}{F_{n-2}}.$$

Hence, if the ratio of $\frac{y}{x}$ be that of $-\frac{a}{b}$ when x and y become infinite, this may ultimately be written

$$\frac{k}{x}f\left(\frac{y}{x}\right),$$

and therefore

$$\text{Perpendicular} = \pm\sqrt{\frac{k}{x}\cdot f\left(\frac{y}{x}\right)},$$

which ultimately vanishes, but x cannot change sign or the perpendicular will become imaginary at one extremity of the asymptote. Hence the line is *only asymptotic at one end* and the curve approaches the asymptote *on opposite sides.*

And in the same way other particular forms may be discussed.

230. **Curvilinear Asymptotes.**

If there be two curves which continually approach each other so that for a common abscissa the limit of the difference of the ordinates is zero, or for a common ordinate the limit of the difference of the abscissae is zero when that common abscissa or common ordinate is infinite, these curves are said to be asymptotic to each other. For example, the curves

$$y=Ax^2+Bx+C+\frac{D}{x},$$

$$y=Ax^2+Bx+C$$

are asymptotic; for the difference of their ordinates for any common abscissa x is $\frac{D}{x}$, a quantity whose limit is zero when x is infinite.

231. **Linear Asymptote obtained by Expansion.**

If it be possible to express the equation of a given curve in the form

$$y=Ax+B+\frac{C}{x}+\frac{D}{x^2}+\ldots,$$

then the line $y=Ax+B$ is clearly asymptotic to the curve. This method of obtaining rectilinear asymptotes is frequently useful.

232. **To find on which side of the Asymptote the Curve lies.**

The sign of C (Art. 231) is useful in determining on which side of the asymptote the curve lies.

Let y be the ordinate of the curve, y' that of the asymptote, then

$$y-y'=\frac{C}{x}+\frac{D}{x^2}+\ldots.$$

If x be taken sufficiently large, the sign of $\frac{C}{x}$ governs the sign of the whole of the right-hand side.

Suppose x and y to be positive, *i.e.*, in the *first quadrant*, then $y-y'$ will have in the limit the same sign as C. If C be *positive*, $y-y'$ will be positive, and the ordinate of the curve will be greater than that of the asymptote, and the curve will therefore approach the asymptote *from above.* Similarly, if C be *negative*, $y-y'$ will be negative, and the curve will approach the asymptote from *below.* And in the same way for portions in the other quadrants.

Ex. *Find the asymptotes of the curve*

$$y^2(x^2-a^2)=x^2(x^2-4a^2).$$

Here $x^2-a^2=0$ gives $x=a$ and $x=-a$, two asymptotes parallel to the axis of y.

Again,
$$y=\pm x\left(\frac{1-\frac{4a^2}{x^2}}{1-\frac{a^2}{x^2}}\right)^{\frac{1}{2}}$$

$$=\pm x\left(1-\frac{4a^2}{x^2}\right)^{\frac{1}{2}}\left(1-\frac{a^2}{x^2}\right)^{-\frac{1}{2}}$$

$$=\pm x\left\{1-\frac{2a^2}{x^2}\ldots\right\}\left\{1+\frac{1}{2}\frac{a^2}{x^2}+\ldots\right\}$$

$$=\pm x\left\{1-\frac{3a^2}{2x^2}+\ldots\right\}$$

$$=\pm x\mp\frac{3a^2}{2x}+\ldots$$

Hence the asymptotes are $\left.\begin{array}{l} y=\pm x \\ x=\pm a \end{array}\right\}.$
and

Again, considering
$$y=x-\frac{3a^2}{2x}+\ldots$$
and
$$y=x,$$

it appears that if x be positive the ordinate of the curve is less than the ordinate of the asymptote, and therefore the curve approaches the line $y=x$ in the positive quadrant from below. Similarly the curve approaches the asymptote $y=-x$ in the fourth quadrant from above.

233. General Investigation.

In order to express the general equation

$$x^n\phi_n\left(\frac{y}{x}\right)+x^{n-1}\phi_{n-1}\left(\frac{y}{x}\right)+x^{n-2}\phi_{n-2}\left(\frac{y}{x}\right)+\ldots=0 \quad (1)$$

in the form

$$y=\mu x+\beta+\frac{\gamma}{x}+\frac{\delta}{x^2}+\ldots, \quad \ldots\ldots\ldots\ldots \quad (2)$$

substitute for y from (2) in (1); then, since the result must be an identity, the coefficient of each power of x will be zero. This will give sufficient equations to determine $\mu, \beta, \gamma, \ldots$.

The result of this substitution is

$$x^n\phi_n(\mu)+x^{n-1}\left|\begin{array}{l}\beta\phi'_n(\mu)\\+\phi_{n-1}(\mu)\end{array}\right.+x^{n-2}\left|\begin{array}{l}\gamma\phi'_n(\mu)\\+\frac{\beta^2}{2!}\phi''_n(\mu)\\+\beta\phi'_{n-1}(\mu)\\+\phi_{n-2}(\mu)\end{array}\right.+\ldots=0$$

which gives us the series of equations

$$\phi_n(\mu)=0,$$

$$\beta\phi'_n(\mu)+\phi_{n-1}(\mu)=0,$$

$$\gamma\phi'_n(\mu)+\frac{\beta^2}{2!}\phi''_n(\mu)+\beta\phi'_{n-1}(\mu)+\phi_{n-2}(\mu)=0,$$

. . . .

Hence $\mu, \beta, \gamma \ldots$ are determined.

234. Polar Co-ordinates.

Let the equation of the curve be

$$r^nf_n(\theta)+r^{n-1}f_{n-1}(\theta)+\ldots+f_0(\theta)=0, \ldots\ldots\ldots \quad (1)$$

or $$u^n f_0(\theta)+u^{n-1}f_1(\theta)+\ldots+f_n(\theta)=0 \qquad (2)$$

To find the directions in which $r=\infty$ or $u=0$ we have

$$f_n(\theta)=0 \qquad (3)$$

Let the roots of this equation be

$$\theta=\alpha,\ \beta,\ \gamma,\ \ldots.$$

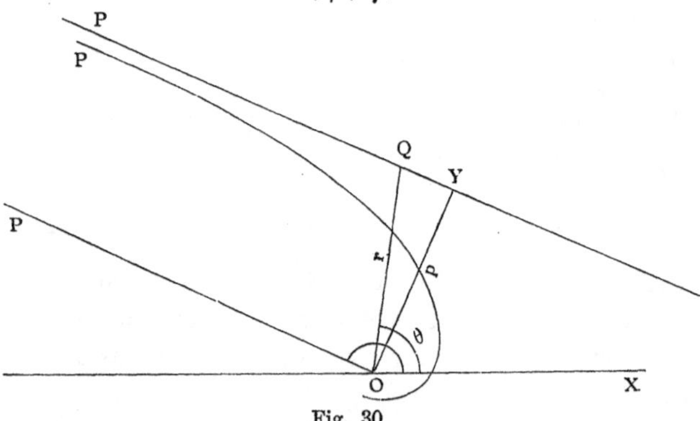

Fig. 30.

Let $X\hat{O}P=\alpha$. Then the radius OP, the curve, and the asymptote meet at infinity towards P. Let $OY(=p)$ be the perpendicular upon the asymptote. Since OY is at right angles to OP it is the polar subtangent, and $p=-\frac{d\theta}{du}$. Let $X\hat{O}Y=\alpha'$, and let Q be any point whose co-ordinates are r, θ upon the asymptote. Then the equation of the asymptote is

$$p=r\cos(\theta-\alpha') \qquad (4)$$

It is clear from the figure that $\alpha'=\alpha-\frac{\pi}{2}$.

To find the value of $-\frac{d\theta}{du}$ when $u=0$ differentiate equation (2), and put $u=0$ and $\theta=\alpha$, and we obtain

$$\left(\frac{du}{d\theta}\right)_{u=0} f_{n-1}(\alpha)+f'_n(\alpha)=0 \qquad (5)$$

Substituting the value of $\left(-\dfrac{d\theta}{du}\right)_{u=0}$ hence deduced for p in equation (4) we have

$$\frac{f_{n-1}(a)}{f'_n(a)}=r\cos\left(\theta-a+\frac{\pi}{2}\right)$$
$$=r\sin(a-\theta).$$

Hence the equations of the asymptotes are

$$r\sin(a-\theta)=\frac{f_{n-1}(a)}{f'_n(a)},$$
$$r\sin(\beta-\theta)=\frac{f_{n-1}(\beta)}{f'_n(\beta)},$$
etc.

COR. The case most often met with is that in which $n=1$, when the equation of the curve is $rf_1(\theta)+f_0(\theta)=0$. Then $f_1(\theta)=0$ gives a, β, γ, etc., and the asymptotes are

$$r\sin(a-\theta)=\frac{f_0(a)}{f'_1(a)}, \text{ etc.}$$

235. **To deduce the Polar Asymptote from the Polar Tangent.**
The same results may be deduced from the equation of a tangent (Art. 182).

The result $u=U\cos(\theta-a)+U'\sin(\theta-a)$ at once reduces to $\dfrac{1}{U'}=r\sin(\theta-a)$, when $U=0$. Putting

$$-\frac{1}{U'}=\frac{f_{n-1}(a)}{f'_n(a)},$$

as found in the last article, we again obtain the equation

$$r\sin(a-\theta)=\frac{f_{n-1}(a)}{f'_n(a)}.$$

Ex. *Find the asymptotes of the curve*

$$r=a\tan\theta \text{ or } r\cos\theta-a\sin\theta=0.$$

Here $\qquad f_1(\theta)=\cos\theta$ and $f_0(\theta)=-a\sin\theta$.

$\cos\theta=0$ gives $\qquad a=\dfrac{\pi}{2},\ \beta=\dfrac{3\pi}{2}$, etc., and $$\frac{f_0(\theta)}{f'_1(\theta)} = \frac{-a\sin\theta}{-\sin\theta} = a.$$

Hence $$\left.\begin{aligned} r\sin\left(\frac{\pi}{2}-\theta\right) &= a \text{ or } r\cos\theta = a \\ r\sin\left(\frac{3\pi}{2}-\theta\right) &= a \text{ or } r\cos\theta = -a \end{aligned}\right\}$$

are the asymptotes.

236. **Circular Asymptotes.**

In many polar equations when θ is increased indefinitely it happens that the equation takes the form of an equation in r, which represents one or more concentric circles. For example, in the curve

$$r = a\frac{\theta}{\theta - 1},$$

which may be written

$$r = a\frac{1}{1-\dfrac{1}{\theta}},$$

it is clear that if θ becomes very large the curve approaches indefinitely near the limiting circle $r = a$. Such a circle is called an *asymptotic circle* of the curve.

EXAMPLES.

Find the asymptotes of the following curves :—

1. $y^3 = x^2(2a - x)$.
2. $y^3 = x(a^2 - x^2)$.
3. $x^3 + y^3 = a^3$.
4. $y(a^2 + x^2) = a^2x$.
5. $axy = x^3 - a^3$.
6. $y^2(2a - x) = x^3$.
7. $x^3 + y^3 = 3axy$.
8. $x^2y + y^2x = a^3$.

9. $x^2y^2 = (a+y)^2(b^2 - y^2)$.
10. $x^2y^2 = a^2y^2 - b^2x^2$.
11. $xy(x-y) - a(x^2 - y^2) = b^3$.
12. $(a^2 - x^2)y^2 = x^2(a^2 + x^2)$.
13. $xy^2 = 4a^2(2a - x)$.
14. $y^2(a - x) = x(b - x)^2$.
15. $x^2y = x^3 + x + y$.
16. $xy^2 + a^2y = x^3 + mx^2 + nx + p$.
17. $x^3 + 2x^2y - xy^2 - 2y^3 + 4y^2 + 2xy + y - 1 = 0$.
18. $x^3 - 2x^2y + xy^2 + x^2 - xy + 2 = 0$.
19. $y(x-y)^3 = y(x-y) + 2$.
20. $x^3 + 2x^2y - 4xy^2 - 8y^3 - 4x + 8y = 1$.
21. $(x+y)^2(x + 2y + 2) = x + 9y - 2$.
22. $3x^3 + 17x^2y + 21xy^2 - 9y^3 - 2ax^2 - 12axy - 18ay^2$
$- 3a^2x + a^2y = 0$.
23. $r\theta^{\frac{1}{2}} = a$.
24. $r\theta = a$.
25. $r \sin n\theta = a$.
26. $r = a \operatorname{cosec} \theta + b$.
27. $r = 2a \sin\theta \tan\theta$.
28. $r \sin 2\theta = a \cos 3\theta$.
29. $r = a + b \cot n\theta$.
30. $r^n \sin n\theta = a^n$.

31. Show that all the asymptotes of the curve $r \tan n\theta = a$ touch the circle $r = \dfrac{a}{n}$.

32. Show that there is an infinite number of asymptotes of the curve $y = (a - x)\tan\dfrac{\pi x}{2a}$, viz.,

$$x = -a, \qquad x = \pm 3a, \qquad x = \pm 5a, \qquad \text{etc.}$$

33. Show that the curve $\theta^2(ar - r^2) = b^2$ has a circular asymptote.

34. Show that there is an infinite series of parallel asymptotes to the curve
$$r = \frac{a}{\theta \sin\theta} + b,$$

and show that their distances from the pole are in Harmonical Progression. Find the circular asymptote.

35. Find the asymptotes of the curve $y=x\dfrac{x^2+a^2}{x^2-a^2}$. Find on which side of the oblique asymptote the curve lies in the positive quadrant. Show also that the hyperbola $x(y-x)=2a^2$ is asymptotic to this cubic curve.

36. Find the asymptotes of the curve $y^2=x^2\dfrac{x+a}{x-a}$, and find on which side the curve approaches these asymptotes.

37. Show that the curve $x=\dfrac{y^3-a^3}{ay}$ has a rectilinear asymptote $y=0$, and a parabolic asymptote $y^2=ax$.

38. Show that the curve $x^2y=x^4+x^3+x^2+x+1$ has a parabolic asymptote whose vertex is at the point $(-\frac{1}{2}, \frac{3}{4})$, and whose latus rectum $=1$.

39. Show that the curve $x^2y=x^3+x^2+x+1$ has a hyperbolic asymptote whose eccentricity $=\dfrac{2}{\sqrt{2+\sqrt{2}}}$.

40. Find the equation of a cubic which has the same asymptotes as the curve $x^3-6x^2y+11xy^2-6y^3+x+y+1=0$, and which touches the axis of y at the origin, and goes through the point (3, 2).

41. Show that the asymptotes of the cubic

$$x^2y-xy^2+xy+y^2+x-y=0$$

cut the curve again in three points which lie on the line $x+y=0$.

42. Find the equation of the conic on which lie the eight points of intersection of the quartic curve

$$xy(x^2-y^2)+a^2y^2+b^2x^2=a^2b^2$$

with its asymptotes.

43. Show that the four asymptotes of the curve

$$(x^2-y^2)(y^2-4x^2)-6x^3+5x^2y+3xy^2-2y^3-x^2+3xy-1=0$$

cut the curve again in eight points which lie on a circle.

44. Form the equation of the cubic curve which has $x=0$ $y=0$, $\frac{x}{a}+\frac{y}{b}=1$ for asymptotes, and cuts its asymptotes in the three points where they intersect the line $\frac{x}{a'}+\frac{y}{b'}=1$, and also passes through the point a, b.

45. Form the equation of a quartic curve which has $x=0$, $y=0$, $y=x$, $y=-x$ for asymptotes, which passes through the point a, b, and cuts its asymptotes again in eight points lying upon the circle $x^2+y^2=a^2$.

46. Form the equation of a quartic curve which has asymptotes $x-y=0$ and $x+y=0$, the curve being supposed to approach each asymptote at one extremity only, but from both sides of that asymptote, and also to touch the axis of y at the origin.

47. Form the equation of a quartic curve with asymptotes $y=0$, $x+y=0$, $x-y=0$, the curve being supposed to approach $y=0$ from opposite sides at the same extremity, but the other two asymptotes from the same side and at opposite extremities in each case. The curve is also to touch the axis of y at the origin and to pass through the point $(2a, a)$.

48. If the equation of a curve be written

$$x^n\phi_n\left(\frac{y}{x}\right)+x^{n-1}\phi_{n-1}\left(\frac{y}{x}\right)+x^{n-2}\phi_{n-2}\left(\frac{y}{x}\right)+\ldots=0$$

and if $\phi_n(\mu_1)=0$, $\phi_n'(\mu_1)=0$, $\phi_{n-1}(\mu_1)=0$, and $\phi'_{n-1}(\mu_1)=0$, show that there are two parallel asymptotes equidistant from the origin, whose equations are

$$y=\mu_1 x \pm\sqrt{-\frac{2\phi_{n-2}(\mu_1)}{\phi_n''(\mu_1)}}.$$

49. Show that the first approximation to the difference of the ordinates of the curve

$$x^n\phi_n\left(\frac{y}{x}\right)+x^{n-1}\phi_{n-1}\left(\frac{y}{x}\right)+x^{n-2}\phi_{n-2}\left(\frac{y}{x}\right)+\ldots=0$$

and its rectilinear asymptote $y=\mu x+\beta$ for a point whose abscissa is x is

$$-\frac{\phi''_n(\mu)[\phi_{n-1}(\mu)]^2-2\phi'_n(\mu)\phi'_{n-1}(\mu)\phi_{n-1}(\mu)+2\phi_{n-2}(\mu)[\phi'_n(\mu)]^2}{2x[\phi'_n(\mu)]^3},$$

assuming that no other asymptote is parallel to this one. Show from this result that the curve at opposite extremities is in general also on opposite sides of the asymptote.

50. Show that the curve

$$(y-2x)^2(y+x)+(y+3x)(y-x)+x=0.$$

has the parabolic asymptote $3y^2-12xy+12x^2+5x=0$.

51. Show, by transforming to the point h, k, that the asymptotes of the general curve of the nth degree

$$(a_0, a_1, \ldots, a_n \between x, y)^n + n(b_0, b_1, \ldots, b_{n-1} \between x, y)^{n-1} + \ldots = 0$$

will all pass through one point if

$$\left\| \begin{matrix} a_0, & a_1, & a_2, & \ldots, & a_{n-1} \\ a_1, & a_2, & a_3, & \ldots, & a_n \\ b_0, & b_1, & b_2, & \ldots, & b_{n-1} \end{matrix} \right\| = 0,$$

and that the co-ordinates of that point are

$$\frac{a_1b_1-a_2b_0}{a_0a_2-a_1^2}, \quad \frac{a_1b_0-a_0b_1}{a_0a_2-a_1^2}.$$

[Professor Cayley uses the notation $(a_0, a_1, \ldots, a_n \between x, y)^n$ for the general binary quantic of the nth degree:

$$a_0x^n+na_1x^{n-1}y+\frac{n(n-1)}{2!}a_2x^{n-2}y^2+\ldots+a_ny^n.]$$

CHAPTER IX.

SINGULAR POINTS.

237. **Concavity. Convexity.**

In the treatment of plane curves the terms concavity and convexity with regard to a point are applied with their ordinary signification. Thus, for example, any arc of a circle is said to be concave to all points within the circle; whilst to a point without the circle the portion lying between that point and the chord of contact of tangents drawn from the point is said to be convex and the remainder of the circumference concave.

238. In general the portion of a curve in the immediate neighbourhood of any specified point lies entirely on one side of the tangent at that point. This is clear from the definition of a tangent, which is considered as the limiting position of a chord. There is an ultimately coincident cross and recross at the point of contact, as shown at the ultimately coincident points P, Q in Fig. 31; so

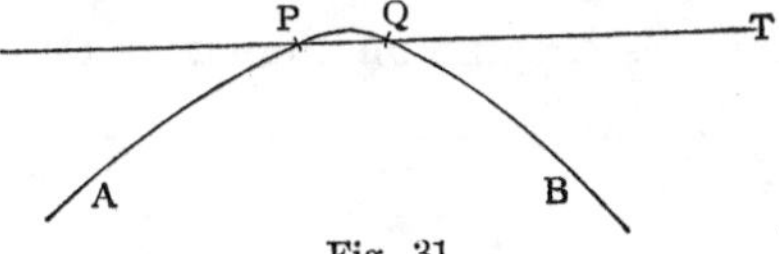

Fig. 31.

that the immediately neighbouring portions AP, QB must in general lie on the same side of the tangent PT.

239. **Point of Inflexion.**

The kind of point discussed in Art. 238 is an ordinary point on a curve. It may however happen that for some point on the curve the tangent, after its cross and recross, *crosses the curve again at a third* ultimately coincident point. Such a point can be seen magnified in Fig. 32.

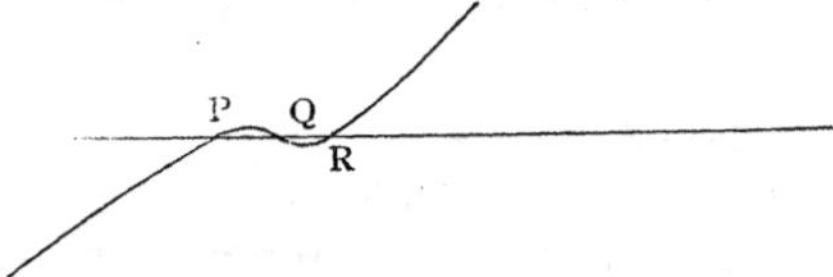

Fig. 32.

In this case it is clear that two successive tangents coincide in position: viz., the limiting positions of the chords PQ, QR. The tangent at such a point is therefore said to be "*stationary*," and the point is called a "*point of contrary flexure*" or a "*point of inflexion*" on the curve. The tangent on the whole crosses its curve at such a point, and the curve changes from being concave to points on one side of the tangent to being convex to the same set of points.

240. **Point of Undulation.**

Again, there may be a point on the curve for which the

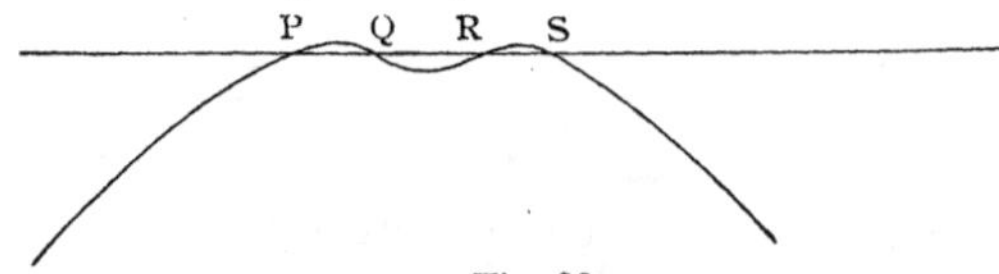

Fig. 33.

tangent crosses its curve in four ultimately coincident points, P, Q, R, S, as seen magnified in Fig. 33, and the point is then called a "*point of undulation*" on the curve. There are now three contiguous tangents coincident, and the tangent on the whole does not cross its curve. And it is clear that singularities of a higher order but of similar character may arise.

241. Analytical Tests. Concavity and Convexity.

It is easy to apply analysis to the investigation of the form of a curve at any particular point.

Let us examine the point x, y on the curve $y=\phi(x)$.

Let P be the point to be considered, P_1 an adjacent point on the curve. Let PN, P_1N_1 be the ordinates of P

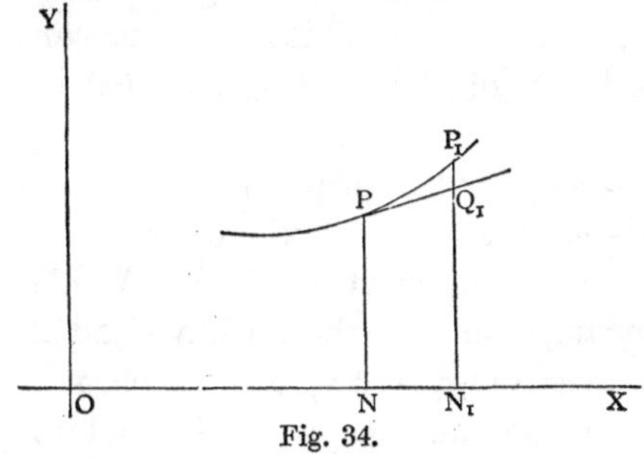

Fig. 34.

and P_1, and suppose P_1N_1 to cut the tangent at P in Q_1. Then $ON=x$, $NP=y=\phi(x)$.

Let $\qquad ON_1=x+h,$

then $\qquad N_1P_1=\phi(x+h)$

$$=\phi(x)+h\phi'(x)+\frac{h^2}{2!}\phi''(x)+\dots, \quad \dots(1)$$

by Taylor's Theorem. Again, the equation of the tangent at P is $\qquad Y-y=\phi'(x)(X-x).$

Putting $\qquad X=x+h$

we obtain $\qquad Y=y+h\phi'(x)=\phi(x)+h\phi'(x), \quad \dots\dots(2)$

which gives the value of N_1Q_1.

Hence the ordinate of the curve exceeds the ordinate of the tangent by

$$N_1P_1 - N_1Q_1 = \frac{h^2}{2!}\phi''(x) + \frac{h^3}{3!}\phi'''(x) + \dots \quad \dots\dots\dots(3)$$

Now, if h be taken sufficiently small, the sign of the right-hand side will be governed by that of its first term; and this term does not change sign with h because it contains an even power of h, viz., the square. Hence *in general*, on whichever side of P the point P_1 be taken, $N_1P_1 - N_1Q_1$ will have the same sign—positive if $\phi''(x)$ be positive, and negative if $\phi''(x)$ be negative; and therefore the element of the curve at P is *convex* or *concave* to the foot of the ordinate of P according as $\phi''(x)$ *is positive or negative.*

We have drawn our figure with the portion of the curve considered above the axis of x. If, however, it had been below, the signs of N_1P_1 and N_1Q_1 would both have been negative and we should have had the contrary result. But observing that $\phi(x)$ is positive for points above the axis of x, and negative for points below, we may obviously state the unrestricted rule that the elementary portion of the curve $y=\phi(x)$ in the neighbourhood of the point (x, y) is *convex or concave to the foot of the ordinate according as* $\phi(x)\phi''(x)$ *or* $y\dfrac{d^2y}{dx^2}$ *is positive or negative.*

242. **Points of Inflexion.**

If $\phi''(x)=0$ at the point under consideration, we have

$$N_1P_1 - N_1Q_1 = \frac{h^3}{3!}\phi'''(x) + \frac{h^4}{4!}\phi''''(x) + \dots,$$

and, as before, the sign of the right-hand side, when h is taken sufficiently small, is governed by the sign of its first term. But this now depends on h^3, and therefore changes sign with h; that is, the ordinate of the curve

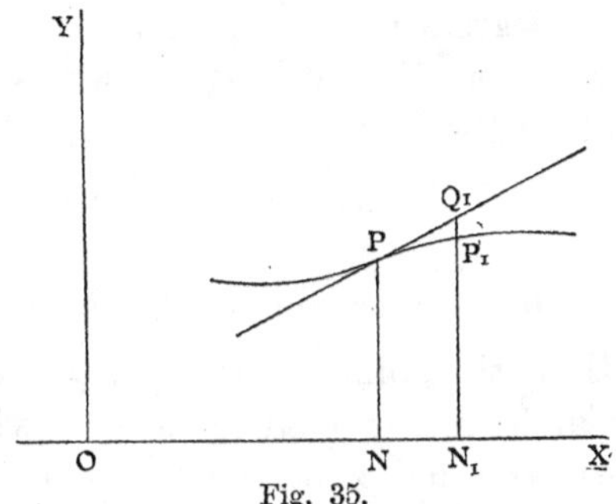

Fig. 35.

is greater than the ordinate of the tangent on one side of P, but less on the other. The tangent now crosses the curve at its point of contact, and the point is of the kind described in Art. 239, and called a *point of inflexion.* A *necessary* condition then for a point of inflexion is that $\phi''(x)$ if not infinite should vanish, and the sign of $\phi'''(x)$ determines the character of the inflexion; for (assuming the element above the axis of x) if $\phi'''(x)$ be positive, $N_1P_1 - N_1Q_1$ changes from negative to positive in passing from negative to positive values of h: *i.e.*, in passing through P the change is from concavity to convexity with regard to the foot of the ordinate. But if $\phi'''(x)$ be negative, the change is from convexity to concavity, and this latter is the case represented in the figure.

243. **Point of Undulation.**

Again, if $\phi'''(x) = 0$ at the same point, and $\phi''''(x)$ do not vanish, the first term in the expansion of $N_1P_1 - N_1Q_1$ depends on h^4, and therefore this expression does not change sign in passing through P. The tangent therefore on the whole does not cross its curve at P. The point is of the kind described in Art. 240 and called a *point of undulation.*

244. **Higher Degrees of Singularity.**

It will now appear that, if by two successive differentiations a result of the form

$$\frac{d^2y}{dx^2}=A(x-a)^{2n}(x-b)^{2m+1}$$

be deduced from the equation to the curve, although $\frac{d^2y}{dx^2}$ vanishes both at the points given by $x=a$ and by $x=b$, yet it only undergoes a change of sign when it passes through $x=b$, the index of the factor $x-b$ being odd. Hence at the points given by $x=a$ there is no ultimate change in the direction of flexure, while at those given by $x=b$ there is a change. The points given by $x=a$ look to the eye like ordinary points on a curve, while those given by $x=b$ resemble points of inflexion, and indeed have been for distinction called by Cramer points of *visible inflexion,** although the singularity is of a higher order than that described in Art. 239, which is the case of $m=0$. If $n=1$, the points given by $x=a$ are points of undulation, such as described in Art. 240. *So that for an Inflexional Point the condition* $\frac{d^2y}{dx^2}=0$, *though necessary, is not sufficient. The complete criterion is that* $\frac{d^2y}{dx^2}$ *should change sign. If* $\frac{d^2y}{dx^2}$ *vanish, but do not change sign, the curve at the point under consideration is undulatory.*

* Dr. Salmon, "Higher Plane Curves," p. 35. Cramer, "Analyse des Lignes Courbes," Geneva.

245. **Case when the Tangent is parallel to the y-axis.** The test of concavity or convexity has been shown to depend upon the sign of $\frac{d^2y}{dx^2}$. In the case, however, of an arc, the tangent to which is parallel to the axis of y, the value of $\frac{dy}{dx}$ and of all subsequent differential coefficients is infinite. But in this case it is obvious that it would be convenient to consider y instead of x for the independent variable, and then the sign of $\frac{d^2x}{dy^2}$ will test the concavity or convexity to the foot of the ordinate drawn from the point under consideration to the axis of y.

Similarly, at a point of inflexion at which the tangent is parallel to the axis of y, $\frac{d^2x}{dy^2}$ must change sign.

And in other cases whenever it is more convenient to use y instead of x for our independent variable, we are of course at liberty to do so with an interchange of the letters x and y in the formula quoted.

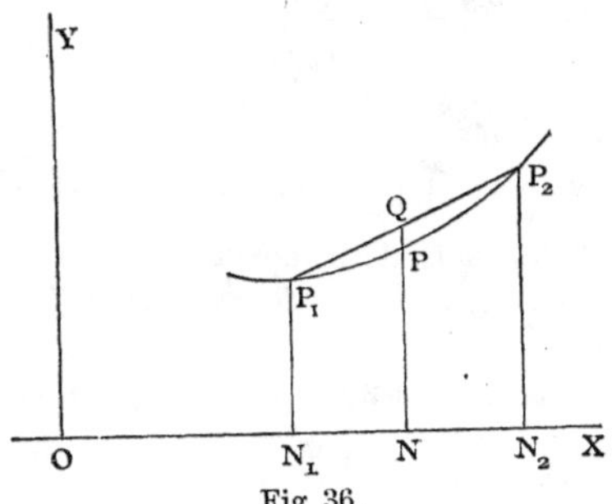

Fig 36.

246. *The test for concavity or convexity may also be investigated as follows:—*

Let P be any point of the curve, co-ordinates x and y. Let the adjacent points on the curve P_1 and P_2 have co-ordinates, $(x-h, y_1)$ and $(x+h, y_2)$ respectively. Let the ordinate of P cut the chord P_1P_2 in Q. Then if h be made infinitesimally small, the portion of the curve in the immediate neighbourhood of P will be convex or concave to N, according as NP is $<$ or $>$ NQ, *i.e.*, as

$$y \text{ is } < \text{ or } > \frac{y_1+y_2}{2}.$$

Now

$$y_2=y+h\frac{dy}{dx}+\frac{h^2}{2!}\frac{d^2y}{dx^2}+\dots,$$

$$y_1=y-h\frac{dy}{dx}+\frac{h^2}{2!}\frac{d^2y}{dx^2}-\dots,$$

so that the criterion depends upon whether

$$y \text{ be } < \text{ or } > y+\frac{h^2}{2!}\frac{d^2y}{dx^2}+\dots,$$

and proceeding to the limit the curve is convex or concave to N according as $\frac{d^2y}{dx^2}$ is positive or negative.

Ex. 1. *Consider the curve $y=2\sqrt{ax}$. Is it convex or concave to the foot of the ordinate?*

Here

$$\frac{d^2y}{dx^2}=-\tfrac{1}{2}\frac{\sqrt{a}}{x^{\frac{3}{2}}},$$

and

$$y\frac{d^2y}{dx^2}=-\frac{a}{x}.$$

Hence $y\frac{d^2y}{dx^2}$ is negative for all positive values of x (and negative values are not admissible), so that the curve in the neighbourhood of any specified point is concave to the foot of the ordinate of that point.

Ex. 2. *Consider the curve $x=y^3+3y^2$. Has it a point of inflexion?*

Here

$$\frac{d^2x}{dy^2}=6(y+1),$$

so that $\frac{d^2x}{dy^2}$ changes sign as y passes through the value $y=-1$. Therefore the point $(2, -1)$ is a point of inflexion on the curve.

247. Convexity and Concavity of a Polar Curve.

Suppose the equation of a curve to be given in polar co-ordinates, and that it is required to find a test of convexity or concavity towards the pole.

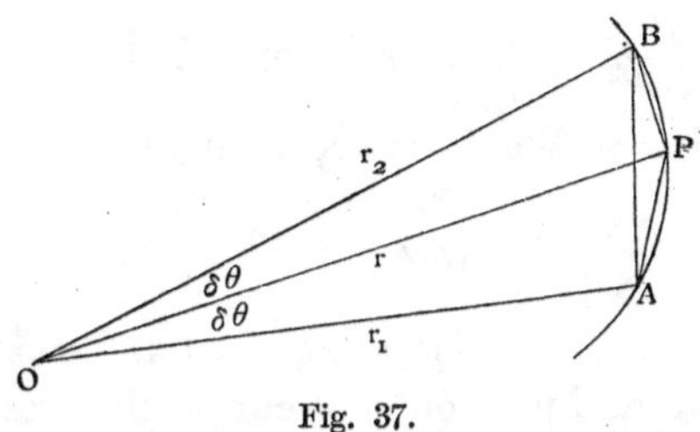

Fig. 37.

Let O be the pole, P the point of the curve to be examined. Let the co-ordinates of P be denoted by r, θ, and let A, B be two points on the curve adjacent to P, and one on each side of it whose co-ordinates are respectively $(r_1, \theta-\delta\theta)$ and $(r_2, \theta+\delta\theta)$. Then the curve in the immediate neighbourhood of P will be concave or convex to O, according as

$$\Delta AOP+\Delta BOP \text{ is } > \text{ or } < \Delta AOB$$

when we proceed to the limit. That is, according as

$$r_1 r\sin\delta\theta+rr_2\sin\delta\theta > \text{ or } < r_1 r_2 \sin 2\delta\theta,$$

or
$$r_1 r+rr_2 > \text{ or } < 2r_1 r_2\cos\delta\theta;$$

i.e., as
$$u_2+u_1 > \text{ or } < 2u\cos\delta\theta,$$

where we have written $r_1=\dfrac{1}{u_1}$, etc.

Now, by Maclaurin's Theorem,

$$u_2=u+\frac{du}{d\theta}\delta\theta+\frac{d^2u}{d\theta^2}\,\frac{\delta\theta^2}{2!}+\ldots,$$

$$u_1=u-\frac{du}{d\theta}\delta\theta+\frac{d^2u}{d\theta^2}\,\frac{\delta\theta^2}{2!}-\ldots,$$

and therefore

$$u_1+u_2=2\left(u+\frac{d^2u}{d\theta^2}\frac{\delta\theta^2}{2!}+\ldots\right),$$

whence we have concavity or convexity to the pole according as $2u+2\frac{d^2u}{d\theta^2}\frac{\delta\theta^2}{2!}+\ldots$ is $>$ or $<$ $2u\left(1-\frac{\delta\theta^2}{2!}+\ldots\right)$, and proceeding to the limit according as

$$u+\frac{d^2u}{d\theta^2} \text{ is } > \text{ or } < 0.$$

248. **Polar Condition for a Point of Inflexion.**

At a point of inflexion the curve changes from concavity to convexity, and therefore the necessary condition is that $u+\frac{d^2u}{d\theta^2}$ should change sign.

Ex. *Find the point of inflexion on the curve* $r=a\theta^{-\frac{1}{2}}$.

Here $$au=\theta^{\frac{1}{2}},$$

therefore $$a\frac{d^2u}{d\theta^2}=-\frac{1}{4}\theta^{-\frac{3}{2}}.$$

Hence, putting $$u+\frac{d^2u}{d\theta^2}=0$$

to find for what value of θ a change of sign can occur, we have

$$\theta^{\frac{1}{2}}-\tfrac{1}{4}\theta^{-\frac{3}{2}}=0,$$
$$\theta^2=\tfrac{1}{4},$$
$$\theta=\pm\tfrac{1}{2}.$$

And the positive value only is admissible, giving

$$\left.\begin{array}{l} r=a\sqrt{2} \\ \theta=\frac{1}{2} \end{array}\right\}$$

as the polar co-ordinates of the point of inflexion.

Multiple Points.

249. **Nature of a Multiple Point.**

A singularity of different nature from those above described occurs on a curve at a point where two branches intersect, as at the point A in the accompanying figure. It will appear from an inspection of the figure that at such a point as the one drawn there are *two* tangents to the

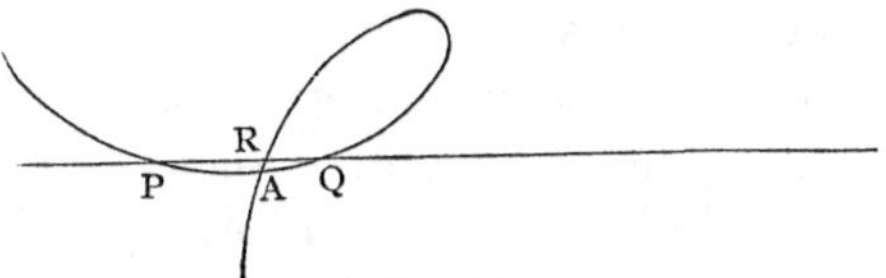

Fig. 38.

curve, one for each branch. Each tangent cuts the curve in two ultimately coincident points, such as P, Q on one branch, and it incidentally intersects the other branch through A in a third point R, ultimately also coinciding with A. Each tangent therefore at such a point intersects the curve in *three* ultimately coincident points at the point of contact; and if the curve be of the nth degree, each tangent will cut the curve again in $n-3$ points real or imaginary. In this respect the tangent at such a point resembles the tangent at a point of inflexion, for (Art. 239) the point of contact of a tangent at a point of inflexion counts for three of the n intersections of the line with the curve.

250. Points through which more than one branch of a curve passes are called "*multiple points*" on the curve. If two branches pass through the point A, as in the above figure, A is called a "*double point.*" If three branches pass through any point, that point is called a "*triple point*" on the curve; and generally, if through any point r branches of the curve pass, that point is referred to as a "*multiple point of the* r^{th} *order*" on the curve. From what has been said with regard to the tangents at a double point it will be obvious that there are r tangents (real or imaginary)

at a multiple point of the rth order, one for each branch. At such a point each of these r tangents cuts its own branch in general in two points, and each of the other branches in one point: *i.e.*, in $r+1$ points altogether, all ultimately coincident with the multiple point. Such a tangent therefore cuts the curve in $n-r-1$ other points real or imaginary. But if at the multiple point there happen to be a point of inflexion on the branch considered, the tangent will cut that branch in three points instead of two at the point of contact, making $r+2$ points of intersection with the curve at the multiple point, and therefore reducing the remaining number of points of intersection to $n-r-2$.

251. **Species of Double Points.**

Consider the case of a double point. The tangents there may be real, coincident, or imaginary.

Case 1. If the tangents be real and not coincident, there are two real branches of the curve passing through the point, and the point is called a *node* or *crunode.*

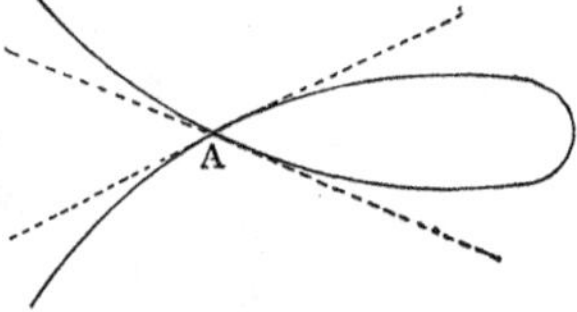

Fig. 39.

Case 2. If the tangents be imaginary, there are no real points on the curve in the immediate neighbourhood of the point considered, and we are unable to travel *along the curve* from such a point in any real direction. Such a point is therefore simply an isolated point, whose co-ordinates satisfy the equation to the curve, and is called a "*conjugate point*" or "*acnode.*"

CASE 3. If the tangents at the double point be coincident, the two branches of the curve will touch at the point considered. The point is then in general of the character called a *cusp* or *spinode.*

252. **Two Species of Cusps.**

There are two kinds of cusps, as shown in the accompanying figures.

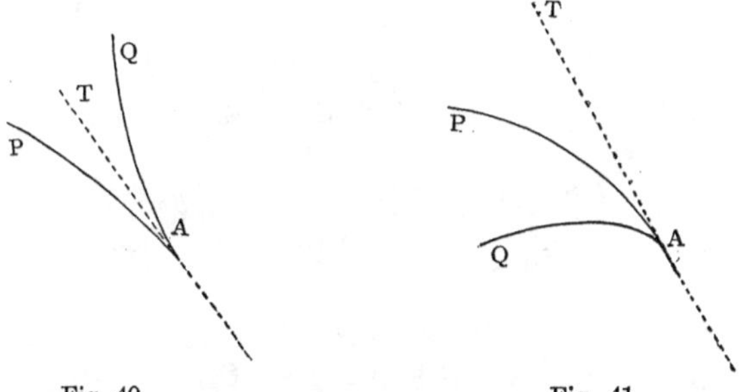

Fig. 40. Fig. 41.

(α) In Fig. 40 the branches PA, QA lie on opposite sides of the tangent at A. This is referred to as a cusp of *the first species* or a *keratoid* cusp (*i.e.*, cusp like *horns*).

(β) In Fig. 41 the branches PA, QA lie on the same side of the tangent at A. This is called a cusp of *the second species* or a *ramphoid* cusp (*i.e.*, cusp like a *beak*).

253. **A Multiple Point can be considered as a Combination of Double Points.**

A *triple* point may obviously be considered as a combination of three double points, for of the three branches intersecting at the point each pair form a double point at their point of intersection. And in general a *multiple* point of the rth order may be considered as the result of the combination of $\frac{r(r-1)}{2}$ double points, since this is the number of ways of combining the r branches two at a time.

254. **To examine the Nature of the Origin.**

If the equation of a curve be rational and algebraic, it may be written in the form

$$\begin{aligned}&a\\&+b_1x+b_2y\\&+c_1x^2+c_2xy+c_3y^2\\&+\dots\\&+k_1x^n+k_2x^{n-1}y+\dots+k_{n+1}y^n=0. \quad\text{(A)}\end{aligned}$$

If this be put into polar co-ordinates it becomes

$$\begin{aligned}&a\\&+r(b_1\cos\theta+b_2\sin\theta)\\&+r^2(c_1\cos^2\theta+c_2\cos\theta\sin\theta+c_3\sin^2\theta)\\&+\dots\\&+r^n(k_1\cos^n\theta+k_2\cos^{n-1}\theta\sin\theta+\dots+k_{n+1}\sin^n\theta)=0. \quad\text{(B)}\end{aligned}$$

Let O be the pole and OA the initial line. Then equation (B) gives the points P_1, P_2, P_3, ..., in which a radius

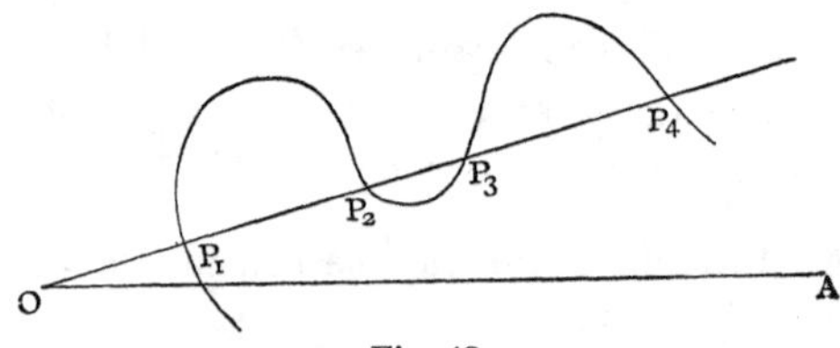

Fig. 42.

vector $OP_1P_2\dots$, making a given angle θ with OA, cuts the curve. The roots of this equation are OP_1, OP_2, OP_3, ...

It is clearly of the nth degree, and therefore has in general n roots. These may, however, become imaginary in pairs.

I. If $a=0$, it will be obvious from either the Cartesian equation (A) or the Polar equation (B) that the curve *passes through the origin O.* In this case one root of the equation (B) is zero, and in the figure $OP_1=0$.

II. In this case, if θ be so chosen as to make $b_1\cos\theta+b_2\sin\theta=0$, a second root of the equation (B) vanishes, and therefore we infer that a straight line making an angle $\tan^{-1}\left(-\frac{b_1}{b_2}\right)$ with the initial line cuts the curve in two contiguous points at the origin, and *therefore is the tangent there.* The Cartesian equation of this line is obvious upon multiplying by r, viz.,

$$b_1x+b_2y=0.$$

Hence if a curve pass through the origin, the *terms of first degree* (if any such exist) *on being equated to zero form the equation of the tangent* at the origin. (See Art. 173.)

III. If $a=0$, $b_1=0$, and $b_2=0$, then in general it is possible to choose θ so that

$$c_1\cos^2\theta+c_2\cos\theta\sin\theta+c_3\sin^2\theta=0,$$

and then three roots of equation (B) will vanish; that is to say, of the pair of lines whose equation is $c_1x^2+c_2xy+c_3y^2=0$ each cuts the curve at the origin in three contiguous points. There are therefore two branches of the curve intersecting at the origin, to each of which a tangent can be drawn, and of the three contiguous points in which it has been seen that each of these tangents cuts the curve two lie on one branch and the other on the remaining

branch. The origin is in this case *a double point on the curve,* and the terms of lowest degree in the equation of the curve, viz., $c_1x^2+c_2xy+c_3y^2$,
when equated to zero form the equation of the tangents at the origin. The *tangent of the angle between these straight lines* is given by

$$\tan\phi=\frac{\sqrt{c_2^2-4c_1c_3}}{c_1+c_3}.$$

If $c_2^2>4c_1c_3$, the tangents are real and not coincident, and there is *a node* at the origin.

If $c_2^2=4c_1c_3$, the tangents are coincident, and the two branches of the curve touch, and there is *in general a cusp* at the origin.

If $c_3^2<4c_1c_3$, there are no real tangents at the origin, although the co-ordinates of the origin satisfy the equation of the curve; there is then a *conjugate point* at the origin.

If $c_1+c_3=0$, the tangents at the origin *intersect at right angles.*

IV. If $a=0$, $b_1=0$, $b_2=0$, $c_1=0$, $c_2=0$, $c_3=0$, the origin is a *triple point* on the curve, and (as shown in III. for the tangents at a double point) the tangents at the origin are $$d_1x^3+d_2x^2y+d_3xy^2+d_4y^3=0.$$

V. And generally, if the lowest terms of an equation are of the rth degree, the origin is a "*multiple point of the rth order*" on the curve, and the terms of the rth degree equated to zero give the r tangents there.

255. To examine the Character of any Specified Point on a Curve.

Results similar to those of the preceding article may be deduced for *any* point on the curve.

Let the straight line $\frac{x-h}{l}=\frac{y-k}{m}=\rho$ be drawn through a given point (h, k) to cut the curve $f(x, y)=0$. Then

$$x=h+l\rho,$$
$$y=k+m\rho.$$

The use of these equations is obviously equivalent to a double transformation of co-ordinates, the first to parallel axes through h, k, the second to polars.

Substituting for x and y in the equation of the curve we obtain $\qquad f(h+l\rho, k+m\rho)=0$
to find the points $P_1, P_2, \ldots$ in which a radius vector through the point h, k cuts the curve.

If this be expanded by the extended form of Taylor's Theorem, the equation becomes

$$f(h, k)+\rho\left(l\frac{\partial}{\partial h}+m\frac{\partial}{\partial k}\right)f+\frac{\rho^2}{2!}\left(l\frac{\partial}{\partial h}+m\frac{\partial}{\partial k}\right)^2 f+\ldots$$
$$+\frac{\rho^n}{n!}\left(l\frac{\partial}{\partial h}+m\frac{\partial}{\partial k}\right)^n f+\ldots=0,$$

which is exactly analogous to equation (B) of Art. 254, and corresponding results follow.

I. If $f(h, k)=0$, one root of the equation for ρ vanishes and the point h, k *lies on the curve* (which is otherwise obvious).

II. In this case, if the ratio $l : m$ be now so chosen that

$$l\frac{\partial f}{\partial h}+m\frac{\partial f}{\partial k}=0,$$

then another root vanishes, and this relation gives the *direction of the tangent*, whose equation is therefore

$$(x-h)\frac{\partial f}{\partial h}+(y-k)\frac{\partial f}{\partial k}=0,$$

as found in Art. 169.

III. But if $\frac{\partial f}{\partial h}=0$ and $\frac{\partial f}{\partial k}=0$, as well as $f(h, k)=0$, then all lines through h, k cut the curve in two contiguous points. But if the ratio $l:m$ be so chosen that

$$l^2\frac{\partial^2 f}{\partial h^2}+2lm\frac{\partial^2 f}{\partial h \partial k}+m^2\frac{\partial^2 f}{\partial k^2}=0,$$

we have in general, as in Art. 254, III., two directions in which a radius vector drawn through (h, k) cuts the curve in three contiguous points. The point (h, k) is a *double point on the curve*, since two branches of the curve pass through this point; and of the three contiguous points in which each of the above-mentioned radii vectores meets the curve, two lie on one branch and one on the other. The equation of the two tangents is

$$(x-h)^2\frac{\partial^2 f}{\partial h^2}+2(x-h)(y-k)\frac{\partial^2 f}{\partial h \partial k}+(y-k)^2\frac{\partial^2 f}{\partial k^2}=0.$$

IV. Further, if $\frac{\partial^2 f}{\partial h^2}=0$, $\frac{\partial^2 f}{\partial h \partial k}=0$, and $\frac{\partial^2 f}{\partial k^2}=0$, in addition to $\frac{\partial f}{\partial h}=0$, $\frac{\partial f}{\partial k}=0$, and $f(h, k)=0$, identically for the same values of h, k, and if on going to terms of the third order we find that all these do *not* identically vanish, the point (h, k) is a *triple point* on the curve.

V. And generally the conditions for the existence of a *multiple point of the* r^{th} *order* at a given point h, k of the curve are that $f(x, y)$ and all its differential coefficients up to those of the $(r-1)^{\text{th}}$ order inclusive should vanish when $x=h$ and $y=k$; and then the equation of the r tangents at that point will be

$$(x-h)^r\frac{\partial^r f}{\partial h^r}+r(x-h)^{r-1}(y-k)\frac{\partial^r f}{\partial h^{r-1}\partial k}+\ldots+(y-k)^r\frac{\partial^r f}{\partial k^r}=0.$$

256. Special Case of Double Point.

Recurring to the case of a double point at a point (h, k), since the equation of the tangents is

$$(x-h)^2\frac{\partial^2 f}{\partial h^2}+2(x-h)(y-k)\frac{\partial^2 f}{\partial h\partial k}+(y-k)^2\frac{\partial^2 f}{\partial k^2}=0,$$

the angle between these tangents is given by

$$\tan\phi=\frac{2\sqrt{\left(\dfrac{\partial^2 f}{\partial h\partial k}\right)^2-\dfrac{\partial^2 f}{\partial h^2}\cdot\dfrac{\partial^2 f}{\partial k^2}}}{\dfrac{\partial^2 f}{\partial h^2}+\dfrac{\partial^2 f}{\partial k^2}},$$

and the point h, k is a *node* or *conjugate* point according as

$$\left(\frac{\partial^2 f}{\partial h\partial k}\right)^2 \text{ is } \gtrless \frac{\partial^2 f}{\partial h^2}\cdot\frac{\partial^2 f}{\partial k^2},$$

and is *in general a cusp* if

$$\left(\frac{\partial^2 f}{\partial h\partial k}\right)^2=\frac{\partial^2 f}{\partial h^2}\cdot\frac{\partial^2 f}{\partial k^2},$$

with the preliminary conditions in each case that

$$f(h, k)=0, \quad \frac{\partial f}{\partial h}=0, \text{ and } \frac{\partial f}{\partial k}=0.$$

We say *in general a cusp;* for it will be seen that in some cases when the above conditions hold the curve *becomes imaginary in the neighbourhood of the point* considered, which must therefore be classed as a conjugate point. In the case of the coincidence of tangents, further investigation is therefore necessary. The mode of procedure is indicated below in the method for the investigation of the character of a cusp.

It appears that $$\left(\frac{\partial^2 f}{\partial x\partial y}\right)^2=\frac{\partial^2 f}{\partial x^2}\cdot\frac{\partial^2 f}{\partial y^2}$$

represents a curve which cuts $f(x, y)=0$ in all its cusps;

and that $$\frac{\partial^2 f}{\partial x^2}+\frac{\partial^2 f}{\partial y^2}=0$$

is a curve which cuts $f(x, y)=0$ in all the double points at which the tangents are at right angles.

257. To search for Double Points.

The rule therefore to search for double points on a curve $f(x, y)=0$ is as follows. Find $\frac{\partial f}{\partial x}$ and $\frac{\partial f}{\partial y}$; equate each to zero and solve. Test whether any of the solutions satisfy the equation of the curve. If so, apply the tests for the character of each of the points denoted, *i.e.*, try whether

$$\left(\frac{\partial^2 f}{\partial x \partial y}\right)^2 \text{ be } \gtreqless \frac{\partial^2 f}{\partial x^2}\cdot\frac{\partial^2 f}{\partial y^2}.$$

258. To discriminate the Species of a Cusp.

METHOD I. Suppose the position of a cusp to have been found by the foregoing rules. Transfer the origin to the cusp. The transformed equation will be of the form

$$(ax+by)^2+u_3+u_4+\ldots=0, \quad \ldots\ldots\ldots\ldots(1)$$

where $ax+by=0$ is the tangent at the origin, and u_3, u_4, ... are homogeneous rational algebraical functions of x and y of the degrees indicated by their respective suffixes.

Let P be the length of the perpendicular drawn from a point x, y of the curve, very near the cusp, upon the tangent $ax+by=0$.

Then $$P=\frac{ax+by}{\sqrt{a^2+b^2}}. \quad \ldots\ldots\ldots\ldots\ldots\ldots(2)$$

If y be eliminated between equations (1) and (2), an equation is obtained giving P in terms of x. It is our object to consider only the two small perpendiculars from points on the curve near the origin, and having a given small abscissa x; hence in comparison with P^2 we reject cubes and all higher powers of P and also all such terms as P^2x, P^2x^2, ... which may arise on substitution.

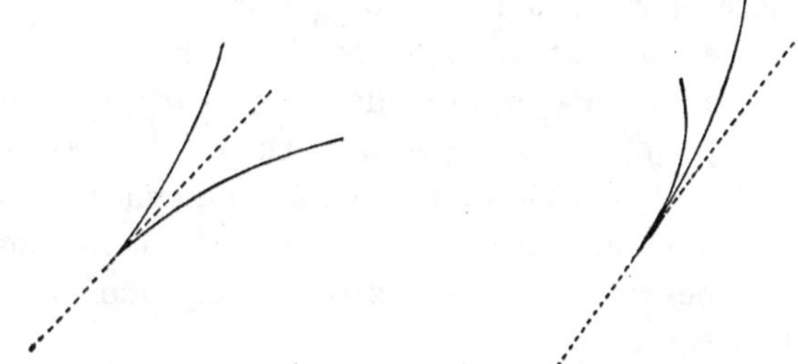

Fig. 43.—Single cusp, first species. Fig. 44.—Single cusp, second species.

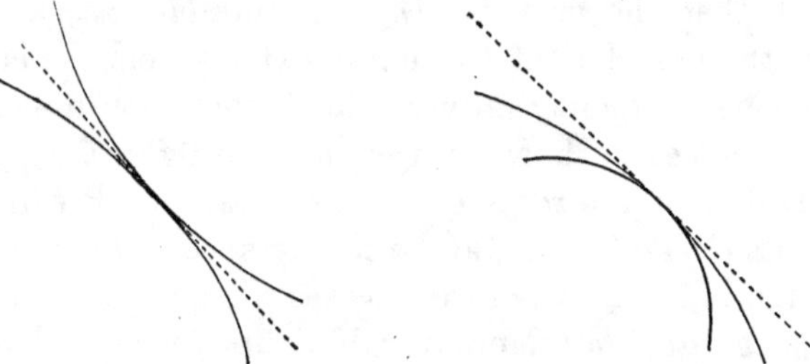

Fig. 45.—Double cusp, first species. Fig. 46.—Double cusp, second species.

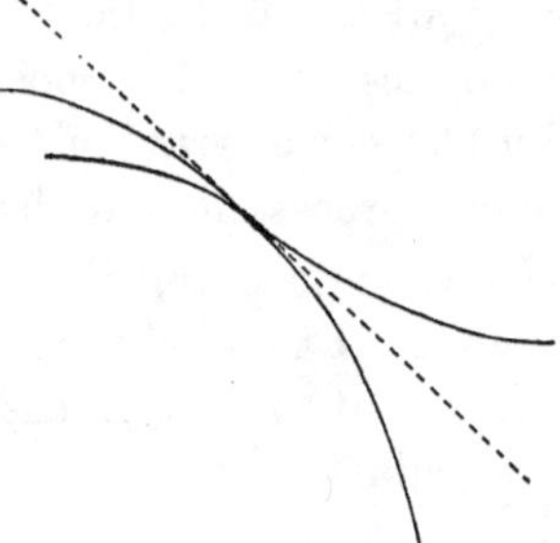

Fig. 47.—Double cusp, change of species. Osculinflexion.

We shall then have a quadratic to determine P. If, when x is made very small, the roots be imaginary, the branches of the curve through the origin are unreal, and therefore there is a conjugate point at the origin. If the roots be real, but *of opposite signs*, the two small perpendiculars lie on opposite sides of the tangent, and there is a *cusp of the first species* at the origin. If the roots be real and *of like sign* the perpendiculars lie on the same side and the cusp *is of the second species*, and the sign of the roots determines on which side of the tangent the cusp lies.

Complete information is also afforded by this method as to whether the cusp is *single or double, i.e.*, as to whether the branches of the curve extend from the cusp towards one extremity only of the tangent, or towards both extremities as shown in the annexed figures.

The reality of the roots of the quadratic for P will in some cases *depend upon*, and in others be *independent of* the sign of x. In the former cases the cusp is *single*; in the latter, *double*. Moreover, if double, we can detect whether the cusp is of the same or of different species towards opposite extremities of the tangent. When the cusp is of different species towards opposite extremities the point is called by Cramer a *point of Osculinflexion*.

In adopting the above process it will clearly be sufficient to put $P=ax+by$, thus dropping the $\sqrt{a^2+b^2}$ for the sake of brevity; the effect of this being to consider a line whose length is proportional to that of the perpendicular instead of the perpendicular itself.

Ex. 1. *Examine the character of the origin on the curve*

$$x^4-4x^2y-2xy^2+4y^2=0.$$

Here the tangent at the origin is $y=0$. According to the rule put $y=P$. The quadratic for P is

$$P^2(4-2x)-4Px^2+x^4=0.$$

The roots of this equation are real or imaginary according as

$$4x^4 \text{ is } > \text{ or } < x^4(4-2x),$$

i.e., according as x is positive or negative. Hence the cusp is "*single*" and lies to the right of the axis of y. Moreover the product of the roots is $\dfrac{x^4}{4-2x}$ and is positive when x is very small, and the roots are therefore of the same sign. The origin is therefore a *single cusp of the second species.* Moreover the sum of the roots is positive, so that the two branches near the origin *lie in the first quadrant.*

Ex. 2. *Examine the character of the curve*

$$x^4 - 3x^2y - 3xy^2 + 9y^2 = 0$$

at the origin. Here $y=0$ is a tangent at the origin. Put $y=P$. The quadratic for P is

$$P^2(9-3x) - 3x^2P + x^4 = 0.$$

The roots are real or imaginary according as $9x^4 - 4(9-3x)x^4$ is positive or negative, *i.e.*, as $-27x^4 + 12x^5$ is positive or negative.

Now, when x is very small, x^5 is *negligible in comparison with* x^4, and therefore the above expression is negative for very small positive or negative values of x. The roots of the equation for P are therefore imaginary, and the origin is a *conjugate point* on the curve.

Ex. 3. *Examine the character of the curve*

$$y = F(x) \pm (x-h)^{\frac{2m+1}{2n}} f(x) \quad \text{.....................(1)}$$

in the neighbourhood of the point $x=h$, $y=F(h)$, *m and n being positive integers.*

By Taylor's Theorem we may write

$$F(x+h) = F(h) + ax + bx^2 + \ldots$$

and $$[f(x+h)]^2 = a_1 + b_1x + \ldots,$$

where a_1 being $[f(h)]^2$ is necessarily positive.

Hence on transforming our origin to the point $\{h, F(h)\}$ we obtain for the transformed equation

$$(y - ax - bx^2 - \ldots)^2 = x^{\frac{2m+1}{n}}(a_1 + b_1x + \ldots). \quad \text{...........(2)}$$

Examining the form of the curve at the origin, there are obviously coincident tangents if $\dfrac{2m+1}{n}$ be > 2.

Put $y - ax = P$, then

$$P^2 - 2P(bx^2 + \ldots) + b^2x^4 - a_1x^{\frac{2m+1}{n}} - \ldots = 0.$$

That the roots of this quadratic are real, if x be positive and small, is obvious from equation (2); also, that the roots are imaginary for small negative values of x. There is therefore a *single cusp extending to the right of the new axis of y.*

Again, the product of the roots $=b^2x^4-a_1x^{\frac{2m+1}{n}}-\ldots$.

If $\frac{2m+1}{n}>4$, this product has the same sign as x^4 when x is taken sufficiently small, and therefore is positive, giving a cusp of the *second species.*

If $\frac{2m+1}{n}<4$, the term $-a_1x^{\frac{2m+1}{n}}$ is the important term in the product and is negative, x being positive. There is therefore in this case a cusp of the *first species.*

We have assumed that the *coefficient* b or $\frac{1}{2!}F'''(h)$ *is not zero.* If however this coefficient vanish, it is easy to make the corresponding change in the subsequent investigation.

Ex. 4. *Examine the nature of the double point on the curve*

$$(x+y)^3-\sqrt{2}(y-x+2)^2=0.$$

Here
$$\left.\begin{aligned}\frac{\partial\phi}{\partial x}&=3(x+y)^2+2\sqrt{2}(y-x+2)=0,\\ \frac{\partial\phi}{\partial y}&=3(x+y)^2-2\sqrt{2}(y-x+2)=0.\end{aligned}\right\}$$

These give
$$\left.\begin{aligned}x+y&=0,\\ y-x+2&=0,\end{aligned}\right\}$$
and

or
$$\left.\begin{aligned}x&=1,\\ y&=-1.\end{aligned}\right\}$$

Now this point obviously lies upon the curve, and there is therefore a multiple point of some description there.

Again,
$$\frac{\partial^2\phi}{\partial x^2}=6(x+y)-2\sqrt{2}=-2\sqrt{2} \text{ at the point } (1,-1),$$
$$\frac{\partial^2\phi}{\partial y^2}=6(x+y)-2\sqrt{2}=-2\sqrt{2},$$
$$\frac{\partial^2\phi}{\partial x\partial y}=6(x+y)+2\sqrt{2}=2\sqrt{2}.$$

Hence at this point
$$\frac{\partial^2\phi}{\partial x^2}\cdot\frac{\partial^2\phi}{\partial y^2}=\left(\frac{\partial^2\phi}{\partial x\partial y}\right)^2,$$

and we have a double point at which the tangents are coincident.

Next, transforming to the point $(1, -1)$ for origin, the equation becomes
$$(x+y)^3 - \sqrt{2}(y-x)^2 = 0.$$
According to the rule we put $y-x=P$. Then rejecting terms in P^3 and P^2x we have $$P^2 - 6x^2\sqrt{2}P - 4x^3\sqrt{2} = 0.$$
The roots are real if $$18x^4 + 4\sqrt{2}x^3 > 0,$$
which is the case if x be very small and positive. There is therefore a *single cusp* at the point $(1, -1)$.

Again, the product of the roots $= -4x^3\sqrt{2}$, and is negative when x is small. This indicates that the cusp is one of the *first species.*

[This curve is obviously only a transformation of the semi-cubical parabola $y^2 = x^3$.]

259. Method II. Another method of discrimination of the species of a cusp depends upon the test for concavity or convexity. Find the two values of $\frac{d^2y}{dx^2}$ (or $\frac{d^2x}{dy^2}$, see Art. 245). If these have *opposite signs* very near to the cusp, the two branches starting from the cusp are *one concave and the other convex* to the foot of the ordinate, and the cusp is of the *first species.* But if the *signs be the same,* the two branches are either *both concave or both convex* to the foot of the ordinate, and the cusp is of the *second species.*

Ex. *Discuss the form of the curve $y = x \pm x^{\frac{3}{2}}$ at the origin.*

Here $$\frac{d^2y}{dx^2} = \pm\frac{3}{4}\frac{1}{\sqrt{x}}.$$

Hence only positive values of x are admissible and the two values of $\frac{d^2y}{dx^2}$ have opposite signs. The origin is therefore a *single cusp of the first species.*

This result is obvious also from the form of the equation to the curve.

260. Singularities of Transcendental Curves.

In addition to the singularities above discussed others occur occasionally in transcendental curves, due to discontinuities in the values of y, $\frac{dy}{dx}$, etc. For instance, if the value of y be discontinuous at a certain point the curve suddenly stops there and the point is called a "*point d'arrêt.*"

Consider the curve $y = a^{\frac{1}{x}}$; $(a > 1)$.
When $x = -\infty$, $y = 1$, and as x increases from $-\infty$ to zero y is always positive and decreases down to zero. As soon, however, as x becomes positive, being still indefinitely small, y suddenly becomes infinitely great, and as x increases to $+\infty$ y gradually diminishes down to unity. The origin is a *point d'arrêt* on this curve, and the shape is that shown in the annexed figure.

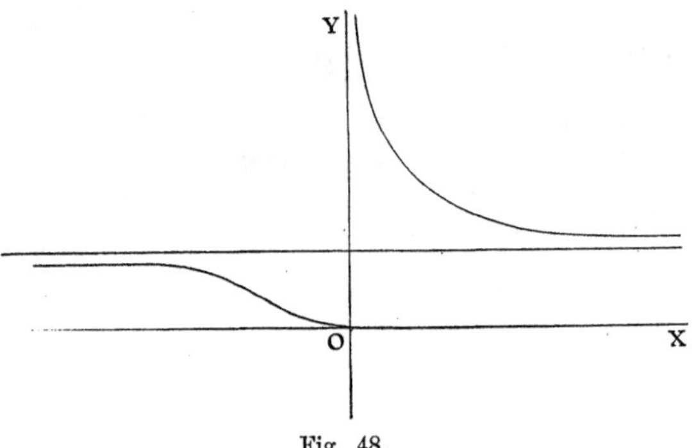

Fig. 48.

Next suppose that the value of y is continuous, but that at a certain point $\frac{dy}{dx}$ becomes discontinuous, so that two branches of the curve meet at certain angle at the same point and stop there. Such a point is called a "*point saillant.*"

261. Branch of Conjugate Points.

It sometimes happens that a curve possesses an infinite series of conjugate points, satisfying the equation to the curve and forming a branch of isolated points. M. Vincent, in a memoir published in vol. xv. of Gergonne's "Annales des Math.," has discussed several such cases, and calls such discontinuous branches by the name *branches pointillées*.

Ex. In tracing the curve $y=x^x$, it is clear that, when $x=\infty$, $y=\infty$; and when $x=1$, $y=1$. Also that as x decreases from ∞ to 1, y also decreases from ∞ to 1. Between $x=1$ and $x=0$ y is less than 1; and when $x=0$, $y=1$ (see Chap. XIII.). There is therefore a continuous branch of the curve, viz., ∞PB, above the axis of x.

Again, whenever x is a fraction with an even denominator there

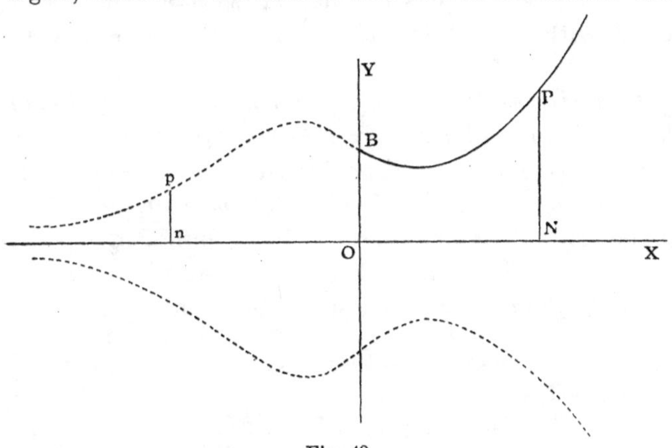

Fig. 49.

are two real values of y, differing only in sign; *e.g.*,

$$(\tfrac{1}{4})^{\frac{1}{4}}=\pm\sqrt{\tfrac{1}{2}},$$

whilst, whenever the denominator of x is odd, there is but one real value for y. There is therefore a branch of conjugate points below the axis forming a discontinuous branch, of the same shape as the continuous branch above the axis.

Next consider what happens when x is negative. Let the coordinates of any point P on the branch in the first quadrant be (x, y), then $ON=x$. Take $On=-x$ along the negative portion of the axis of x, then, if p be the corresponding point on the curve, we have

$$pn=(-x)^{-x}, \quad PN=x^{x},$$

and therefore

$$pn \,.\, PN=(-1)^{x},$$

which may be $=1$, -1, or imaginary, according to the particular value of x. Hence, when the ordinate pn is real, its magnitude is inverse to that of the corresponding ordinate PN. Hence on this curve we have two infinite series of conjugate points, as shown in the figure.

For an account of M. Vincent's memoir and criticisms upon it see Dr. Salmon's "Higher Plane Curves," p. 275, or a paper by Mr. D. F. Gregory, "Camb. Math. Journal," vol. i., pp. 231, 264.

262. **Maclaurin's Theorem with regard to Cubics.**

We conclude the present chapter with an important theorem with regard to cubic curves, which is due to Maclaurin.

If a radius vector OPQ be drawn through a point of inflexion (O) of a cubic, cutting the curve again in P and Q, to show that the locus of the extremities of the harmonic means, between OP and OQ, is a straight line.

If the origin be taken at the point of inflexion and the tangent at the point of inflexion as the axis of y, the equation of the cubic must assume the form

$$y^3+xu=0 \text{.........................} \quad (1)$$

where u is the most general expression of the second and lower degrees, viz.,

$$ax^2+2hxy+by^2+2gx+2fy+c,$$

for it is clear that the axis of y cuts this curve in three points ultimately coincident with the origin.

The equation (1) when put into polars takes the form

$$Lr^2+Mr+N=0,$$

where

$$L=\sin^3\theta+(a\cos^2\theta+2h\sin\theta\cos\theta+b\sin^2\theta)\cos\theta,$$
$$M=(2g\cos\theta+2f\sin\theta)\cos\theta,$$
$$N=c\cos\theta.$$

If r_1, r_2 be the roots of this quadratic, and ρ the harmonic mean between them, we have

$$\frac{2}{\rho}=\frac{1}{r_1}+\frac{1}{r_2}=-\frac{M}{N}=-\frac{2g\cos\theta+2f\sin\theta}{c},$$

which shows that the Cartesian Equation of the locus of the extremity of the harmonic mean is the straight line

$$gx+fy+c=0.$$

263. It is obvious from Art. 187 that the equation of the polar conic of the cubic (1) with regard to the origin is

$$x(2gx+2fy)+2cx=0,$$

or

$$x(gx+fy+c)=0.$$

Hence the polar conic of a point of inflexion on a cubic breaks up into two straight lines, one of which is the tangent at the point of inflexion, and the other the locus of the extremities of the harmonic means of the radii vectores through the point of inflexion. It appears from this that only three tangents can be drawn from a point of inflexion on a cubic to the curve, viz., one to each of the points in which the line $gx+fy+c=0$ meets the curve, and consequently also that their three points of contact lie in a straight line.

264. **If a Cubic have three real points of Inflexion they are Collinear.**

It follows immediately from Maclaurin's Theorem above proved that if A and B be two points of inflexion on a cubic, the line AB produced will cut the curve in a third point C, which is also a point of inflexion on the cubic. For if B, B_1, B_2 be the three ultimately coincident points on the cubic, which lie in a straight line (B being a point of inflexion), let AB, AB_1, AB_2 cut the curve in C, C_1, C_2, and let AH, AH_1, AH_2 be the harmonic means between AB, AC; AB_1, AC_1; AB_2, AC_2 respectively, then H, H_1, H_2 lie in a straight line by Maclaurin's Theorem, and B, B_1, B_2 lie in a straight line; therefore by a theorem in conic sections C, C_1, C_2 also lie in a straight line, and they are ultimately coincident points. C is therefore a point of inflexion.

EXAMPLES.

1. Write down the equations of the tangents at the origin for each of the following curves :—

$$(\alpha) \qquad y + c = c \cosh \frac{x}{c}.$$

$$(\beta) \qquad y = a \tan \frac{x}{b}.$$

$$(\gamma) \qquad y^2 = x \log(1 + x).$$

$$(\delta) \qquad x^3 + y^3 = 3axy.$$

2. Show that the curve $y = e^x$ is at every point convex to the foot of the ordinate of that point.

3. Show that for the cubical parabola

$$a^2y = (x - b)^3$$

there is a point of inflexion whose abscissa is b.

4. Show that for the semicubical parabola

$$ay^2 = x^3$$

the origin is a cusp of the first species.

5. Show that the origin is a cusp of the first species on the curve

$$a(y - x)^2 = x^3.$$

6. Show that on the curve

$$(ay - x^2)^2 = bx^3$$

there is a cusp of the first species at the origin, and a point of inflexion whose abscissa is $\frac{9}{64}b$.

7. Show that there are points of inflexion at the origin on each of the curves

$$(\alpha)\ \ y = x \cos \frac{x}{a}.$$

$$(\beta)\ \ y = a \tan \frac{x}{b}.$$

$$(\gamma)\ \ y = x^2 \log (1 - x).$$

8. Show that the curves

$$y^2 = x^2 \sin \frac{x}{a}, \qquad y^2 = x^2 \tan \frac{x}{a}$$

have cusps of the first kind at the origin.

9. Show that at the origin on the curve

$$y^2 = bx \sin \frac{x}{a}$$

there is a node or a conjugate point according as a and b have like or unlike signs.

10. Show that there is a point of inflexion on the curve

$$y = e^{x^{\frac{1}{3}}}$$

at the point $(8, e^2)$.

11. Show that the Trident curve

$$axy + a^3 = x^3$$

has a point of inflexion at the point in which it cuts the axis of x, and show that the tangent at the point of inflexion makes with the axis of x an angle $\tan^{-1}3$.

12. Show that the curve

$$b(ay - x^2)^2 = x^5$$

has a cusp of the second species at the origin.

13. Show that, if n be greater than 2, the curve

$$b^{n-4}(ay - x^2)^2 = x^n$$

has a cusp at the origin of the first or second species according as n is less or greater than 4.

14. Find the two points of inflexion of the curve

$$\frac{y}{c} = \frac{x^2}{9a^2} + \left(\frac{x-a}{a}\right)^{\frac{1}{3}}$$

and draw figures showing the characters of the inflexions.

15. Show that every point in which the curve of sines

$$\frac{y}{a} = \sin\frac{x}{b}$$

cuts the axis of x is a point of inflexion on the curve.

16. Show that the points of inflexion on the cubic

$$y = \frac{a^2x}{x^2 + a^2}$$

are given by $x = 0$ and $x = \pm a\sqrt{3}$.

Show that these three points of inflexion lie on the straight line

$$x = 4y.$$

17. Find by polars the points of inflexion on the curve

$$2x(x^2 + y^2) = a(2x^2 + y^2).$$

18. Show that the curve $au = \theta^n$ has a point of inflexion where

$$au = \{n(1-n)\}^{\frac{n}{2}}.$$

19. Show that if the origin be a point of inflexion on the curve

$$u_1 + u_2 + u_3 + \ldots = 0$$

u_2 will contain u_1 for a factor.

20. Show that there is a point of inflexion at the origin on the cubic

$$y = axy + by^2 + cx^3.$$

21. Show that there is a point of undulation at the origin on the curve

$$y = ax^4 + bx^2y^2 + cy^4.$$

22. Show that the origin is a triple point on the curve

$$x^4 + y^4 = axy^2,$$

and that there is a cusp of the first species there.

23. Show that there are two points of inflexion at the origin on each of the curves

$$x^2y^2 = a^2(x^2 - y^2),$$
$$r^2 = a^2 \cos 2\theta.$$

24. Show that the abscissae of the points of inflexion on the curve

$$y^n = f(x)$$

are roots of the equation

$$\frac{n-1}{n}\{f'(x)\}^2 = f(x)f''(x).$$

25. Show that the abscissae of the points of inflexion on the curve

$$y = e^{-\lambda x} \tan \mu x$$

are given by

$$2\mu \sec^2 \mu x(\mu \tan \mu x - \lambda) + \lambda^2 \tan \mu x = 0.$$

26. Show that the curve

$$y = \frac{x^3 + ax^2 + a^3}{x^2 - a^2}$$

has but one point of inflexion, and that its abscissa is

$$-a\frac{\sqrt[3]{3}+1}{\sqrt[3]{3}-1}.$$

27. Find the positions of the points of inflexion on the curve

$$12y = x^4 - 16x^3 + 42x^2 + 12x + 1.$$

28. Prove that the curve

$$y = be^{-\left(\frac{x}{a}\right)^n}$$

has a point of inflexion given by

$$x = a\sqrt[n]{\frac{n-1}{n}}.$$

29. Prove that the point $\left(-2, \frac{-2}{e^2}\right)$ is a point of inflexion on the curve

$$y = xe^x.$$

30. Show that there are two points of inflexion on the cubic

$$x^3 + y^3 = a^3$$

at the points $(a, 0)$, $(0, a)$ respectively.

31. In the curve $$x^3 + y^3 = ax^2$$
show that there is a cusp of the first kind at the origin, and a point of inflexion where $x = a$.

32. In the curve $y^2 = (x - a)(x - b)(x - c)$
show that if $a = b$ there is a node, cusp, or conjugate point at $x = a$ according as a is $>$, $=$, or $<$ c. Also show that the points of inflexion have for their abscissae $x = \dfrac{4c - a}{3}$. Hence show that the points of inflexion on this curve are real or imaginary according as the curve has a conjugate point or a node.

33. Show that for the curve
$$r = a(1 - \cos\theta)$$
there is a cusp of the first kind at the origin.

34. Show that the curve
$$r^2\cos^2\theta = a^2\cos 2\theta$$
has a double point at the origin.

35. Show that the curve
$$r = a\sin n\theta$$
has a multiple point at the origin of order n or $2n$ according as n is odd or even.

36. Show that the curve
$$r = \frac{a\theta^2}{1 + \theta^2}$$
has a cusp of the first kind at the pole.

37. Show that if the cubic
$$xy^2 + ey = ax^3 + bx^2 + cx + d$$
have a centre, then will $b = 0$ and $d = 0$ and the centre is at the origin.

In this case show also that the origin is a point of inflexion on the curve.

38. Show that there is a conjugate point on the locus
$$x^3 + y^3 + 3cxy = c^3$$
at the point $(-c, -c)$. Trace the curve.

39. Show that the curve

$$x^5 + y^5 = 5ax^2y^2$$

has two cusps of the first species at the origin, and that $x + y = a$ is an asymptote.

40. Show that a cubic curve cannot have more than one double point, and cannot have a triple point.

Examine the case of the curve

$$2(x^3 + y^3) - 3(3x^2 + y^2) + 12x = 4$$

and show that there are apparently two nodes at (1, 1) and at (2, 0) respectively. Explain this result.

41. Show that the curve

$$by^2 = x^3 \sin^2 \frac{x}{a}$$

has a cusp of the first species at the origin and is symmetrical with regard to the axis of x. Show also that it has an infinite series of conjugate points lying at equal distances from each other along the negative portion of the axis of x.

42. Show that the curve

$$y - x = \log_e \frac{ey^2}{4x}$$

has a node at the point (1, 2).

43. Show that the curve

$$(x^2 + y^2)^2 = a(3x^2y - y^3)$$

has a triple point at the origin, and that the angles between the branches through the origin are equal.

44. Show that the curve

$$(x^2 + y^2)^{\frac{5}{2}} = 4axy(x^2 - y^2)$$

has a multiple point of the eighth order at the origin, and that the curve consists of eight equal loops.

45. Show that for the Cissoid

$$y^2 = \frac{x^3}{2a - x}$$

the origin is a cusp of the first species.

46. Show that for the Conchoid

$$x^2y^2 = (a+y)^2(b^2 - y^2),$$

if b be $> a$ there is a node at $x=0, y= -a$, and if $b = a$ there is a cusp at the same point.

47. Show that at the point $(-1, -2)$ there is a cusp of the first species on the curve

$$x^3 + 2x^2 + 2xy - y^2 + 5x - 2y = 0.$$

48. Examine the singularities of the curve

$$x^4 - 4ax^3 - 2ay^3 + 4a^2x^2 + 3a^2y^2 - a^4 = 0.$$

There are nodes at the points $(0, a)$, $(a, 0)$, $(2a, a)$. Find the directions of the tangents at these points.

49. Show that the curve

$$x^4 - 2x^2y - xy^2 - 2x^2 - 2xy + y^2 - x + 2y + 1 = 0$$

has a single cusp of the second kind at the point $(0, -1)$.

50. Examine the character of the curve

$$ay^4 - ax^2y^2 + x^5 + x^4y = 0$$

in the immediate neighbourhood of the origin.

51. Show that at each of the four points of intersection of the curve

$$(ax)^{\frac{2}{3}} + (by)^{\frac{2}{3}} = (a^2 - b^2)^{\frac{2}{3}}$$

with the axes there is a cusp of the first species.

52. Show that the origin is a conjugate point on the curve

$$x^4 - ax^2y + axy^2 + a^2y^2 = 0.$$

53. Show that at the origin there is a single cusp of the second species on the curve $x^4 - 2ax^2y - axy^2 + a^2y^2 = 0$.

54. Show that the curve

$$y^2 = 2x^2y + x^3y + x^3$$

has a single cusp of the first species at the origin.

55. Show that the curve

$$y^2 = 2x^2y + x^4y + x^4$$

has a double keratoid cusp at the origin.

56. Show that the curve

$$y^2 = 2x^2y + x^4y - 2x^4$$

has a conjugate point at the origin.

CHAPTER X.

CURVATURE.

CURVATURE.

265. Angle of Contingence.

Let PQ be an arc of a curve. Suppose that between P and Q there is no point of inflexion or other singularity, but that the bending is continuously in one direction. Let LPR and MQ be the tangents at P and Q, intersect-

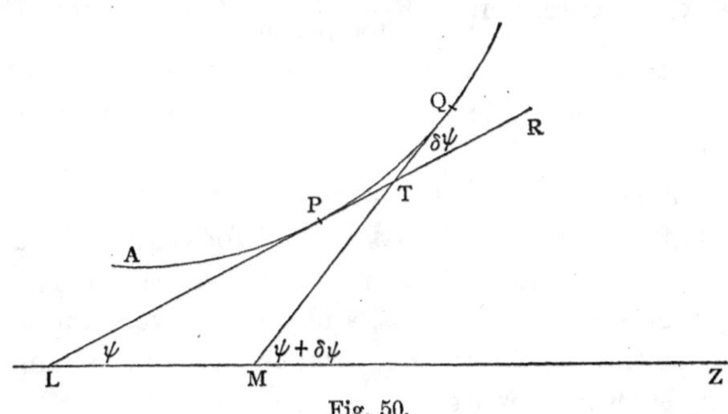

Fig. 50.

ing at T and cutting a given fixed straight line LZ in L and M. Then the angle RTQ is called the *angle of contingence* of the arc PQ.

The angle of contingence of any arc is therefore the

difference of the angles which the tangents at its extremities make with any given fixed straight line. It is also obviously the angle *turned through* by a line which rolls along the curve from one extremity of the arc to the other.

266. **Measure of Curvature.**

It is clear that the *whole bending* or *curvature* which the curve undergoes between P and Q is greater or less according as the angle of contingence RTQ is greater or less. The fraction $\dfrac{\text{angle of contingence}}{\text{length of arc}}$ is called the *average bending* or *average curvature* of the arc. We shall define the *curvature* of a curve in the immediate neighbourhood of a given point to be *the rate of deflection* from the tangent at that point. And we shall take as a measure of this rate of deflection at the given point the limit of the expression $\dfrac{\text{angle of contingence}}{\text{length of arc}}$ when the length of the arc measured from the given point and therefore also the angle of contingence are indefinitely diminished.

That this is a proper measure of the rate of deflection is obvious from the consideration that, for a given length of arc, the deflection is greater or less as the angle of contingence is greater or less, and for a given angle of contingence the deflection is greater or less as the length of the arc is less or greater.

267. **Curvature of a Circle.**

In the case of the circle the curvature is the same at

every point and is measured by the RECIPROCAL OF THE RADIUS.

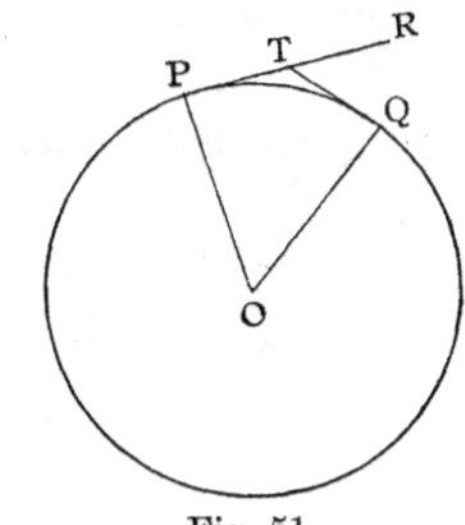

Fig. 51.

For let r be the radius, O the centre. Then

$$R\hat{T}Q = P\hat{O}Q = \frac{\text{arc } PQ}{r},$$

the angle being supposed measured in circular measure.

Hence $\dfrac{\text{angle of contingence}}{\text{length of arc}} = \dfrac{1}{r}$,

and this is true whether the limit be taken or not. Hence the "curvature" of a circle at any point is measured by the reciprocal of the radius.

268. **Circle of Curvature.**

If three contiguous points be taken on a curve, a circle may be drawn to pass through those three points. Let them be P, Q, R. Then, when the points are indefinitely close together, PQ and QR are ultimately tangents both to the curve and to the circle. Hence at the point of ultimate coincidence the curve and the circle have the *same angle of contingence*, viz., the angle RQZ (see Fig. 52). Moreover, the *arcs* PR of the circle and the curve differ by a small quantity of order higher than their own, and therefore *may be considered equal in the limit* (see Art. 36). Hence the curvatures of this circle and of the curve at the point of contact are equal. It is therefore convenient to describe the curvature of a curve at a given point by reference to a circle thus drawn, the reciprocal of the radius being a correct measure of the rate

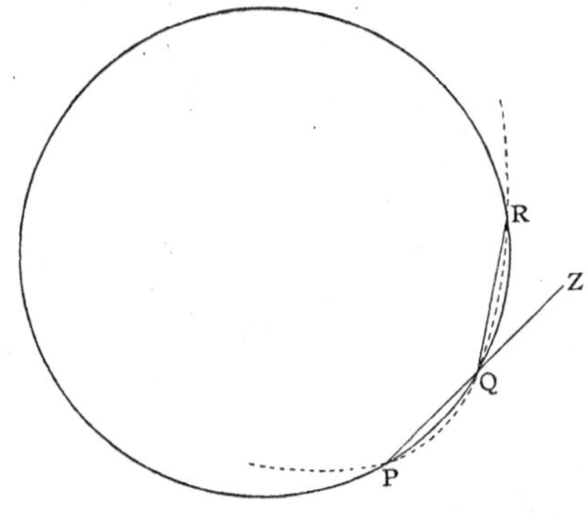

Fig. 52.

of bend. We shall therefore consider such a circle to exist for each point of a curve and shall speak of it as the **circle of curvature** of that point. Its radius and centre will be called the **radius** and **centre of curvature** respectively, and a chord of this circle drawn through the point of contact in any direction will be referred to as the **chord of curvature** in that direction.

269. **Formula for Radius of Curvature.**

Referring to the figure of Art. 265, let the arc AP measured from some fixed point A on the curve up to P be called s, and AQ, $s+\delta s$; let the angle $PLZ=\psi$, and $QMZ=\psi+\delta\psi$. Then the angle of contingence $RTQ=\delta\psi$ and the measure of the curvature $=Lt\frac{\delta\psi}{\delta s}=\frac{d\psi}{ds}$. If therefore the radius of curvature be called ρ we have

$$\frac{1}{\rho}=\frac{d\psi}{ds}, \text{ or } \rho=\frac{ds}{d\psi}. \quad\ldots\ldots\ldots\ldots\ldots\ldots(\text{A})$$

270. This formula may also be arrived at thus. Let PQ and QR (Fig. 52) be considered equal chords, and therefore when we proceed to the limit the elementary arcs PQ and QR may be considered equal. Call each δs, and the angle $R\hat{Q}Z=\delta\psi$.

Now the radius of the circum-circle of the triangle PQR is $\dfrac{PR}{2\sin PQR}$.

Hence $\rho=Lt\dfrac{PR}{2\sin PQR}=Lt\dfrac{2\delta s}{2\sin\delta\psi}=Lt\dfrac{\delta s}{\delta\psi}\cdot\dfrac{\delta\psi}{\sin\delta\psi}=\dfrac{ds}{d\psi}$.

Also, it is clear that the lines which bisect at right angles the chords PQ, QR intersect at the circum-centre of PQR, *i.e.*, in the limit the centre of curvature of any point on a curve may be considered as the *point of intersection of the normal at that point with the normal at a contiguous and ultimately coincident point.*

271. The formula (A) is useful in the case in which the equation of the curve is given in its intrinsic form, *i.e.*, when the equation is given as a relation between s and ψ (Art. 291). For example, that relation for a catenary is known to be $s=c\tan\psi$; whence we deduce at once

$$\rho=\frac{ds}{d\psi}=c\sec^2\psi,$$

and the rate of its deflection at any point is measured by

$$\frac{1}{\rho}=\frac{\cos^2\psi}{c}=\frac{c}{s^2+c^2}.$$

272. **Transformations.**

This formula however must be transformed so as to suit each of the systems of co-ordinates in which it is usual to express the equation of a curve. These transformations we proceed to perform.

We have the equations

$$\cos\psi=\frac{dx}{ds},\quad \sin\psi=\frac{dy}{ds}.$$

Hence, differentiating each of these with respect to s,

$$-\sin\psi\frac{d\psi}{ds}=\frac{d^2x}{ds^2},\quad \cos\psi\frac{d\psi}{ds}=\frac{d^2y}{ds^2},$$

whence

$$\frac{1}{\rho}=\frac{-\dfrac{d^2x}{ds^2}}{\dfrac{dy}{ds}}=\frac{\dfrac{d^2y}{ds^2}}{\dfrac{dx}{ds}}, \quad \text{......................(B)}$$

and by squaring and adding

$$\frac{1}{\rho^2}=\left(\frac{d^2x}{ds^2}\right)^2+\left(\frac{d^2y}{ds^2}\right)^2. \quad \text{................(C)}$$

These formulae (B) and (C) are only suitable for the case in which both x and y are known functions of s.

273. **Cartesian Formula. Explicit Functions.**

Again, since $\tan\psi=\dfrac{dy}{dx}$,

we have $$\sec^2\psi\frac{d\psi}{dx}=\frac{d^2y}{dx^2}$$

by differentiation with regard to x.

Now $$\frac{d\psi}{dx}=\frac{d\psi}{ds}\cdot\frac{ds}{dx}=\frac{1}{\rho\cos\psi};$$

therefore $$\sec^3\psi\cdot\frac{1}{\rho}=\frac{d^2y}{dx^2},$$

and $$\sec^2\psi=1+\tan^2\psi=1+\left(\frac{dy}{dx}\right)^2;$$

therefore $$\rho=\frac{\left\{1+\left(\dfrac{dy}{dx}\right)^2\right\}^{\frac{3}{2}}}{\dfrac{d^2y}{dx^2}}. \quad \text{.............(D)}$$

This important form of the result is adapted to the evaluation of the radius of curvature when the equation of the curve is given in Cartesian co-ordinates, y being an explicit function of x.

274. **Cartesians. Implicit Functions.**

We may throw this into another shape specially adapted to Cartesian curves, in which neither variable can be expressed explicitly as a function of the other.

Thus, if $\phi(x, y)=0$ be the equation to the curve, we have

$$\frac{\partial\phi}{\partial x}+\frac{\partial\phi}{\partial y}\cdot\frac{dy}{dx}=0,$$

and

$$\frac{\partial^2\phi}{\partial x^2}+2\frac{\partial^2\phi}{\partial x\partial y}\cdot\frac{dy}{dx}+\frac{\partial^2\phi}{\partial y^2}\left(\frac{dy}{dx}\right)^2+\frac{\partial\phi}{\partial y}\cdot\frac{d^2y}{dx^2}=0.$$

Hence, substituting for $\frac{dy}{dx}$ and $\frac{d^2y}{dx^2}$ in formula (D),

$$\rho=\frac{-\left[1+\left\{-\dfrac{\frac{\partial\phi}{\partial x}}{\frac{\partial\phi}{\partial y}}\right\}^2\right]^{\frac{3}{2}}\dfrac{\partial\phi}{\partial y}}{\dfrac{\partial^2\phi}{\partial x^2}+2\dfrac{\partial^2\phi}{\partial x\partial y}\left\{-\dfrac{\frac{\partial\phi}{\partial x}}{\frac{\partial\phi}{\partial y}}\right\}+\dfrac{\partial^2\phi}{\partial y^2}\left\{-\dfrac{\frac{\partial\phi}{\partial x}}{\frac{\partial\phi}{\partial y}}\right\}^2}$$

or

$$\rho=-\frac{\left\{\left(\frac{\partial\phi}{\partial x}\right)^2+\left(\frac{\partial\phi}{\partial y}\right)^2\right\}^{\frac{3}{2}}}{\frac{\partial^2\phi}{\partial x^2}\left(\frac{\partial\phi}{\partial y}\right)^2-2\frac{\partial^2\phi}{\partial x\partial y}\cdot\frac{\partial\phi}{\partial x}\cdot\frac{\partial\phi}{\partial y}+\frac{\partial^2\phi}{\partial y^2}\left(\frac{\partial\phi}{\partial x}\right)^2}\text{. ..(E)}$$

275. A curve is frequently defined by giving the two Cartesian co-ordinates x, y in terms of a third variable,

e.g., the equation of a cycloid is most conveniently expressed as
$$x=a(\theta+\sin\theta),$$
$$y=a(1-\cos\theta).$$

Formula (D) is very easily modified to meet the requirements of this case.

Let
$$\left.\begin{array}{l}x=F(t)\\ y=f(t)\end{array}\right\}\text{ be the equations of the curve.}$$

Then
$$\frac{dy}{dx}=\frac{\dfrac{dy}{dt}}{\dfrac{dx}{dt}}=\frac{f'(t)}{F'(t)},$$

and
$$\frac{d^2y}{dx^2}=\frac{d}{dt}\cdot\left(\frac{dy}{dx}\right)\cdot\frac{dt}{dx}$$
$$=\frac{\dfrac{d^2y}{dt^2}\cdot\dfrac{dx}{dt}-\dfrac{d^2x}{dt^2}\cdot\dfrac{dy}{dt}}{\left(\dfrac{dx}{dt}\right)^3}$$
$$=\frac{f''(t)\,.\,F'(t)-f'(t)\,.\,F''(t)}{\{F'(t)\}^3},$$

and formula (D) becomes
$$\rho=\frac{\left\{\left(\dfrac{dx}{dt}\right)^2+\left(\dfrac{dy}{dt}\right)^2\right\}^{\frac{3}{2}}}{\dfrac{d^2y}{dt^2}\cdot\dfrac{dx}{dt}-\dfrac{d^2x}{dt^2}\cdot\dfrac{dy}{dt}}=\frac{\{[F'(t)]^2+[f'(t)]^2\}^{\frac{3}{2}}}{f''(t)\,.\,F'(t)-f'(t)\,.\,F''(t)}. \qquad \text{.......(F)}$$

Ex. In the above-mentioned case of *the cycloid*
$$\frac{dx}{d\theta}=a(1+\cos\theta),\qquad \frac{d^2x}{d\theta^2}=-a\sin\theta.$$
$$\frac{dy}{d\theta}=a\sin\theta,\qquad \frac{d^2y}{d\theta^2}=a\cos\theta,$$

and by formula (F)
$$\rho=\frac{a\{(1+\cos\theta)^2+\sin^2\theta\}^{\frac{3}{2}}}{\cos\theta(1+\cos\theta)+\sin^2\theta}=\frac{8a\cos^3\dfrac{\theta}{2}}{2\cos^2\dfrac{\theta}{2}}=4a\cos\frac{\theta}{2}.$$

276. Curvature at the Origin.

When the curve passes through the origin the values of $\frac{dy}{dx}(=p)$ and $\frac{d^2y}{dx^2}(=q)$ at the origin may be deduced by substituting for y the expression $px+\frac{qx^2}{2!}+\dots$ (the expansion of y by Maclaurin's Theorem) and equating coefficients of like powers of x in the identity obtained. The radius of curvature at the origin may then be at once deduced from the formula $\rho=\frac{(1+p^2)^{\frac{3}{2}}}{q}$ [Formula (D)].

Ex. Let the curve be

$$\begin{aligned} &ax+by \\ &+a'x^2+2h'xy+b'y^2 \\ &+\dots\dots \quad =0. \end{aligned}$$

Putting
$$y=px+\frac{q}{2!}x^2+\dots$$

we have
$$\left.\begin{array}{r} a \\ +bp \end{array}\right| x+ \left.\begin{array}{r} a' \\ +2h'p \\ +b'p^2 \\ +\frac{bq}{2} \end{array}\right| x^2+\dots \equiv 0,$$

therefore
$$a+bp=0,$$

and
$$a'+2h'p+b'p^2+\frac{bq}{2}=0,$$

etc.,

giving
$$p=-\frac{a}{b} \text{ and } q=-2\frac{a'+2h'p+b'p^2}{b},$$

whence
$$\rho=\frac{(1+p^2)^{\frac{3}{2}}}{q}=-\frac{1}{2}\,\frac{(a^2+b^2)^{\frac{3}{2}}}{a'b^2-2h'ab+b'a^2}.$$

This result of course might be deduced at once from formula (E).

277. It will be noticed that, if the lowest terms of the equation be of the second degree, we should get a quadratic equation giving two values for p, and consequently also two values for q. These indicate the two values of

ρ corresponding to the two branches of the curve passing through the origin.

Ex. *Find the radii of curvature at the origin for the curve*

$$y^2-3xy+2x^2-x^3+y^4=0.$$

Substituting $px+\frac{q}{2!}x^2+\ldots$ for y we have

$$\left.\begin{array}{r}p^2\\-3p\\+2\end{array}\right|x^2+\left.\begin{array}{r}pq\\-\frac{3}{2}q\\-1\end{array}\right|x^3+\ldots\equiv 0,$$

whence
$$p^2-3p+2=0,$$
$$pq-\tfrac{3}{2}q-1=0,$$
etc.,

whence $p=1$ or 2,

and $q=-2$ or 2,

and therefore $\rho=\dfrac{(1+p^2)^{\frac{3}{2}}}{q}=\dfrac{2^{\frac{3}{2}}}{-2}=-\sqrt{2}=-1\cdot414\ldots,$

or $=\dfrac{5^{\frac{3}{2}}}{2}=\dfrac{5}{2}\sqrt{5}=5\cdot590\ldots.$

The difference of sign introduced by the q indicates that the two branches passing through the origin bend in opposite directions.

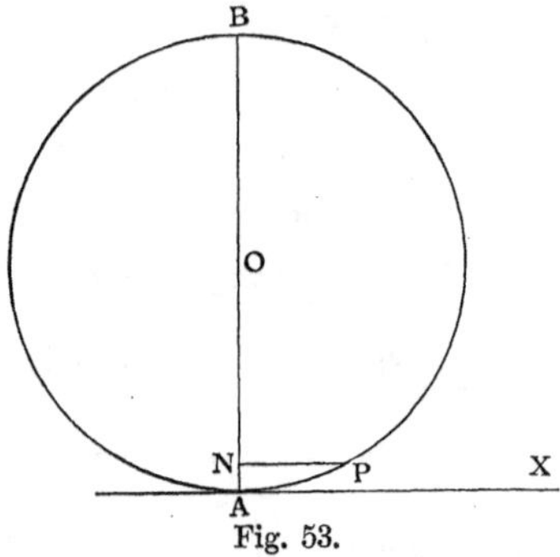

Fig. 53.

278. Newtonian Method.

The Newtonian method of finding the curvature of the curve at the origin is instructive and interesting. Suppose the axes taken so that the axis of x is a tangent to the curve at the point A, and the axis of y, viz., AB, is therefore the normal. Let APB be the circle of curvature, P the point adjacent to and ultimately coincident with A in which the curve and the circle intersect. Then

$$PN^2 = AN \cdot NB,$$

or

$$NB = \frac{PN^2}{AN}.$$

Now in the limit

$$NB = AB = \text{twice the radius of curvature.}$$

Hence

$$\rho = Lt\frac{1}{2}\frac{PN^2}{AN} = Lt\frac{x^2}{2y}. \qquad \text{(G)}$$

Similarly, if the axis of y be the tangent at the origin, we have

$$\rho = Lt\frac{y^2}{2x}.$$

Ex. *Find the radius of curvature at the origin for the curve*

$$2x^4 + 3y^4 + 4x^2y + xy - y^2 + 2x = 0.$$

In this case the *axis of y* is a tangent at the origin, and therefore we shall endeavour to find $Lt\frac{y^2}{2x}$.

Dividing by x

$$2x^3 + 3y^2 \cdot \frac{y^2}{x} + 4xy + y - \frac{y^2}{x} + 2 = 0.$$

Now, at the origin $Lt\frac{y^2}{x} = 2\rho$, $x=0$, $y=0$, and the equation becomes

$$-2\rho + 2 = 0,$$

or

$$\rho = 1.$$

279. The same method may be applied when the tangent to the curve at the origin does not coincide with one of the axes; but as the method of Art. 276 is very simple we leave the investigation as an exercise to the student.

Ex. Establish in the above manner the result of the Example in Art. 276.

280. Formula for Pedal Equations.

To find a formula for the radius of curvature adapted to pedal equations. Let O be the pole and C the centre

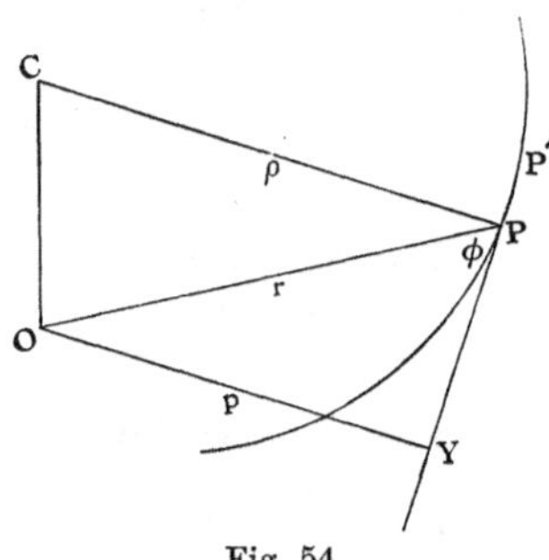

Fig. 54.

of curvature corresponding to the point P on the curve, P' a contiguous point on the curve ultimately coincident with P, the normals at P and P' intersecting at C (Art. 270). Let OY, the perpendicular on the tangent at P, $=p$. Then

$$OC^2=r^2+\rho^2-2r\rho\cos OPC$$
$$=r^2+\rho^2-2r\rho\sin\phi$$
$$=r^2+\rho^2-2\rho p,$$

since $$\sin\phi=\frac{p}{r}.$$

Again, for the contiguous point P', r becomes $r+\delta r$ and p becomes $p+\delta p$, while ρ and OC remain the same. Hence $$OC^2=(r+\delta r)^2+\rho^2-2\rho(p+\delta p);$$
therefore by subtraction
$$(r+\delta r)^2-r^2=2\rho\delta p,$$

or in the limit $$\rho=Lt\left(r\frac{\delta r}{\delta p}+\frac{\delta r^2}{2\delta p}\right)=r\frac{dr}{dp}. \quad \text{.............(H)}$$

Ex. In the equation $p^2 = Ar^2 + B$, which represents any epi- or hypo-cycloid [Chap. VII., Ex. 36], we have

$$p = Ar\frac{dr}{dp},$$

and therefore $\rho \propto p$.

The equiangular spiral, in which $p \propto r$, is included as the case in which $B = 0$.

281. **Polar Curves.**

We shall next reduce the formula to a shape suited for application to curves given by their polar equations.

We proved in Art. 181

$$\frac{1}{p^2} = u^2 + \left(\frac{du}{d\theta}\right)^2.$$

Hence

$$-\frac{1}{p^3}\frac{dp}{d\theta} = \left(u + \frac{d^2u}{d\theta^2}\right)\frac{du}{d\theta},$$

or

$$\frac{dp}{du} = -p^3\left(u + \frac{d^2u}{d\theta^2}\right).$$

Now

$$\rho = \frac{rdr}{dp} \text{ and } r = \frac{1}{u};$$

therefore

$$\rho = -\frac{1}{u^3}\frac{du}{dp} = \frac{1}{p^3u^3\left(u + \dfrac{d^2u}{d\theta^2}\right)},$$

or

$$\rho = \frac{\left\{u^2 + \left(\dfrac{du}{d\theta}\right)^2\right\}^{\frac{3}{2}}}{u^3\left(u + \dfrac{d^2u}{d\theta^2}\right)}. \quad \text{.....................(I)}$$

282. This may easily be put in the r, θ form thus:—

Since

$$u = \frac{1}{r},$$

we have

$$\frac{du}{d\theta} = -\frac{1}{r^2}\frac{dr}{d\theta},$$

and therefore $$\frac{d^2u}{d\theta^2}=\frac{2}{r^3}\left(\frac{dr}{d\theta}\right)^2-\frac{1}{r^2}\frac{d^2r}{d\theta^2};$$

therefore $$\rho=\frac{\left\{\frac{1}{r^2}+\frac{1}{r^4}\left(\frac{dr}{d\theta}\right)^2\right\}^{\frac{3}{2}}}{\frac{1}{r^3}\left\{\frac{1}{r}+\frac{2}{r^3}\left(\frac{dr}{d\theta}\right)^2-\frac{1}{r^2}\frac{d^2r}{d\theta^2}\right\}}$$

$$=\frac{\left\{r^2+\left(\frac{dr}{d\theta}\right)^2\right\}^{\frac{3}{2}}}{r^2+2\left(\frac{dr}{d\theta}\right)^2-r\frac{d^2r}{d\theta^2}}\text{................(J)}$$

283. **Tangential-Polar Form.**

In Art. 195 it was proved that

$$\rho=p+\frac{d^2p}{d\psi^2},\text{ ....(K)}$$

giving us a formula for the radius of curvature suitable for p, ψ equations.

Ex. It is known that the general p, ψ equation of all epi- and hypo-cycloids can be written in the form

$$p=A\sin B\psi \text{ (Chap. VII., Ex. 36).}$$

Hence $$\rho=A\sin B\psi-AB^2\sin B\psi,$$

and therefore $$\rho\propto p,$$

thus again proving the result of the Example in Art. 280.

284. **Point of Inflexion.**

At a point of inflexion the radius of curvature is infinite. This is geometrically obvious from the fact that it is the radius of a circle which passes through three collinear points. We may hence deduce various forms of the condition for a point of inflexion; thus if

$$\rho=\infty,$$

we get $$\frac{d\psi}{ds}=0 \text{ from (A),}$$

$$\frac{d^2y}{dx^2}=0 \text{ from (D)},$$

$$\frac{\partial^2\phi}{\partial x^2}\cdot\left(\frac{\partial\phi}{\partial y}\right)^2-2\frac{\partial^2\phi}{\partial x\partial y}\cdot\frac{\partial\phi}{\partial x}\cdot\frac{\partial\phi}{\partial y}+\frac{\partial^2\phi}{\partial y^2}\cdot\left(\frac{\partial\phi}{\partial x}\right)^2=0 \text{ from (E)},$$

$$u+\frac{d^2u}{d\theta^2}=0 \text{ from (I)},$$

$$r^2+2\left(\frac{dr}{d\theta}\right)^2-r\frac{d^2r}{d\theta^2}=0 \text{ from (J)},$$

some of which have already been established otherwise.

EXAMPLES.

1. Apply formula A to the curves
$$s=a\psi,\quad s=a\sin\psi,\quad s=a\tan\psi.$$
2. Apply formula D to the curves
$$y^2=4ax,\quad y=c\cosh\frac{x}{c}.$$
3. Apply formula E to the curve
$$ax+by+a'x^2+2h'xy+b'y^2+\ldots$$
to find the radius of curvature at the origin.
4. Apply formula F to the ellipse
$$\left.\begin{aligned}x&=a\cos\theta\\ y&=b\sin\theta\end{aligned}\right\}.$$
5. Apply formula H to the curves
$$p^2=ar,\quad ap=r^2,\quad p=\frac{r^{m+1}}{a^m}.$$
6. Apply formula I to the reciprocal spiral
$$au=\theta.$$

285. Centre of Curvature.

The Cartesian co-ordinates of the centre of curvature may be found thus :—

Let Q be the centre of curvature corresponding to the point P of the curve. Let OX be the axis of x; O the origin; x, y the co-ordinates of P; $\bar{x}, \bar{y}$ those of Q; ψ the angle the tangent makes with the axis of x. Draw PN, QM perpendiculars upon the x-axis and PR a perpendi-

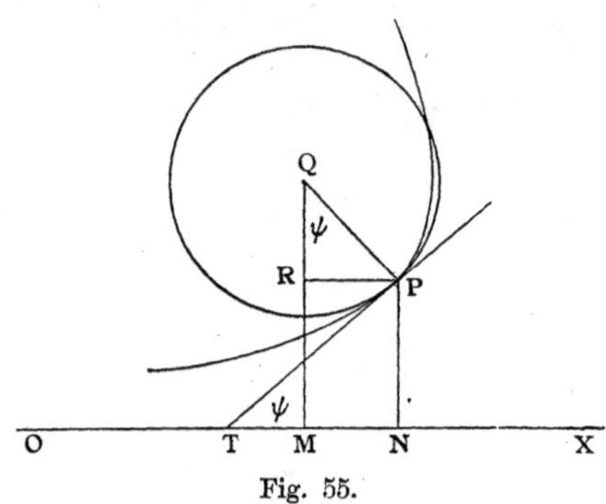

Fig. 55.

cular upon QM. Then

$$\bar{x} = OM = ON - RP$$
$$= ON - QP \sin\psi$$
$$= x - \rho \sin\psi,$$

and

$$\bar{y} = MQ = NP + RQ$$
$$= y + \rho \cos\psi.$$

Now $$\tan\psi = \frac{dy}{dx};$$

therefore $$\sin\psi = \frac{\frac{dy}{dx}}{\sqrt{1+\left(\frac{dy}{dx}\right)^2}},$$

and $$\cos\psi = \frac{1}{\sqrt{1+\left(\frac{dy}{dx}\right)^2}}.$$

Also $$\rho = \frac{\left\{1+\left(\frac{dy}{dx}\right)^2\right\}^{\frac{3}{2}}}{\frac{d^2y}{dx^2}}.$$

Hence
$$\bar{x}=x-\frac{\frac{dy}{dx}\left\{1+\left(\frac{dy}{dx}\right)^2\right\}}{\frac{d^2y}{dx^2}} \qquad \ldots\ldots\ldots\ldots\ldots\ldots (\alpha)$$

$$\bar{y}=y+\frac{1+\left(\frac{dy}{dx}\right)^2}{\frac{d^2y}{dx^2}} \qquad \ldots\ldots\ldots\ldots\ldots\ldots (\beta)$$

INVOLUTES AND EVOLUTES.

286. DEF. The locus of the centres of curvature of all points of a given plane curve is called the *evolute* of that curve. If the evolute itself be regarded as the original curve, a curve of which it is the evolute is called an *involute*.

The equation of the evolute of a given curve may be found by eliminating x and y between equations (α), (β) of the last article and the equation of the curve.

Ex. *To find the locus of the centres of curvature of the parabola*
$$y=\frac{x^2}{4a}.$$

Here
$$\frac{dy}{dx}=\frac{x}{2a}, \quad \frac{d^2y}{dx^2}=\frac{1}{2a}.$$

Hence
$$\bar{x}=x-\frac{\frac{dy}{dx}\left\{1+\left(\frac{dy}{dx}\right)^2\right\}}{\frac{d^2y}{dx^2}}=-\frac{x^3}{4a^2},$$

$$\bar{y}=y+\frac{1+\left(\frac{dy}{dx}\right)^2}{\frac{d^2y}{dx^2}}=2a+\frac{3x^2}{4a};$$

whence
$$(\bar{y}-2a)^3=\frac{27x^6}{64a^3}=\frac{27a\bar{x}^2}{4}.$$

Hence the equation of the evolute is
$$4(y-2a)^3=27ax^2.$$

287. **Evolute touched by the Normals.**

Let P_1, P_2, P_3 be contiguous points on a given curve, and let the normals at P_1, P_2 and at P_2, P_3 intersect at Q_1, Q_2 respectively. Then in the limit when P_2, P_3 move along the curve to ultimate coincidence with P_1 the limiting positions of Q_1, Q_2 are the centres of curvature corresponding to the points P_1, P_2 of the curve. Now Q_1 and Q_2 both lie on the normal at P_2, and therefore it

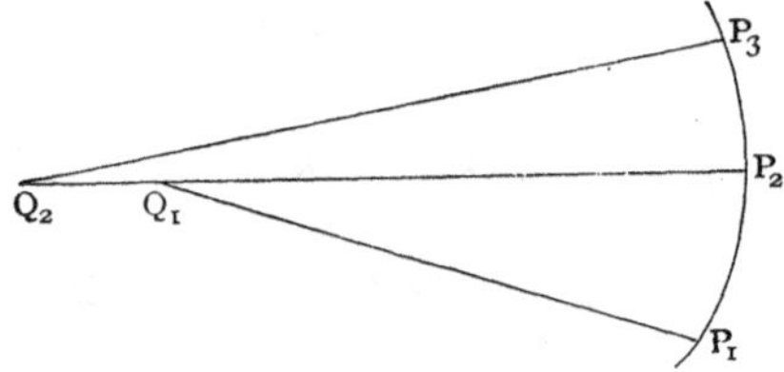

Fig. 56.

is clear that the normal is a tangent to the locus of such points as Q_1, Q_2, *i.e.*, each of the normals of the original curve is *a tangent to the evolute;* and it will be seen in the chapter on Envelopes (Art. 313) that in general the best method of investigating the equation of the evolute of any proposed curve is to consider it as the *envelope of the normals* of that curve.

288. **There is but one Evolute, but an infinite number of Involutes.**

Let *ABCD*... be the original curve on which the successive points *A*, *B*, *C*, *D*, ... are indefinitely close to each other. Let *a*, *b*, *c*, ... be the successive points of intersection of normals at *A*, *B*, *C*, ... and therefore the centres of curvature of those points. Then looking at *ABC*... as the original curve, *abcd*... is its *evolute.* And regarding $abcd\ldots$ as the original curve, $ABCD\ldots$ is *an involute.*

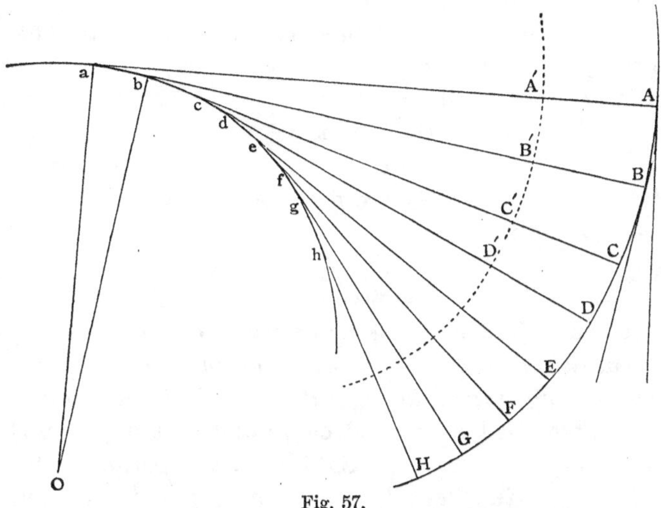

Fig. 57.

If we suppose any equal lengths AA', BB', CC', ... to be taken along each normal, as shown in the figure, then a new curve is formed, viz., $A'B'C'\ldots$, which may be called a *parallel* to the original curve, having the same normals as the original curve and therefore having the same evolute. It is therefore clear that if any curve be given it can have but *one evolute*, but an infinite number of curves may have the same evolute, and therefore any curve may have an *infinite number of involutes.* The involutes of a given curve thus form a system of *parallel curves.*

289. **Involutes traced out by the several points of a string unwound from a curve.**

Since a is the centre of the circle of curvature for the point A (Fig. 57), $aA = aB$

$$= bB + \text{elementary arc } ab \text{ (Art. 36).}$$

Hence $\quad aA - bB = \text{arc } ab.$

Similarly $\quad bB - cC = \text{arc } bc,$

$$cC - dD = \text{arc } cd,$$

$$\text{etc.,}$$

$$fF - gG = \text{arc } fg.$$

Hence by addition

$$aA - gG = \text{arc } ab + \text{arc } bc + \ldots + \text{arc } fg$$
$$= \text{arc } ag.$$

Hence the *difference between the radii of curvature at two points of a curve is equal to the length of the corresponding arc of the evolute.* Also, if the evolute $abc\ldots$ be regarded as a rigid curve and a string be unwound from it, being kept tight, then the *points of the unwinding string describe a system of parallel curves, each of which is an involute of the curve abcd...,* one of them coinciding with the original curve $ABC\ldots$. It is from this property that the names involute and evolute are derived.

290. Radius of Curvature of the Evolute.

It is easy to find an expression for the radius of curvature at that point of the evolute which corresponds to any given point of the original curve.

Let O (Fig. 57) be the centre of curvature for the point a of the evolute. The angle $\delta\psi'$ between the normals at a, b

= the angle between the tangents at a, b

= the angle between the tangents at A, B to the original curve

= $\delta\psi$.

And if s' be the arc of the evolute measured from some fixed point up to a, and ρ' the radius of curvature of the evolute at a, and ρ that of the original curve at A, we have, rejecting infinitesimals of order higher than the first,

$$\delta s' = \text{arc } ab = \delta\rho,$$

and therefore

$$\rho' = Lt\frac{\delta s'}{\delta\psi'} = Lt\frac{\delta\rho}{\delta\psi} = \frac{d\rho}{d\psi} = \frac{d^2 s}{d\psi^2},$$

s being the arc of the original curve measured from some fixed point up to A, and ψ the angle which the tangent at A makes with some fixed straight line.

INTRINSIC EQUATION.

291. The relation between the length of the arc (s) of a given curve, measured from a given fixed point on the curve, and the angle between the tangents at its extremities (ψ) has been aptly styled by Dr. Whewell the *Intrinsic Equation* of the curve. For many curves this relation takes a very elegant form. The name seems specially suitable to a relation between such quantities as these, depending as it does upon no external system of co-ordinates. The method of obtaining the intrinsic equation from the Cartesian or polar relation is dependent in general upon processes of integration. If the equation of the curve be given as $y=f(x)$, the axis of x being supposed a tangent at the origin, and the length of the arc being measured from the origin, we have

$$\tan\psi = f'(x), \qquad (1)$$

and

$$\frac{ds}{dx} = \sqrt{1+[f'(x)]^2}. \qquad (2)$$

If s be determined by integration from (2) and x eliminated between the result and equation (1), the required relation between s and ψ will be obtained.

Ex. 1. *Intrinsic equation of a circle.*

If ψ be the angle between the initial tangent at A and the tangent at the point P, and a the radius of the circle, we have

$$P\hat{O}A = P\hat{T}X = \psi,$$

and therefore $$s = a\psi.$$

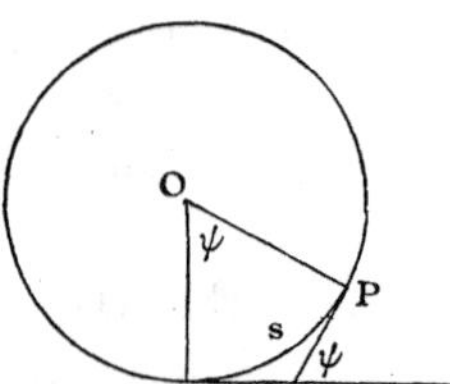

Fig. 58.

Ex. 2. *In the case of the catenary whose equation is*

$$y = c \cosh \frac{x}{c}$$

the intrinsic equation is $$s = c \tan \psi.$$

For $$\tan \psi = \frac{dy}{dx} = \sinh \frac{x}{c},$$

and $$\frac{ds}{dx} = \sqrt{1 + \sinh^2 \frac{x}{c}} = \cosh \frac{x}{c},$$

and therefore $$s = c \sinh \frac{x}{c},$$

the constant of integration being chosen so that x and s vanish together, whence $$s = c \tan \psi.$$

Examples.

1. Show that the cycloid $$\left.\begin{aligned} x &= a(\theta + \sin \theta) \\ y &= a(1 - \cos \theta) \end{aligned}\right\}$$

has for its intrinsic equation $$s = 4a \sin \psi.$$

2. Show that the epi- or hypo-cycloid given by

$$\left.\begin{aligned} x &= (a+b)\cos\theta - b\cos\frac{a+b}{b}\theta \\ y &= (a+b)\sin\theta - b\sin\frac{a+b}{b}\theta \end{aligned}\right\}$$

has an intrinsic equation of the form

$$s = A\sin B\psi.$$

292. Intrinsic Equation of the Evolute.

Let $s = f(\psi)$ be the equation of the given curve. Let s' be the length of the arc of the evolute measured from some fixed point A to any other point Q. Let O and P

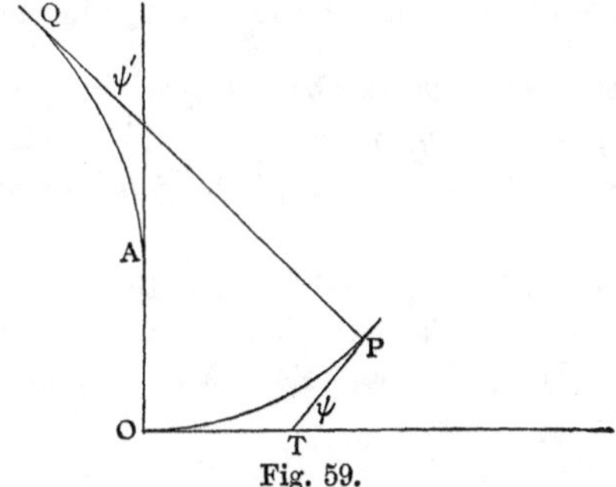

Fig. 59.

be the points on the original curve corresponding to the points A, Q on the evolute, ρ_0, ρ the radii of curvature at O and P; ψ' the angle the tangent QP makes with OA produced, and ψ the angle the tangent PT makes with the tangent at O.

Then $\psi' = \psi$, and $\quad s' = \rho - \rho_0 = \dfrac{ds}{d\psi} - \rho_0,$

or $\quad s' = f'(\psi') - \rho_0,$

the intrinsic equation of the evolute.

293. Intrinsic Equation of an Involute.

With the same figure, if the curve AQ be given by the equation $\quad s' = f(\psi'),$

we have $\rho = s' + \rho_0, \quad \rho = \dfrac{ds}{d\psi}, \text{ and } \psi = \psi',$

whence $$s = \int \{ f(\psi') + \rho_0 \} d\psi'.$$

294. Evolutes of Cycloids or Epi- and Hypo-Cycloids.

If we apply the result of Art. 292 to the intrinsic equation $s = A \sin B\psi$, we get for the equation of the evolute $$s' = AB \cos B\psi' - \rho_0,$$
or, dropping the dashes,
$$s = AB \cos B\psi,$$
if s be supposed measured from the point where $\psi = \dfrac{\pi}{2B}$.

This proves that the *evolute of an epi- or hypo-cycloid is a similar epi- or hypo-cycloid.* Also, the case in which $B = 1$ shows that the *evolute of a cycloid is an equal cycloid.*

[For further information on Intrinsic Equations the student is referred to Boole, "Differential Equations," p. 263, and to "Camb. Phil. Trans.," Vol. VIII, p. 689, and Vol. IX., p. 150.]

CONTACT.

295. First, consider the point P at which two curves cut. It is clear that in general each has its own tangent at that point, and that if the curves be of the mth and n^{th} degrees respectively, they will cut in $mn - 1$ other points real or imaginary.

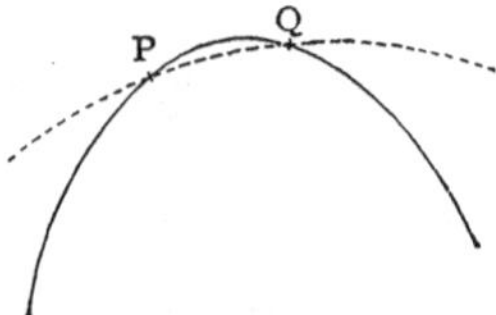

Fig. 60.

Next, suppose one of these other points (say Q) to

move along one of the curves up to coincidence with P. The curves now cut in two ultimately coincident points at P, and therefore have a common tangent. There is then said to be contact *of the first order.* It will be observed that at such a point the curves *do not on the whole cross each other.*

Again, suppose another of the mn points of intersection (viz., R) to follow Q along one of the curves to coincidence with P. There are now three contiguous points on each curve common, and therefore the curves

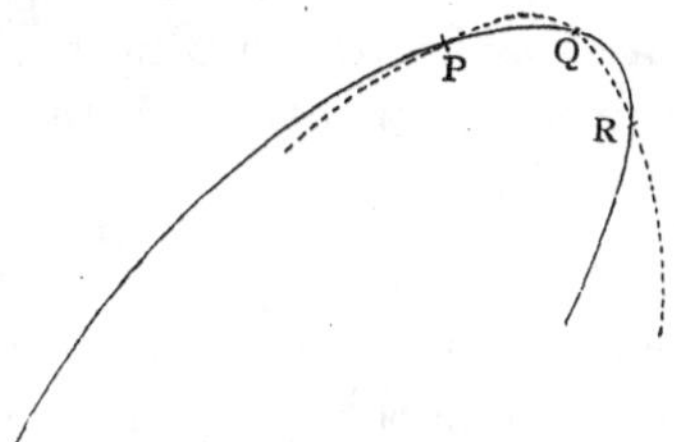

Fig. 61.

have two contiguous tangents common, namely, the ultimate position of the chord PQ and the ultimate position of the chord QR. Contact of this kind is said to be *of the second order*, and the curves on the whole *cross each other.*

Finally, if other points of intersection follow Q and R up to P, so that ultimately k points of intersection coincide at P, there will be $k-1$ contiguous common tangents at P, and the contact is said to be of the $(k-1)^{\text{th}}$ order. And if k be odd and the contact of an *even order* the curves *will cross*, but if k be even and the contact therefore of an *odd order* they *will not cross.*

296. Closest Degree of Contact of the Conic Sections with a Curve.

The simplest curve which can be drawn so as to pass

through	two	given points is a	straight line,
do.	three	do.	circle,
do.	four	do.	parabola,
do.	five	do.	conic.

Hence, if the points be contiguous and ultimately coincident points on a given curve, we can have respectively the

Straight Line of Closest Contact (or tangent), having contact of the *first order* and cutting the curve in *two* ultimately coincident points, and therefore *not in general crossing* its curve; the

Circle of Closest Contact, having contact of the *second order* and cutting the curve in *three* ultimately coincident points, and therefore *in general crossing* its curve (this is the circle already investigated as the circle of curvature); the

Parabola of Closest Contact, having contact of the *third order* and cutting the curve in *four* ultimately coincident points, and therefore in general *not crossing;* and the

Conic of Closest Contact, having contact of the *fourth order* and cutting the curve in *five* ultimately coincident points, and therefore in general *crossing.*

We say *in general:* for take for instance the "circle of closest contact" at a given point on a conic section. Now, a circle and a conic section intersect in four points real or imaginary, and since in our case three of these are real and coincident, the circle of closest contact cuts the curve again in some one real fourth point. But *it may happen* as in the case in which the three ultimately coincident points are at an end of one of the axes of the conic *that the fourth point is coincident with the other three,* in which case the circle of closest contact has a contact of higher order than usual, viz., of the *third* order, cutting the curve in four ultimately coincident points, and therefore on the whole *not crossing* the curve. The student should draw for himself figures of the circle of closest contact at various points of a conic section, remembering from Geometrical Conics that the common chord of the circle and conic, and the tangent at the point of contact make equal angles with either axis. The conic which has the closest possible contact is said to *osculate* its curve at the point of contact and is called the *osculating conic.* Thus the circle of curvature is called the *osculating circle,* the parabola of closest contact is called the *osculating parabola,* and so on.

297. **Analytical Conditions for Contact of a given order.**

We may treat this subject analytically as follows.

Let
$$\left.\begin{aligned} y&=\phi(x) \\ y&=\psi(x) \end{aligned}\right\}$$
be the equations of two curves which cut at the point $P(x, y)$.

Consider the values of the respective ordinates at the points P_1, P_2 whose common abscissa is $x+h$.

Let $MN=h$,

Then $NP_1=\phi(x+h)$,

$NP_2=\psi(x+h)$

and $$P_2P_1 = NP_1 - NP_2 = \phi(x+h) - \psi(x+h)$$
$$= [\phi(x) - \psi(x)] + h[\phi'(x) - \psi'(x)]$$
$$+ \frac{h^2}{2!}[\phi''(x) - \psi''(x)] + \ldots.$$

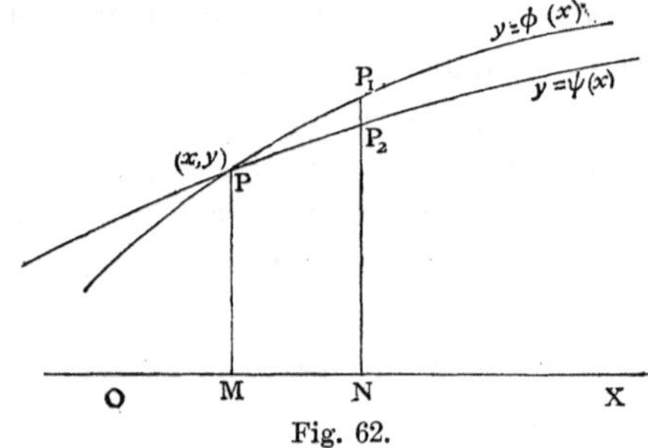

Fig. 62.

If the expression for P_2P_1 be equated to zero, the roots of the resulting equation for h will determine the points at which the curves cut.

If $\phi(x) = \psi(x)$, the equation has one root zero and the curves cut at P.

If also $\phi'(x) = \psi'(x)$ for the same value of x, the equation has two roots zero and the curves cut in *two* contiguous points at P, and therefore have a common tangent. The contact is now of the *first order*.

If also $\phi''(x) = \psi''(x)$ for the same value of x, the equation for h has three roots zero and the curves cut in *three* ultimately coincident points at P. There are now two contiguous tangents common, and the contact is said to be of the *second order;* and so on.

Similarly for curves given by their polar equations, if $r = f(\theta)$, $r = \phi(\theta)$ be the two equations, there will be $n+1$ equations to be satisfied for the same value of θ in order that for that value there may be contact of the nth order, viz.,

$$f(\theta)=\phi(\theta),\ f'(\theta)=\phi'(\theta),\ f''(\theta)=\phi''(\theta),\ \ldots,\ f^n(\theta)=\phi^n(\theta).$$

298. **Osculating Circle.**

The circle of curvature may now be investigated as the circle which has contact of the second order with a given curve at a given point.

Suppose $$y=f(x) \quad \ldots\ldots\ldots\ldots\ldots\ldots(1)$$
to be the equation of the curve.

Let $$(x-\bar{x})^2+(y-\bar{y})^2=\rho^2 \ldots\ldots\ldots\ldots\ldots\ldots(2)$$
be the equation of the circle of curvature.

By differentiating (2) we have

$$x-x+(y-\bar{y})\frac{dy}{dx}=0, \quad \ldots\ldots\ldots\ldots\ldots\ldots(3)$$

and differentiating again

$$1+\left(\frac{dy}{dx}\right)^2+(y-\bar{y})\frac{d^2y}{dx^2}=0. \quad \ldots\ldots\ldots\ldots\ldots\ldots(4)$$

Now the x, y, $\frac{dy}{dx}$, $\frac{d^2y}{dx^2}$ of equations (2), (3), (4) refer to the circle. But, since there is to be contact of the second order with the curve $y=f(x)$ at the point (x, y), $\frac{dy}{dx}$ and $\frac{d^2y}{dx^2}$ *have the same value as when deduced from the equation to the curve, i.e.*, we may write $f'(x)$ for $\frac{dy}{dx}$ and $f''(x)$ for $\frac{d^2y}{dx^2}$.

From equation (4)

$$y-\bar{y}=-\frac{1+\left(\frac{dy}{dx}\right)^2}{\frac{d^2y}{dx^2}}=-\frac{1+\{f'(x)\}^2}{f''(x)},$$

whence $$x-\bar{x}=\frac{\dfrac{dy}{dx}\left\{1+\left(\dfrac{dy}{dx}\right)^2\right\}}{\dfrac{d^2y}{dx^2}}=\frac{f'(x)[1+\{f'(x)\}^2]}{f''(x)},$$

and by squaring and adding

$$\rho=\pm\frac{\left\{1+\left(\dfrac{dy}{dx}\right)^2\right\}^{\frac{3}{2}}}{\dfrac{d^2y}{dx^2}}=\pm\frac{[1+\{f'(x)\}^2]^{\frac{3}{2}}}{f''(x)},$$

such a sign being given to the radical as will make ρ positive, *i.e.*, if $\dfrac{d^2y}{dx^2}$ be positive we must choose the $+$ sign for the numerator, and if $\dfrac{d^2y}{dx^2}$ be negative we must choose the $-$ sign.

The values of $\bar{x}$ and $\bar{y}$ are the same as those found geometrically in Art. 285, viz.,

$$\bar{x}=x-\frac{\dfrac{dy}{dx}\left\{1+\left(\dfrac{dy}{dx}\right)^2\right\}}{\dfrac{d^2y}{dx^2}},$$

$$\bar{y}=y+\frac{1+\left(\dfrac{dy}{dx}\right)^2}{\dfrac{d^2y}{dx^2}}.$$

299. **Conic having Third Order Contact at a given point.**

The locus of the centres of all conics having third order contact with a given curve at a given point (*i.e.*, cutting the curve in four ultimately coincident points) is a *straight line* which passes through the point of contact.

Let P be a point on the curve and C the centre of one of the conics having third order contact with the given

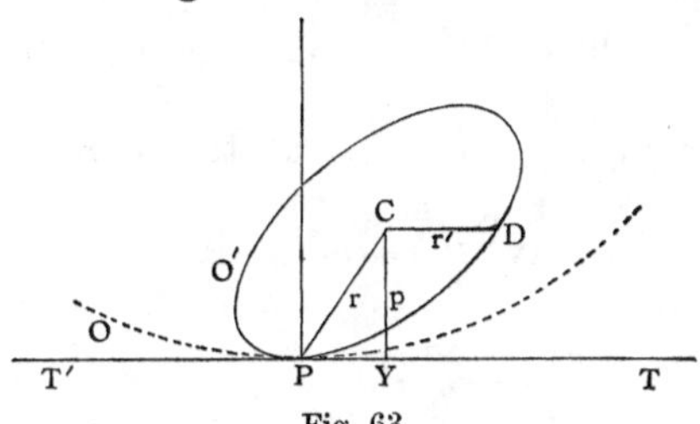

Fig. 63.

curve at P. Let CD be the semiconjugate to CP and CY a perpendicular on the tangent at P.

Let $CP=r$, $CD=r'$, $CY=p$, and let PC make an angle ϕ with the normal at P.

Then we have $$r^2+r'^2=a^2+b^2,$$
and $$pr'=ab,$$
and therefore $$rdr+r'dr'=0;$$
and for a conic $$\rho=\frac{CD^3}{ab}=\frac{r'^3}{ab};$$ (See Ex. 18, p. 301)

therefore $$\frac{d\rho}{ds}=\frac{3r'^2}{ab}\cdot\frac{dr'}{ds}=-\frac{3r'}{ab}\cdot\frac{rdr}{ds}$$
$$=-\frac{3r}{p}\cdot\frac{dr}{ds}=3\frac{\sin\phi}{\cos\phi},$$

for $\frac{dr}{ds}=\cos CPT'=-\sin\phi$, the arcs of the curve and of the conic being measured from the points O and O' up to P, and $$\frac{p}{r}=\cos\phi;$$

therefore $$\frac{d\rho}{ds}=3\tan\phi,$$

and $\tan\phi=\frac{1}{3}\frac{d\rho}{ds}$, where $\frac{d\rho}{ds}$ is found for *one of the conics.*

But since the conic and the curve have contact of the third order they have the same tangent, the same $\frac{dr}{d\theta}$, the same $\frac{d^2r}{d\theta^2}$, and the same $\frac{d^3r}{d\theta^3}$ at the point of contact. They therefore also have *the same* ρ and *the same* $\frac{d\rho}{ds}$, for ρ depends on $\frac{d^2r}{d\theta^2}$ and $\frac{d\rho}{ds}$ on $\frac{d^3r}{d\theta^3}$.

Hence the value of ϕ found above is the same for all the conics, and depends only upon the shape of the curve at the point of contact. The locus of all such centres is therefore a straight line through the point of contact inclined in front of the normal at an angle $\tan^{-1}\left(\frac{1}{3}\frac{d\rho}{ds}\right)$, where $\frac{d\rho}{ds}$ *is found from the curve.*

300. **Osculating Conic.**

We can now pick out the particular conic which has *fourth order* contact with the given curve at the given point.

Let O be the centre of curvature of the point considered and C the required centre of the conic of closest contact. Let P_1 be a point on the curve adjacent to the given point P. Join CP, CP_1 and draw P_1N at right angles to CP.

Let $\quad O\hat{P}C=\phi, \quad O\hat{P}_1C=\phi+\delta\phi, \quad PC=R.$

Then $\quad P\hat{E}P_1=P\hat{O}E+\phi,$

and also $\quad =P_1\hat{C}E+\phi+\delta\phi,$

whence $\quad P\hat{O}E=P_1\hat{C}E+\delta\phi.$

Also, neglecting infinitesimals of higher order than the first, $\quad PP_1=\delta s,$

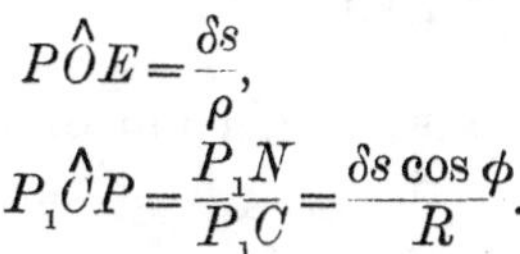

$$P\hat{O}E = \frac{\delta s}{\rho},$$

and
$$P_1\hat{C}P = \frac{P_1N}{P_1C} = \frac{\delta s \cos\phi}{R}.$$

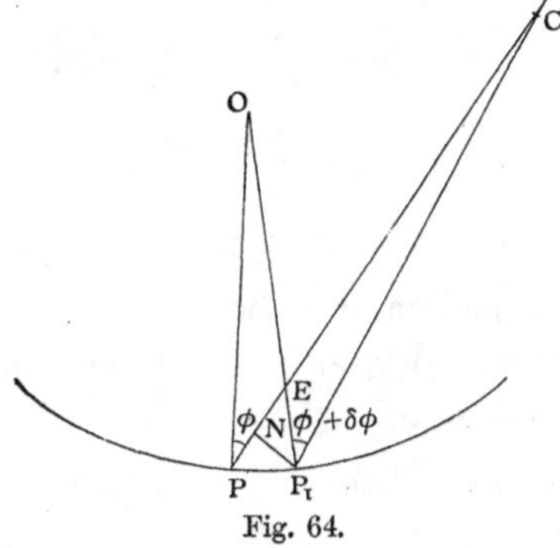

Fig. 64.

Hence
$$\frac{\delta s}{\rho} = \frac{\delta s \cos\phi}{R} + \delta\phi,$$

or, proceeding to the limit,

$$\left.\begin{aligned} \frac{\cos\phi}{R} &= \frac{1}{\rho} - \frac{d\phi}{ds}, \\ \text{where} \quad \phi &= \tan^{-1}\frac{1}{3}\frac{d\rho}{ds} \end{aligned}\right\}.$$

And since the contact is of the fourth order, $\dfrac{d\phi}{ds}$ is the same for *the curve as for the conic,* and may therefore be *supposed derived from the equation of the curve.*

These equations determine the position of C.

301. **Tangent and Normal as Axes. Co-ordinates of a Point near the Origin in terms of the Arc.**

When the tangent and normal at any point of a curve are taken as the axes of x and y it is sometimes requisite to express the co-ordinates of a point on the curve near the origin in terms of the length of the arc measured from the origin up to that point.

Assume $$x = a + a_1 s + a_2 \frac{s^2}{2!} + a_3 \frac{s^3}{3!} + \dots,$$
$$y = b + b_1 s + b_2 \frac{s^2}{2!} + b_3 \frac{s^3}{3!} + \dots,$$

the letters $a, a_1, \dots, b, b_1 \dots$ denoting constants whose values are to be determined, and s being the length of arc. Then, when $s=0$, x and y both vanish, and therefore

$$a = b = 0.$$

Again, by Maclaurin's Theorem

$$\left.\begin{aligned} a_1 &= \left(\frac{dx}{ds}\right)_0 = (\cos\psi)_0 = 1, \\ b_1 &= \left(\frac{dy}{ds}\right)_0 = (\sin\psi)_0 = 0, \end{aligned}\right\} \text{[the suffix zero denoting the values at the origin]}$$

$$\left.\begin{aligned} a_2 &= \left(\frac{d^2x}{ds^2}\right)_0 = -\left(\sin\psi\frac{d\psi}{ds}\right)_0 = -\left(\frac{\sin\psi}{\rho}\right)_0 = 0, \\ b_2 &= \left(\frac{d^2y}{ds^2}\right)_0 = \left(\cos\psi\frac{d\psi}{ds}\right)_0 = \left(\frac{\cos\psi}{\rho}\right)_0 = \frac{1}{\rho}, \end{aligned}\right\}$$

$$\left.\begin{aligned} a_3 &= \left(\frac{d^3x}{ds^3}\right)_0 = -\left(\frac{\cos\psi}{\rho^2} - \frac{\sin\psi}{\rho^2}\frac{d\rho}{ds}\right)_0 = -\frac{1}{\rho^2}, \\ b_3 &= \left(\frac{d^3y}{ds^3}\right)_0 = \left(-\frac{\sin\psi}{\rho^2} - \frac{\cos\psi}{\rho^2}\frac{d\rho}{ds}\right)_0 = -\frac{1}{\rho^2}\frac{d\rho}{ds}, \end{aligned}\right\}$$

etc.,

whence $$x = s - \frac{s^3}{6\rho^2} + \dots,$$

$$y = \frac{s^2}{2\rho} - \frac{s^3}{6\rho^2}\frac{d\rho}{ds} - \dots.$$

EXAMPLES.

1. Prove that in the case of the equiangular spiral whose intrinsic equation is

$$s = a(e^{m\psi} - 1),$$
$$\rho = mae^{m\psi}.$$

2. For the tractrix $s = c \log \sec \psi$ prove that $\rho = c \tan \psi$.

3. Show that in the curve

$$y = x + 3x^2 - x^3$$

the radius of curvature at the origin $= \cdot 4714\ldots$, and that at the point (1, 3) it is infinite.

4. Show that in the curve

$$y^2 - 3xy - 4x^2 + x^3 + x^4y + y^5 = 0$$

the radii of curvature at the origin are

$$\frac{85}{2}\sqrt{17} \text{ and } 5\sqrt{2}.$$

5. Show that the radii of curvature of the curve

$$y^2 = x^2\frac{a + x}{a - x}$$

for the origin $= \pm a\sqrt{2}$,

and for the point $(-a, 0)$ $= \dfrac{a}{4}$.

6. Show that the radii of curvature at the origin for the curve

$$x^3 + y^3 = 3axy$$

are each $= \dfrac{3a}{2}$.

7. Prove that the chord of curvature parallel to the axis of y for the curve

$$y = a \log \sec \frac{x}{a}$$

is of constant length.

8. Prove that for the curve

$$s = m(\sec^3\psi - 1),$$
$$\rho = 3m \tan \psi \sec^3\psi,$$

and hence that $$3m\frac{dy}{dx}\frac{d^2y}{dx^2}=1.$$

Also, that this differential equation is satisfied by the semicubical parabola $$27my^2=8x^3.$$

9. Prove that for the curve

$$s=a\log\cot\left(\frac{\pi}{4}-\frac{\psi}{2}\right)+a\frac{\sin\psi}{\cos^2\psi},$$

$$\rho=2a\sec^3\psi;$$

and hence that $$\frac{d^2y}{dx^2}=\frac{1}{2a},$$

and that this differential equation is satisfied by the parabola

$$x^2=4ay.$$

10. Show that for the curve in which $s=ae^{\frac{x}{c}}$

$$c\rho=s(s^2-c^2)^{\frac{1}{2}}.$$

11. Show that the curve for which $s=\sqrt{8ay}$ (the cycloid) has for its intrinsic equation

$$s=4a\sin\psi.$$

Hence prove $$\rho=4a\sqrt{1-\frac{y}{2a}}.$$

12. Prove that the curve for which $y^2=c^2+s^2$ (the catenary) has for its intrinsic equation

$$s=c\tan\psi.$$

Hence prove $\rho=\frac{y^2}{c}=$ the part of the normal intercepted between the curve and the x-axis.

13. For the parabola $$y^2=4ax,$$

prove $$\bar{x}=2a+3x,$$

$$\bar{y}=-2\frac{x^{\frac{3}{2}}}{a^{\frac{1}{2}}},$$

$$\rho=2\frac{SP^{\frac{3}{2}}}{a^{\frac{1}{2}}},$$

SP being the focal distance of the point of the parabola whose co-ordinates are (x, y).

14. Show that the circles of curvature of the parabola $y^2 = 4ax$ for the ends of the latus rectum have for their equations

$$x^2 + y^2 - 10ax \pm 4ay - 3a^2 = 0,$$

and that they cut the curve again in the points $(9a, \mp 6a)$.

15. Show that the evolute of the parabola $y^2 = 4ax$ is the semicubical parabola

$$27ay^2 = 4(x - 2a)^3,$$

and that the length of the evolute from the cusp to the point where it meets the parabola $= 2a(3\sqrt{3} - 1)$.

16. For the ellipse $\dfrac{x^2}{a^2} + \dfrac{y^2}{b^2} = 1,$

prove

$$\left.\begin{aligned} \bar{x} &= \frac{a^2 - b^2}{a^4} x^3 \\ \bar{y} &= \frac{b^2 - a^2}{b^4} y^3 \end{aligned}\right\}.$$

Hence show that the equation of the evolute is

$$(ax)^{\frac{2}{3}} + (by)^{\frac{2}{3}} = (a^2 - b^2)^{\frac{2}{3}},$$

and prove that the whole length of the evolute

$$= 4\left(\frac{a^2}{b} - \frac{b^2}{a}\right).$$

17. Show that in a parabola the radius of curvature is twice the part of the normal intercepted between the curve and the directrix.

18. Prove that in an ellipse, centre C, the radius of curvature at any point P is given by

$$\rho = \frac{CD^3}{ab} = \frac{a^2b^2}{p^3} = \frac{(rr')^{\frac{3}{2}}}{ab},$$

where a, b are the semi-axes, r, r' are the focal distances of P, p the perpendicular from the centre on the tangent at P, and CD the semi-diameter conjugate to CP.

19. Show that in any conic

$$\rho = \frac{(\text{normal})^3}{(\text{semi-latus-rectum})^2}$$

20. Apply the polar formula for radius of curvature to show that the radius of the circle

$$r = a\cos\theta \text{ is } \frac{a}{2}.$$

21. Show that for the cardioide $r = a(1+\cos\theta)$

$$\rho = \frac{4a}{3}\cos\frac{\theta}{2}; \ i.e., \propto \sqrt{r}.$$

Also deduce the same result from the pedal equation of the curve, viz.,

$$p\sqrt{2a} = r^{\frac{3}{2}}.$$

22. Show that at the points in which the Archimedean spiral $r = a\theta$ intersects the reciprocal spiral $r\theta = a$ their curvatures are in the ratio 3 : 1.

23. For the equiangular spiral $r = ae^{m\theta}$ prove that the centre of curvature is at the point where the perpendicular to the radius vector through the pole intersects the normal.

24. Prove that for the curve

$$r = a\sec 2\theta,$$

$$\rho = -\frac{r^4}{3p^3}.$$

25. For any curve prove the formula

$$\rho = \frac{r}{\sin\phi\left(1+\frac{d\phi}{d\theta}\right)}$$

where
$$\tan\phi = \frac{rd\theta}{dr}.$$

Deduce the ordinary formula in terms of r and θ.

26. Show that the chord of curvature through the pole for the curve

$$p = f(r)$$

is given by
$$\text{chord} = 2p\frac{dr}{dp} = 2\frac{f(r)}{f'(r)}.$$

27. Show that the chord of curvature through the pole of the cardioide $r = a(1+\cos\theta)$ is $\frac{4}{3}r$.

28. Show that the chord of curvature through the pole of the equiangular spiral $r = ae^{m\theta}$ is $2r$.

29. Show that the chord of curvature through the pole of the curve $r^m = a^m \cos m\theta$ is $\dfrac{2r}{m+1}$.

Examine the cases when $m = -2, -1, -\frac{1}{2}, \frac{1}{2}, 1, 2$.

30. Show that the radius of curvature of the curve

$$r = a \sin n\theta$$

at the origin is $\dfrac{na}{2}$.

31. Show that for the curve

$$x^m + y^m = k^m$$

we may write ρ in the form

$$\frac{k}{m-1} \frac{\left(\cos^{4\frac{1-m}{m}}\phi + \sin^{4\frac{1-m}{m}}\phi\right)^{\frac{3}{2}}}{\cos^{2\frac{1-2m}{m}}\phi \sin^{2\frac{1-2m}{m}}\phi}, \text{ where } x = k\cos^{\frac{2}{m}}\phi.$$

Examine the cases $m = 2, \frac{2}{3}, 1$.

32. For the rectangular hyperbola

$$xy = k^2,$$

prove that $\rho = \dfrac{r^3}{2k^2}$,

r being the central radius vector of the point considered.

33. For the curve $r^m = a^m \cos m\theta$,

prove that $\rho = \dfrac{a^m}{(m+1)r^{m-1}}$.

Examine the particular cases of a rectangular hyperbola, lemniscate, parabola, cardioide, straight line, circle.

34. Show that the co-ordinates of the centre of curvature of any curve may be written

$$\left\{ \frac{1}{2}\frac{\dfrac{d^2r^2}{dy^2}}{\dfrac{d^2x}{dy^2}}, \ \frac{1}{2}\frac{\dfrac{d^2r^2}{dx^2}}{\dfrac{d^2y}{dx^2}} \right\}.$$

35. If A be the area of the portion of a curve included between the curve, two radii of curvature, and the evolute, prove

$$\rho = 2\frac{dA}{ds}.$$

36. Show that the evolute of an equiangular spiral is an equal equiangular spiral.

37. Given the pedal equation of a curve, viz., $p = f(r)$; show that the pedal equation of its evolute may be found by eliminating p and r between this equation and the equations

$$r'^2 = \rho^2 + r^2 - 2\rho p, \qquad (\alpha)$$

$$p'^2 = r^2 - p^2. \qquad (\beta)$$

Again, that if the equation $p' = f(r')$ of a curve be given, the general differential equation of its involutes may be obtained by eliminating p', r' between this equation and the equations (α), (β).

38. Show that the curve whose equation is

$$p^2 = r^2 - a^2$$

is an involute of a circle, and that its intrinsic equation is

$$s = a\frac{\psi^2}{2}.$$

39. Show that the evolute of the epi- or hypo-cycloid denoted by

$$p^2 = Ar^2 + B$$

is another epi- or hypo-cycloid denoted by

$$p^2 = Ar^2 + B\left(1 - \frac{1}{A}\right).$$

40. Show that the pedal equation of the evolute of the curve

$$r^m = a^m \sin m\theta$$

is obtained by eliminating r between

$$r'^2 = \frac{a^{2m} + (m^2 - 1)r^{2m}}{(m+1)^2 r^{2m-2}}$$

and

$$p'^2 = r^2\frac{a^{2m} - r^{2m}}{a^{2m}}.$$

41. Show that the intrinsic equation of the evolute of a parabola is

$$s = 2a(\sec^3\psi - 1).$$

42. If x, y be the co-ordinates of a point P of a curve OP passing through the origin O, then the radius of curvature at O

$$=\tfrac{1}{2}Lt\frac{x^2+y^2}{x\sin\alpha-y\cos\alpha},$$

where $y=x\tan\alpha$ is the equation of the tangent at the origin.

Hence show that the radius of curvature of the curve

$$x^4+y^2=2a(x+y)$$

at the origin is $2a\sqrt{2}$.

43. Show that the curvature at any point of the pedal of an epi- or hypo-cycloid is $\dfrac{p(r^4+a^2p^2)}{r^6}$,

where a is the radius of the fixed circle and r and p refer to the pedal curve. [SIDNEY COLL., CAMB.]

44. If r, p, ρ be respectively the radius vector, perpendicular from the origin on the tangent and the radius of curvature at any point of a curve, prove that the radius of curvature at the corresponding point of the reciprocal polar with regard to the origin is

$$\frac{k^2r^3}{p^3\rho},$$

where k^2 is the constant of reciprocation.

Hence show that the reciprocal of a circle is a conic with the origin as focus.

45. If r, p, ρ be the same as in the last question, show that the radius of curvature at the corresponding point of the inverse with regard to the origin is $\dfrac{k^2\rho}{2p\rho-r^2}$,

k^2 being the constant of inversion.

46. Show that the parabola whose axis is parallel to the axis of y, and which has the closest possible contact with the curve

$$a^{n-1}y=x^n$$

at the point (a, a), has for its equation

$$n(n-1)x^2=2ay+2n(n-2)ax-(n-1)(n-2)a^2.$$

47. Show that the locus of the centre of the rectangular hyperbola, having contact of the third order with the conic

$$Ax^2 + By^2 = 1,$$

has for its equation $$x^2 + y^2 = \left(\frac{1}{A} + \frac{1}{B}\right) \sqrt{Ax^2 + By^2}.$$

48. Show that the locus of the centres of the rectangular hyperbolae, having contact of the third order with the parabola

$$y^2 = 4ax,$$

is the equal parabola $y^2 + 4a(x + 2a) = 0$.

49. If the equation to a curve passing through the origin be

$$u_1 + u_2 + u_3 + \ldots = 0,$$

where u_n is a homogeneous function of x, y of n dimensions, show that the general equation to all conics having the same curvature at the origin as the given curve is

$$u_1 + u_2 + (lx + my)u_1 = 0.$$

Thence find the circle of curvature.

50. Show that the circle of curvature at the origin for the curve $$x + y = ax^2 + by^2 + cx^3$$ is $$(a + b)(x^2 + y^2) = 2x + 2y.$$

51. If a right line move in any manner in a plane, the centres of curvature of the paths described by the different points in it in any position lie on a conic.

52. If, on the tangent at each point of a curve, a constant length be measured from the point of contact, prove that the normal to the locus of the points so found passes through the corresponding centre of curvature of the given curve.

[BERTRAND.]

53. If through each point of a curve a line of given length be drawn, making a constant angle with the normal to the curve, the normal to the locus of the extremity of this line passes through the corresponding centre of curvature of the proposed curve. [BERTRAND.]

54. If on the tangent at each point of a curve a constant length c be measured from the point of contact, show that the radius of curvature of the curve locus of its extremity is given by

$$\rho' = \frac{(\rho^2 + c^2)^{\frac{3}{2}}}{\rho^2 + c^2 - c\dfrac{d\rho}{d\psi}},$$

where ρ and ψ refer to the corresponding point of the original curve.

55. If through each point of a curve a line of given length c be drawn, making a constant angle a with the normal at that point, the radius of curvature of the locus of its extremity is given by

$$\rho' = \frac{(\rho^2 + c^2 - 2\rho c \cos a)^{\frac{3}{2}}}{\rho^2 + c^2 - 2\rho c \cos a - c \sin a\dfrac{d\rho}{d\psi}},$$

where ρ and ψ refer to the corresponding point of the original curve.

56. If on each tangent to a given curve a length be measured from the point of contact equal to the radius of curvature there, the centre of curvature at any point on the locus of the extremity of the measured length is at the centre of curvature of the corresponding point of the original curve.

57. If accented letters refer to a point on a curve and unaccented letters to the corresponding point on the involute, prove

$$x = x' \mp \rho\frac{dx'}{ds'},$$

$$y = y' \mp \rho\frac{dy'}{ds'}.$$

Show how, by means of these equations and

$$s' \mp \rho = l,$$

the equation of an involute of a given curve may be found; s' being supposed known in terms of the co-ordinates of the extremities of the arc.

58. Show that the equation of the involute of the catenary

$$y = c \cosh \frac{x}{c}$$

which begins at the point where $x=0$, $y=c$,
is the Tractrix

$$x = c \cosh^{-1}\frac{c}{y} - \sqrt{c^2 - y^2}.$$

59. If a straight line be drawn through the pole perpendicular to the radius vector of a point on the equiangular spiral $r = ae^{\theta \cot \alpha}$ to meet the corresponding tangent, show that the distance between the point of intersection and the point of contact of the tangent is equal to the arc of the curve measured from the pole to the point of contact. Hence prove that the locus of this point of intersection is one of the involutes of the spiral, and show that it is an equal equiangular spiral.

60. An equiangular spiral has contact of the second order with a given curve at a given point; prove that its pole lies on a certain circle, and that, if the contact be the closest possible, the distance of the pole from the point of contact is

$$\frac{\rho}{\sqrt{1 + \left(\frac{d\rho}{ds}\right)^2}}.$$

[MATH. TRIPOS.]

61. If the tangent and normal to a curve at any point be taken as the axes of x and y respectively, and if s be the distance, measured along the arc, of a point very near to the origin, show that the Cartesian co-ordinates of that point are approximately

$$x = s - \frac{s^3}{6\rho^2} + \frac{s^4}{8\rho^3}\frac{d\rho}{ds}\ldots,$$

$$y = \frac{s^2}{2\rho} - \frac{s^3}{6\rho^2}\frac{d\rho}{ds} - \frac{s^4}{24\rho^3}\left(1 - 2\left|\frac{d\rho}{ds}\right|^2 + \rho\frac{d^2\rho}{ds^2}\right)\ldots,$$

the values of ρ, $\frac{d\rho}{ds}$, and $\frac{d^2\rho}{ds^2}$ being those at the origin.

62. If a line be drawn parallel to the common tangent of a curve and its circle of curvature, and so near to it as to intercept on the curve a small arc of length s measured from the point of contact, of the first order of small quantities, show that the distance between the two points on the same side of the common normal in which the line cuts the curve and the circle of curvature is $\frac{s^2}{6\rho}\frac{d\rho}{ds}$, *i.e.*, is of the second order of small quantities, the values of ρ and $\frac{d\rho}{ds}$ being those at the point of contact; and again, if a line be drawn parallel to the common normal, the distance between the points of intersection with the curve and the circle is $\frac{s^3}{6\rho^2}\frac{d\rho}{ds}$ and is of the third order of small quantities.

63. Prove that the circle

$$\sqrt{2}(x^2+y^2+2)=3(x+y)$$

has contact of the third order with the conic

$$5x^2-6xy+5y^2=8.$$

64. Show that for the portion of the curve

$$a^5y^2=x^7$$

very near the origin the shape of the evolute is approximately given by $$1225x^3y^2=16a^5.$$

65. A line is drawn through the origin meeting the cardioide $r=a(1-\cos\theta)$ in the points P, Q, and the normals at P and Q meet in C. Show that the radii of curvature at P and Q are proportional to PC and QC.

66. If PQ be an arc not containing a point of maximum or minimum curvature, the circles of curvature at P, Q will lie one entirely within the other.

67. If in the plane curve $\phi(x, y)=0$, we have at any point $\frac{\partial\phi}{\partial x}=0$, $\frac{\partial\phi}{\partial y}=0$, $\frac{\partial^2\phi}{\partial x^2}=0$, prove that the curvature of one of the branches of the curve which passes through that point is

$$\frac{1}{3}\frac{\partial^3\phi}{\partial x^3}\left(\frac{\partial^2\phi}{\partial x\partial y}\right)^{-1}.$$

[CAIUS COLL., CAMB.]

68. If θ be the angle between the normal at any point P of a plane curve $\phi(x, y)=0$, and the line drawn from P to the centre of the chord parallel and indefinitely near to the tangent at P, prove that

$$\cos\theta=\frac{bp^2-2hpq+aq^2}{\sqrt{p^2+q^2}\sqrt{\{(b^2+h^2)p^2-2(a+b)hpq+(a^2+h^2)p^2\}}},$$

where $p=\frac{\partial\phi}{\partial x}$, $q=\frac{\partial\phi}{\partial y}$, $a=\frac{\partial^2\phi}{\partial x^2}$, $h=\frac{\partial^2\phi}{\partial x\partial y}$, and $b=\frac{\partial^2\phi}{\partial y^2}$.

69. A curve is such that any two corresponding points of its evolute and an involute are at a constant distance. Prove that the line joining the two points is also constant in direction.

CHAPTER XI.

ENVELOPES.

302. Families of Curves.

If in the equation $\phi(x, y, c)=0$ we give any arbitrary numerical values to the constant c, we obtain a number of equations representing a certain family of curves; and any member of the family may be specified by the particular value assigned to the constant c. The quantity c, which is constant for the same curve but different for different curves, is called the *parameter* of the family.

303. Envelope. Definition.

Let all the members of the family of curves $\phi(x, y, c)=0$ be drawn which correspond to a system of infinitesimally close values of the parameter, supposed arranged in order of magnitude. We shall designate as consecutive curves any two curves which correspond to two consecutive values of c from the list. Then the locus of the ultimate points of intersection of consecutive members of this family of curves is called the ENVELOPE of the family.

304. The Envelope touches each of the Intersecting Members of the Family.

It is easy to show that the envelope touches every

curve of the system. For, let A, B, C represent three consecutive members of the family. Let P be the point of intersection of A and B, and Q that of B and C.

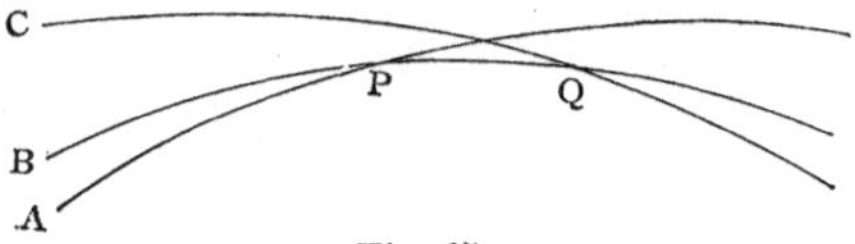

Fig. 65.

Now, by definition, P and Q are points on the envelope. Thus the curve B and the envelope have two contiguous points common, and therefore have ultimately a common tangent, and therefore touch each other. Similarly, the envelope may be shown to touch any other curve of the system.

305. **To find the Equation of an Envelope.**

To find the equation of the envelope of the family of curves of which $\phi(x, y, c)=0$ is the typical equation.

Let
$$\left.\begin{aligned}\phi(x, y, c)=0,\\ \phi(x, y, c+\delta c)=0,\end{aligned}\right\}\quad\text{(A)}$$
be two consecutive members of the family. Expanding the latter we have
$$\phi(x, y, c)+\delta c\frac{\partial}{\partial c}\phi(x, y, c)+\ldots=0.$$
Hence in the limit, when δc is infinitesimally small, we obtain
$$\frac{\partial}{\partial c}\phi(x, y, c)=0$$
as the equation of a curve passing through the ultimate point of intersection of the curves (A).

If we eliminate c between the equations
$$\phi(x, y, c)=0$$
and
$$\frac{\partial}{\partial c}\phi(x, y, c)=0$$

we obtain the locus of that point of intersection for all values of the parameter c. That is, we obtain the equations of the envelope of the family of curves of which $\phi(x, y, c)=0$ is the type.

The polar curves $\phi(r, \theta, c)$ may be treated in the same manner.

Ex. *Find the envelope of the system of straight lines of which $y=cx+\frac{a}{c}$ is the type, c being the parameter and (a) constant for all lines of the system.*

Here $$\phi(x, y, c)=y-cx-\frac{a}{c}=0,$$

and $$\frac{\partial}{\partial c}\phi(x, y, c)=-x+\frac{a}{c^2}=0,$$

therefore $$c=\pm\sqrt{\frac{a}{x}};$$

whence $$y=\pm x\sqrt{\frac{a}{x}}+\frac{a}{\pm\sqrt{\frac{a}{x}}}=\pm 2\sqrt{ax},$$

or $$y^2=4ax,$$

a parabola, which is therefore the envelope. In other words, every straight line, obtained by giving any arbitrary special value to c in the equation $y=cx+\frac{a}{c}$, touches the parabola $y^2=4ax$.

306. The Envelope of $A\lambda^2+2B\lambda+C=0$ is $B^2=AC$.

If A, B, C be any functions of x and y, and the equation of any curve be $A\lambda^2+2B\lambda+C=0$,

λ being an arbitrary parameter, the envelope of all such curves is $B^2=AC$.

For we have to eliminate λ between

$$A\lambda^2+2B\lambda+C=0$$

and $$2A\lambda+2B=0,$$

and the result is clearly $B^2=AC$.

The result of the example of Art. 305 may be obtained in this way; for the equation $$y=mx+\frac{a}{m}$$
may be written $m^2x-my+a=0$,
and therefore the envelope is $y^2=4ax$.

307. Another Mode of Establishing the Rule.

The equation $A\lambda^2+2B\lambda+C=0$ may be regarded as a quadratic equation to find the values of λ for the two particular members of the family which pass through a given point (x, y). Now, if (x, y) be supposed to be a point on the envelope, these members will be coincident. Hence for such values of x, y the quadratic for λ must have two equal roots, and the locus of such points is therefore $$B^2=AC.$$

The envelope of the system $\phi(x, y, c)=0$ might be considered in a similar manner. And it is proved in Theory of Equations that if $f(c)=0$ is a rational algebraic equation for c, the condition that it should have a pair of equal roots is obtained by eliminating c between the equations
$$f(c)=0,$$
$$f'(c)=0,$$
a result agreeing with that of Art. 305.

EXAMPLES.

1. Show that the envelope of the line $\frac{x}{a}+\frac{y}{b}=1$, where $ab=c^2$, a constant, is $4xy=c^2$.

2. Show that the envelope of the line $lx+my+1=0$, where the parameters l, m are connected by the quadratic relation
$$al^2+2hlm+bm^2+2gl+2fm+c=0,$$
is the conic $Ax^2+2Hxy+By^2+2Gx+2Fy+C=0$,

A, B, C, F, G, H being minors of the determinant $\begin{vmatrix} a, & h, & g \\ h, & b, & f \\ g, & f, & c \end{vmatrix}$.

308. Case of Two Parameters.

Next, suppose the typical equation of the family of curves to involve two parameters α, β connected by a given equation. Then two courses are open to us. We may *eliminate one of the parameters* by means of the connecting equation and thus reduce the problem to that solved in Art. 305, *or*, as is frequently better from considerations of symmetry, *consider one of the parameters capable of independent variation and the other dependent upon it.* We then proceed as follows.

Let $$\phi(x, y, \alpha, \beta)=0 \text{ ....(1)}$$
be the typical equation of the curves whose envelope is to be investigated, and $f(\alpha, \beta)=0$ ....(2)
the relation connecting α and β.

Then, supposing α the independent parameter, we have

$$\frac{\partial\phi}{\partial\alpha}+\frac{\partial\phi}{\partial\beta}\cdot\frac{d\beta}{d\alpha}=0, \text{....................(3)}$$

where $$\frac{\partial f}{\partial\alpha}+\frac{\partial f}{\partial\beta}\cdot\frac{d\beta}{d\alpha}=0. \text{....................(4)}$$

We thus have four equations and three quantities to eliminate, viz., α, β, $\frac{d\beta}{d\alpha}$. The result of elimination is the equation of the envelope.

The parameters α, β, connected by the relation $f(\alpha, \beta)=0$, may be regarded as the co-ordinates of a parametric point which lies on the curve $f(x, y)=0$.

309. Indeterminate Multipliers.

The equations (3) and (4) may be written

$$\frac{\partial\phi}{\partial\alpha}d\alpha+\frac{\partial\phi}{\partial\beta}d\beta=0 \text{ (Art. 139)},$$

$$\frac{\partial f}{\partial\alpha}d\alpha+\frac{\partial f}{\partial\beta}d\beta=0.$$

The result of eliminating $d\alpha$, $d\beta$ between these equations

is
$$\frac{\dfrac{\partial\phi}{\partial\alpha}}{\dfrac{\partial f}{\partial\alpha}}=\frac{\dfrac{\partial\phi}{\partial\beta}}{\dfrac{\partial f}{\partial\beta}}.$$

Call each of these ratios λ. We then have

$$\frac{\partial\phi}{\partial\alpha}=\lambda\frac{\partial f}{\partial\alpha}, \qquad\text{..............(5)}$$

$$\frac{\partial\phi}{\partial\beta}=\lambda\frac{\partial f}{\partial\beta}. \qquad\text{..............(6)}$$

This quantity λ is called an "Indeterminate Multiplier."

It remains to eliminate α, β, and λ between equations (1), (2), (5), and (6).

This method is peculiarly adapted to the case in which

$$\phi(x, y, \alpha, \beta)\equiv\phi_1(x, y, \alpha, \beta)-a_1=0,$$

and
$$f(\alpha, \beta)\equiv f_1(\alpha, \beta)-a_2=0,$$

where ϕ_1 and f_1 are homogeneous in α and β, and of the pth and q^{th} degrees respectively, a_1 and a_2 being absolute constants.

Multiply equation (5) by α and (6) by β, and add. Then by Euler's Theorem

$$pa_1=qa_2\lambda,$$

so that in such cases λ is easily found.

Ex. *Find the envelope of* $\frac{x}{a}+\frac{y}{b}=1$, *where a and b are connected by the relation*
$$a^2+b^2=c^2,$$
c being an absolute constant; i.e., the envelope of a line of constant length which slides with its extremities upon two fixed rods at right angles to each other.

Here
$$\frac{x}{a^2}da+\frac{y}{b^2}db=0,$$
$$a\,da+b\,db=0,$$

and therefore
$$\frac{x}{a^2}=\lambda a,$$

$$\frac{y}{b^2}=\lambda b.$$

Multiplying by a and b respectively, and adding,

$$\frac{x}{a}+\frac{y}{b}=\lambda(a^2+b^2),$$

or $$1=\lambda c^2.$$

Hence $$\left.\begin{aligned}a^3&=c^2x,\\ b^3&=c^2y,\end{aligned}\right\}$$

and since $$a^2+b^2=c^2$$

we have $$(c^2x)^{\frac{2}{3}}+(c^2y)^{\frac{2}{3}}=c^2,$$

or $$x^{\frac{2}{3}}+y^{\frac{2}{3}}=c^{\frac{2}{3}}.$$

310. **Case of Three Parameters connected by Two Equations.**

Next, suppose the equation of a curve to contain *three* parameters connected by *two* equations.

Let the equation of the curve be

$$\phi(x, y, \alpha, \beta, \gamma)=0, \quad\text{..............(1)}$$

and let

$$\left.\begin{aligned}f_1(\alpha, \beta, \gamma)&=0,\\ f_2(\alpha, \beta, \gamma)&=0,\end{aligned}\right\}\quad\begin{aligned}&\text{...........(2)}\\ &\text{...........(3)}\end{aligned}$$

be the two connecting equations. Then we have

$$\frac{\partial\phi}{\partial\alpha}d\alpha+\frac{\partial\phi}{\partial\beta}d\beta+\frac{\partial\phi}{\partial\gamma}d\gamma=0, \quad\text{..............(4)}$$

$$\frac{\partial f_1}{\partial\alpha}d\alpha+\frac{\partial f_1}{\partial\beta}d\beta+\frac{\partial f_1}{\partial\gamma}d\gamma=0, \quad\text{..............(5)}$$

$$\frac{\partial f_2}{\partial\alpha}d\alpha+\frac{\partial f_2}{\partial\beta}d\beta+\frac{\partial f_2}{\partial\gamma}d\gamma=0. \quad\text{..............(6)}$$

The result of eliminating $d\alpha$, $d\beta$, $d\gamma$ between these three equations is

$$\begin{vmatrix}\frac{\partial\phi}{\partial\alpha}, & \frac{\partial\phi}{\partial\beta}, & \frac{\partial\phi}{\partial\gamma}\\ \frac{\partial f_1}{\partial\alpha}, & \frac{\partial f_1}{\partial\beta}, & \frac{\partial f_1}{\partial\gamma}\\ \frac{\partial f_2}{\partial\alpha}, & \frac{\partial f_2}{\partial\beta}, & \frac{\partial f_2}{\partial\gamma}\end{vmatrix}=0. \quad\text{..............(7)}$$

If α, β, γ be eliminated between the four equations (1), (2), (3) and (7), the result will be the equation of the envelope.

It is to be noted that the same determinant would arise from the elimination of the "indeterminate multipliers" λ_1 and λ_2 from the equations

$$\frac{\partial\phi}{\partial\alpha}+\lambda_1\frac{\partial f_1}{\partial\alpha}+\lambda_2\frac{\partial f_2}{\partial\alpha}=0, \quad\text{.............}(8)$$

$$\frac{\partial\phi}{\partial\beta}+\lambda_1\frac{\partial f_1}{\partial\beta}+\lambda_2\frac{\partial f_2}{\partial\beta}=0, \quad\text{.............}(9)$$

$$\frac{\partial\phi}{\partial\gamma}+\lambda_1\frac{\partial f_1}{\partial\gamma}+\lambda_2\frac{\partial f_2}{\partial\gamma}=0, \quad\text{...........}(10)$$

and it is often advantageous to use these latter equations in place of (4), (5), (6), involving $d\alpha$, $d\beta$, $d\gamma$.

The result of eliminating $\alpha, \beta, \gamma, \lambda_1, \lambda_2$ between the six equations (1), (2), (3), (8), (9), (10) will then be the equation to the envelope.

311. The *general* investigation of the envelope of a curve whose equation contains r parameters connected by $r-1$ equations proceeds in exactly the same way, and is the result of the elimination of the r parameters and $r-1$ indeterminate multipliers between $2r$ equations.

312. **Converse Problem. Given the Family and the Envelope to find the relation between the Parameters.**

Suppose we are given the equation of a curve

$$\phi(x, y, \alpha, \beta)=0 \quad\text{..................}\quad (1)$$

containing two parameters. Suppose also the envelope given, viz.,

$$F(x, y)=0. \quad\text{.......................}\quad (2)$$

Required the relation between α and β.

Eliminate y between (1) and (2). We obtain an equation of the form $f(x, \alpha, \beta)=0$, ... (3)

giving the abscissa of the point of contact of the curve with its envelope. Since the curve touches its envelope, equation (3) must also be true for a contiguous value of x, viz., $x+\delta x$ (unless the tangent at the point of contact be parallel to the axis of y, in which case we could have eliminated x between (1) and (2) and proceeded in the same way with y). Hence

$$\left.\begin{aligned} f(x, a, b)&=0, \quad \text{..................} \quad (4)\\ f(x+\delta x, a, b)&=0. \quad \text{..................} \quad (5)\end{aligned}\right\}$$

The latter may be expanded in powers of δx, when it becomes

$$f(x, a, b)+\frac{\partial f}{\partial x}\delta x+\ldots=0, \quad \text{..............} \quad (6)$$

and therefore in the limit

$$\frac{\partial f}{\partial x}=0. \quad \text{.........................} \quad (7)$$

If, then, x be eliminated between

$$f(x, \alpha, \beta)=0,$$

$$\frac{\partial}{\partial x}f(x, \alpha, \beta)=0,$$

we obtain the relation sought.

It will be observed that this is precisely the same process as finding the envelope of

$$\phi(x, y, \alpha, \beta)=0,$$

considering α, β *as the current co-ordinates and* x, y *as parameters connected by the relation*

$$F(x, y)=0.$$

Ex. *Given that* $x^{\frac{2}{3}}+y^{\frac{2}{3}}=c^{\frac{2}{3}}$ *is the envelope of* $\frac{x}{a}+\frac{y}{b}=1$, *find the necessary relation between* a *and* b.

We have
$$\frac{dx}{x^{\frac{1}{3}}}+\frac{dy}{y^{\frac{1}{3}}}=0,$$

$$\frac{dx}{a}+\frac{dy}{b}=0;$$

therefore
$$x^{\frac{1}{3}}=\lambda a,$$

$$y^{\frac{1}{3}}=\lambda b.$$

Hence $$\frac{x}{a}=\lambda x^{\frac{2}{3}},\ \frac{y}{b}=\lambda y^{\frac{2}{3}},$$

and by addition $$1=\lambda c^{\frac{2}{3}}.$$

This gives $$a=c^{\frac{2}{3}}x^{\frac{1}{3}},\ b=c^{\frac{2}{3}}y^{\frac{1}{3}},$$

and by squaring and adding

$$a^2+b^2=c^2,$$

the relation required. (See Ex., Art. 309.)

313. Evolutes considered as Envelopes.

The evolute of a curve has been defined as the locus of the centre of curvature, and it has been shown (Art. 287) that the centre of curvature is the ultimate point of intersection of two consecutive normals. Hence the evolute is the *envelope of the normals* to a curve. It is from this point of view that the equation of the evolute of a given curve is in general most easily obtained.

Ex. *To find the evolute of the ellipse* $\frac{x^2}{a^2}+\frac{y^2}{b^2}=1$.

The equation of the normal at the point whose eccentric angle is ϕ is
$$\frac{ax}{\cos\phi}-\frac{by}{\sin\phi}=a^2-b^2. \quad\text{.........................}\quad (1)$$

We have to find the envelope of this line for different values of the parameter ϕ.

Differentiating with regard to ϕ,

$$ax\frac{\sin\phi}{\cos^2\phi}+by\frac{\cos\phi}{\sin^2\phi}=0, \quad\text{..............................}\quad (2)$$

or $$\frac{\sin^3\phi}{by}+\frac{\cos^3\phi}{ax}=0.$$

Hence $$\frac{\sin\phi}{-\sqrt[3]{by}}=\frac{\cos\phi}{\sqrt[3]{ax}}=\frac{1}{\sqrt{(ax)^{\frac{2}{3}}+(by)^{\frac{2}{3}}}}. \quad\text{............}\quad (3)$$

Substituting these values of $\sin\phi$ and $\cos\phi$ in equation (1) we obtain, after reduction, $$(ax)^{\frac{2}{3}}+(by)^{\frac{2}{3}}=(a^2-b^2)^{\frac{2}{3}}.$$

314. Pedal Curves as Envelopes.

It has already been pointed out (Art. 197) that if circles be described on radii vectores of a given curve as diameters they all touch the *first positive pedal* of the curve with regard to the origin. It is obvious, therefore, that the problem of finding the first positive pedal of a given curve is identical with that of finding the *envelope of circles described on the radii vectores as diameters.*

Again, the *first negative pedal* is the envelope of a *straight line drawn through any point of the curve and at right angles to the radius vector to the point.*

Ex. 1. *Find the first positive pedal of the circle* $r=2a\cos\theta$ *with regard to the origin.*

Let d, α be the polar co-ordinates of any point on the circle, then

$$d=2a\cos\alpha.$$

Again, the equation of a circle on the radius vector d for diameter is

$$r=d\cos(\theta-\alpha), \quad\text{.................................}\quad (1)$$

or

$$r=2a\cos\alpha\cos(\theta-\alpha). \quad\text{.........................}\quad (2)$$

Here α is the parameter.

Differentiating with regard to α,

$$-\sin\alpha\cos(\theta-\alpha)+\cos\alpha\sin(\theta-\alpha)=0,$$

whence

$$\sin(\theta-2\alpha)=0,$$

or

$$\alpha=\frac{\theta}{2}. \quad\text{.....................}\quad (3)$$

Substituting this value of α in equation (2)

$$r=2a\cos^2\frac{\theta}{2},$$

or

$$r=a(1+\cos\theta),$$

the equation of a cardioide.

Ex. 2. *Find the equation of the first negative pedal of the cardioide* $r=a(1+\cos\theta)$ *with regard to the origin.*

Here we have to find the envelope of the line

$$x\cos\alpha+y\sin\alpha=d,$$

where d, α are the polar co-ordinates of any point on the cardioide; *i.e.*, where

$$d=a(1+\cos\alpha).$$

The equation of the line is therefore

$$x \cos \alpha + y \sin \alpha = a(1 + \cos \alpha),$$

or

$$(x - a) \cos \alpha + y \sin \alpha = a,$$

a line which, from its form, is easily seen to be a tangent to

$$(x - a)^2 + y^2 = a^2,$$

or

$$r = 2a \cos \theta,$$

which is therefore its envelope.

EXAMPLES.

1. Find the equation of the curve whose tangent is of the form $y = mx + m^4$, m being independent of x and y.

2. Find the envelope of the curves

$$\frac{a^2 \cos \theta}{x} - \frac{b^2 \sin \theta}{y} = \frac{c^2}{a}$$

for different values of θ.

3. Find the envelope of the family of trajectories

$$y = x \tan \theta - \tfrac{1}{2} g \frac{x^2}{u^2 \cos^2 \theta},$$

θ being the arbitrary parameter.

4. Find the envelopes of straight lines drawn at right angles to tangents to a given parabola and passing through the points in which those tangents cut

(1) the axis of the parabola,
(2) a fixed line parallel to the directrix.

5. Find the envelope of straight lines drawn at right angles to normals to a given parabola and passing through the points in which those normals cut the axis of the parabola.

6. A series of circles have their centres on a given straight line, and their radii are proportional to the distances of their corresponding centres from a given point in that line. Find the envelope.

7. Find the envelopes of the line

$$\frac{x}{a}+\frac{y}{b}=1$$

under the following conditions :—

$$(1)\quad a+b=k,$$

$$(2)\quad a^n+b^n=k^n,$$

$$(3)\quad a^m b^n=k^{m+n},$$

k being a constant in each case.

8. P is a point which moves along a given straight line. PM, PN are perpendiculars on the co-ordinate axes supposed rectangular. Find the envelope of the line MN.

9. A straight line has its extremities on two fixed straight lines and forms with them a triangle of constant area. Find its envelope.

10. Find the envelope of the line $y=mx-2am-am^3$ for different values of m; *i.e.*, find the equation of the evolute of the parabola $y^2=4ax$.

11. Show that the envelope of the family of curves

$$A\lambda^3+3B\lambda^2+3C\lambda+D=0,$$

where λ is the arbitrary parameter and A, B, C, D are functions of x and y, is

$$(BC-AD)^2=4(BD-C^2)(AC-B^2).$$

12. Show that the envelope of the family of curves

$$A\cos^n\theta+B\sin^n\theta=C,$$

where θ is the arbitrary parameter and A, B, C are functions of x and y, is

$$A^{\frac{2}{2-n}}+B^{\frac{2}{2-n}}=C^{\frac{2}{2-n}}.$$

13. Show that the envelope of the lines whose equations are

$$x\sec^2\theta+y\,\mathrm{cosec}^2\theta=c$$

is a parabola touching the axes of co-ordinates.

14. Find the envelopes of the systems of coaxial ellipses whose semiaxes a and b are connected by the equations

$$(1) \qquad a+b=k,$$
$$(2) \qquad \sqrt{a}+\sqrt{b}=\sqrt{k},$$
$$(3) \qquad a^m+b^m=k^m,$$
$$(4) \qquad ab=k^2,$$

k being a constant in each case.

15. Find the envelopes of the parabolas which touch the co-ordinate axes and are such that the distances (α, β) from the origin to the points of contact are connected by the relations

$$(1) \qquad \alpha+\beta=k,$$
$$(2) \qquad \alpha^m+\beta^m=k^m,$$
$$(3) \qquad \alpha\beta=k^2,$$

k being a constant in each case.

16. Show that the system of conics obtained by varying λ in the equation

$$\frac{x^2}{a^2}+2\lambda\frac{xy}{ab}+\frac{y^2}{b^2}=1-\lambda^2$$

have for their envelope the parallelogram whose sides are

$$x=\pm a, \quad y=\pm b.$$

17. Find the envelope of the line which joins the feet of the two perpendiculars from any point of a circle upon a given pair of perpendicular diameters.

18. Show that the envelope of straight lines which join the extremities of a pair of conjugate diameters of an ellipse is a similar ellipse.

19. Show that if PM, PN be perpendiculars from any point P of the curve $y=mx^3$ upon the axes the envelope of MN is

$$27y+4mx^3=0.$$

20. Find the envelope of circles described on the radii vectores of an ellipse drawn from the centre as diameters.

21. Show that the envelope of a circle whose centre lies on the parabola $y^2 = 4ax$ and which passes through its vertex is

$$2ay^2 + x(x^2 + y^2) = 0.$$

22. Show that the envelope of a circle whose centre lies on the parabola $y^2 = 4ax$ and whose radius = the abscissa of the centre is made up of the tangent at the vertex and a circle with centre at the focus.

23. If a lamina rotate in its own plane about any fixed point in that plane, show that the directions of motion at any instant of any given curve of points in the lamina have for their envelope the first negative pedal of that curve with regard to the fixed point.

Examine the particular cases of a straight line and a circle.

24. Two particles move along parallel straight lines, the one with uniform velocity and the other with the same initial velocity but with uniform acceleration. Show that the line joining them always touches a fixed hyperbola.

25. A series of circles is described having their centres on an equilateral hyperbola and passing through its centre. Show that the locus of their ultimate points of intersection is a lemniscate.

26. Prove that the equation of the normal to the curve

$$x^{\frac{2}{3}} + y^{\frac{2}{3}} = a^{\frac{2}{3}}$$

may be written in the form

$$y \cos \phi - x \sin \phi = a \cos 2\phi.$$

Hence show that the evolute of the curve is

$$(x+y)^{\frac{2}{3}} + (x-y)^{\frac{2}{3}} = 2a^{\frac{2}{3}}.$$

27. Show that the envelope of the lines

$$x \cos m\alpha + y \sin m\alpha = a(\cos n\alpha)^{\frac{m}{n}},$$

where α is the arbitrary parameter, is

$$r^{\frac{n}{m-n}} = a^{\frac{n}{m-n}} \cos \frac{n}{m-n}\theta.$$

28. If O be the pole and P any point of the curve

$$r = a \cos m\theta,$$

and if with O for pole and P for vertex a similar curve be described, the envelope of all such curves is

$$r^{\frac{1}{2}} = a^{\frac{1}{2}} \cos \frac{m\theta}{2}.$$

29. If O be the pole and P any point of the curve

$$r^m = a^m \cos m\theta,$$

and if with O for pole and P for vertex a curve similar to

$$r^n = a^n \cos n\theta$$

be described, the envelope of all such curves is

$$r^{\frac{mn}{m+n}} = a^{\frac{mn}{m+n}} \cos \frac{mn}{m+n} \theta.$$

30. If O be the pole and Y the foot of the perpendicular from O on any tangent to the curve

$$r^m = a^m \cos m\theta,$$

and if with O for pole and Y for vertex a curve similar to

$$r^n = a^n \cos n\theta$$

be described, the envelope of all such curves is

$$r^p = a^p \cos p\theta, \text{ where } p = \frac{mn}{m+n+mn}.$$

31. If a point on the circumference of a given circle be taken as pole, and circles be described on radii vectores of the given circle as diameters, the envelope of these circles is a cardioide.

32. Show that the envelope of all cardioides on radii vectores of the circle $r = a \cos \theta$ for axes, and having their cusps at the pole, is $$r^{\frac{1}{3}} = a^{\frac{1}{3}} \cos \tfrac{1}{3}\theta.$$

33. Show that the envelope of all cardioides described on radii vectores of the cardioide $r = a(1 + \cos \theta)$ for axes, and having their cusps at the pole, is

$$r^{\frac{1}{4}} = (2a)^{\frac{1}{4}} \cos \frac{\theta}{4}.$$

34. On radii vectores of $r^{2n}=a^{2n}\cos 2n\theta$ as axes curves similar to it are described, the curves being all concentric. Show that the envelope of all these is

$$r^n=a^n\cos n\theta.$$

35. Prove that the pedal equation of the envelope of the line

$$x\cos 2\theta+y\sin 2\theta=2a\cos\theta$$

is

$$p^2=\tfrac{4}{3}(r^2-a^2).$$

36. Prove that the pedal equation of the envelope of the line

$$x\cos m\theta+y\sin m\theta=a\cos n\theta$$

is

$$m^2r^2=(m^2-n^2)p^2+n^2a^2.$$

37. Two central radii vectores of a circle of radius a rotate from coincidence in a given initial position with uniform angular velocities ω and ω'. Show that the pedal equation of the envelope of a line joining their extremities is

$$(\omega+\omega')^2r^2=4\omega\omega'p^2+(\omega-\omega')^2a^2.$$

38. The envelope of polars with respect to the circle

$$x^2+y^2=2ax$$

of points which lie on the circle

$$x^2+y^2=2bx$$

is

$$\{(a-b)x+ab\}^2=b^2\{(x-a)^2+y^2\}.$$

39. A square slides with two of its adjacent sides passing through fixed points. Show that its remaining sides touch a pair of fixed circles, one diagonal passes through a fixed point, and that the envelope of the other is a circle.

40. An equilateral triangle moves so that two of its sides pass through two fixed points. Prove that the envelope of the third side is a circle.

41. Prove that the envelope of the circles obtained by varying the arbitrary parameter a in the equation

$$c^2(y-a)^2+(cx-a^2)^2=(a^2+c^2)^2$$

consists of a straight line and a circle.

42. Two points are taken on an ellipse on the same side of the major axis and such that the sum of their abscissae is equal to the semi-major axis. Show that the line joining them envelopes a parabola which goes through the extremities of the minor axis and whose latus rectum is equal to that of the ellipse.

43. Given the centre and directrices of an ellipse, show that the envelope of the normals at the ends of the latera recta is

$$27y^4 \pm 256cx^3 = 0.$$

44. Prove that the envelope of a circle which passes through a fixed point F and subtends a constant angle at another fixed point F' is a limaçon.

45. Find the envelope of a parabola of which the directrix and one point are given.

46. Find the condition between a and b that the envelope of the line

$$\frac{x}{a} + \frac{y}{b} = 1$$

may be the curve

$$x^p y^q = k^{p+q}.$$

47. S is a fixed point, and with any point P of a curve for centre and with radius $PS + k$ a circle is described. Show that the envelopes for different values of k consist of two sets of parallel curves, one set being circles; and find what the original curve must be that both sets may be circles.

48. Rays emanate from a luminous point O and are reflected at a plane curve. OY is the perpendicular from O on the tangent at any point P, and OY is produced to a point Q, such that $YQ = OY$. Show that the caustic curve is the evolute of the locus of Q. Show that the caustic curve may also be regarded as the evolute of the envelope of a circle whose centre is P and radius OP.

[If a ray of light in the plane of a given bright curve be incident upon the curve the reflected ray and the incident ray make equal angles with the normal to the curve at the point of incidence, and the reflected ray lies in the plane of the curve. If a given system of rays be incident upon the curve, the envelope of the reflected rays is called the caustic by reflection.]

49. Parallel rays are incident on a bright semicircular wire (radius a) and in its plane. Show that the caustic curve is the epicycloid formed by a point attached to a circle of radius $\frac{a}{4}$ rolling upon the circumference of a circle of radius $\frac{a}{2}$.

50. Rays emanate from a point on the circumference of a reflecting circular arc. Show that the caustic after reflection is a cardioide.

51. Show that if rays emanate from the pole of an equiangular spiral and are reflected by the curve the caustic is a similar equiangular spiral.

CHAPTER XII.

CURVE TRACING.

315. Nature of the Problem. Cartesian Equations.

If, in the Cartesian equation of any algebraic curve, various values of x be assigned, we obtain a number of equations whose roots give the corresponding values of the ordinates. The real roots of these equations can always be either found exactly or approximated closely to by methods explained in the Theory of Equations. We can by this means, laborious though it will in most cases be, find as many points as we like which satisfy the given equation of the curve; and by joining these points by a curved line drawn freely through them we can form a fairly good idea as to its shape. The experience, however, which we have gained in previous chapters will in general obviate any necessity of resort to the usually tedious process of approximating to the roots of equations of high degree; and we propose to give a list of suggestions for guidance in curve tracing which in most cases will enable us to form, without much difficulty, a sufficiently exact notion of the character of the curve represented by any specified equation.

316. Order of Procedure.

1. A glance will suffice to detect *symmetry* in a curve. If no odd powers of y occur, the curve is symmetrical with respect to the axis of x. Similarly for symmetry about the axis of y. If all the powers of both x and y which occur be even, the curve is symmetrical about both axes, as, for instance, in the case of the ellipse $\frac{x^2}{a^2}+\frac{y^2}{b^2}=1$. Again, if on changing the signs of x and y the equation of the curve remain unchanged, there is symmetry in opposite quadrants, as in the case of the hyperbola $xy=k^2$. The origin is then said to be a centre of the curve.

If the curve be not symmetrical with regard to either axis, consider whether any obvious transformation of co-ordinates could make it so.

2. Notice whether the curve passes through the origin; also the points *where it crosses the co-ordinate axes;* or, in fact, any points whose co-ordinates present themselves as obviously satisfying the equation to the curve.

3. *What asymptotes are there?* First find those parallel to the co-ordinate axes; next, the oblique ones (Art. 210). These results point out in what directions the curve extends to infinity.

Find also *on which side of each asymptote the curve lies* (Art. 232).

4. If the curve pass through the origin, equate to zero the terms of lowest degree. These terms will give the *tangent or tangents at the origin* (Art. 254), and thus tell the direction in which the curve passes through the origin. A more complete method of finding the *shape of the curve near to and at a great distance from the origin* is to follow in Art. 320.

5. If there be a *node, cusp,* or *conjugate point* at the origin, or a multiple point of higher order than the second, take note of the fact. If there be a cusp, *test its species* (Art. 258).

6. Find *what other multiple points* the curve has (Art. 257), and ascertain the position and character of each.

7. *Find* $\frac{dy}{dx}$; and for what points it *vanishes* or *becomes infinite.* These results will indicate the points at which the tangent is parallel or perpendicular to the axis of x. The direction of the tangent at other points may also be ascertained if desirable.

8. Find, if convenient, the *points of inflexion.*

9. A straight line will cut a curve of the nth degree *in n points* real or imaginary, and imaginary intersections *occur in pairs.* These facts are often useful in detecting a false notion of the shape of a curve.

10. If we can solve the equation for one of the variables, say y, in terms of the other, x, it will be frequently found that radicals occur in the solution, and that the range of admissible values of x which give real values for y *is thereby limited.* The *existence of loops* upon a curve is frequently detected thus.

11. It sometimes happens that the equation is much simplified upon *reduction to the polar form.* This is especially the case when the origin is a multiple point on the curve.

317. It is not necessary of course in every case to take all the steps indicated above, or to keep to the order laid down, but the student is advised in any curve he may attempt to trace to note down the result of each inves-

tigation he may make. For instance, he should remark the absence just as much as the existence of symmetry, asymptotes, or singular points, and the total information gained will generally be sufficient to give a tolerably good diagram of the curve.

318. We add a few examples to illustrate the points enumerated.

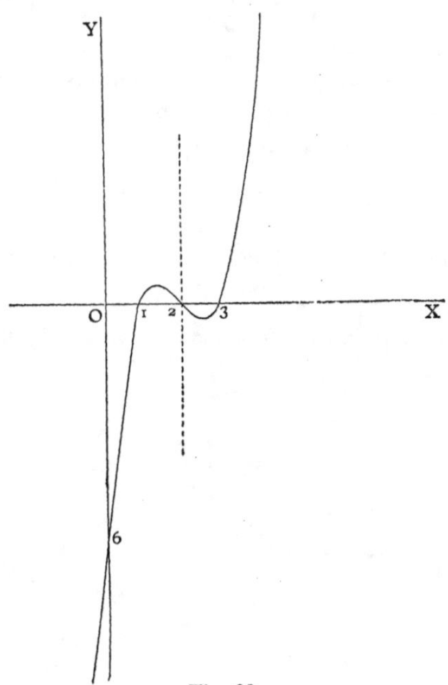

Fig. 66.

I. *To trace the curve* $y=(x-1)(x-2)(x-3)$.

(*a*) This curve is not symmetrical about either axis; but if the origin be transferred to the point (2, 0) the equation becomes

$$y=x(x^2-1),$$

showing symmetry in opposite quadrants when referred to the new axes, and that the tangent at the new origin is inclined at an angle 135° to the axis of x.

(β) Recurring to the original equation,

If $\quad y=0, \quad x=1,\ 2,$ or 3;
If $\quad x=0, \quad y=-6$;
If $\quad x=\infty, \quad y=\infty$;
If $\quad x=-\infty, \quad y=-\infty$.
When $\quad x$ is >3 $\quad y$ is positive,
$x<3$ but >2 $\quad y$ is negative,
$x<2$ but >1 $\quad y$ is positive,
$x<1$ $\quad y$ is negative.

(γ) The curve does not go through the origin, and, although extending to infinity, it has no rectilineal asymptote.

(δ) Since $\quad y=x^3-6x^2+11x-6$

we have
$$\frac{dy}{dx}=3x^2-12x+11,$$

which vanishes when $\quad x=2\pm\frac{1}{\sqrt{3}}$.

(ϵ) Also $\frac{d^2y}{dx^2}=6(x-2)$, which shows that there is a point of inflexion at the point where $x=2$.

The shape of the curve is therefore that shown in Fig. 66.

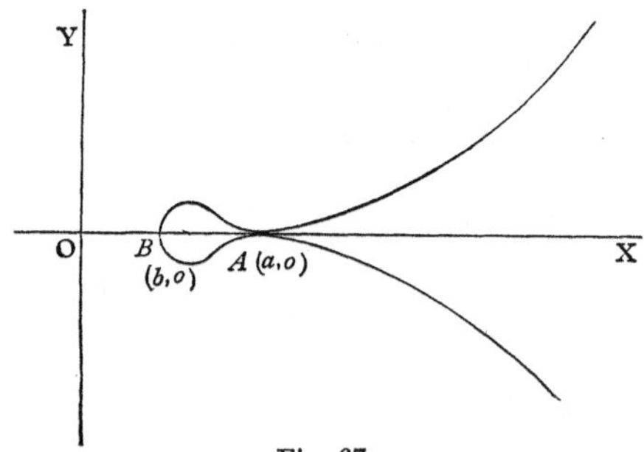

Fig. 67.

II. *To trace the curve*

$$y=\pm\frac{(x-a)^2\sqrt{x-b}}{a^{\frac{3}{2}}}.$$

Case 1. Suppose $\quad a>b$ (Fig 67).

(α) The curve is symmetrical with regard to the axis of x.

(β) While $x < b$, y is imaginary,
and y is real for all values of x from b to ∞, and the curve meets the axis of x when $x=a$ and when $x=b$.

(γ) $\frac{dy}{dx}=0$ when $x=a$, and $=\infty$ when $x=b$, so that the curve touches the axis x at the point $(a, 0)$, and cuts it at right angles at $(b, 0)$.

(δ) There is no asymptote; but, when $x=\infty$, y and $\frac{dy}{dx}$ are both ∞ in the limit, the curve ultimately taking the shape of

$$y=\pm\frac{x^{\frac{5}{2}}}{a^{\frac{3}{2}}}.$$

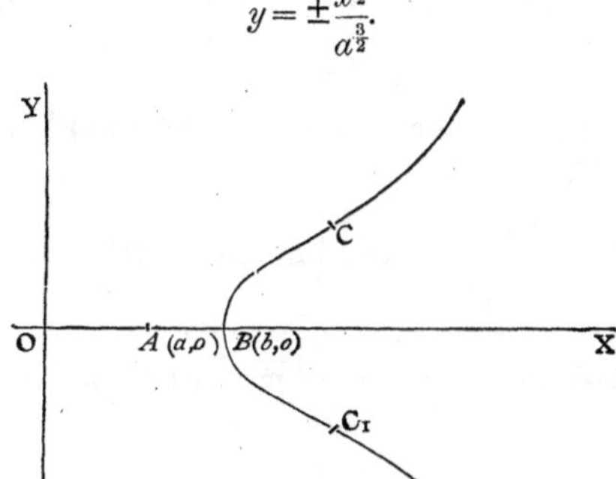

Fig. 68.

CASE 2. Next consider $a < b$ (Fig. 68).

(α') There is in this case also symmetry about the axis of x.

(β') The equation to the curve is satisfied by the point $(a, 0)$, but by no other point in its vicinity, for if x be $< b$, y is imaginary except when $x=a$. The point $(a, 0)$ is therefore a conjugate point.

(γ') Moreover $\frac{dy}{dx}=\infty$ when $x=b$, and the curve cuts the axis of x at right angles at this point.

(δ') Also, when $x=\infty, \frac{dy}{dx}=\infty$; so the curve in departing from $(b, 0)$ (the point B in Fig. 68) must bend towards the positive direction of the axis of x, and, finally, $\frac{dy}{dx}$ again becomes infinite, showing that there must be a point of inflexion at some point C between B and ∞. Its exact position is of course given by the equation

$$\frac{d^2y}{dx^2}=0.$$

The shapes of the curves in the two cases are given in Figs. 67 and 68 respectively.

EXAMPLES.

1. Trace the curve $\qquad y=x^2(x-1),$
showing that its tangent is parallel to the axis of x at the origin and at the point $x=\frac{2}{3}$.

2. Trace $\qquad y=x^3,$
and show that there is a point of inflexion at the origin. This curve is called the *cubical parabola.*

3. Trace the curve $\qquad y^2=x^3,$
showing that there is a cusp of the first kind at the origin. This curve is called a *semicubical parabola.*

4. Trace the curve $\quad ay^2=(x-a)(x-b)(x-c),$
where a, b, c are in descending order of magnitude, and examine the cases
(1) $a=b$.
(2) $b=c$.
(3) $a=b=c$.

III. *To trace the curve*

$$y=\frac{x^3+ax^2+a^3}{x^2-a^2},$$

a being positive.

α. There is no symmetry about either axis and the curve does not pass through the origin.

β. The curve cuts the axis of y at the point $(0, -a)$ and the axis of x at the point given by the real root of

$$x^3+ax^2+a^3=0.$$

(It is clear that two roots of this equation are imaginary, for the sum of the squares of the reciprocals of its roots is negative.) Also, the real root is obviously negative and numerically greater than a.

γ. When x is $>a$, $\qquad y$ is positive.
When x lies between a and $-a$, y is negative.

When x is $<-a$, y is positive until x passes the negative root above referred to, and then is negative afterwards.

δ. The asymptotes parallel to the axes are $x=\pm a$. To find the oblique asymptote

$$y=x\frac{1+\frac{a}{x}+\frac{a^3}{x^3}}{1-\frac{a^2}{x^2}}=x\left(1+\frac{a}{x}+\frac{a^3}{x^3}\right)\left(1+\frac{a^2}{x^2}+\ldots\right),$$

or
$$y=x+a+\frac{a^2}{x}+\ldots.$$

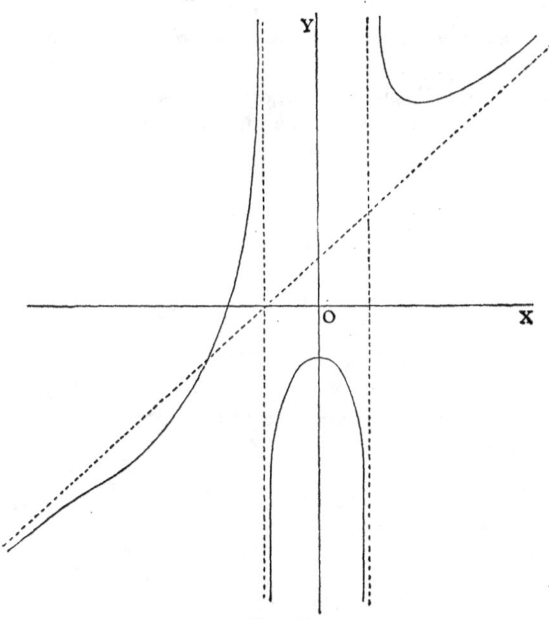

Fig. 69.

Hence $y=x+a$ is the oblique asymptote, and, if x be positive, the ordinate of the curve is obviously greater than that of the asymptote, and the curve lies above the oblique asymptote. If x be negative, the curve lies below it.

Y

$$\frac{dy}{dx}=\frac{x(x^3-3a^2x-4a^3)}{(x^2-a^2)^2},$$

which gives $\frac{dy}{dx}=0$, when $x=0$ or when $x^3-3a^2x-4a^3=0$, which clearly has a positive root lying between $x=2a$ and $x=3a$, and which can be shown to have only this one real root. Also, $\frac{dy}{dx}=\infty$ only when $x=\pm a$.

ζ. A point of inflexion lies between $x=-5a$ and $x=-6a$ (Ex. 26, Chap. X.).

The shape is therefore that given in Fig. 69.

IV. *To trace the curve* $y^2+2x^3y+x^7=0$.

α. The curve is not symmetrical about either axis and there are no asymptotes.

β. The curve passes through the origin, but cuts neither axis again.

γ. There is a cusp at the origin, the equation of the tangent being

$$y=0.$$

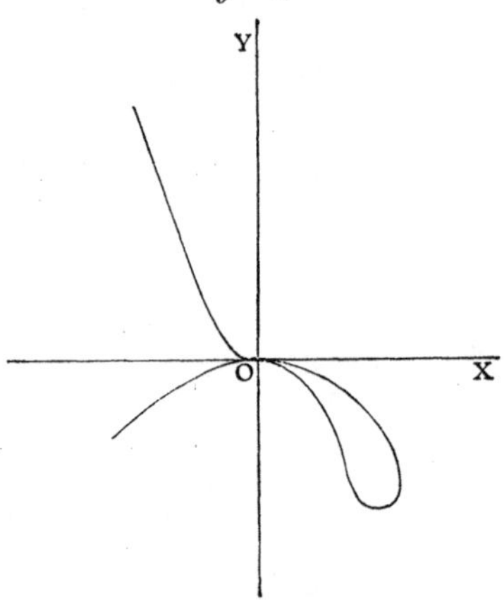

Fig. 70.

Proceeding according to Art. 258 the quadratic for P is

$$P^2+2Px^3+x^7=0,$$

an equation whose roots are real if x be very small, positive or negative; for the criterion for real roots is that x^6-x^7 should be >0.

This condition is fulfilled until x is > 1, when P or y becomes imaginary.

Moreover, the product of the roots $= x^7$ and is positive or negative according as x is positive or negative. There is therefore a double cusp at the origin, and on the positive side of the axis of y it is of the second species, while on the negative side it is of the first species. The point is therefore a point of oscul-inflexion (Fig. 47).

δ. $$y = -x^3 \pm x^3 \sqrt{1-x},$$

so that $\frac{dy}{dx} = \infty$ if $x = 1$. Also, one value of $\frac{dy}{dx}$ is zero when $x = \frac{48}{49}$. The shape of the curve is now readily seen to be that shown in Fig. 70.

319. The following curve illustrates a particular artifice which may be occasionally employed, namely to express the ordinate of the curve as the sum or difference of the ordinates of two known or easily traceable curves.

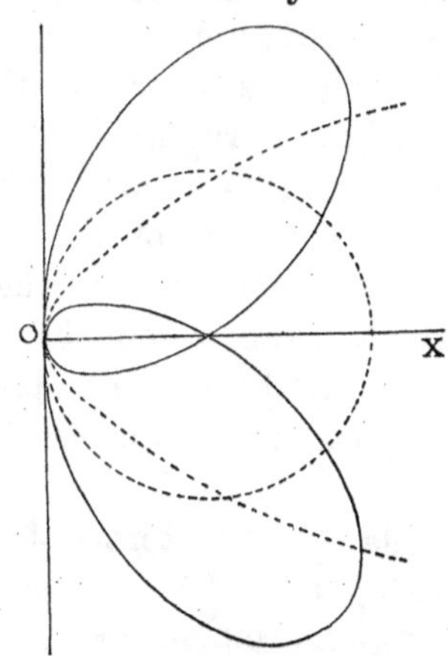

Fig. 71.

V. *To trace* $\quad (x^2 + y^2 - 3ax)^2 = 4ax^2(2a - x).$

Here $$y^2 = 2ax - x^2 \pm 2\sqrt{ax}\sqrt{2ax - x^2} + ax$$
$$= (\sqrt{2ax - x^2} \pm \sqrt{ax})^2;$$

therefore $$y = \pm\sqrt{2ax - x^2} \pm \sqrt{ax},$$

or $$y = \pm y_1 \pm y_2,$$

where y_1 and y_2 are corresponding ordinates of the circle $x^2+y^2=2ax$ and of the parabola $y^2=ax$. Hence the ordinate of the curve is the sum or difference of the corresponding ordinates of these curves. The circle and the parabola are shown by dotted lines in the accompanying figure, and the resultant curve by the continuous line.

EXAMPLES.

1. Trace the curve $\quad (x+y+1)^2=(1-x)^5,$
showing that there is a cusp of the first species at $(1, -2)$; also that all chords parallel to the axis of y are bisected by the line

$$x+y+1=0.$$

2. Trace the curve $\quad r=a\sec\theta\pm a\cos\theta,$
the radius vector being the sum or difference of the radii vectores of a straight line and a circle.

320. **Newton's Diagram of Squares.**

When a curve whose equation is algebraic and rational passes through the origin it is frequently desirable to ascertain the shape of the curve in the immediate neighbourhood of the origin more accurately than can be predicted from a mere knowledge of the direction of the tangents, and also to form some idea of the limiting form of the curve at a great distance from the origin.

The following is a graphical method of determining what terms of an equation are to be retained or rejected in such cases:—

Let Ax^py^q, Bx^ry^s be any two terms of the equation of the curve; and let us suppose them to be such that they are of the same order of magnitude. Take a pair of co-ordinate axes and mark down the positions of the points (p, q) (r, s), which we shall call P and R respectively. Then, since x^py^q and x^ry^s are of the same order of magnitude, x^{p-r} and y^{s-q} are also of the same order, and therefore the order of x is that of $y^{\frac{s-q}{p-r}}$

Now $\frac{s-q}{r-p}=\tan\theta$, where θ is the angle which the line PR makes with OX. So that the order of x is that of $y^{-\tan\theta}$, and therefore the order of the term Ax^py^q is that of $y^{q-p\tan\theta}$. Now $q-p\tan\theta$ = the intercept OA made

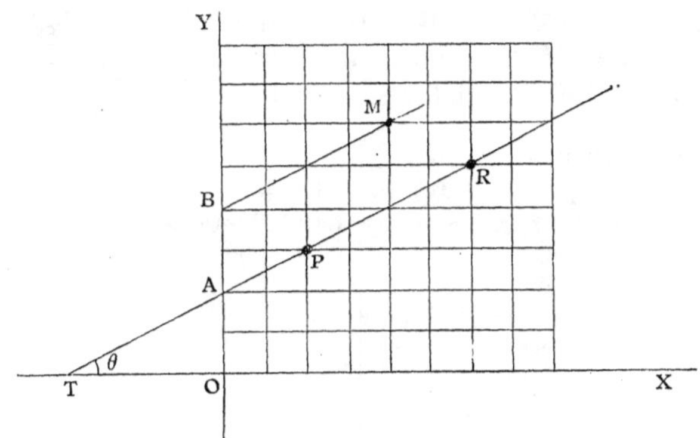

Fig. 72.

by the line PR upon OY, so that the order of the terms Ax^py^q and Bx^ry^s is that of y^{OA} and is measured by the intercept OA.

Consider next any other term Cx^my^n in the equation. Let its graphical point (m, n) be denoted by M in the figure. Then the order of this term is that of

$$y^{n-m\tan\theta} \text{ or } y^{OB},$$

the line MB being drawn parallel to RP, cutting off the intercept OB on the axis of y. OB therefore graphically marks the order of this term, which may therefore be rejected in tracing near the origin in comparison with the terms denoted by the points P and R if OB be greater than OA; and in tracing the curve at a great distance from the origin it may be rejected if OB be less than OA.

Thus if all the terms of the equation be represented graphically by the series of points P, Q, R, S ... in the manner above described, and if when any two, say P and R, are chosen all the other points lie on the side of the line PR, remote from the origin, they may all be rejected in tracing the portion of the curve in the immediate proximity of the origin; but if they all lie on the origin side of the line PR they may all be rejected in tracing the curve at an infinite distance from the origin.

Ex. If the equation be

$$x^2y^3+2xy+3x^4y+x^2y^2+y^4=0,$$

the points A, B, C, D, E represent the 1st, 2nd, etc. terms respect-

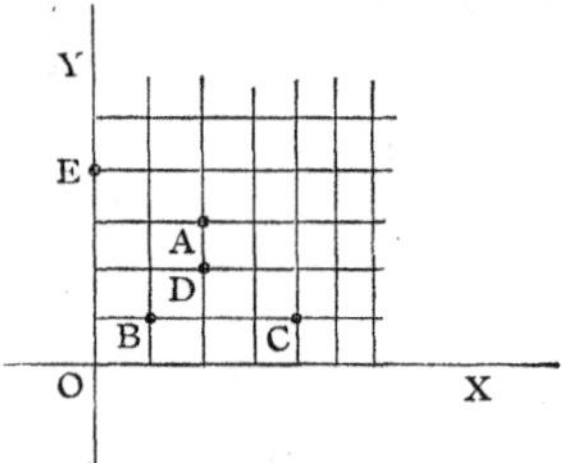

Fig. 73.

ively, and a glance at the diagram will show that the

the second and third
and the second and fifth

are groups which may be taken together in tracing near the origin,

whilst the first and third
and the first and fifth

are groups which may be taken together in approximating to the form of the curve at an infinite distance from the origin.

321. The above method is a modification of the one adopted in such cases by Newton, and is known as Newton's Parallelogram. A further slight variation on the same method is due to De Gua, and is known as De

Gua's "Analytical Triangle." [De Gua's "Usage de l'Analyse de Descartes," Paris, 1740.]

VI. *To trace* $x^5+y^5-5a^2x^2y=0.$

a. Newton's diagram shows at once that near the origin the first and third of these terms, or the second and third, may be taken together, whilst at a great distance from the origin the first and

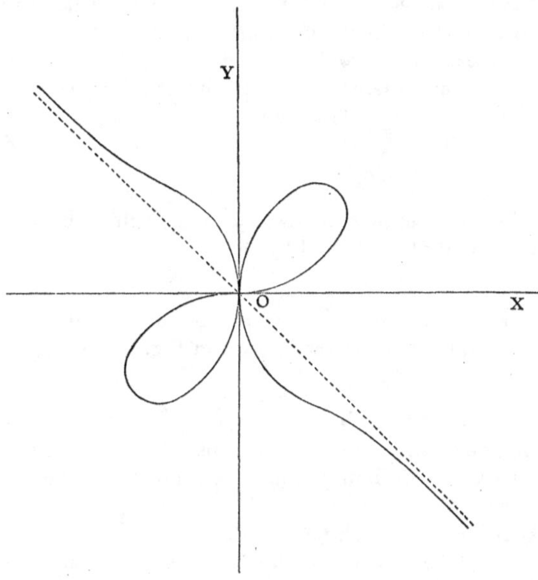

Fig. 74.

second may be taken together. This indicates that at the origin the curve assumes the parabolic forms

$$y^2=\pm a\sqrt{5}x,$$
$$x^3=5a^2y,$$

and that at infinity it approximates to the straight line $x+y=0$, which is obviously the only asymptote.

β. Moreover, the equation may be written

$$y = -x\left(1 - 5a^2\frac{y}{x^3}\right)^{\frac{1}{5}}$$
$$= -x + \frac{a^2y}{x^2} + \ldots$$
$$= -x - \frac{a^2}{x} + \ldots,$$

when in the limit $y = -x =$ a very large quantity.
Hence again $y = -x$ is an asymptote, but we gain the additional information that if x be negative and very large the ordinate of the curve is greater than the ordinate of the asymptote.

γ. Since when the signs of x and y are both changed the equation remains of the same form there is symmetry in opposite quadrants.

δ. Since $$\frac{dy}{dx} = \frac{x(2a^2y - x^3)}{y^4 - a^2x^2},$$
we have $$\frac{dy}{dx} = 0$$
at the points where the curve is intersected by the cubical parabola $2a^2y = x^3$ (which is easily traced), and by the axis of y; and
$$\frac{dy}{dx} = \infty,$$
where the curve is cut by either of the parabolas $y^2 = \pm ax$. The form of the equation is therefore that shown in Fig. 74.

EXAMPLES.

1. Trace $$x^5 + y^5 = 5ax^2y^2,$$
showing that at the origin there are two cusps of the first species, an asymptote $x + y = a$, two infinite branches below the asymptote, and a loop in the first quadrant.

2. Show that the curve $y^6 - a^2x^2y^2 + x^6 = 0$
consists of four equal loops, one in each of the four quadrants and lying entirely within the circle $r = a$.

322. Polar Equations. Order of Procedure.

In tracing a curve from its polar equation it is advisable to follow some such routine as the following:—

1. If possible *form a table* of corresponding values of r and θ which satisfy the equation of the curve. Consider both positive and negative values of θ.

2. Obtain the *value of* $\tan\phi$, Art. 178. This will indicate the direction of the tangent at any point. The length of the polar subtangent is often useful, Art. 179.

3. Examine whether any values of θ exist which give an infinite value of r. If so, find whether the curve has *asymptotes* in such directions (Art. 234) and find their equations.

4. Examine whether there be an *asymptotic circle* (Art. 236).

5. Find the positions of the *points of inflexion* (Art. 248).

6. It will frequently be obvious from the equation of the curve that the values of r or θ are *confined between certain limits*. If such exist they should be ascertained.

E.g., if $r=a\sin n\theta$ it is clear that r must lie in magnitude between the limits 0 and a, and the curve lie wholly within the circle $r=a$.

323. Curves of the Classes $r=a\sin n\theta$, $r\sin n\theta=a$.

VII. *To trace* $r=a\sin 5\theta$.

a. We have the following table of corresponding values of r and θ:

Values of θ	0	Intermed. Values.	$\frac{\pi}{10}$	Intermed. Values.	$\frac{2\pi}{10}$	Intermed. Values.	$\frac{3\pi}{10}$	Intermed. Values.	$\frac{4\pi}{10}$	Intermed. Values.
Corresponding Values of r	0	Pos. and Incr.	a	Pos. and Decr.	0	Neg.	$-a$	Neg.	0	Pos.
Values of θ	$\frac{5\pi}{10}$		$\frac{6\pi}{10}$		$\frac{7\pi}{10}$		$\frac{8\pi}{10}$	etc.		
Corresponding Values of r	a	Pos.	0	Neg.	$-a$	Neg.	0	etc.		

β. r is never greater than a, and there is no asymptote.

γ. $\tan\phi = \frac{1}{5}\tan 5\theta$, and therefore vanishes whenever r vanishes and $=\infty$ whenever $r=\pm a$. The curve therefore consists of a series of similar loops as shown in Fig. 75, all being arranged symmetrically about the origin and lying entirely within a circle whose centre is at the pole and radius a.

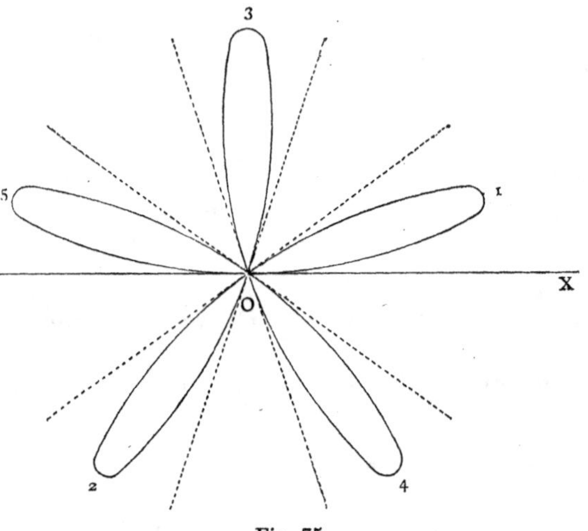

Fig. 75.

324. Any other curve of the class

$$r = a\sin n\theta$$

may be traced in a similar manner.

We annex a figure of the curve

$$r = a\sin 6\theta \text{ (Fig. 76).}$$

It will be noticed for this class of curves that if *n be odd there are n loops,* whilst if *n be even there are 2n loops.* This will be easily seen from the *order of description* of the loops, which we have denoted by the numerals 1, 2, 3 ... in the figures.

325. Curves of the class

$$r \sin n\theta = a$$

are *inverse* to the above species, and their forms are

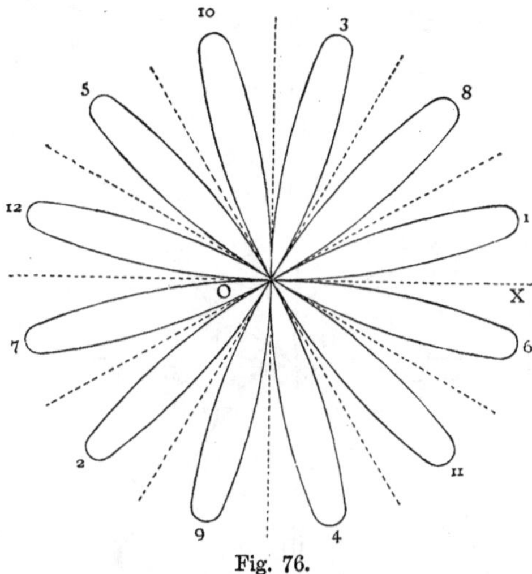

Fig. 76.

therefore obvious, going to ∞ along a radial asymptote whenever the radius of the companion curve $r = a \sin n\theta$ vanishes, and touching $r = a \sin n\theta$ at the extremity of each loop. We give in illustration a tracing of the curves

$$\left. \begin{array}{ll} & r = a \sin 4\theta \\ \text{and} & r \sin 4\theta = a \end{array} \right\},$$

with the asymptotes of the latter, in one figure (Fig. 77).

326. **Class $r^m = a^n \cos n\theta$.**

The class of curves of which

$$r^m = a^n \cos n\theta$$

is the type embraces, as has been previously noticed, several important and well known curves. For instance, we get Bernoulli's lemniscate ($n=2$), the circle ($n=1$), the cardioide ($n=\frac{1}{2}$), the parabola ($n=-\frac{1}{2}$), the straight line ($n=-1$), the rectangular hyperbola ($n=-2$).

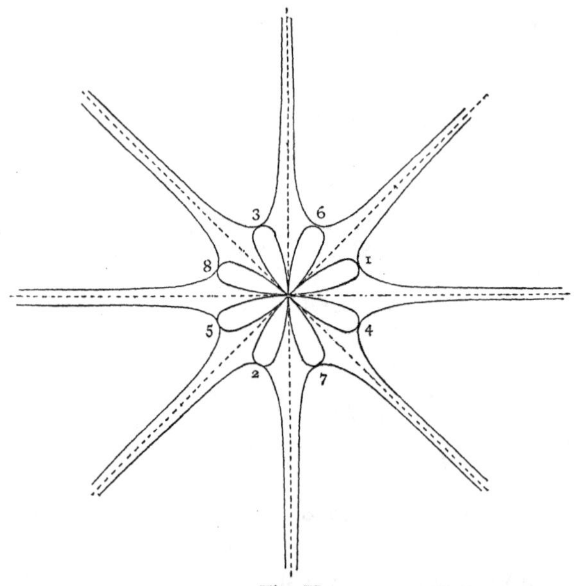

Fig. 77.

VIII. *To trace* $r^2=a^2\cos 2\theta$ (*Bernoulli's Lemniscate*).

α. Negative values of $\cos 2\theta$ give imaginary values of r. Hence the only real portions of the curve lie in the two quadrants bounded by $\theta=-\frac{\pi}{4}$ and $\theta=+\frac{\pi}{4}$, and by $\theta=\frac{3\pi}{4}$ and $\theta=\frac{5\pi}{4}$.

β. $$r=0 \text{ when } \theta=\pm\frac{\pi}{4} \text{ or } \frac{3\pi}{4} \text{ or } \frac{5\pi}{4},$$

and $$=\pm a \text{ when } \theta=0 \text{ or } \pi.$$

γ. Since the only power of r occurring is even, the curve is symmetrical about the origin. Again, since the equation is unaltered by writing $-\theta$ for θ the curve is obviously symmetrical about the initial line.

Also, r increases from $\theta=-\frac{\pi}{4}$ to 0 and decreases again from $\theta=0$ to $\frac{\pi}{4}$ and is nowhere infinite or in fact greater than a.

The curve therefore consists of two similar loops as shown in Fig. 78.

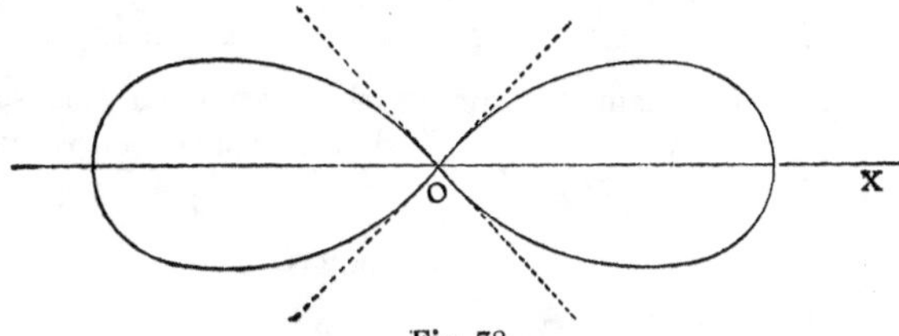

Fig. 78.

Other curves of this species may be treated in a similar manner. It will be easily seen that if n be fractional $\left(=\frac{p}{q}\right)$, the curve will have p portions arranged symmetrically about the origin.

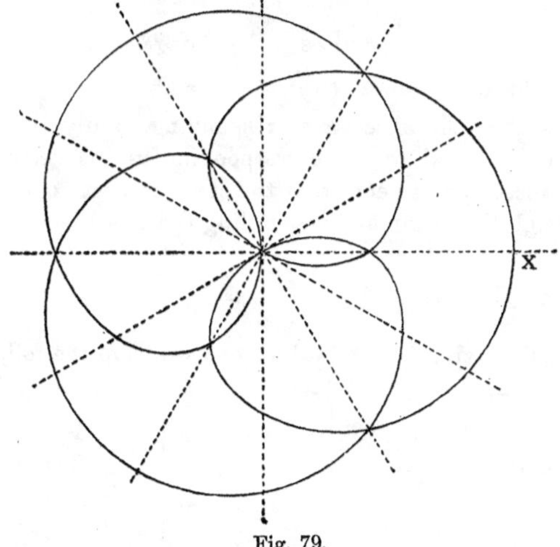

Fig. 79.

For example, in the curve

$$r^{\frac{3}{5}}=a^{\frac{3}{5}}\cos\frac{3}{5}\theta$$

we have the following scheme of values for r and θ:

θ	0	$\frac{5\pi}{6}$	$\frac{10\pi}{6}$	$\frac{15\pi}{6}$	$\frac{20\pi}{6}$	$\frac{25\pi}{6}$	$\frac{30\pi}{6}$	etc.
r	a	0	$-a$	0	a	0	$-a$	etc.

whence we obtain a figure with three equal loops, the whole lying within a circle whose radius is a and centre at the origin (Fig. 79).

EXAMPLES.

1. Trace the curves

$$r=a\cos 2\theta, \qquad r\cos 2\theta=a,$$

$$r=a\cos 3\theta, \qquad r=a\cos 4\theta.$$

2. Trace

$$r^3=a^3\cos 3\theta, \qquad r^3\cos 3\theta=a^3,$$

$$r^{\frac{1}{3}}=a^{\frac{1}{3}}\cos \tfrac{1}{3}\theta, \qquad r^{\frac{1}{3}}\cos \tfrac{1}{3}\theta=a^{\frac{1}{3}},$$

$$r^{\frac{3}{4}}=a^{\frac{3}{4}}\cos \tfrac{3}{4}\theta, \qquad r^{\frac{3}{4}}\cos \tfrac{3}{4}\theta=a^{\frac{3}{4}}.$$

3. Trace the curve $y^2(x^2+a^2)=x^2(a^2-x^2)$. [I. C. S., 1885.]

Show that the abscissa corresponding to any given central radius vector is equal to the corresponding radius vector in Bernoulli's Lemniscate, and hence that the curve consists of two loops passing through the origin and resembling those of the Lemniscate.

IX. *To trace* $r=\dfrac{a\theta}{1+\theta}$.

α. By giving a set of values to θ we have the following table:

Values of θ in Circular Measure	∞	4	3	2	1	$\frac{1}{2}$	$\frac{1}{4}$	0	$-\frac{1}{4}$	$-\frac{1}{2}$
Values of r	a	$\frac{4a}{5}$	$\frac{3a}{4}$	$\frac{2a}{3}$	$\frac{a}{2}$	$\frac{a}{3}$	$\frac{a}{5}$	0	$-\frac{a}{3}$	$-a$

Values of θ in Circular Measure	$-\frac{3}{4}$	-1	$-\frac{5}{4}$	$-\frac{4}{3}$	$-\frac{3}{2}$	-2	-3	-4	-10	$-\infty$
Values of r	$-3a$	∞	$5a$	$4a$	$3a$	$2a$	$\frac{3a}{2}$	$\frac{4a}{3}$	$\frac{10a}{9}$	a

β. Since we may write the equation

$$r=\frac{a}{1+\frac{1}{\theta}},$$

when θ becomes very large, either positively or negatively, the form of the curve approximates to that of an asymptotic circle $r=a$, which it approaches both from within and without.

γ. Art. 234 shows that $\quad r\sin(\theta+1)+a=0$
is an asymptote to the curve. This line touches the asymptotic circle and is shown by the dotted straight line in the figure.

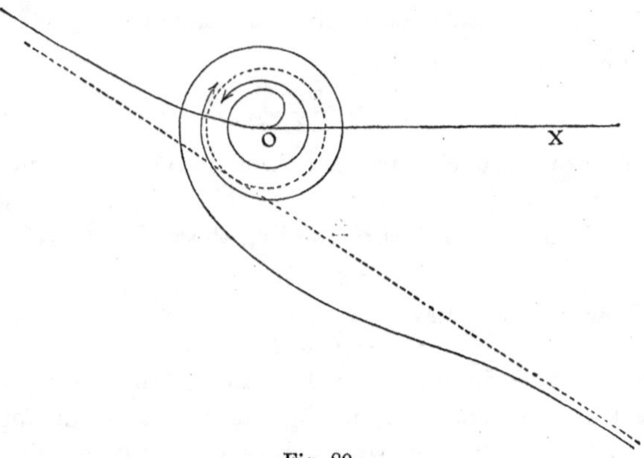

Fig. 80.

δ. The points of inflexion (Art. 248) are given by the equation

$$\theta^3+\theta^2+2=0,$$

an equation which has one real root which lies between $\theta=-1$ and $\theta=-2$. The curve is therefore that shown in Fig. 80.

EXAMPLES.

1. Trace $$r=\frac{a\theta^2}{\theta^2+1},$$
showing that it lies entirely within the circle $r=a$, which is an asymptotic circle; also, that there is a cusp of the first species at the origin.

2. Trace $$r=\frac{a\theta^2}{\theta^2-1}.$$
Show that there are two linear asymptotes and an asymptotic circle; also a cusp of the first species at the origin and a point of inflexion when $\theta^2=3$.

EXAMPLES.

1. Show that the curve
$$y^2=x^2\frac{a^2+x^2}{a^2-x^2}$$
consists of two branches each passing through the origin and extending to infinity, and that the whole curve is contained between two asymptotes parallel to the axis of y.

2. Show that the curve
$$y^2=x^2\frac{x^2-4a^2}{x^2-a^2}$$
has two infinite branches passing through the origin and lying between the asymptotes $x=\pm a$, and that there are in addition two other infinite branches resembling those of the hyperbola
$$x^2-y^2=4a^2.$$

3. Show that the curve
$$x^3+y^3=a^3$$
consists of one infinite branch running to the asymptote $x+y=0$ at each end and cutting the axes at right angles at the points $(a, 0)$, $(0, a)$ at which there are points of inflexion.

4. Show that the curve
$$x^3+y^3=3axy$$
consists of one infinite branch running to the asymptote

$x+y+a=0$ at each end and lying on the upper side of that line. Also, that the axes of co-ordinates are tangents at the origin, and that there is a loop in the first quadrant. This curve is called the Folium of Descartes.

5. Trace the curves

$$(\alpha)\quad x^3+y^3=a^2x.$$

$$(\beta)\quad x^3+y^3=2ax^2.$$

$$(\gamma)\quad ay^2=x(a^2-x^2).$$ [R. M. A., Nov., 1883.]

6. Show that the curve

$$ay^2=x^2y+x^3$$ [R. M. A., July, 1880.]

has a cusp of the first species at the origin and an asymptote $x+y=a$ cutting the curve at $\left(\frac{a}{2}, \frac{a}{2}\right)$. Trace the curve.

7. Trace the curves

$$(\alpha)\quad ay^2-2axy+x^3=0.$$ [R. M. A., Nov., 1880.]

$$(\beta)\quad y^3+axy+bx^2=0.$$ [R. M. A., Nov., 1881.]

a and b both being positive quantities.

8. Trace $$xy^2=4a^2(2a-x).$$

Show that this curve may be constructed thus: take a semicircle APB whose diameter is AB, produce MP the ordinate of P so that $$MP:MQ=AM:AB,$$ then the given curve is the locus of Q. [This curve is called the Witch and was discussed by Maria Gaetana Agnesi, Professor of Mathematics at Bologna, 1748.]

9. Trace the curve $y^2(2a-x)=x^3$. [Cissoid of Diocles.] Show that this curve arises from the following geometrical construction. AB is the diameter of a semicircle APB, BT the tangent at B, APT a straight line through A cutting the semicircle and the straight line in P and T; then, if Q be taken on this line so that $AQ=PT$, the locus of Q is the Cissoid.

10. Trace $$\left(\frac{y}{x+a}\right)^3 = \frac{x-a}{x+2a},$$
and show that the oblique asymptote cuts the curve at an angle $\tan^{-1}8$. [R. M. A., Nov., 1882.]

11. Trace $$2x(x^2+y^2) = a(2x^2+y^2)$$
and find by polars the co-ordinates of the points of inflexion. [R. M. A., June, 1883.]

12. Trace $$y(a^2+x^2) = a^2x,$$
showing that there are points of contrary flexure where $x=0$ or $\pm a\sqrt{3}$, that the tangent is parallel to the axis of x where $x = \pm a$, and that the axis of x is an asymptote.

13. Trace $$x^2y^2 = a^2(x^2-y^2),$$
showing that the curve lies entirely between its asymptotes $y = \pm a$, and that its tangents at the origin are $y = \pm x$.

14. Trace the curve
$$(x^2-a^2)(y^2-b^2) = a^2b^2.$$

15. Trace $$x^4 = a^2(x^2-y^2).$$

16. Trace $$(y^2-a^2)^2 = x^2(x^2-2a^2).$$

17. Trace $$axy = x^3 - a^3.$$ (The Trident.)

18. Trace the curve
$$x^4 - 2mx^2y^2 + y^4 = a^4$$
when m is respectively greater than, equal to, and less than unity, and also when m is zero. [LONDON, 1880.]

19. Trace $$y^2 = x^2\frac{x+a}{x-a}.$$

20. Trace $$y^2 = x^2\frac{x^2+a^2}{x^2-a^2}.$$

21. Trace $$x(x+y)^2 = a(x-y)^2.$$ [I. C. S., 1879.]

22. Trace $$x^3 = y(x-a)^2.$$ [OXFORD, 1876.]

23. Trace $\left(\frac{x}{y-a}\right)^2 = \frac{y-a}{y+a}$. [H. C. S., 1881.]

24. Find the multiple points on the curve

$$2(x^4+y^4)+5x^2y^2+4a^4=6a^2(x^2+y^2)$$

and the directions of the tangents at those points.

[H. C. S., 1881.]

Also trace the curve.

25. Trace the curve $x^3+y^3+3cxy=a^3$,
and prove that as c diminishes to a the ultimate form of the loop is that of an ellipse whose eccentricity $=\sqrt{\frac{2}{3}}$.

[MATH. TRIPOS.]

26. Trace $(x-y)^2(x+y)(2x+y)=a^2y^2$. [CAMB., 1879.]

27. Trace the curve $r=a(1+\cos\theta)$. (Cardioide.)

28. Trace $r=a+b\cos\theta$. (The Limaçon of Pascal.)

29. Trace $r=a(2\cos\theta\pm 1)$. (The Trisectrix.)

30. Trace the following spirals :—

(α) $r=a\theta$. (Spiral of Archimedes.)

(β) $r\theta=a$. (The Hyperbolic or Reciprocal Spiral.)

(γ) $r^2\theta=a^2$. (The Lituus.)

(δ) $r=ae^{m\theta}$. (The Logarithmic or Equiangular Spiral.)

Show that in each case there is an infinite number of convolutions round the pole, and that $r\sin\theta=a$ is an asymptote to (β) and the initial line an asymptote to (γ).

31. Trace the curves

$$r=a\cos 5\theta,\quad r\cos 5\theta=a,\quad r=a\cos\tfrac{1}{5}\theta.$$

32. Trace the curves

$$r^{\frac{2}{3}}=a^{\frac{2}{3}}\cos\tfrac{2}{3}\theta,\quad r^{\frac{2}{3}}=a^{\frac{2}{3}}\sec\tfrac{2}{3}\theta,\quad r^{\frac{2}{5}}=a^{\frac{2}{5}}\cos\tfrac{2}{5}\theta.$$

What is the relation between them? [CAMB., 1876.]

33. Trace the curve $\theta = \frac{a}{r-a}$,
showing that a line parallel to the initial line at a distance a above it is an asymptote. Show also that there is an asymptotic circle $r=a$.

34. Trace $r = a\frac{\theta + \sin\theta}{\theta - \sin\theta}$.
Show that this curve has an asymptotic circle; also that as each branch of the curve comes from infinity it approaches the asymptotic circle from the outside on one side of the initial line and from the inside upon the other.

35. Trace $r = 2a\frac{\sin^2\theta}{\cos\theta}$ (The Cissoid)
from the polar equation.

36. Trace $r = a\frac{\theta - \alpha}{\theta + \alpha}$. [R. M. A., July, 1880.]

37. Trace $r\theta^2 = \tan\theta$, from $\theta = 0$ to $\theta = 2\pi$. [OXFORD, 1876.]

38. Trace $r^3 \sin 3(\theta - \alpha) = \sin\theta - \sin\alpha$. [CAMB., 1879.]

39. Trace the "curve of sines"
$$y = b\sin\frac{x}{a}.$$

40. Trace $y = e^{-\lambda x}\tan\mu x$.

41. Trace $r = \frac{a}{\theta^5 - 1}$
for positive values of θ. [TRIN. COLL. CAMB., 1873.]

42. Trace $r = \frac{a}{1 - \sin 2\theta}$. [OXFORD.]

43. Trace $(x+a)^2(y-a) + (y+a)^2(x-a) = 0$. [OXFORD.]

44. Trace $$y^2 = c^2\frac{(x+a)(x+b)}{(x-a)(x-b)}.$$

45. Trace $$\left.\begin{aligned} x &= a(1-\cos\theta) \\ y &= a\theta \end{aligned}\right\}.$$

(The companion to the Cycloid.)

46. Trace $y = c\cosh\frac{x}{c}$. (The Catenary.)

47. Trace $y = x^2 + \cosh x$.

48. Trace $x^2y^2 = (a+y)^2(b^2-y^2)$,

or $r = a\,\mathrm{cosec}\,\theta \pm b$.

(The Conchoid of Nicomedes).

49. Trace $\{y^2+(a+x)^2\}\{y^2+(a-x)^2\} = a^2b^2$,
examining the cases

(1) $a<b$.
(2) $a=b$. (Lemniscate of Bernoulli.)
(3) $a>b$. (Cassini's Ovals.)

50. Trace $y^4 + x^2y^2 + 2y^3 + x^3 = 0$. [CRAMER.]

51. Trace
$$r = a(\cos\alpha\cos\theta - \tfrac{1}{3}\cos 3\alpha\cos 3\theta + \tfrac{1}{5}\cos 5\alpha\cos 5\theta - \ldots).$$
[MATH. TRIPOS, 1878.]

52. Trace $y = e^{-x^2}$. (The Probability Curve.)

53. Trace the curves

(α) $y^4 - axy^2 + x^4 = 0$.
(β) $a^3y^2 - 2abx^2y - x^5 = 0$.
(γ) $y^5 + ax^4 - b^2xy^2 = 0$. [CRAMER.]

54. Trace $x^4 - ax^2y + by^3 = 0$. [DE GUA.]

55. Trace (α) $x^5 + y^5 = 2a^3xy$.
(β) $x^5 + y^5 = xy(a^2x + b^2y)$. [FROST.]

APPLICATION TO THE EVALUATION OF SINGULAR FORMS AND MAXIMA AND MINIMA VALUES.

CHAPTER XIII.

UNDETERMINED FORMS.

327. In Chap. I. it was explained that a function may involve an independent variable in such a manner that its value for a certain assigned value of the variable cannot be found by a direct substitution of that value. And in such cases the function is said to assume a "*Singular*," "*Undetermined*," "*Illusory*," or "*Indeterminate*" form.

328. It is proposed in the present chapter to consider more fully the method of evaluation of the true limiting values of such quantities when the independent variable is made to approach indefinitely near its assigned value.

329. **List of Forms occurring.**

Several cases are to be considered, viz., when, upon substitution of the assigned value of the independent variable, the function reduces to one of the forms

$$\frac{0}{0}, \quad 0 \times \infty, \quad \frac{\infty}{\infty}, \quad \infty - \infty, \quad 0^0, \quad \infty^0, \quad \text{or } 1^\infty.$$

It is frequently easy to treat these cases by algebraical

or trigonometrical methods without having recourse to the Differential Calculus, though the latter is required for a general discussion of such forms.

By far the most important case to consider is that in which the function takes the form $\frac{0}{0}$; for, in the first place, it is the one which most frequently occurs; and, secondly, any of the other forms may be made to depend upon this one by some special artifice.

330. **Algebraical Treatment.**

Suppose the function to take the form $\frac{0}{0}$ when the independent variable x ultimately coincides with its assigned value a. Put $x=a+h$ and *expand both numerator and denominator* of the function. It will now become apparent that the reason why both numerator and denominator vanish is that some power of h is a common factor of each. This should now be *divided out.* Finally, put $h=0$ so that x becomes $=a$, and the true limiting value of the function will be apparent.

In the particular case in which x is to become *zero* the expansion of numerator and denominator in powers of x should be at once proceeded with without any preliminary substitution for x.

In the case in which x is to become infinite, put $x=\frac{1}{y}$, so that when x becomes $=\infty$ y becomes $=0$.

The method thus explained will be better understood by examining the mode of solution of the following examples.

Ex. 1. *Find* $$Lt_{x=0}\frac{a^x-b^x}{x}.$$

Here numerator and denominator both vanish if x be put equal to 0. We therefore expand a^x and b^x by the exponential theorem. Hence

$$Lt_{x=0}\frac{a^x-b^x}{x}$$

$$=Lt_{x=0}\frac{\left\{1+x\log_e a+\frac{x^2}{2!}(\log_e a)^2+\dots\right\}-\left\{1+x\log_e b+\frac{x^2}{2!}(\log_e b)^2+\dots\right\}}{x}$$

$$=Lt_{x=0}\left\{\log_e a-\log_e b+\frac{x}{2!}(\overline{\log_e a}|^2-\overline{\log_e b}|^2)+\dots\right\}$$

$$=\log_e a-\log_e b=\log_e\frac{a}{b}.$$

Ex. 2. *Find* $$Lt_{x=1}\frac{x^7-2x^5+1}{x^3-3x^2+2}.$$

This is of the form $\frac{0}{0}$ if we put $x=1$. Therefore we put $x=1+h$ and expand. We thus obtain

$$Lt_{x=1}\frac{x^7-2x^5+1}{x^3-3x^2+2}=Lt_{h=0}\frac{(1+h)^7-2(1+h)^5+1}{(1+h)^3-3(1+h)^2+2}$$

$$=Lt_{h=0}\frac{(1+7h+21h^2+\dots)-2(1+5h+10h^2+\dots)+1}{(1+3h+3h^2+\dots)-3(1+2h+h^2)+2}$$

$$=Lt_{h=0}\frac{-3h+h^2+\dots}{-3h+\dots}$$

$$=Lt_{h=0}\frac{-3+h+\dots}{-3+\dots}$$

$$=\frac{-3}{-3}=1.$$

It will be seen from these examples that in the process of expansion it is only necessary in general *to retain a few of the lowest powers of h.*

Ex. 3. *Find* $$Lt_{x=0}\left(\frac{\tan x}{x}\right)^{\frac{1}{x^2}}.$$

Since $$\frac{\tan x}{x}=\frac{1}{\cos x}\cdot\frac{\sin x}{x}$$

we have $Lt_{x=0}\dfrac{\tan x}{x}=1.$

Hence the form assumed by $\left(\dfrac{\tan x}{x}\right)^{\frac{1}{x^2}}$ is 1^∞ when we put $x=0$.

Expand $\sin x$ and $\cos x$ in powers of x. This gives

$$Lt_{x=0}\left(\frac{\tan x}{x}\right)^{\frac{1}{x^2}}=Lt_{x=0}\left(\frac{x-\dfrac{x^3}{3!}+\dots}{x-\dfrac{x^3}{2!}+\dots}\right)^{\frac{1}{x^2}}$$

$$=Lt_{x=0}\left(1+\frac{x^2}{3}+\text{higher powers of } x\right)^{\frac{1}{x^2}}$$

$$=Lt_{x=0}\left(1+\frac{x^2l}{3}\right)^{\frac{1}{x^2}},$$

where l is a series in ascending powers of x whose first term (and therefore whose limit when $x=0$) is unity. Hence

$$Lt_{x=0}\left(\frac{\tan x}{x}\right)^{\frac{1}{x^2}}=Lt_{x=0}\left\{\left(1+\frac{x^2l}{3}\right)^{\frac{3}{lx^2}}\right\}^{\frac{l}{3}}=e^{\frac{1}{3}}, \text{ by Art. 21.}$$

Ex. 4. *Find* $$Lt_{x=1}x^{\frac{1}{1-x}}.$$

This expression is of the form 1^∞. Put

$$1-x=y,$$

and therefore, if $x=1$, $y=0$;

therefore $$\text{Limit required}=Lt_{y=0}(1-y)^{\frac{1}{y}}=e^{-1} \text{ (Art. 21).}$$

Ex. 5. $$Lt_{x=\infty}x(a^{\frac{1}{x}}-1).$$ This is of the form $\infty\times 0$.

Put $$x=\frac{1}{y},$$

therefore, if $x=\infty$, $y=0$, and

$$\text{Limit required}=Lt_{y=0}\frac{a^y-1}{y}=\log_e a \text{ (Art. 22).}$$

EXAMPLES.

Find the values of the following limits :

1. $Lt_{x=0}\frac{a^x-1}{b^x-1}$.

2. $Lt_{x=1}\frac{x^{\frac{3}{2}}-1}{x^{\frac{5}{2}}-1}$.

3. $Lt_{x=1}\frac{x^m-1}{x^n-1}$.

4. $Lt_{x=0}\frac{(1+x)^{\frac{1}{n}}-1}{x}$.

5. $Lt_{x=1}\frac{x^4+x^3-x^2-5x+4}{x^3-x^2-x+1}$.

6. $Lt_{x=1}\frac{x^5-2x^3-4x^2+9x-4}{x^4-2x^3+2x-1}$.

7. $Lt_{x=0}\frac{e^x-e^{-x}}{x}$.

8. $Lt_{x=0}\frac{e^x+e^{-x}-2}{x^2}$.

9. $Lt_{x=0}\frac{x\cos x-\log(1+x)}{x^2}$.

10. $Lt_{x=0}\frac{xe^x-\log(1+x)}{x^2}$.

11. $Lt_{x=0}\frac{x-\sin x\cos x}{x^3}$.

12. $Lt_{x=0}\frac{\sin^{-1}x-x}{x^3\cos x}$.

13. $Lt_{x=0}\frac{\cosh x-\cos x}{x\sin x}$.

14. $Lt_{x=0}\frac{\sin^{-1}x}{\tan^{-1}x}$.

15. $Lt_{x=0}\frac{\sin^{-1}x-\sinh x}{x^5}$.

16. $Lt_{x=0}\frac{x\cos^3 x-\log(1+x)-\sin^{-1}\frac{x^2}{2}}{x^3}$.

17. $Lt_{x=0}\frac{2\sin x+\tanh^{-1}x-3x}{x^5}$.

18. $Lt_{x=0}\frac{e^x\sin x-x-x^2}{x^2+x\log(1-x)}$.

19. $Lt_{x=0}\frac{x^3e^{\frac{x^4}{4}}-\sin^{\frac{3}{2}}x^2}{x^7}$.

20. $Lt_{x=0}\left(\frac{\tan x}{x}\right)^{\frac{1}{x}}$.

21. $Lt_{x=0}\left(\frac{\tan x}{x}\right)^{\frac{1}{x^3}}$.

22. $Lt_{x=0}\left(\frac{\sin x}{x}\right)^{\frac{1}{x}}$.

23. $Lt_{x=0}\left(\frac{\sin x}{x}\right)^{\frac{1}{x^2}}$.

24. $Lt_{x=0}\left(\frac{\sin x}{x}\right)^{\frac{1}{x^3}}$.

25. $Lt_{x=0}(\text{covers }x)^{\frac{1}{x}}$.

26. $Lt_{x=\frac{\pi}{2}}(\text{cosec }x)^{\tan^2 x}$.

331. APPLICATION OF THE DIFFERENTIAL CALCULUS. John Bernoulli * was the first to make use of the processes of the Differential Calculus in the determination of

* "Acta Eruditorum," 1704.

the true values of functions assuming singular forms. We propose now to discuss each singularity in order.

332. I. Form $\frac{0}{0}$.

Consider a curve passing through the origin and defined by the equations

$$\left.\begin{aligned} x &= \phi(t), \\ y &= \psi(t). \end{aligned}\right\}$$

Let x, y be the co-ordinates of a point P on the curve very near the origin, and suppose a to be the value of t

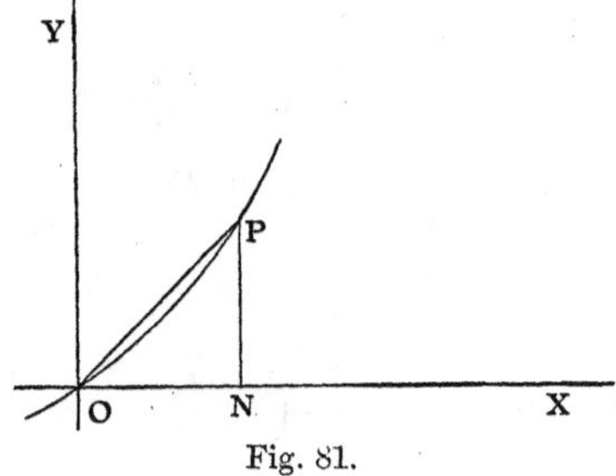

Fig. 81.

corresponding to the origin, so that $\phi(a)=0$ and $\psi(a)=0$.

Then ultimately we have

$$Lt\frac{y}{x}=Lt\tan PON=\text{the value of } \frac{dy}{dx} \text{ at the origin};$$

and

$$\frac{dy}{dx}=\frac{\frac{dy}{dt}}{\frac{dx}{dt}}=\frac{\phi'(t)}{\psi'(t)}.$$

Hence

$$Lt_{t=a}\frac{\phi(t)}{\psi(t)}=Lt_{t=a}\frac{\phi'(t)}{\psi'(t)};$$

and if $\frac{\phi'(t)}{\psi'(t)}$ be not of the form $\frac{0}{0}$ when t takes its assigned value a, we therefore obtain

$$Lt_{t=a}\frac{\phi(t)}{\psi(t)}=\frac{\phi'(a)}{\psi'(a)}.$$

But, if $\frac{\phi'(t)}{\psi'(t)}$ be also of undetermined form, we may repeat the process and say

$$Lt_{t=a}\frac{\phi'(t)}{\psi'(t)}=Lt_{t=a}\frac{\phi''(t)}{\psi''(t)}=\text{etc.},$$

proceeding in this manner until we arrive at a fraction such that when the value a is substituted for t its numerator and denominator *do not both vanish,* and thus obtaining an intelligible result—zero, finite, or infinite.

333. Another Proof of the Method.

We may arrive at the same result in another way, thus:—

Let $\frac{\phi(x)}{\psi(x)}$ take the form $\frac{0}{0}$ when x approaches and ultimately coincides with the value a. Let $x=a+h$. Then by Taylor's Theorem

$$\frac{\phi(x)}{\psi(x)}=\frac{\phi(a)+h\phi'(a+\theta h)}{\psi(a)+h\psi'(a+\theta_1 h)}=\frac{\phi'(a+\theta h)}{\psi'(a+\theta_1 h)},$$

for $\phi(a)=0$ and $\psi(a)=0$ by supposition. Hence in the limit when $x=a$ (and therefore $h=0$), we have

$$Lt_{x=a}\frac{\phi(x)}{\psi(x)}=Lt_{h=0}\frac{\phi'(a+\theta h)}{\psi'(a+\theta_1 h)}=\frac{\phi'(a)}{\psi'(a)}.$$

If it should happen that $\phi'(a)$ and $\psi'(a)$ are both zero, we can, as before, repeat the process of differentiating the numerator and denominator before substitution for x.

Ex. 1. $Lt_{\theta=0}\frac{\sin\theta-\theta}{\theta^3}$.

Here $\phi(\theta)=\sin\theta-\theta$, and $\psi(\theta)=\theta^3$, which both vanish when θ vanishes.

$$\phi'(\theta) = \cos\theta - 1, \text{ and } \psi'(\theta) = 3\theta^2,$$

and both of these expressions vanish with θ.

Differentiating again

$$\phi''(\theta) = -\sin\theta, \text{ and } \psi''(\theta) = 6\theta,$$

and still both expressions vanish with θ. We must therefore differentiate again

$$\phi'''(\theta) = -\cos\theta, \text{ and } \psi'''(\theta) = 6,$$

whence $\phi'''(0) = -1$, and $\psi'''(0) = 6$;

therefore $Lt_{\theta=0} \dfrac{\sin\theta - \theta}{\theta^3} = -\dfrac{1}{6}$.

Ex. 2. $Lt_{\theta=0} \dfrac{e^\theta - e^{-\theta} + 2\sin\theta - 4\theta}{\theta^5}$ $\left[\text{Form } \frac{0}{0}\right]$

$= Lt_{\theta=0} \dfrac{e^\theta + e^{-\theta} + 2\cos\theta - 4}{5\theta^4}$ $\left[\text{Form } \frac{0}{0}\right]$

$= Lt_{\theta=0} \dfrac{e^\theta - e^{-\theta} - 2\sin\theta}{20\theta^3}$ $\left[\text{Form } \frac{0}{0}\right]$

$= Lt_{\theta=0} \dfrac{e^\theta + e^{-\theta} - 2\cos\theta}{60\theta^2}$ $\left[\text{Form } \frac{0}{0}\right]$

$= Lt_{\theta=0} \dfrac{e^\theta - e^{-\theta} + 2\sin\theta}{120\theta}$ $\left[\text{Form } \frac{0}{0}\right]$

$= Lt_{\theta=0} \dfrac{e^\theta + e^{-\theta} + 2\cos\theta}{120} = \dfrac{1}{30}$.

334. The proposition of Art. 332 may also be treated as follows.

Let $\phi(a) = 0$ and $\psi(a) = 0$, and let the pth differential coefficient of $\phi(x)$ and the qth of $\psi(x)$ be the first which do not vanish when x is put equal to a. Then by Taylor's Theorem, putting $x = a + h$,

$$\phi(x) = \phi(a) + h\phi'(a) + \ldots + \frac{h^{p-1}}{(p-1)!}\phi^{p-1}(a) + \frac{h^p}{p!}\phi^p(a+\theta h)$$

$$= \frac{h^p}{p!}\phi^p(a+\theta h).$$

Similarly

$$\psi(x) = \frac{h^q}{q!}\psi^q(a+\theta_1 h).$$

Hence $$Lt_{x=a}\frac{\phi(x)}{\psi(x)}=\frac{q!}{p!}Lt_{h=0}h^{p-q}\frac{\phi^p(a+\theta h)}{\psi^q(a+\theta_1 h)}$$
$$=\frac{q!}{p!}\frac{\phi^p(a)}{\psi^q(a)}Lt_{h=0}h^{p-q}.$$

Now, if $p>q$, $Lt_{h=0}h^{p-q}=0$.

If $p<q$, $Lt_{h=0}h^{p-q}=\infty$.

If $p=q$, $Lt_{x=a}\frac{\phi(x)}{\psi(x)}=\frac{\phi^p(a)}{\psi^p(a)}$;

so that the limit is 0, $\frac{\phi^p(a)}{\psi^p(a)}$, or ∞, according as p is $>$, $=$, or $<q$.

335. II. Form $0\times\infty$.

Let $\phi(a)=0$ and $\psi(a)=\infty$, so that $\phi(x)\psi(x)$ takes the form $0\times\infty$ when x approaches and ultimately coincides with the value a.

Then $$Lt_{x=a}\phi(x)\psi(x)=Lt_{x=a}\frac{\phi(x)}{\dfrac{1}{\psi(x)}},$$

and since $$\frac{1}{\psi(a)}=\frac{1}{\infty}=0,$$

the limit may be supposed to take the form $\frac{0}{0}$, and may be treated like Form I.

Ex. 1. $$Lt_{\theta=0}\theta\cot\theta=Lt_{\theta=0}\frac{\theta}{\tan\theta}=Lt_{\theta=0}\frac{1}{\sec^2\theta}=1.$$

Ex. 2. $$Lt_{x=\infty}x\sin\frac{a}{x}=Lt_{x=\infty}\frac{\sin\dfrac{a}{x}}{\dfrac{1}{x}}=Lt_{\frac{a}{x}=0}a\frac{\sin\dfrac{a}{x}}{\dfrac{a}{x}}=a.$$

336. III. Form $\frac{\infty}{\infty}$.

Let $\phi(a)=\infty$, $\psi(a)=\infty$, so that $\frac{\phi(x)}{\psi(x)}$ takes the form $\frac{\infty}{\infty}$ when x approaches indefinitely near the value a.

The artifice adopted in this case is to write

$$\frac{\phi(x)}{\psi(x)}=\frac{\dfrac{1}{\psi(x)}}{\dfrac{1}{\phi(x)}}.$$

Then since $\dfrac{1}{\psi(a)}=\dfrac{1}{\infty}=0$, and $\dfrac{1}{\phi(a)}=\dfrac{1}{\infty}=0$, we may consider this as taking the form $\dfrac{0}{0}$, and therefore we may apply the preceding rule.

$$Lt_{x=a}\frac{\phi(x)}{\psi(x)}=Lt_{x=a}\frac{\dfrac{1}{\psi(x)}}{\dfrac{1}{\phi(x)}}=Lt_{x=a}\frac{\dfrac{\psi'(x)}{[\psi(x)]^2}}{\dfrac{\phi'(x)}{[\phi(x)]^2}}$$

$$=Lt_{x=a}\left[\frac{\phi(x)}{\psi(x)}\right]^2\frac{\psi'(x)}{\phi'(x)}.$$

Therefore $\quad Lt_{x=a}\dfrac{\phi(x)}{\psi(x)}=\left[Lt_{x=a}\dfrac{\phi(x)}{\psi(x)}\right]^2 Lt_{x=a}\dfrac{\psi'(x)}{\phi'(x)}$ (Art. 12).

Hence, *unless* $Lt_{x=a}\dfrac{\phi(x)}{\psi(x)}$ *be zero or infinite*, we have

$$1=\left\{Lt_{x=a}\frac{\phi(x)}{\psi(x)}\right\}\left\{Lt_{x=a}\frac{\psi'(x)}{\phi'(x)}\right\},$$

or $$Lt_{x=a}\frac{\phi(x)}{\psi(x)}=Lt_{x=a}\frac{\phi'(x)}{\psi'(x)}.$$

If, however, $Lt_{x=a}\dfrac{\phi(x)}{\psi(x)}$ be zero, then

$$Lt_{x=a}\frac{\phi(x)+\psi(x)}{\psi(x)}=1,$$

and therefore, by the former case (the limit being neither zero nor infinite), $\quad =Lt_{x=a}\dfrac{\phi'(x)+\psi'(x)}{\psi'(x)}.$

Hence, subtracting unity from each side,

$$Lt_{x=a}\frac{\phi(x)}{\psi(x)}=Lt_{x=a}\frac{\phi'(x)}{\psi'(x)}.$$

Finally, in the case in which

$$Lt_{x=a}\frac{\phi(x)}{\psi(x)}=\infty, \quad Lt_{x=a}\frac{\psi(x)}{\phi(x)}=0,$$

and therefore by the last case

$$=Lt_{x=a}\frac{\psi'(x)}{\phi'(x)};$$

therefore $\quad Lt_{x=a}\dfrac{\phi(x)}{\psi(x)}=Lt_{x=a}\dfrac{\phi'(x)}{\psi'(x)}.$

This result is therefore proved true in all cases.

337. *If any function become infinite for any finite value of the independent variable, then all its differential coefficients will also become infinite for the same value.* An algebraical function only becomes infinite by the vanishing of some factor in the denominator. Now, the process of differentiating never removes such a factor, but raises it to a higher power in the denominator. Hence all differential coefficients of the given function will contain that vanishing factor in the denominator, and will therefore become infinite when such a value is given to the independent variable as will make that factor vanish.

It is obvious too that the circular functions which admit of infinite values, viz., $\tan x$, $\cot x$, $\sec x$, $\operatorname{cosec} x$, are really fractional forms, and become infinite by the vanishing of a sine or cosine *in the denominator*, and therefore these follow the same rule as the above.

The rule is also true for the logarithmic function

$\log(x-a)$ when $x=a$, or for the exponential function $b^{\frac{1}{x-a}}$ when $x=a$, b being supposed greater than unity.*

338. From the above remarks it will appear that if $\phi(a)$ and $\psi(a)$ become infinite so also *in general* will $\phi'(a)$ and $\psi'(a)$. Hence at first sight it would appear that the formula $Lt_{x=a}\dfrac{\phi'(x)}{\psi'(x)}$ is no better than the original form $Lt_{x=a}\dfrac{\phi(x)}{\psi(x)}$. But it *generally* happens that the limit of the expression $\dfrac{\phi'(x)}{\psi'(x)}$, when $x=a$, can be more easily evaluated.

Ex. 1. *Find* $Lt_{\theta=\frac{\pi}{2}}\dfrac{\log\left(\theta-\frac{\pi}{2}\right)}{\tan\theta}$ *which is of the form* $\dfrac{\infty}{\infty}$.

Following the rule of differentiating numerator for new numerator, and denominator for new denominator, we may write the above limit

$$=Lt_{\theta=\frac{\pi}{2}}\frac{\dfrac{1}{\theta-\frac{\pi}{2}}}{\sec^2\theta},$$

which is still of the form $\dfrac{\infty}{\infty}$. But it can be written

$$=Lt_{\theta=\frac{\pi}{2}}\frac{\cos^2\theta}{\theta-\frac{\pi}{2}}\ \left(\text{which is of the form } \frac{0}{0}\right)$$

$$=Lt_{\theta=\frac{\pi}{2}}\frac{-2\cos\theta\sin\theta}{1}=0.$$

Ex. 2. *Evaluate* $Lt_{x=\infty}\dfrac{x^n}{e^x}$, *which is of the form* $\dfrac{\infty}{\infty}$.

* For further discussion of this point the student is referred to Professor De Morgan's "Diff. and Int. Calculus."

$$Lt_{x=\infty}\frac{x^n}{e^x}=Lt_{x=\infty}\frac{nx^{n-1}}{e^x}=\dots$$

$$=Lt_{x=\infty}\frac{n!}{e^x}=\frac{n!}{\infty}=0.$$

It is obvious that the same result is true when n is fractional.

Ex. 3. *Evaluate $Lt_{x=0}x^m(\log x)^n$, m and n being positive.*

This is of the form $0\times\infty$, but may be written

$$Lt_{x=0}\left\{\frac{\log x}{x^{-\frac{m}{n}}}\right\}^n \qquad \left[\text{Form } \frac{\infty}{\infty}\right]$$

and by putting $x^{\frac{m}{n}}=e^{-y}$ this expression is reduced to

$$Lt_{y=\infty}\left\{\frac{-\frac{n}{m}y}{e^y}\right\}^n=0 \text{ as in Ex. 2.}$$

339. IV. Form $\infty-\infty$.

Next, suppose $\phi(a)=\infty$ and $\psi(a)=\infty$, so that $\phi(x)-\psi(x)$ takes the form $\infty-\infty$, when x approaches and ultimately coincides with the value a.

Let $$u=\phi(x)-\psi(x)=\psi(x)\left\{\frac{\phi(x)}{\psi(x)}-1\right\}.$$

From this method of writing the expression it is obvious that unless $Lt_{x=a}\frac{\phi(x)}{\psi(x)}=1$ the limit of u becomes $\psi(a)\times$ (a quantity which does not vanish); and therefore the limit sought is ∞.

But if $Lt_{x=0}\frac{\phi(x)}{\psi(x)}=1$, the problem is reduced to the evaluation of an expression which takes the form $\infty\times 0$, a form which has been already discussed (II.).

Ex. $Lt_{x=0}\left(\frac{1}{x}-\cot x\right)=Lt_{x=0}\frac{1}{x}(1-x\cot x)$

$$=Lt_{x=0}\frac{\sin x-x\cos x}{x\sin x}\ \left(\text{which is of the form } \frac{0}{0}\right)$$

$$=Lt_{x=0}\frac{x\sin x}{\sin x+x\cos x}\ \left(\text{which is of the same form still}\right)$$

$$=Lt_{x=0}\frac{\sin x+x\cos x}{2\cos x-x\sin x}=0.$$

340. V. Forms 0^0, ∞^0, 1^∞.

Let $y=u^v$, u and v being functions of x; then

$$\log_e y = v\log_e u.$$

Now $\log_e 1=0$, $\log_e\infty=\infty$, $\log_e 0=-\infty$; and therefore when the expression u^v takes one of the forms 0^0, ∞^0, 1^∞, $\log y$ takes the undetermined form $0\times\infty$. The rule is therefore to *take the logarithm and proceed as in Art.* 335.

Ex. 1. *Find $Lt_{x=0}x^x$, which takes the undetermined form 0^0.*

$$Lt_{x=0}\log_e x^x = Lt_{x=0}\frac{\log_e x}{\frac{1}{x}} = Lt_{x=0}\frac{\frac{1}{x}}{-\frac{1}{x^2}} = Lt_{x=0}(-x)=0,$$

whence $\quad Lt_{x=0}x^x=e^0=1.$

Ex. 2. *Find $Lt_{x=\frac{\pi}{2}}(\sin x)^{\tan x}$. This takes the form 1^∞.*

$$Lt_{x=\frac{\pi}{2}}(\sin x)^{\tan x} = Lt_{x=\frac{\pi}{2}}e^{\tan x\log\sin x},$$

and
$$Lt_{x=\frac{\pi}{2}}\tan x\log\sin x = Lt_{x=\frac{\pi}{2}}\frac{\log\sin x}{\cot x} = Lt_{x=\frac{\pi}{2}}\frac{\cot x}{-\operatorname{cosec}^2 x}$$
$$= Lt_{x=\frac{\pi}{2}}(-\sin x\cos x)=0,$$

whence $\quad$ required limit $=e^0=1.$

A slightly different arrangement of the work is exemplified here.

341. The following example is worthy of notice, viz.,

$$Lt_{x=a}\{1+\phi(x)\}^{\psi(x)},$$

given that $\quad \phi(a)=0,\quad \psi(a)=\infty,\quad Lt_{x=a}\phi(x)\psi(x)=m.$

We can write the above in the form

$$Lt_{x=a}\left[\{1+\phi(x)\}^{\frac{1}{\phi(x)}}\right]^{\phi(x)\,.\,\psi(x)},$$

which is clearly e^m by Art. 21, Chap. I.

It will be observed that many examples take this form, such, for example, as $Lt_{x=0}\left(\frac{\tan x}{x}\right)^{\frac{1}{x^2}}$ on p. 364, and Exs. 20 to 26 on p. 365.

342. $\frac{dy}{dx}$ of doubtful value at a Multiple Point.

Since $\frac{\partial u}{\partial x}=0$ and $\frac{\partial u}{\partial y}=0$ at any multiple point on the curve $u=0$, it will be apparent that at such a point the value of $\frac{dy}{dx}$ as derived from the formula

$$\frac{dy}{dx}=-\frac{\frac{\partial \phi}{\partial x}}{\frac{\partial \phi}{\partial y}}$$

will be of the undetermined form $\frac{0}{0}$.

The rule of Art. 332 may be applied to find the true limiting values of $\frac{dy}{dx}$ for such cases, but it is generally better to proceed otherwise.

If the multiple point be at the origin, the equations of the tangents at that point can be at once written down by inspection and the required values of $\frac{dy}{dx}$ thus found.

If the multiple point be not at the origin, the equation of the curve should be transformed to parallel axes through the multiple point and the problem is then solved as before.

Ex. *Consider the value of $\frac{dy}{dx}$ at the origin for the curve*

$$x^4+ax^2y+bxy^2+y^4=0.$$

The tangents at the origin are obviously

$$x=0, \qquad y=0, \qquad ax+by=0,$$

making with the axis of x angles whose tangents are respectively

$$\infty, \quad 0, \quad -\frac{a}{b},$$

which are therefore the required values of $\frac{dy}{dx}$.

EXAMPLES.

Investigate the following limiting forms :—

1. $Lt_{x=0}\dfrac{\log(1-x^2)}{\log\cos x}$.

2. $Lt_{x=1}\dfrac{2x^3-3x^2+1}{3x^5-5x^3+2}$.

3. $Lt_{x=\frac{\pi}{4}}\dfrac{1-\tan x}{1-\sqrt{2}\sin x}$.

4. $Lt_{x=1}\dfrac{1+\cos\pi x}{\tan^2\pi x}$.

5. $Lt_{x=a}\log\left(2-\dfrac{x}{a}\right)\cot(x-a)$.

6. $Lt_{x=0}\dfrac{\log_{\sin x}\cos x}{\log_{\sin\frac{x}{2}}\cos\frac{x}{2}}$.

7. $Lt_{\theta=0}\dfrac{\cot\theta\tan^{-1}(m\tan\theta)-m\cos^2\dfrac{\theta}{2}}{\sin^2\dfrac{\theta}{2}}$.

8. $Lt_{x=0}(\cos x)^{\cot^2 x}$.

9. $Lt_{x=1}(1-x^2)^{\frac{1}{\log(1-x)}}$.

10. $Lt_{x=0}(\log x)^{\log(1-x)}$.

11. $Lt_{x=\infty}\dfrac{Ax^n+Bx^{n-1}+Cx^{n-2}+\ldots}{ax^m+bx^{m-1}+cx^{m-2}+\ldots}$ according as n is $>$, $=$, or $<$ m.

12. $Lt_{x=0}x^{-x^m}$, m being positive.

13. $Lt_{x=\infty}\left(\dfrac{ax+1}{ax-1}\right)^x$.

14. $Lt_{x=\frac{\pi}{2}}\left(\dfrac{1+\cos x}{1-\cos x}\right)^{\frac{1}{\cos x}}$.

15. $Lt_{x=0}\{\cot(45° - x)\}^{\cot x}$.

16. $Lt_{x=\infty}\left(\dfrac{a_1^{\frac{1}{x}} + a_2^{\frac{1}{x}} + a_3^{\frac{1}{x}} + \ldots + a_n^{\frac{1}{x}}}{n}\right)^{nx}$.

17. $Lt_{x=0}\dfrac{2x^2 - 2e^{x^2} + 2\cos x^{\frac{3}{2}} + \sin^3 x}{x^4}$.

18. $Lt_{x=1}\dfrac{\sqrt{1+x} - \sqrt{1+x^2}}{\sqrt{1-x} - \sqrt{1-x^2}}$.

19. $Lt_{x=\infty}a^x \sin\dfrac{b}{a^x}$. $\left\{\begin{array}{l}\text{(i.) If } a \text{ be} > 1.\\ \text{(ii.) If } a \text{ be} < 1.\end{array}\right.$

20. $Lt_{x=0}\dfrac{\operatorname{cosec} x - \cot x}{x}$.

21. $Lt_{x=0}\dfrac{\sqrt{a^2 + ax + x^2} - a\sqrt{1 + \dfrac{x}{a}}}{\log\cos\dfrac{x}{a}}$.

22. $Lt_{x=a}\dfrac{e^{\frac{a-x}{a}}\sin^{-1}\dfrac{a-x}{a} + \log\dfrac{x}{a} - \dfrac{1}{2}\left(1 - \dfrac{x}{a}\right)^2}{\left(\dfrac{3x}{a} - \dfrac{3x^2}{a^2} + \dfrac{x^3}{a^3}\right)^{-1} - 1}$.

23. $Lt_{x=0}\left(\dfrac{a}{x} - \cot\dfrac{x}{a}\right)$.

24. $Lt_{x=0}\dfrac{\sqrt{a^2 + ax + x^2} - \sqrt{a^2 - ax + x^2}}{\sqrt{a+x} - \sqrt{a-x}}$.

25. $Lt_{x=0}\dfrac{\log(1 + x + x^2) + \log(1 - x + x^2)}{\sec x - \cos x}$.

26. $Lt_{x=0}\dfrac{x\sin(\sin x) - \sin^2 x}{x^6}$.

27. $Lt_{x=0}\dfrac{(1+x)^{\frac{1}{x}} - e}{x}$.

28. $Lt_{x=0}\dfrac{(1+x)^{\frac{1}{x}}-e+\dfrac{ex}{2}-\dfrac{11ex^2}{24}}{x^3}$.

29. $Lt_{x=\infty}\left[x\left(1+\dfrac{1}{x}\right)^x-ex^2\log\left(1+\dfrac{1}{x}\right)\right]$.

30. $Lt_{x=y=a}\dfrac{(x-y)a^n+(y-a)x^n+(a-x)y^n}{(x-y)(y-a)(a-x)}$.

[Put $x=a+h$, $y=a+k$, and expand in powers of h and k, and finally, after reduction, put $h=0$, $k=0$.]

31. $Lt_{x=y=1}\dfrac{\log x+\log y}{x+y-2}$.

32. Show that *generally*, if a function of two independent variables take one of the singular forms $\dfrac{0}{0}$, etc., for certain values of the variables, its value is truly indeterminate.

33. Given $$x^3+y^3+a^3=3axy,$$ find the values of $\dfrac{dy}{dx}$ when $x=y=a$.

34. Find the values of $\dfrac{dy}{dx}$ at the origin for the curve
$$x^3+y^3=3axy.$$

35. For the curve $\quad x^2y^2=(a^2-y^2)(b+y)^2$
find the values of $\dfrac{dy}{dx}$ at the point $(0, -b)$.

36. For the curve $\quad x^4+ax^2y=ay^3$
find the values of $\dfrac{dy}{dx}$ when $x=0$.

37. Prove $\quad Lt_{x=0}\dfrac{a^x-1}{x^n\sin x}\left(\dfrac{b\sin x-\sin bx}{\cos x-\cos bx}\right)^n=\left(\dfrac{b}{3}\right)^n\log a.$

38. Prove $$Lt_{x=0}\frac{\dfrac{d^{n+1}u}{dx^{n+1}}}{\dfrac{d^{n-1}u}{dx^{n-1}}}=n^2-m^2,$$

where $u=\dfrac{\cos my}{\cos y}$ and $x=\sin y$.

39. Find $Lt_{\theta=0}\dfrac{d^2y}{dx^2}$, where $y=\dfrac{\theta}{\sin\theta}$ and $\theta=\cos^{-1}(1-x)$.

[I. C. S., 1884.]

40. If $y=(\sin^{-1}x)^2$, prove that
$$Lt_{x=0}\frac{\dfrac{d^{n+2}y}{dx^{n+2}}}{\dfrac{d^ny}{dx^n}}=n^2.$$

41. Prove that $Lt_{x=\infty}\dfrac{a^{x^m}}{b^{x^n}}$ is zero or infinite according as n is greater or less than m, a and b being both greater than unity.

42. Prove $Lt_{x=a}\left(2-\dfrac{x}{a}\right)^{\tan\frac{\pi x}{2a}}=e^{\frac{2}{\pi}}$.

43. Prove $Lt_{x=a}\sqrt{a^2-x^2}\cot\left\{\dfrac{\pi}{2}\sqrt{\dfrac{a-x}{a+x}}\right\}=\dfrac{4a}{\pi}$.

44. Find $Lt_{x=0}(\cos ax)^{\operatorname{cosec}^2 bx}$.

45. Find $Lt\dfrac{a^x\sin bx-b^x\sin ax}{\tan bx-\tan ax}$, $\left\{\begin{array}{l}(1)\ \text{If } x=0.\\(2)\ \text{If } a=b.\end{array}\right.$

46. Find $Lt_{x=1}\dfrac{x^{\frac{3}{2}}-1+(x-1)^{\frac{3}{2}}}{(x^2-1)^{\frac{3}{2}}-x+1}$.

47. Find $Lt_{x=a}\dfrac{\sqrt{x}-\sqrt{a}+\sqrt{x-a}}{\sqrt{x^2-a^2}}$.

48. Find $Lt_{x=\frac{\pi}{2}}(\sin x)^{\tan x}$.

49. Prove that if, when x is infinite, $\phi(x)=\infty$, then will

$$Lt_{x=\infty}\frac{\phi(x)}{x}=Lt\{\phi(x+1)-\phi(x)\},$$

and also that $Lt_{x=\infty}\{\phi(x)\}^{\frac{1}{x}}=Lt\frac{\phi(x+1)}{\phi(x)}$.

[TODHUNTER'S DIFF. CALC.]

50. Prove that $Lt_{x=\infty}\left\{\frac{x^x}{x!}\right\}^{\frac{1}{x}}=e$. [TODHUNTER'S DIFF. CALC.]

51. Prove $$Lt_{n=\infty}\frac{1^m+2^m+3^m+\ldots+n^m}{n^{m+1}}=\frac{1}{m+1},$$

m being positive.

52. Prove $Lt_{h=0}h\{a^m+\overline{a+h}|^m+\overline{a+2h}|^m+\ldots+\overline{a+(n-1)h}|^m\}$

$$=\frac{b^{m+1}-a^{m+1}}{m+1},$$

where $h=\frac{b-a}{n}$, and a, b are any given quantities.

CHAPTER XIV.

MAXIMA AND MINIMA—ONE INDEPENDENT VARIABLE.

343. Elementary Algebraical Methods.

Examples frequently occur in elementary algebra and geometry in which it is required to find whether any limitations exist to the admissible values of certain functions for *real* values of the variable or variables upon which they depend.

For example, the function x^2-4x+9 may be written in the form $$(x-2)^2+5,$$ from which it is at once apparent that the least admissible value of the expression is 5, the value which it assumes when $x=2$. For the square of a real quantity is essentially positive, and therefore any value of x other than 2 will give a greater value than 5 to the expression considered.

As a second illustration let us investigate whether any limitation exists to the values of the expression

$$\frac{x^2-x+1}{x^2+x+1}$$

for real values of x.

Putting $$\frac{x^2-x+1}{x^2+x+1}=y,$$

we have $$x^2(1-y)-x(1+y)+1-y=0,$$

an equation whose roots are real only when

$$(1+y)^2>4(1-y)^2,$$

i.e., when $(3y-1)(3-y)$ is positive;

i.e., when y lies between the values 3 and $\frac{1}{3}$. It appears therefore that the given expression *always lies in value between* 3 *and* $\frac{1}{3}$. Its maximum value is therefore 3 and its minimum $\frac{1}{3}$.

344. **Method of Projection.**

Ex. *Suppose it be required to determine geometrically the greatest triangle inscribed in a given ellipse.*

It is obvious from elementary considerations that if the ellipse be projected orthogonally into a circle the greatest triangle inscribed in the given ellipse must project into the greatest triangle inscribed in a circle; and such a triangle is equilateral and the tangent to the circle at each angular point is parallel to the opposite side. This property of parallelism is a projective property, and therefore holds for the greatest triangle inscribed in the given ellipse.

Moreover

$$\frac{\text{Area of greatest triangle inscribed in the ellipse}}{\text{Area of ellipse}}$$

$$=\frac{\text{Area of equilateral triangle inscribed in a circle}}{\text{Area of the circle}}$$

$$=\frac{3\sqrt{3}}{4\pi}.$$

Hence the area of the greatest triangle inscribed in an

ellipse whose semiaxes are a, b is

$$\frac{3\sqrt{3}}{4}ab.$$

Examples.

1. Show algebraically that the expression $x+\frac{1}{x}$ cannot lie between 2 and -2 for real values of x. Illustrate this geometrically by tracing the hyperbola $xy-x^2=1$.

2. Prove that, if x be real, $\frac{x^2-4x+9}{x^2+4x+9}$ must lie between 5 and $\frac{1}{5}$.

3. Show that, if x be real, $\frac{x+a}{x-a}\cdot\frac{x+b}{x-b}$ cannot lie between the values $$-\left(\frac{\sqrt{a}+\sqrt{b}}{\sqrt{a}-\sqrt{b}}\right)^2 \text{ and } -\left(\frac{\sqrt{a}-\sqrt{b}}{\sqrt{a}+\sqrt{b}}\right)^2.$$

4. Show that the triangle of greatest area with given base and vertical angle is isosceles.

5. Show that the greatest chord passing through a point of intersection of two given circles is that which is drawn parallel to the line joining the centres.

6. If A, B be two given points on the same side of a given straight line and P be a point in the line, then $AP+BP$ will be least when AP and BP are equally inclined to the straight line.

7. Show that the triangle of least perimeter inscribable in a given triangle is the pedal triangle.

8. If A, B, C be the angular points of a triangle and P any other point, then $AP+BP+CP$ will be a minimum when each of the angles at P is 120°. [AP is a normal to the ellipse with foci B, C and passing through P.]

9. The diagonals of a maximum parallelogram inscribed in an ellipse are conjugate diameters of the ellipse.

10. If the sum of two varying positive quantities be constant show that their product is greatest when the quantities are equal. Extend this to the case of any number of positive quantities.

The General Problem.

345. Suppose x to be any independent variable capable of assuming *any real value whatever*, and let $\phi(x)$ be any given function of x. Let the curve $y=\phi(x)$ be represented in the adjoining figure, and let A, B, C, D, ... be those points on the curve at which the tangent is parallel to one of the co-ordinate axes.

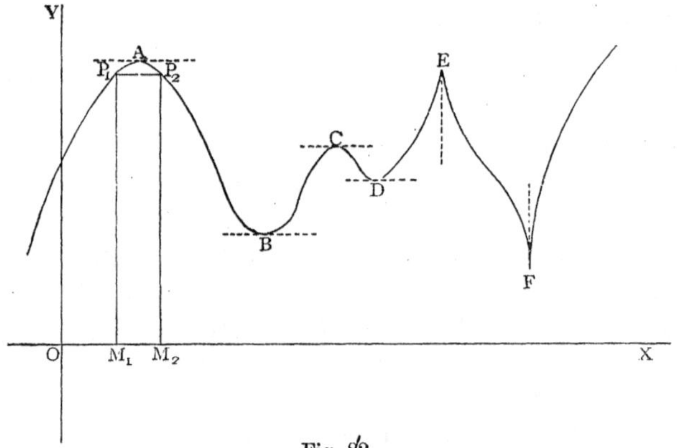

Fig. 82.

Suppose an ordinate to travel from left to right along the axis of x. Then it will be seen that as the ordinate passes such points as A, C, or E it *ceases to increase and begins to decrease;* whilst when it passes through B, D, or F it *ceases to decrease and begins to increase.* At each of the former set of points the ordinate is said to have a **maximum** value, whilst at the latter it is said to have a **minimum** value.

346. **Points of Inflexion.**

On inspection of Fig. 83 it will be at once obvious that at such points of inflexion as G or H, where the tangent is parallel to one of the co-ordinate axes, there is neither a maximum nor a minimum ordinate. Near G, for instance, the ordinate increases up to a certain value NG, and then as it passes through G it continues to increase without any prior *sensible* decrease.

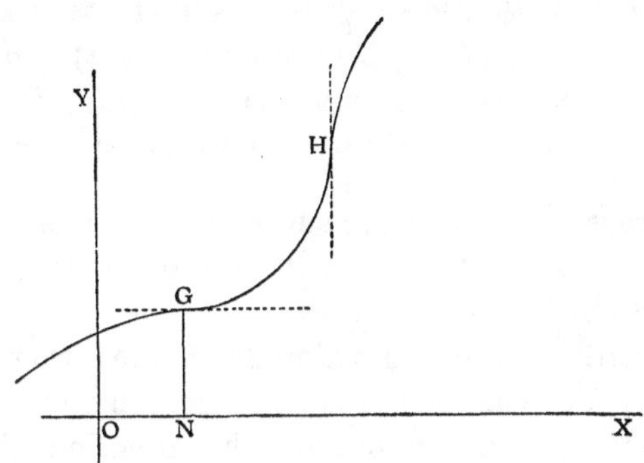

Fig. 83.

This point may however be considered as a combination of two such points as A and B in Fig. 82, the ordinate

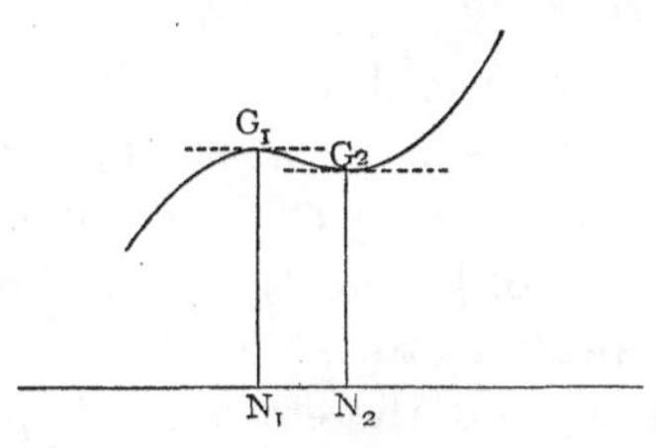

Fig. 84.

increasing up to a certain value N_1G_1, then decreasing through an indefinitely small and negligible interval to N_2G_2, and then increasing again as shown in the magnified figure (Fig. 84), the points G_1, G_2 being ultimately coincident.

347. We are thus led to the following definition:—

DEF. *If, while the independent variable x increases continuously, a function dependent upon it, say $\phi(x)$, increases through any finite interval however small until $x=a$ and then decreases, $\phi(a)$ is said to be a* MAXIMUM *value of $\phi(x)$. And if $\phi(x)$ decrease to $\phi(a)$ and then increase, both decrease and increase being through a finite interval, then $\phi(a)$ is said to be a* MINIMUM *value of $\phi(x)$.*

348. Properties of Maxima and Minima Values. Criteria. The following statements will now be obvious from the figures 82 and 83:—

(α) According to the definition given, the term maximum value does not mean the absolutely greatest nor minimum the absolutely least value of the function discussed. Moreover there may be *several maxima* values and *several minima* values of the same function, some greater and some less than others, as in the case of the ordinates at A, B, C, ... (Fig. 82).

(β) Between two equal values of a function *at least one maximum or one minimum must lie;* for whether the function be increasing or decreasing as it passes the value [M_1P_1 in Fig. 82] it must, if continuous, respectively decrease or increase again at least once before it attains its original value, and therefore must pass through at least one maximum or minimum value in the interval.

(γ) For a similar reason it is clear that between two maxima at least one minimum must lie; and between two minima at least one maximum must lie. In other words, maxima and minima values must occur **alternately.** Thus we have a maximum at A, a minimum at B, a maximum at C, etc.

(δ) In the immediate neighbourhood of a maximum or minimum ordinate two contiguous ordinates are equal, one on each side of the maximum or minimum ordinate; and these may be considered as ultimately coincident with the maximum or minimum ordinate. Moreover as the ordinate is ceasing to increase and beginning to decrease its rate of variation is itself in general an infinitesimal. This is expressed by saying that at a maximum or minimum the function discussed has a **stationary** value. This principle is of much use in the geometrical treatment of maxima and minima problems.

(ϵ) At all points, such as A, B, C, D, E, ..., at which maxima and minima ordinates occur the *tangent is parallel to one or other of the co-ordinate axes.* At points like A, B, C, D the value of $\frac{dy}{dx}$ vanishes, whilst at the cuspidal points E, F, $\frac{dy}{dx}$ becomes infinite. The positions of maxima and minima ordinates are therefore given by the roots of the equations

$$\left.\begin{array}{l}\phi'(x)=0\\ \phi'(x)=\infty\end{array}\right\}.$$

(ζ) That $\frac{dy}{dx}=0$, or $\frac{dy}{dx}=\infty$, are not in themselves *sufficient* conditions for the existence of a maximum or minimum value is clear from observing the points G, H of Fig. 83, at which the tangent is parallel to one of the co-ordinate axes, but at which the ordinate has not a maximum or minimum value. But in passing a *maximum* value of the ordinate the angle ψ which the tangent makes with OX changes from acute to obtuse (Fig.

85), and therefore $\tan \psi$, or $\frac{dy}{dx}$, changes from *positive to negative;* while in passing a *minimum* value ψ changes from obtuse to acute (Fig. 86), and therefore $\frac{dy}{dx}$ changes from *negative to positive.*

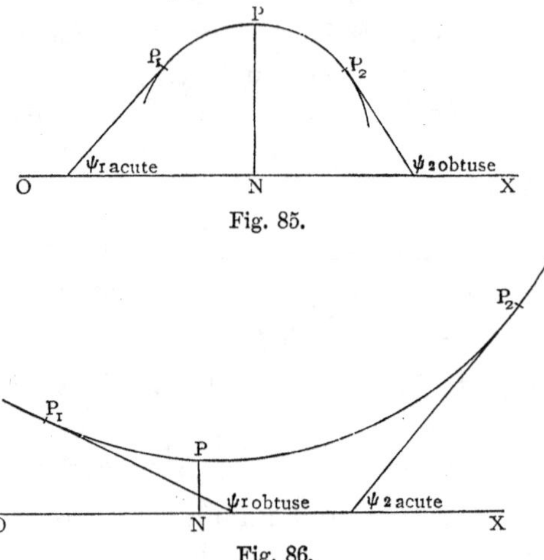

Fig. 85.

Fig. 86.

We can therefore make the following rule for the **detection and discrimination** of maxima and minima values. First *find* $\frac{dy}{dx}$ and by equating it to zero find for what values of x *it vanishes;* also observe if any values of x will make it become *infinite.* Then test for each of these values whether the sign of $\frac{dy}{dx}$ *changes from* $+$ *to* $-$ *or from* $-$ *to* $+$ as x increases through that value. If the former be the case y has a **maximum** value for that value of x; but if the latter, a

minimum. If no change of sign take place the point is a **point of inflexion** at which the tangent is parallel to one of the co-ordinate axes.

349. **Criteria for the discrimination of Maxima and Minima Values. Another Method of Investigation.**

The same criteria may be deduced at once from the aspect of $\frac{dy}{dx}$ as a *rate-measurer*. For $\frac{dy}{dx}$ is positive or negative according as y is an increasing or a decreasing function. Now, if y have a *maximum* value it is ceasing to increase and beginning to decrease, and therefore $\frac{dy}{dx}$ must be changing from *positive to negative;* and if y have a *minimum* value it is ceasing to decrease and beginning to increase, and therefore $\frac{dy}{dx}$ must be changing from *negative to positive.* Moreover, since a change from positive to negative, or *vice versa,* can only occur by passing through one of the values zero or infinity, we must search for the maximum and minimum values among those corresponding to the values of x given by $\phi'(x)=0$ or by $\phi'(x)=\infty$.

Further, since $\frac{dy}{dx}$ must be increasing when it changes from negative to positive, $\frac{d^2y}{dx^2}$ must then be positive; and similarly, when $\frac{dy}{dx}$ changes from positive to negative $\frac{d^2y}{dx^2}$ must be negative, so we arrive at another form of the criterion for maxima and minima values, viz., that there will be a maximum or minimum according as the value of x which makes $\frac{dy}{dx}$ zero or infinite gives $\frac{d^2y}{dx^2}$ a *negative or a positive sign.*

EXAMPLES.

1. Find the maximum and minimum values of y where

$$y=(x-1)(x-2)^2.$$

Here
$$\frac{dy}{dx}=(x-2)^2+2(x-1)(x-2)$$
$$=(x-2)(3x-4).$$

Putting this expression $=0$ we obtain for the values of x which give possible maxima or minima values

$$x=2 \text{ and } x=\frac{4}{3}.$$

To test these: we have,

if x be a little less than 2, $\frac{dy}{dx}=(-)(+)=\text{negative}$,

if x be a little greater than 2, $\frac{dy}{dx}=(+)(+)=\text{positive}$.

Hence there is a change of sign, viz., from negative to positive as x passes through the value 2, and therefore $x=2$ gives y a *minimum* value.

Again, if x be a little less than $\frac{4}{3}$, $\frac{dy}{dx}=(-)(-)=\text{positive}$,

and if x be a little greater than $\frac{4}{3}$, $\frac{dy}{dx}=(-)(+)=\text{negative}$,

showing that there is a change of sign in $\frac{dy}{dx}$, viz., from positive to negative, and therefore $x=\frac{4}{3}$ gives a *maximum* value for y.

Otherwise:
$$\frac{dy}{dx}=(x-2)(3x-4),$$

so that when $\frac{dy}{dx}$ is put $=0$ we obtain $x=2$ or $\frac{4}{3}$.

And
$$\frac{d^2y}{dx^2}=6x-10,$$

so that, when $x=2$, $\frac{d^2y}{dx^2}=2$,

a positive quantity, showing that, when $x=2$, y assumes a minimum value, whilst, when
$$x=\frac{4}{3},\quad \frac{d^2y}{dx^2}=-2,$$
which is negative, showing that, for this value of x, y assumes a maximum value.

2. If
$$\frac{dy}{dx}=(x-a)^{2n}(x-b)^{2p+1},$$
where n and p are positive integers, show that $x=a$ gives neither maximum nor minimum values of y, but that $x=b$ gives a minimum.

It will be clear from this example that neither maxima nor minima values can arise from the vanishing of such factors of $\frac{dy}{dx}$ *as have even indices.*

3. Show that $\frac{x^2-7x+6}{x-10}$ has a maximum value when $x=4$ and a minimum when $x=16$.

4. If
$$\frac{dy}{dx}=x(x-1)^2(x-3)^3,$$
show that $x=0$ gives a maximum value to y
and $x=3$ gives a minimum.

5. *To show that a triangle of maximum area inscribed in any oval curve is such that the tangent at each angular point is parallel to the opposite side.*

If PQR be a maximum triangle inscribed in the oval, its vertex P lies between the vertices L, M of two equal triangles LQR, MQR inscribed in the oval. Now, the chord LM is parallel to QR and

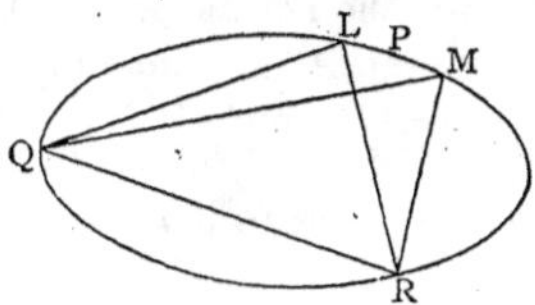

Fig. 87.

the tangent at P is the limiting position of the chord LM, which proves the proposition.

It follows that, if the oval be an ellipse, the medians of the triangle are diameters of the curve, and therefore the centre of gravity of the triangle is at the centre of the ellipse.

6. Show that the sides of a triangle of minimum area circumscribing any oval curve are bisected at the points of contact; and hence that, if the oval be an ellipse, the centre of gravity of such a triangle coincides with the centre of the ellipse.

7. *To find the path of a ray of light from a point A in one medium to a point B in another medium supposing the path to be such that the least possible time is occupied in passing from A to B, and that the velocity of propagation of light changes from v to v′ on passing the boundary separating the media.* [FERMAT'S PROBLEM.]

We shall, for simplicity, consider A and B to lie in the plane of the paper, and the separating surface of the media to be cylindrical with its generators perpendicular to the plane of the paper.

Let OPP' be the section of the separating surface by the plane of the paper, and let APB, $AP'B$ be two contiguous paths from A to B.

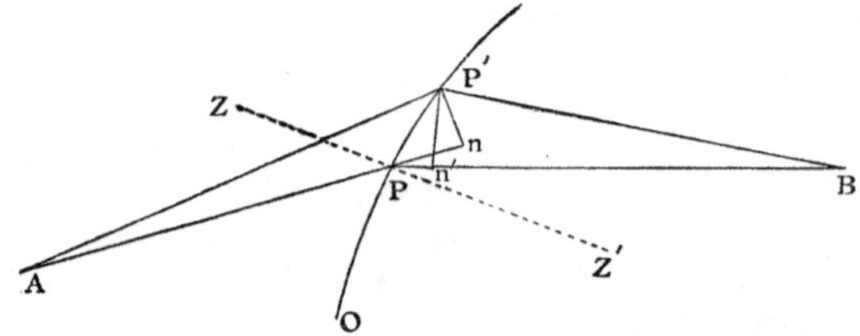

Fig. 88.

Then, if the times in these two paths be equal, the quickest path lies between them. Let fall perpendiculars $P'n$, $P'n'$ from P' upon AP and BP, and draw the normal ZPZ' at the point P.

Then, since the time in APB = time in $AP'B$,

$$\frac{AP}{v}+\frac{PB}{v'}=\frac{AP'}{v}+\frac{BP'}{v'},$$

or in the limit

$$\frac{Pn}{v}=\frac{Pn'}{v'},$$

whence

$$Lt\frac{\sin nPZ'}{\sin n'PZ'}=Lt\frac{Pn}{Pn'}=\frac{v}{v'};$$

and therefore, if in the limit the incident ray AP and the refracted ray PB make angles i, i' respectively with the normal at P, we obtain

$$\frac{\sin i}{\sin i'}=\frac{v}{v'},$$

thus proving Snell's well known law of refraction.

350. **Analytical Investigation.**

We now proceed to investigate the conditions for the existence of maxima and minima values from a purely analytical point of view.

It appears from the definition given of maxima and minima values that as x increases or decreases from the value a through any small but finite interval h, if $\phi(x)$ be always less than $\phi(a)$, then $\phi(a)$ is a maximum value of $\phi(x)$; and that if $\phi(x)$ be always greater than $\phi(a)$, then $\phi(a)$ is a minimum value of $\phi(x)$.

We shall assume in the present article that none of the derived functions we find it necessary to employ become infinite or discontinuous for the particular values discussed of the independent variable. We then have by Lagrange's modification of Taylor's Theorem

$$\left.\begin{aligned}\phi(x+h)-\phi(x)&=h\phi'(x)+\frac{h^2}{2!}\phi''(x+\theta h)\\ \text{and}\quad \phi(x-h)-\phi(x)&=-h\phi'(x)+\frac{h^2}{2!}\phi''(x-\theta h)\end{aligned}\right\}\ldots\ldots\ldots \text{(A)}$$

And when h is made sufficiently small the sign of the right-hand side of each equation, and therefore also of the left-hand side, is ultimately dependent upon that of $h\phi'(x)$, that being the term of lowest degree in h.

Hence $\qquad \phi(x+h)-\phi(x)$

and $\qquad \phi(x-h)-\phi(x)$

have in general *opposite signs*.

For a maximum or minimum value, however, it has been explained above that these expressions must, when h is taken small enough, have the *same sign*. It is therefore necessary that $\phi'(x)$ should *vanish*, so that the lowest terms of the right-hand sides of the equations (A) should depend upon an even power of h. $\phi'(x)=0$ is therefore an *essential* condition for the occurrence of a maximum or minimum value. Let the roots of this equation be $a, b, c, \ldots$.

Consider the root $x=a$.

We may now replace equations (A) by the two equations

$$\left.\begin{aligned}\phi(a+h)-\phi(a)&=\frac{h^2}{2!}\phi''(a)+\frac{h^3}{3!}\phi'''(a+\theta_1 h)\\ \phi(a-h)-\phi(a)&=\frac{h^2}{2!}\phi''(a)-\frac{h^3}{3!}\phi'''(a-\theta_1 h)\end{aligned}\right\}\ldots\ldots\ldots \text{(B)}$$

It is obvious now as before that the term $\frac{h^2}{2!}\phi''(a)$, being that of lowest degree, governs the sign of the right and therefore also of the left side of each of equations (B); *i.e.*, in general the signs of

$$\left.\begin{aligned}&\phi(a+h)-\phi(a)\\ \text{and}\qquad &\phi(a-h)-\phi(a)\end{aligned}\right\}$$

are the same as that of $\phi''(a)$. Hence if $\phi''(a)$ be *negative* $\phi(a+h)$ and $\phi(a-h)$ are both $<\phi(a)$, and therefore $\phi(a)$ is a *maximum* value of $\phi(x)$; while if $\phi''(a)$ be *positive* both $\phi(a+h)$ and $\phi(a-h)$ are $>\phi(a)$, and therefore $\phi(a)$ is a *minimum* value of $\phi(x)$. But if it should happen that $\phi''(a)$ vanishes, equations (B) are replaced by

$$\left.\begin{aligned}\phi(a+h)-\phi(a)&=\frac{h^3}{3!}\phi'''(a)+\frac{h^4}{4!}\phi''''(a+\theta_2 h)\\ \phi(a-h)-\phi(a)&=-\frac{h^3}{3!}\phi'''(a)+\frac{h^4}{4!}\phi''''(a-\theta_2 h)\end{aligned}\right\},\ldots\ldots \text{(C)}$$

and therefore when h is sufficiently small

$$\left.\begin{aligned}&\phi(a+h)-\phi(a)\\ &\phi(a-h)-\phi(a)\end{aligned}\right\}$$

are of opposite signs, and therefore there cannot be a maximum or minimum value of $\phi(x)$ when $x=a$ unless

$\phi'''(a)$ also vanish, in which case the sign of the right side of each equation depends upon that of $\phi''''(a)$. And, as before, if this be negative we have a maximum value and if positive a minimum.

Similarly, if several successive differential coefficients vanish when x is put equal to a, it appears that for a maximum or minimum value it is essential that the first not vanishing should be of an *even order*, and that if that differential coefficient be *negative* when $x = a$ a *maximum* value of $\phi(x)$ is indicated, but if *positive* a *minimum.*

EXAMPLES.

1. Determine for what values of x the function

$$\phi(x) = 12x^5 - 45x^4 + 40x^3 + 6$$

acquires maximum or minimum values.

Here $$\phi'(x) = 60(x^4 - 3x^3 + 2x^2).$$

Putting this $=0$ we obtain $x=0$, $x=1$, $x=2$.

Again $$\phi''(x) = 60(4x^3 + 9x^2 + 4x).$$

If $x=1$, $\phi''(x)$ is negative and therefore we have a maximum value; if $x=2$, $\phi''(x)$ is positive and therefore this value of x gives a minimum value for $\phi(x)$. If $x=0$, $\phi''(x)$ vanishes, so we must proceed further.

Now $$\phi'''(x) = 60(12x^2 - 18x + 4),$$

which does not vanish when $x=0$, so $x=0$ gives neither a maximum nor a minimum.

2. Show that $x=0$ gives a maximum value, and $x=1$ a minimum, for the function $$\frac{x^3}{3} - \frac{x^2}{2}.$$

3. Show that $x=0$ gives a maximum and $x=1$ a minimum for

$$\frac{x^5}{5} - \frac{x^4}{4}.$$

4. Show that the expression $\sin^3\theta \cos\theta$ attains a maximum value when $\theta = 60°$.

5. Illustrate geometrically the statement of Art. 350 that in general $\phi(x+h) - \phi(x)$ and $\phi(x-h) - \phi(x)$ are of opposite sign.

Implicit Functions.

351. In the case in which the quantity y, whose maximum and minimum values are the subject of investigation, appears as an implicit function of x, and cannot readily be expressed explicitly, we may proceed as follows:—

Let the connecting relation between x and y be

$$\phi(x, y)=0, \qquad (1)$$

then

$$\frac{\partial\phi}{\partial x}+\frac{\partial\phi}{\partial y}\frac{dy}{dx}=0. \qquad (2)$$

But for a maximum or minimum value of y, $\frac{dy}{dx}=0$, and therefore $\frac{\partial\phi}{\partial x}=0$.

The values of y deduced from the equations

$$\left.\begin{aligned}\phi(x, y)&=0\\ \frac{\partial\phi}{\partial x}&=0\end{aligned}\right\} \qquad (3)$$

therefore include the required maxima and minima.

Differentiating equation (2) we have

$$\frac{\partial^2\phi}{\partial x^2}+\frac{\partial^2\phi}{\partial x\partial y}\cdot\frac{dy}{dx}+\left(\frac{\partial^2\phi}{\partial x\partial y}+\frac{\partial^2\phi}{\partial y^2}\cdot\frac{dy}{dx}\right)\frac{dy}{dx}+\frac{\partial\phi}{\partial y}\cdot\frac{d^2y}{dx^2}=0, \qquad (4)$$

and, remembering that $\frac{dy}{dx}=0$, this reduces to

$$\frac{d^2y}{dx^2}=-\frac{\frac{\partial^2\phi}{\partial x^2}}{\frac{\partial\phi}{\partial y}}. \qquad (5)$$

Substituting the values of x and y derived from equations (3) we can test the sign of $\frac{d^2y}{dx^2}$, and thus discriminate between the maxima and minima values.

The case in which this test fails, viz., when $\frac{\partial^2 \phi}{\partial x^2}=0$ for the values of x and y deduced by equations (3), is complicated owing to the complex nature of the general formulae for $\frac{d^3y}{dx^3}$ and $\frac{d^4y}{dx^4}$.

Ex. Find the maximum and minimum ordinates of the curve

$$x^3+y^3=3axy.$$

Here
$$(x^2-ay)+(y^2-ax)\frac{dy}{dx}=0, \dots\dots\dots\dots\dots\dots\dots \quad (1)$$

and $\frac{dy}{dx}=0$ gives
$$x^2=ay.$$

Combining this with the equation to the curve we obtain

$$y^3=2axy;$$

i.e.,
$$y=0 \text{ or } y^2=2ax.$$

$y=0$ gives
$$x=0,$$

whilst $y^2=2ax$ and $x^2=ay$ give
$$y^4=4a^3y,$$

which presents the additional solution

$$y=a\sqrt[3]{4},$$
$$x=a\sqrt[3]{2}.$$

Hence the points at which maxima or minima ordinates may exist have for their co-ordinates (0, 0) and $(a\sqrt[3]{2}, a\sqrt[3]{4})$.

Now
$$\frac{\partial^2\phi}{\partial x^2}=6x \text{ and } \frac{\partial\phi}{\partial y}=3(y^2-ax),$$

and therefore at the point
$$x=a\sqrt[3]{2},$$
$$y=a\sqrt[3]{4},$$

$$\frac{d^2y}{dx^2}=-\frac{\frac{\partial^2\phi}{\partial x^2}}{\frac{\partial\phi}{\partial y}}=-\frac{2x}{y^2-ax}=\frac{-2a\sqrt[3]{2}}{2a^2\sqrt[3]{2}-a^2\sqrt[3]{2}}=-\frac{2}{a},$$

and is negative, and therefore at this point y has a maximum value.

At the point $x=0$, $y=0$, the formulae for $\frac{dy}{dx}$ and $\frac{d^2y}{dx^2}$, both become indeterminate, and we have to investigate their true values.

Differentiating equation (1) we have

$$2x - 2a\frac{dy}{dx} + 2y\left(\frac{dy}{dx}\right)^2 + (y^2 - ax)\frac{d^2y}{dx^2} = 0,$$

$$2 + 2\left(\frac{dy}{dx}\right)^3 + \left(6y\frac{dy}{dx} - 3a\right)\frac{d^2y}{dx^2} + (y^2 - ax)\frac{d^3y}{dx^3} = 0.$$

And when x and y both vanish these give

$$\frac{dy}{dx} = 0 \text{ and } \frac{d^2y}{dx^2} = \frac{2}{3a},$$

showing that the ordinate y has for this point a minimum value.

SEVERAL DEPENDENT VARIABLES.

352. Suppose the quantity u, whose maxima and minima values are the subject of investigation, to be a function of n variables x, y, z, etc., but that by virtue of $n-1$ relations between them there is but one variable independent, say x. We may now, from the $n-1$ equations, theoretically find the $n-1$ dependent variables y, z, ... in terms of x, and suppose that by substitution u is expressed as a function of the one independent variable x. The methods of the preceding articles can now be applied. It is often, however, inconvenient, even if possible, actually to eliminate the $n-1$ dependent variables y, z, etc., and it is not necessary that this should be immediately done.

Suppose, for instance, $u = \phi(x, y, z)$ a function such as the one discussed, x the independent variable, y and z dependent variables connected with x and y by the relations

$$F_1(x, y, z) = 0,$$
$$F_2(x, y, z) = 0.$$

Then, putting $\dfrac{du}{dx} = 0$ for a maximum or minimum, we

have
$$\frac{du}{dx}=\frac{\partial\phi}{\partial x}+\frac{\partial\phi}{\partial y}\cdot\frac{dy}{dx}+\frac{\partial\phi}{\partial z}\cdot\frac{dz}{dx}=0, \ldots\ldots\ldots\ldots(1)$$
$$\frac{\partial F_1}{\partial x}+\frac{\partial F_1}{\partial y}\cdot\frac{dy}{dx}+\frac{\partial F_1}{\partial z}\cdot\frac{dz}{dx}=0, \ldots\ldots\ldots\ldots(2)$$
$$\frac{\partial F_2}{\partial x}+\frac{\partial F_2}{\partial y}\cdot\frac{dy}{dx}+\frac{\partial F_2}{\partial z}\cdot\frac{dz}{dx}=0, \ldots\ldots\ldots\ldots(3)$$
and eliminating $\frac{dy}{dx}$ and $\frac{dz}{dx}$,
$$\begin{vmatrix} \frac{\partial\phi}{\partial x}, & \frac{\partial\phi}{\partial y}, & \frac{\partial\phi}{\partial z} \\ \frac{\partial F_1}{\partial x}, & \frac{\partial F_1}{\partial y}, & \frac{\partial F_1}{\partial z} \\ \frac{\partial F_2}{\partial x}, & \frac{\partial F_2}{dy}, & \frac{\partial F_2}{\partial z} \end{vmatrix}=0, \ldots\ldots\ldots\ldots(4)$$
an equation in x, y, z which, with $u=\phi(x, y, z)$, $F_1=0$, and $F_2=0$, will serve to find x, y, z and u.

Again, by differentiating equations (1), (2), (3), and eliminating $\frac{dy}{dx}, \frac{dz}{dx}, \frac{d^2y}{dx^2}, \frac{d^2z}{dx^2}$ we may deduce the value of $\frac{d^2u}{dx^2}$ and test its sign for the values of x, y, z found.

Ex. A Norman window consists of a rectangle surmounted by a semicircle. Given the perimeter, show that, when the quantity of light admitted is a maximum, the radius of the semicircle must equal the height of the rectangle.

[Todhunter's Diff. Calc., p. 214, Ex. 30.]

Let y be the height and $2x$ the breadth of the rectangle, then the area of the window is given by
$$A=\tfrac{1}{2}\pi x^2+2xy,$$
and this is to be a maximum.

For the perimeter we have
$$P=2y+2x+\pi x=\text{constant}.$$

Choose x to be the independent variable. Then we have, since A is a maximum,
$$\frac{dA}{dx}=0=\pi x+2y+2x\frac{dy}{dx},$$
and since P is constant
$$\frac{dP}{dx}=0=2\frac{dy}{dx}+2+\pi.$$
Eliminating $\frac{dy}{dx}$ we have
$$\pi x+2y=x(\pi+2),$$
or
$$x=y=\frac{P}{\pi+4},$$
and therefore the radius of the semicircle is equal to the height of the rectangle.

To test whether this result gives a maximum value to A we have
$$\frac{d^2A}{dx^2}=\pi+4\frac{dy}{dx}+2x\frac{d^2y}{dx^2},$$
and
$$\frac{d^2P}{dx^2}=0=2\frac{d^2y}{dx^2};$$
therefore
$$\frac{d^2A}{dx^2}=\pi+2(-2-\pi)=-\pi-4,$$
and is therefore negative.

Hence the relation found, viz., $x=y$, indicates a *maximum* value of the area.

353. In the solution of such questions as the foregoing it is frequently unnecessary to employ any test for the discrimination between the maxima and minima, since it is often sufficiently obvious from geometrical considerations which results give the maxima values and which give the minima.

354. **Function of a Function.**

Suppose $z=f(x)$, where x is capable of assuming all possible values, and let $y=F(z)$; then it appears that since
$$\frac{dy}{dx}=\frac{dy}{dz}\cdot\frac{dz}{dx}=F'(z)f'(x),$$

the vanishing of either of the factors $f'(x)$ or $F'(z)$ will give $\frac{dy}{dx}=0$, and therefore y may have maxima or minima either for solutions of $F'(z)=0$ or for such values of x as make $f'(x)=0$, and which therefore make z a maximum or minimum. Moreover, if z be not capable of assuming all possible values, it may happen that some of the roots of $F'(z)=0$ are excluded by reason of their not lying within the limits to which z is restricted. Several such problems have been discussed at length in the "Cambridge Mathematical Journal," Vol. III., p. 237.

Ex. 1. To find the maxima and minima values of the perpendicular from the centre of an ellipse upon a tangent.

If r and r' be conjugate semi-diameters, a and b the semi-axes, and p the perpendicular from the centre on the tangent at the point whose radius vector is r, we have

$$r^2+r'^2=a^2+b^2,$$

$$pr'=ab,$$

giving

$$\frac{a^2b^2}{p^2}=a^2+b^2-r^2.$$

Differentiating with respect to r

$$\frac{a^2b^2}{p^3}\frac{dp}{dr}=r,$$

and putting

$$\frac{dp}{dr}=0,$$

we obtain

$$r=0,$$

a result which is inadmissible, since r is restricted to lie between the limits a and b. It appears therefore at first sight as if the ordinary criteria had failed to determine the true maxima and minima values of r. We should remember, however, that since r is restricted to lie between certain values it will not do for an independent variable, and we should therefore have substituted the value of r from the equation of the curve in terms of θ, which is susceptible of all values and therefore suitable for an independent variable. We should thus have

$$\frac{a^2b^2}{p^3}\frac{dp}{d\theta}=r\frac{dr}{d\theta},$$

and the vanishing of $\frac{dr}{d\theta}$ indicates that the maximum and minimum values of p are to be sought at the same values of θ for which the maximum and minimum values of r occur; *i.e.*, obviously when $r=a$ and when $r=b$. This result was of course apparent *a priori* from the form of the relation between p and r.

Ex. 2. The orbits of the earth and Venus being assumed circular and co-planar, to investigate in what position Venus appears brightest.

The brightness of a planet varies directly as the area of its phase, and inversely as the square of the distance of the planet from the earth.

Let E and S be the earth and the sun and V the centre of Venus, the plane of the paper being the plane of motion.

Let PVP', QVQ' be diametral planes of the planet, perpendicular to the lines EV and SV, and let ZVZ' be the diameter perpendicular to the plane of motion. Draw QN at right angles to PP'. Let c be the planet's radius and x, a, r the lengths of EV, ES, and SV

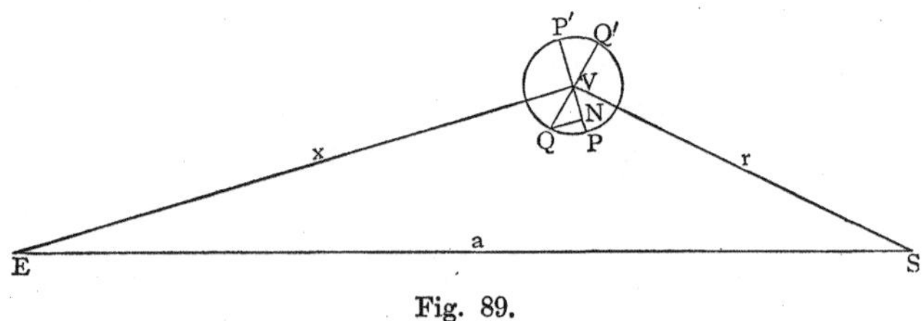

Fig. 89.

respectively. The hemispherical portion QPQ' is illuminated by the sun's rays, whilst PQP' is the portion exposed to view from the earth. The illuminated portion visible is therefore bounded by the line $ZQZ'PZ$, whose projection upon the plane $PZP'Z'$ is a crescent-shaped area bounded by a semicircle and a semi-ellipse, the greatest breadth being PN. The area of this crescent is

$$\tfrac{1}{2}\pi c^2 - \tfrac{1}{2}\pi c \,.\, c \cos NVQ,$$

and therefore $$\propto 1 - \cos NVQ.$$

The brightness therefore $\propto \dfrac{1-\cos NVQ}{EV^2}$ or $\dfrac{1+\cos EVS}{EV^2}$.

Now $$\cos EVS = \frac{x^2+r^2-a^2}{2xr},$$

whence brightness $\propto \dfrac{(x+r)^2-a^2}{x^3}$ or $\dfrac{1}{x}+\dfrac{2r}{x^2}+\dfrac{r^2-a^2}{x^3}$.

This expression has its maximum and minimum values,

(1) when x is a maximum or a minimum,

i.e., when $$x = a \pm r;$$

(2) when $$\frac{1}{x^2}+\frac{4r}{x^3}+\frac{3(r^2-a^2)}{x^4}=0.$$

This second relation gives

$$x^2+4rx+3(r^2-a^2)=0,$$

or $$x=\sqrt{3a^2+r^2}-2r,$$

the negative root being inadmissible.

We have now to inquire whether this value of x lies between the greatest and least of the admissible values of x, viz., $a \pm r$.

Now $$\sqrt{3a^2+r^2}-2r > a-r$$

if $$r < a,$$

and $$\sqrt{3a^2+r^2}-2r < a+r$$

if $$r > \frac{a}{4}.$$

For the inferior planets, Venus and Mercury, whose mean distances from the sun are respectively $\cdot 7a$ and $\cdot 39a$ roughly, r obviously lies within the prescribed limits. To distinguish between the maxima and minima, we observe that when the earth and planet are in conjunction, *i.e.*, when $x=a-r$, the brightness $=0$, and is obviously a minimum. Hence $x=\sqrt{3a^2+r^2}-2r$ gives a maximum, and $x=a+r$ a minimum. It is easy to deduce hence that, for the position of maximum brightness,

$$2\tan E = \tan\frac{V}{2},$$

an equation due to Halley, and

$$3a\cos^2 E + 4r\cos E - 4a = 0,$$

which determines the angle E.

[See GODFRAY'S ASTRONOMY, 2nd Ed., p. 287.]

355. Other Maxima and Minima; Singularities.

The accompanying figure (Fig. 90) is intended to illustrate some points with regard to maxima and minima which we have not at present considered.

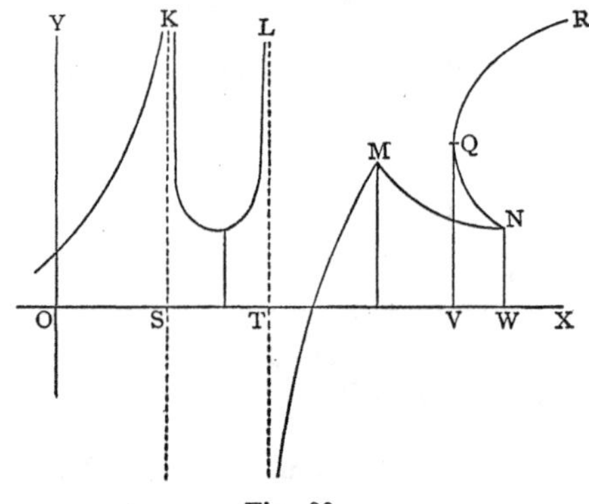

Fig. 90.

At S there is an asymptote parallel to the y-axis. The curve $y = \phi(x)$ approaches the asymptote at each side towards the same extremity. Here $y = \infty$ and $\frac{dy}{dx} = \infty$, but $\frac{dy}{dx}$ changes sign in crossing the asymptote, and there is an *infinite maximum* ordinate at S.

At T there is another asymptote parallel to the y-axis, but in crossing the asymptote the curve reappears at the opposite extremity and $\frac{dy}{dx}$ does not change sign; there is therefore neither a maximum nor a minimum at T.

At M there is a "*point saillant*" giving a discontinuity in the value of $\frac{dy}{dx}$. The ordinate at such a point is a maximum or a minimum. In the case in the figure we have a maximum ordinate.

At R the curve has a "*point d'arrêt*" and a maximum ordinate, though $\frac{dy}{dx}$ does not vanish or become infinite.

At N there is a *cusp*, but $\frac{dy}{dx}$ is neither zero nor infinite. Yet the ordinate at N is the smallest in its immediate neighbourhood, and therefore a minimum. It is to be noticed, however, that in travelling along the branch MN the value of x does not *pass through* OW, and therefore the ordinary theory does not apply.

At such points as Q, $\frac{dy}{dx}=\infty$ and changes sign, and yet obviously the value of y is not a maximum or minimum. As in the last case, it should be observed that in travelling along the branch NQR the value of x does not *pass through* the value OV, but recedes to it from W to V and then increases again. We notice, however, that this result may be written as $\frac{dx}{dy}=0$, and that $\frac{dx}{dy}$ changes sign at Q, indicating a *maximum or minimum value of the abscissa* x.

For further information upon this subject the student is referred to Professor De Morgan's "Diff. and Int. Calculus."

EXAMPLES.

1. Find the maximum and minimum values of

$$2x^3 - 15x^2 + 36x + 6.$$

2. Show that the expression

$$(x-2)(x-3)^2$$

has a maximum value when $x = \frac{7}{3}$, and a minimum value when $x = 3$.

3. Show algebraically that the greatest value of

$$x(a-x)$$

is $\frac{a^2}{4}$, and illustrate the result geometrically.

4. Show that the expression

$$x^3 - 3x^2 + 6x + 3$$

has neither a maximum nor a minimum value.

5. Investigate the maximum and minimum values of the expression $\quad 3x^5 - 25x^3 + 60x.$

6. For a certain curve

$$\frac{dy}{dx} = (x-1)(x-2)^2(x-3)^3(x-4)^4;$$

discuss the character of the curve at the points $x = 1$, $x = 2$, $x = 3$, $x = 4$.

7. Find the positions of the maximum and minimum ordinates of the curve for which

$$\frac{dy}{dx} = (x-2)^3(2x-3)^4(3x-4)^5(4x-5)^6.$$

8. Find for what values of x the expression

$$(x-1)^4(x+3)^5$$

has maximum or minimum values.

9. Find algebraically the limits between which the expression

$$ax+\frac{b}{x}$$

must or must not lie for real values of x. Illustrate your result by a sketch of the curve

$$y=ax+\frac{b}{x}.$$

10. Investigate algebraically the maximum and minimum values of the expression $\dfrac{x^2-4x+2}{2x-7}$ for real values of x. Illustrate your answer geometrically.

11. Investigate the maximum and minimum values of the expression $2x^3-21x^2+60x+30.$

12. Find the minimum ordinate and the point of inflexion on the curve $x^3-axy+b^3=0.$

13. Find the maximum and minimum ordinates of the curve

$$(y-c)^2=(x-a)^6(x-b).$$

14. Show that the curve $y=xe^x$ has a minimum ordinate where $x=-1$.

15. Show that the values of x for which $e^{x\sin x}$ has maximum or minimum values may be determined graphically as the abscissae of the points of intersection of the straight line

$$y=-x,$$

with the curve of tangents $y=\tan x$.

16. Show that the expression

$$a+(x-b)^{\frac{2}{3}}+(x-b)^{\frac{4}{3}}$$

has a minimum value when $x=b$.

17. Find the minimum value of

$$\frac{a^2}{\sin^2 x}+\frac{b^2}{\cos^2 x}.$$

18. Show that $$\sin^p\theta \cos^q\theta$$
attains a maximum value when
$$\theta = \tan^{-1}\sqrt{\frac{p}{q}}.$$

19. Show that the function
$$x \sin x + \cos x + \cos^2 x$$
continually diminishes as x increases from 0 to $\frac{\pi}{2}$.

20. Show that $\sqrt[e]{e}$ is a maximum value of $\left(\frac{1}{x}\right)^x$.

21. Show trigonometrically that the greatest and least values of the expression $$a \sin x + b \cos x$$
are $$\sqrt{a^2+b^2} \text{ and } -\sqrt{a^2+b^2}.$$

22. Show by trigonometry that the greatest and least values of the function $a \cos^2\theta + 2h \sin\theta \cos\theta + b \sin^2\theta$
are respectively $$\frac{a+b}{2} \pm \sqrt{\left(\frac{a-b}{2}\right)^2 + h^2}.$$

23. Find in an elementary manner the maximum and minimum values of the expression
$$(a \sin\theta + b \cos\theta)(a \cos\theta + b \sin\theta).$$

24. If $$y = 2x - \tan^{-1}x - \log\{x + \sqrt{1+x^2}\},$$
show that y continually increases as x changes from zero to positive infinity.

25. If $$z = \frac{a^2}{x} + \frac{b^2}{y},$$
where $$x + y = a,$$
show that z has a minimum value when
$$x = \frac{a^2}{a+b},$$
and a maximum when $$x = \frac{a^2}{a-b}.$$

26. Given that $$\frac{x}{a}+\frac{y}{b}=1,$$
show that the maximum value of xy is $\frac{ab}{4}$ and that the minimum value of x^2+y^2 is $\frac{a^2b^2}{a^2+b^2}$.

27. Show that the area of the greatest rectangle inscribed in a given ellipse and having its sides parallel to the axes of the ellipse is to that of the ellipse as $2:\pi$.

28. Show that the maximum and minimum values of
$$x^2+y^2,$$
where $$ax^2+2hxy+by^2=1$$
are given by the roots of the quadratic
$$\left(a-\frac{1}{r^2}\right)\left(b-\frac{1}{r^2}\right)=h^2.$$
Hence find the area of the conic denoted by the first equation.

29. PSP', QSQ' are focal chords of a conic intersecting at right angles. Find the positions of the chords when $PP'+QQ'$ has a maximum or minimum value.

30. Divide a given number a into two parts, such that the product of the pth power of one and the qth power of the other shall be as great as possible.

31. Show that if a number be divided into two factors, such that the sum of their squares is a minimum, the factors are each equal to the square root of the given number.

32. Into how many equal parts must the number ne be divided so that their continued product may be a maximum; n being a positive integer and e the base of the Napierian Logarithms?

33. What fraction exceeds its pth power by the greatest number possible?

34. Given the length of an arc of a circle, find the radius of the circle when the corresponding segment has a maximum or minimum area. [PAPPUS ALEXANDRINUS.]

35. The centres of two spheres, radii r_1, r_2, are at the extremities of a straight line of length $2a$, on which a circle is described. Find a point in the circumference from which the greatest amount of spherical surface is visible.

36. In the line joining the centres of two spheres find a point such that the sum of the spherical surfaces visible therefrom may be a maximum. [EDUCATIONAL TIMES.]

37. AC and BD are parallel straight lines, and AD is drawn. Show how to draw a straight line COE, cutting AD and BD in O and E respectively, so that the sum of the triangles EOD, COA may be a minimum. [VIVIANI.]

38. A person wishes to divide a triangular field into two equal parts by a straight fence. Show how it is to be done so that the fence may be of the least expense.

39. If four straight rods be freely hinged at their extremities the greatest quadrilateral they can form is inscribable in a circle.

40. A tree in the form of a frustum of a cone is n feet long, and its greater and less diameters are a and b feet respectively. Show that the greatest beam of square section that can be cut out of it is $\dfrac{na}{3(a-b)}$ feet long.

41. If the polar diameter of the earth be to the equatorial as $229:230$, show that the greatest angle made by a body falling to the earth with a perpendicular to the surface is about $14' \, 59''$, and that the latitude is $45° \, 7' \, 29''$.

42. The resistance to a steamer's motion in still water varies as the nth power of the velocity. Find the rate at which the

steamer must be propelled against a tide running at a knots an hour so as to consume the least amount of fuel in a given journey.

43. Show that the volume of the greatest cylinder which can be inscribed in a cone of height b and semivertical angle α is $\frac{4}{27}\pi b^3 \tan^2\alpha$.

44. Show that the height of the cone of greatest convex surface which can be inscribed in a given sphere is to the radius of the sphere as 4 : 3.

45. Show that the chord of a given curve which passes through a given point and cuts off a maximum or minimum area is bisected at the point.

46. Two particles move uniformly along the axes of x and y with velocities u and v respectively. They are initially at distances a and b respectively from the origin, and the axes are inclined at an angle ω. Show that the least distance between the particles is $\dfrac{(av - bu)\sin\omega}{(u^2 + v^2 - 2uv\cos\omega)^{\frac{1}{2}}}$.

47. Find the area of the greatest triangle which can be inscribed in a given parabolic segment having for its base the bounding chord of the segment.

48. For a maximum or minimum parabola circumscribing a given triangle ABC, show that the sum of the perpendiculars from ABC upon the axis is algebraically zero.

49. In a submarine telegraph cable the speed of signalling varies as $x^2 \log \frac{1}{x}$ where x is the ratio of the radius of the core to that of the covering. Show that the greatest speed is attained when this ratio is $1 : \sqrt{e}$.

50. S is the focus of an ellipse of eccentricity e, and E is a fixed point on the major axis, and P is any point on the curve. Show that when PE is a minimum $SP = \frac{SE}{e}$.

51. Find the maximum value of

$$(x-a)^2(x-b), \begin{cases} (1) \text{ when } a > b. \\ (2) \text{ when } a < b. \end{cases}$$

What happens if $a = b$? Illustrate your answers by diagrams of the curve
$$y = (x-a)^2(x-b)$$
in the three different cases.

52. An open tank is to be constructed with a square base and vertical sides so as to contain a given quantity of water. Show that the expense of lining it with lead will be least if the depth is made half of the width.

53. If two variables x and y are connected by the relation $ax^2 + by^2 = ab$, show that the maximum and minimum values of the function $x^2 + y^2 + xy$ will be the values of u given by the equation
$$4(u-a)(u-b) = ab.$$

54. If SP and SQ be two focal distances in an ellipse inclined to each other at the given angle 2α, find the greatest and least values of the area of the triangle PSQ.

55. SQ is a focal radius vector in a given ellipse inclined at a given angle α to SA, where A is the vertex nearest to the focus S. Find the angle ASP, where SP is another focal radius, such that the area of the triangle PSQ may be a maximum.

56. Find the point P on the parabola $y^2 = 4ax$ such that the perpendicular on the tangent at P from a given point on the axis distant h from the vertex may be the least possible. What is the geometrical meaning of the result?

57. Find the area and position of the maximum triangle

having a given angle which can be inscribed in a given circle, and prove that the area cannot have a minimum value.

58. From a fixed point A on the circumference of a circle of radius c the perpendicular AY is let fall on the tangent at P. Prove that the maximum area of the triangle APY is

$$\tfrac{3}{8}c^2\sqrt{3}.$$

59. If a parallelogram be inscribed in an ellipse the greatest possible value of its perimeter is equal to twice the diagonal of the rectangle described on the axes.

60. O is a fixed point without a circle, A one of the extremities of the diameter through O, OQQ' a chord through O. Find its position when the area of the triangle QAQ' is a maximum. Does it ever become a minimum?

61. Describe the equilateral triangle of greatest area, each of whose sides passes through a given fixed point.

62. A length l of wire is cut into two portions which are bent into the shapes of a circle and a square respectively. Show that if the sum of the areas be the least possible the side of the square is double the radius of the circle.

63. Find the least isosceles triangle which can be described about an ellipse with its base parallel to one of the axes, and show that its sides are parallel to those of the greatest isosceles triangle which can be inscribed in the same ellipse with its vertex at one extremity of the other axis.

64. Obtain the maximum and minimum values of the volume of a right circular cone whose vertex is at a given point and whose base is a plane section of a given sphere; and point out the difference of the cases of the point being within or without the sphere.

65. Prove that a chord of constant inclination to the arc of

a closed curve divides the area most unequally when it is a chord of curvature.

66. Show that the normal chord to the parabola $y^2=4ax$ which cuts off the least arc is normal where $y=\frac{\text{Lat. Rect.}}{\sqrt{3}}$ and is inclined to the axis at an angle $\tan^{-1}\frac{2}{\sqrt{3}}$.

67. When the product of two perpendicular radii vectores of a curve is a maximum or a minimum, show that they make supplementary angles with the tangents at their extremities.

68. Two perpendicular lines intersect on a parabola, one passing through the focus. Show that the triangle formed by them with the directrix has its least values when the focal distances of the right angle and the vertex of the parabola include an angle of 36° or of 108°.

69. Show that when the angle between the tangent to a curve and the radius vector of the point of contact has a maximum or minimum value the radius of curvature at that point is given by $\rho=\frac{r^2}{p}$.

70. Show how to find the co-ordinates of the points on a curve given in Cartesians at which the curvature is a maximum or a minimum.

APPENDIX.

APPENDIX.

GEOMETRICAL PROPERTIES OF THE CYCLOID.

356. DEF. *When a circle rolls in a plane along a given straight line, the locus traced out by any point on the circumference of the rolling circle is called a* CYCLOID.

357. Description of the Curve.

The nature of the motion shows that there is an infinite number of cusps arranged at equal distances along the given straight line. It is usual to confine the name

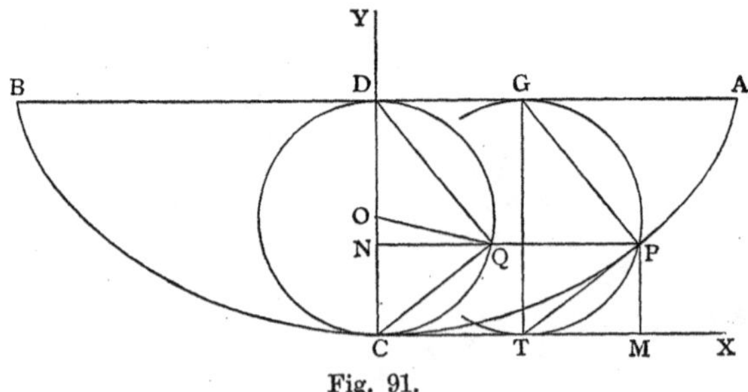

Fig. 91.

cycloid to the portion of the curve lying between two consecutive cusps.

Let A, B be two consecutive cusps, ACB the arc of the cycloid lying between them. The line AB along which the circle rolls is called the *base*. Let GPT be the rolling circle, G the point of contact, GT the diameter through G, and P the point attached to the circumference, which by its motion traces the cycloid. The circle GPT is called the *generating circle*. Let C be the point of the curve at greatest distance from AB; this point is called the *vertex*. Let CX be the tangent at C, and CY the normal, obviously bisecting the base AB in the point D. We shall take these lines as co-ordinate axes. It is clear that the curve is symmetrical about CY.

358. **Tangent and Normal.**

Since a circle may be considered as the limit of an inscribed regular polygon with an indefinitely large number of sides, the circle GPT may be supposed to be for the instant turning about an angular point of this polygon situated at G. Hence the motion of the point P is instantaneously perpendicular to the line PG, which is therefore the direction of the normal at P. Moreover, since this motion is in the direction of PT, PT is the tangent at P to the locus of P.

359. **Equations of the Cycloid.**

Let DQC be the circle described upon DC for diameter and let O be its centre. Draw PM, PN perpendicular to CX and CY respectively, the latter cutting the circle DQC in Q. Join DQ, OQ, CQ.

Now, since the circle rolls *without sliding* along the line AB, every point of the circle comes successively into contact with the straight line, so that the length of AD is half of the circumference of the circle, and the portion $GA = \text{arc } GP = \text{arc } DQ$. Hence the remainder $DG = \text{arc } CQ$.

Now, $PQCT$, $PQDG$ are parallelograms; whence, if a be the radius of the generating circle and θ the angle COQ,
$$PQ = DG = \text{arc } CQ = a\theta.$$
Hence, if x, y be co-ordinates of P,
$$\left.\begin{aligned} x &= CM = NQ + QP = a(\theta + \sin\theta) \\ y &= CN = CO - NO = a(1 - \cos\theta) \end{aligned}\right\} \text{ ....} \quad \text{(A)}$$

From these equations the Cartesian equation may be at once obtained by eliminating θ; the result being
$$x = a \operatorname{vers}^{-1}\frac{y}{a} + \sqrt{2ay - y^2}, \text{ ....} \quad \text{(B)}$$
but from the form of the result the equation is not so useful as the two equations marked (A).

360. **Length of the arc CP.**

Since
$$\left.\begin{aligned} x &= a(\theta + \sin\theta) \\ y &= a(1 - \cos\theta) \end{aligned}\right\},$$
we obtain
$$\left.\begin{aligned} dx &= a(1 + \cos\theta)d\theta \\ dy &= a\sin\theta d\theta \end{aligned}\right\},$$
squaring and adding
$$\begin{aligned} ds^2 = dx^2 + dy^2 &= 2a^2(1 + \cos\theta)d\theta^2 \\ &= 4a^2\cos^2\frac{\theta}{2}d\theta^2, \end{aligned}$$
or
$$ds = 2a\cos\frac{\theta}{2}d\theta,$$
and upon integration
$$s = 4a\sin\frac{\theta}{2}, \text{ ....} \quad \text{(C)}$$
the constant of integration vanishing if s be measured from C, so that s and θ vanish together.

Again, since $\text{chord } CQ = 2a \sin\frac{\theta}{2}$,

we have $\text{arc } CP = 2 \text{ chord } CQ.$ .... (D)

Further, since $y = 2a \sin^2\frac{\theta}{2}$,

$$s = 4a\sqrt{\frac{y}{2a}} = \sqrt{8ay}. \quad \text{........ (E)}$$

361. **Geometrical Proofs.**

These results may be established by geometry as follows:—

Let TPG be any position of the generating circle, G being the point of contact, GT the diameter through G, and P the tracing point. Let the circle roll through an infinitesimal distance till the point of contact comes to G'. Let the circle in rolling turn through an infinitesimal angle equal to POQ, OQ being a radius of the circle, and let P come to P'. Then QP' is parallel and equal to GG', and therefore to the arc QP. PP' is ultimately the

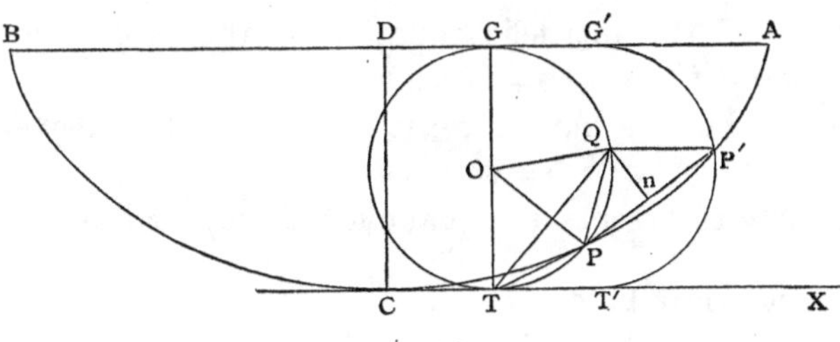

Fig. 92.

tangent at P and therefore ultimately in a straight line with TP. Draw Qn at right angles to PP'; then Tn and TQ are ultimately equal and Pn is therefore the increase in the chord TP in rolling from G to G'. Moreover PP' is ultimately the increase of arc, and since in the limit $QP' = \text{arc } QP = \text{chord } QP$, and Qn is drawn perpendicularly to PP', n is the middle point of PP', and therefore the rate of growth of the arc CP is double that of the chord TP, and they begin their growth together at C. Hence arc $CP = 2$ chord TP.

362. Intrinsic Equation.

If in Fig. 91 $PTX = \psi$, we have $\psi = \frac{\theta}{2}$; whence the intrinsic equation of the cycloid is $s = 4a \sin \psi$.

363. Radius of Curvature.

The formula of Art. 270 gives

$$\rho = \frac{ds}{d\psi} = 4a \cos \psi = 4a \cos \frac{\theta}{2} = 2PG,$$

i.e., radius of curvature $= 2$. normal.

364. Evolute.

By Art. 292 the intrinsic equation of the evolute of the curve $s = f(\psi)$ is $s = f'(\psi)$.

Applying this, we have for the evolute of the above cycloid

$$s = 4a \cos \psi,$$

which clearly represents an equal cycloid (see Art. 294).

365. Geometrical Proofs.

These results may also be established geometrically as follows:—

Let AD be half the base and CD the axis of a given cycloid APC. Produce CD to F, making DF equal to CD, and through F draw FE parallel to DA. Through any point G on the base draw TGG' parallel to CD and cutting the tangent at C in T and the line FE in G'.

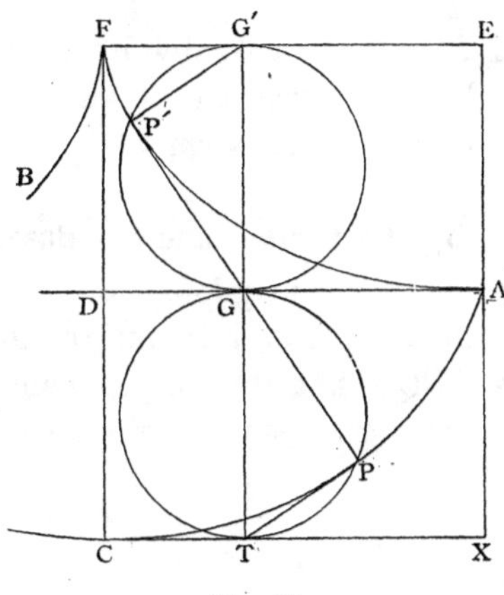

Fig. 93.

On GT and $G'G$ as diameters describe circles, the former cutting the cycloid in the tracing point P. Join PT, PG and produce PG to meet the circle $GP'G'$ in P' and join $P'G'$. Then obviously the arc $G'P' =$ arc $PT = DG = FG'$, and therefore the point P' lies on a cycloid, equal to the original cycloid, with cusp at F and vertex at A. Moreover $P'G$ is a tangent to this cycloid and $P'G'$ a normal. The cycloid FA is therefore the envelope of the normals of the cycloid AC and therefore its *evolute*; and P' is the *centre of curvature* corresponding to the point P on the original cycloid.

If, therefore, a string of length equal to the arc $FP'A$ have one extremity attached to a fixed point at F the other end, when the string is unwound from the curve $FP'A$, will trace out the cycloidal arc APC. Thus a heavy particle may be made to *oscillate along a cycloidal arc*, by allowing the suspending string to wrap alternately upon two rigid cycloidal cheeks such as FA, FB.

Moreover, since PP' is obviously by its construction bisected at G, the radius of curvature at any point of a cycloid is *double the length of the normal.*

366. **Area bounded by the Cycloid and its Base.**

Let PGP', $QG'Q'$ be two contiguous normals. Then G, G' are their middle points, and therefore ultimately the elementary area $GPQG'$ is treble the elementary area $P'GG'Q'$. Hence, summing all such elements, the area

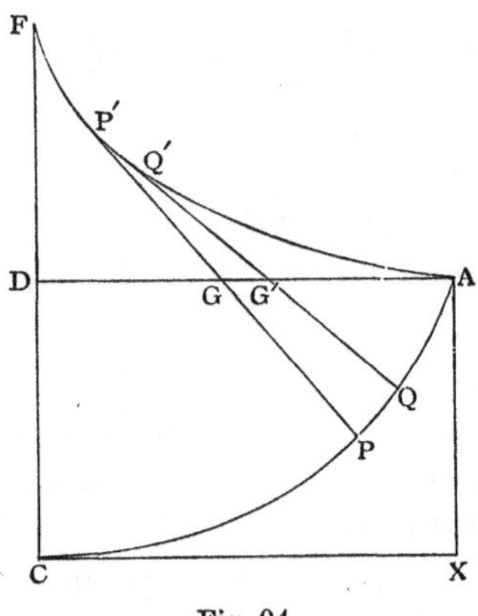

Fig. 94.

$APCD$ is treble the area $ADFP'$; *i.e.*, the area of the cycloid is three-fourths of the circumscribing rectangle, for the area of $ADFP'$ is equal to the area $CXAP$.

Now the length of AD = half the circumference of the circle

$$= \pi a.$$

Hence the

$$\text{rectangle } AXCD = \pi a \cdot 2a = 2\pi a^2,$$

and therefore the

$$\text{semicycloidal area } APCD = \tfrac{3}{4} \cdot 2\pi a^2 = \tfrac{3}{2}\pi a^2,$$

and the area bounded by the whole cycloid and its base $= 3\pi a^2$, and is therefore *three times the area of the generating circle.*

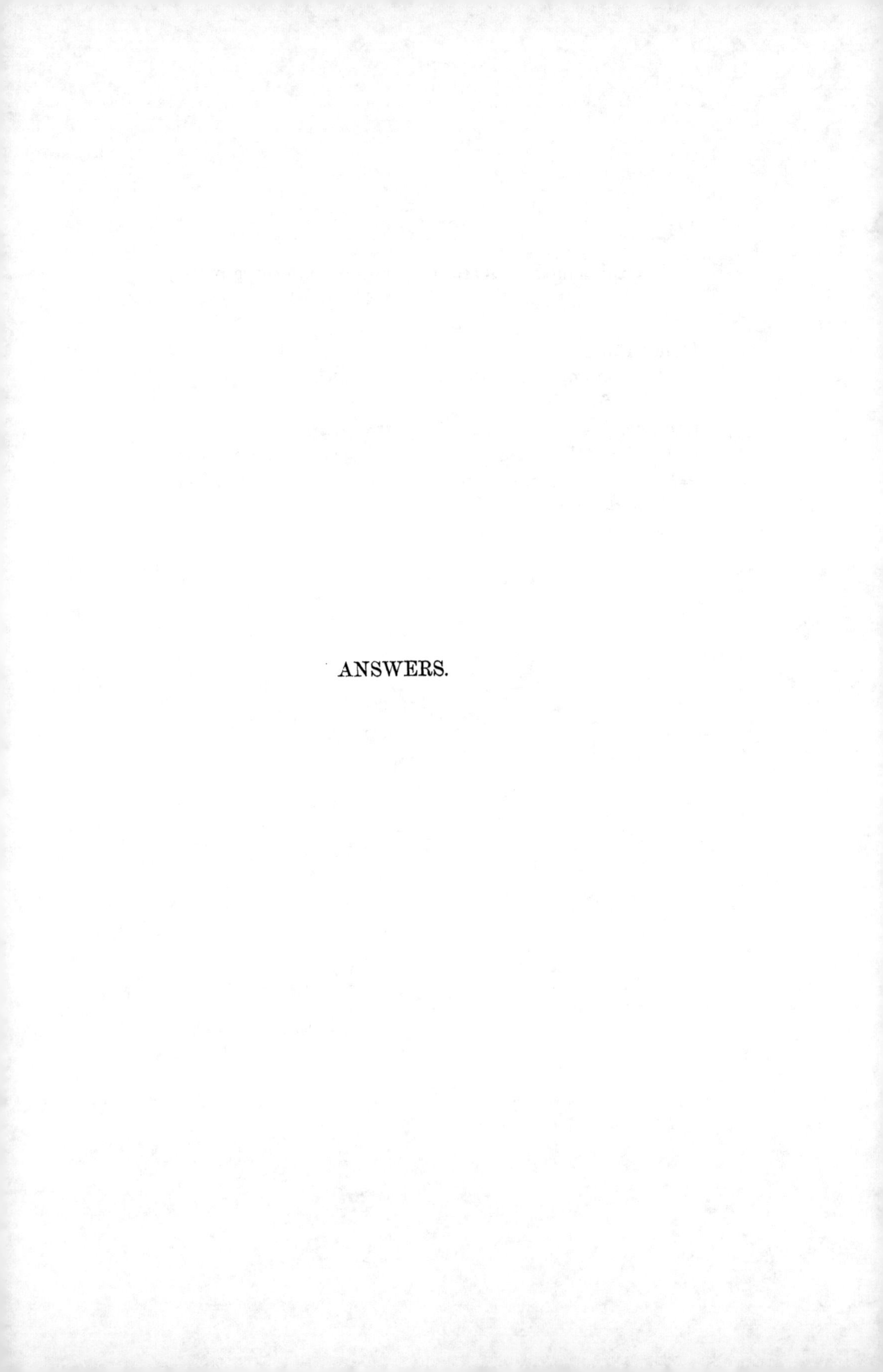

ANSWERS.

ANSWERS TO THE EXAMPLES.

CHAPTER I.

PAGE 11.

1. (i.) ∞; (ii.) $\frac{1}{a}$; (iii.) ∞.
2. (i.) $\frac{1}{2}$; (ii.) 2.
3. ∞.
4. $\pm\frac{b}{a}$.
5. $\frac{1}{2}$.
6. a.
7. $3a^2$.
8. (i.) $\frac{b}{a}$; (ii.) $\frac{a}{b}$.
9. $\frac{1}{2}$.

CHAPTER II.

PAGE 30.

1. $Xx + Yy = c^2$.
2. $\frac{Xx}{a^2} + \frac{Yy}{b^2} = 1$.
3. $Y - y = y(X - x)$.
4. $x(Y - y) = X - x$.
5. $\cos^2 x(Y - y) = X - x$.
6. $(1 + x^2)(Y - y) = X - x$.

PAGE 32.

1. $\sec^2 x$.
2. $\frac{1}{1 + x^2}$.
3. $-\frac{\cos x}{\sin^2 x}$.
4. $-\frac{1}{x\sqrt{x^2 - 1}}$.

PAGE 36.

1. $3x^2$.
2. $\sqrt{\frac{a}{x}}$.
3. $\frac{x}{\sqrt{a^2 + x^2}}$.
4. e^x.
5. $\frac{e^{\sqrt{x}}}{2\sqrt{x}}$.
6. $a^{\sin x}\cos x \log_e a$.
7. $\frac{1}{x}a^{\log x}\log_e a$.
8. $\frac{3x^2}{1 + x^6}$.
9. $-\tan x$.
10. $\frac{2}{\sin 2x}$.
11. $x^x(\log_e x + 1)$.

12. $x^{\sin x}\left\{\cos x \cdot \log x + \frac{\sin x}{x}\right\}$.

13. $(\sin x)^x\{\log \sin x + x \cot x\}$.

14. $\frac{(\sin x)^{\sqrt{x}}}{\sqrt{x}}(\log\sqrt{\sin x} + x \cot x)$.

17. $(0, 0)$ and $\left(2a, -\frac{4a^3}{3b^2}\right)$.

18. $\left(\pm\frac{a^2}{\sqrt{a^2+b^2}}, \pm\frac{b^2}{\sqrt{a^2+b^2}}\right)$.

PAGE 48.

1. $\log \sin x + x \cot x$.

2. $\frac{a^2-2x^2}{\sqrt{a^2-x^2}}$.

3. $\frac{c}{x}e^{\frac{x}{c}}\left(1-\frac{c}{x}\right)$.

4. $\frac{-a^3}{x^2\sqrt{a^2-x^2}}$.

5. $a^{\frac{1+\sin x}{2}}\cos x \cdot \log_e a$.

6. $e^{\sqrt{u}}\frac{x}{\sqrt{u}}\cot v(\sin w)^w(\log \sin w + w \cot w)$.

CHAPTER III.

PAGE 65.

1. $\frac{1}{2\sqrt{x}}$.

2. $-\frac{1}{2}x^{-\frac{3}{2}}$.

3. $\frac{b}{c}$.

4. $1-\frac{1}{x^2}$.

5. e^x.

6. $\cos x$.

7. $-\sin x$.

8. $b\cos(a+bx)$.

9. $bnx^{n-1}\cos(a+bx^n)$.

10. $\frac{\cos x}{2\sqrt{x}}$.

11. $\frac{\cos x}{2\sqrt{\sin x}}$.

12. $\frac{\cos\sqrt{x}}{4\sqrt{x\sin\sqrt{x}}}$.

13. $pqx^{q-1}\cos x^q \sin^{p-1}x^q$.

14. $\frac{2x}{\sqrt{1-x^4}}$.

15. $\frac{\pi}{\sqrt{1-x^2}}$.

16. $\frac{1}{x[1+(\log x)^2]}$.

17. $\frac{\pi}{180}\cos x°$.

18. $\log x + 1$.

19. $\frac{e^x}{x}\log(ex^x)$.

20. $\cos e^x \cdot e^x \cdot \log x + \frac{\sin e^x}{x}$.

21. $\frac{\log\sqrt{\cot x}}{\cosh x} - \frac{2\tan^{-1}e^x}{\sin 2x}$.

22. $(x+a)^{m-1}(x+b)^{n-1}[(m+n)x+mb+na]$.

23. $\frac{x^2+2x-2}{(x+1)^2}$.

24. $\frac{1}{n}(a+x)^{\frac{1-n}{n}}$.

25. $\frac{2x}{n}(a^2+x^2)^{\frac{1-n}{n}}$.

26. $\frac{\sinh x}{2\sqrt{\cosh x}}$.

27. $\tanh x$.

28. $\operatorname{sech} 2x$.

29. $\frac{2}{\sqrt{2-x^2}}$.

30. $-\frac{2\operatorname{cosec} 2x}{\sqrt{2\log\cot x - (\log\cot x)^2}}$.

31. $\frac{\cos x}{1+\sin^2 x}$.

32. $-\frac{1}{1+x^2}$.

33. $-\frac{1}{x\sqrt{x^2-1}}$.

34. $\dfrac{x^2 - 2x^{\frac{3}{2}} - 2x^{\frac{1}{2}} + 1}{2x^{\frac{1}{2}}(1+x)(1+x^2)}$. 35. $\sin^{m-1}x\cos^{n-1}x(m\cos^2 x - n\sin^2 x)$.

36. $\dfrac{(\sin^{-1}x)^{m-1}(\cos^{-1}x)^{n-1}}{\sqrt{1-x^2}}(m\cos^{-1}x - n\sin^{-1}x)$.

37. $\cos(e^x\log x)e^x\log\left(xe^{\frac{1}{x}}\right)\sqrt{1-(\log x)^2} - \sin(e^x\log x)\dfrac{\log x^{\frac{1}{x}}}{\sqrt{1-(\log x)^2}}$.

38. $-\dfrac{1}{(1-x)^{\frac{1}{2}}(1+x)^{\frac{3}{2}}}$.

39. $-\dfrac{3x+x^3}{(1+x^2)^{\frac{3}{2}}}$.

40. $\dfrac{x^4 - 2a^2x^2 + 4a^4}{(x^2-a^2)^{\frac{3}{2}}(x^2-4a^2)^{\frac{1}{2}}}$.

41. $-\dfrac{2+2x-x^2}{2(1-x)^{\frac{1}{2}}(1+x+x^2)^{\frac{3}{2}}}$.

42. $\dfrac{2(1-x^2)}{1+x^2+x^4}$.

43. $\log\left(\dfrac{e^{\frac{1}{x}}}{a}\right)$.

44. $\dfrac{2}{\sqrt{1-x^2}}$.

45. $\left(\dfrac{x}{n}\right)^{nx}\left\{n\left(\log\dfrac{ex}{n}\right)^2 + \dfrac{1}{x}\right\}$.

46. $\dfrac{ab}{a^2 + x^2\left(\tan^{-1}\dfrac{x}{a}\right)^2}\left\{\tan^{-1}\dfrac{x}{a} + \dfrac{ax}{a^2+x^2}\right\}$.

47. $\dfrac{\cos^{-1}x - x\sqrt{1-x^2}}{(1-x^2)^{\frac{3}{2}}}$. 48. $\dfrac{a\sin(a\operatorname{cosec}^{-1}x)}{x\sqrt{x^2-1}}$. 49. $-\dfrac{\sqrt{b^2-a^2}}{b+a\cos x}$.

50. $2e^{\tan^{-1}x}\left\{\dfrac{\log\sec x^3}{1+x^2} + 3x^2\tan x^3\right\}$.

51. $e^{ax}\left\{a\cos(b\tan^{-1}x) - \dfrac{b}{1+x^2}\sin(b\tan^{-1}x)\right\}$.

52. $\dfrac{xa^{cx}(2+cx\log_e a)}{1+a^{2cx}x^4}$.

53. $\dfrac{x\log_a e\sin(\log_a\sqrt{a^2+x^2})}{(a^2+x^2)\cos^2(\log_a\sqrt{a^2+x^2})}$.

54. $\dfrac{2}{1-x^4}$.

55. $\dfrac{6}{1-x^4}$.

56. $\dfrac{1}{x\log x}$.

57. $\dfrac{1}{x\log x\log^2 x\log^3 x\ldots\log^{n-1}x}$.

58. $\dfrac{1}{a+b\cos x}$.

59. $\dfrac{1}{\sqrt{1-x^2}} - \dfrac{1}{2\sqrt{x-x^2}}$.

60. $\dfrac{2}{\sqrt{x}(1+4x)}$.

61. $\dfrac{x^2-1}{x^2-4}$.

62. $x^x\log ex - x^{\frac{1}{x}-2}\log\dfrac{x}{e}$.

63. $x^{x^x} \,.\, x^x\left\{(\log x)^2 + \log x + \dfrac{1}{x}\right\}$.

64. $x^{e^x} \,.\, e^x\left\{\log x + \dfrac{1}{x}\right\}$.

65. $e^{x^x} \,.\, x^x(\log x + 1)$.

66. $e^x \,.\, e^{e^x}$.

67. $10^x \,.\, 10^{10^x}(\log_e 10)^2$.

68. $(\sin x)^{\cos x}\left(\dfrac{\cos^2 x}{\sin x} - \sin x \log \sin x\right) - (\cos x)^{\sin x}\left(\dfrac{\sin^2 x}{\cos x} - \cos x \log \cos x\right)$.

69. $-(\cot x)^{\cot x} \operatorname{cosec}^2 x \log e \cot x - (\coth x)^{\coth x} \operatorname{cosech}^2 x \log e \coth x$.

70. $\dfrac{\sqrt{x}}{1+x^{\frac{3}{2}}} \dfrac{a^{cx}x^{\sin x}}{1+a^{2cx}x^{2\sin x}} \log\left(a^{c}x^{\cos x} \,.\, e^{\frac{\sin x}{x}}\right) + \tan^{-1}(a^{cx}x^{\sin x}) \dfrac{1-2x^{\frac{3}{2}}}{2x^{\frac{1}{2}}\left(1+x^{\frac{3}{2}}\right)^2}$.

71. $\dfrac{1}{\sqrt{1-e^{2\tan^{-1}x}}} \cdot \dfrac{e^{\tan^{-1}x}}{1+x^2}$.

72. $$\frac{\left(\sin\frac{m}{x} + \cos\frac{m}{x}\right)\left(1 - \sin\frac{m}{x} + \cos\frac{m}{x}\right)\frac{m}{x^2}}{2\sqrt{\left(1+\cos\frac{m}{x}\right)\left(1-\sin\frac{m}{x}\right)}}.$$

73. $$\frac{\sqrt{1-x^2} - 2\sqrt{x}}{4\sqrt{x}\sqrt{1-x^2}\sqrt{\sqrt{x}+\cos^{-1}x}\,(1+\sqrt{x}+\cos^{-1}x)}.$$

74. $y\left[2xe^{x^2}\cos e^{x^2} \log\dfrac{1+\sqrt{x}}{1+2\sqrt{x}} - \dfrac{\sin e^{x^2}}{2\sqrt{x}(1+\sqrt{x})(1+2\sqrt{x})}\right]$.

75. $-y\cot x(1+2\operatorname{cosec}^2 x \log \cos x)$.

76. $-y\left\{\dfrac{\log \cot^{-1}x}{x^2} + \dfrac{1}{x(1+x^2)\cot^{-1}x}\right\}$.

77. $\left(1+\dfrac{1}{x}\right)^x\left\{\log\dfrac{x+1}{x} - \dfrac{1}{x+1}\right\} + x^{\frac{1}{x}-1}\{x+1-\log x\}$.

78. $\dfrac{b}{a} \dfrac{x^2+y^2-ay}{(x^2+y^2)\sec^2\frac{y}{b} - bx}$.

79. $\cos x \cos 2x \cos^2 y e^{\cos^2 x}$.

80. $-\dfrac{ax+hy}{hx+by}$.

81. $\dfrac{n}{2x}\left[\dfrac{bx^n}{(a+bx^n)^{\frac{1}{2}}\{(a+bx^n)^{\frac{1}{2}} - a^{\frac{1}{2}}\}} - 1\right]$.

82. $\dfrac{y\tan x + \log \sin y}{\log \cos x - x\cot y}$.

83. $x(3+2\tan\log x + \tan^2\log x)$.

84. $\dfrac{y(x-y)}{x(x+y)}$.

85. $\dfrac{y^2}{x - xy\log x}$.

86. $\dfrac{y\log y}{x\log x} \dfrac{1+x\log x\log y}{1-x\log y}$.

87. $\dfrac{y\{(a+bx)y - bx^2\}}{x(y-x)(a+bx)}$.

88. $-\dfrac{ax+hy+g}{hx+by+f}$.

89. $\dfrac{y}{x}$.

90. $\dfrac{6x^2(1+y^2)\tan x^3 \,.\, e^{\tan^{-1}y}}{1+y^2 - \log \sec^2 x^3 \,.\, e^{\tan^{-1}y}}$.

91. $\dfrac{\log_{10}e}{2x^2}$.

92. $$\frac{(1+a^2\cos^2 bx)(x^2+ax+a^2)^{n-1}\left[n(2x+a)\log\cot\frac{x}{2} - \operatorname{cosec} x(x^2+ax+a^2)\right]}{-ab\sin bx}.$$

93. $\dfrac{ab}{a^2+b^2}\left(a^2\cot\dfrac{x}{2} - b^2\tan\dfrac{x}{2}\right)$.

94. $x^{\sin^{-1}x}\left(\log x + \dfrac{\sqrt{1-x^2}}{x}\sin^{-1}x\right)$.

95. $\frac{1}{2}$. 96. $\frac{1}{x^4}\frac{\sqrt{1+x^2}+\sqrt{1-x^2}}{\sqrt{1+x^2}-\sqrt{1-x^2}}$. 97. $\frac{2}{x}$. 98. $-\frac{1}{2}$. 99. 1.

100. $2\dfrac{n(1+x^2)\tan^{-1}x\log\tan^{-1}x+x}{(1+x^2)\tan^{-1}x(\sqrt{x}\cos\sqrt{x}-3\sin\sqrt{x})}x^{\frac{2n+3}{2}}$.

110. $\dfrac{e^{-xz}}{4z^3+x^2}\left[\dfrac{8z^4+5x^5}{2xz\sqrt{x^2z-1}}-(5z^4+4x^5)\sec^{-1}x\sqrt{z}\right]$.

CHAPTER IV.

PAGE 82.

1. $\dfrac{(-1)^n n!}{a-b}\left\{\dfrac{a}{(x-a)^{n+1}}-\dfrac{b}{(x-b)^{n+1}}\right\}$.

2. $\dfrac{(-1)^n n!}{7}\left\{\dfrac{1}{(x-3)^{n+1}}-\dfrac{3^{n+1}}{(3x-2)^{n+1}}\right\}$.

3. $\frac{3}{4}\sin\left(x+\frac{n\pi}{2}\right)-\frac{3^n}{4}\sin\left(3x+\frac{n\pi}{2}\right)$.

4. $e^{2x}.\,2^{n-1}\left\{1+2^{\frac{n}{2}}\cos\left(2x+\frac{n\pi}{4}\right)\right\}$.

5. $\dfrac{e^{ax}}{4}\left[3(a^2+b^2)^{\frac{n}{2}}\sin\left(bx+n\tan^{-1}\dfrac{b}{a}\right)-(a^2+9b^2)^{\frac{n}{2}}\sin\left(3bx+n\tan^{-1}\dfrac{3b}{a}\right)\right]$.

6. $(-1)^{n-1}n!\left[\dfrac{(n+2)(n+1)}{2(x-1)^{n+3}}+\dfrac{3(n+1)}{(x-1)^{n+2}}+\dfrac{4}{(x-1)^{n+1}}-\dfrac{4}{(x-2)^{n+1}}\right]$.

7. $\dfrac{(-1)^n n!}{2a}\left\{\dfrac{1}{(x-a)^{n+1}}-\dfrac{1}{(x+a)^{n+1}}\right\}$.

8. $\dfrac{(-1)^n n!}{a^{n+2}}\sin(n+1)\theta\sin^{n+1}\theta$, where $x=a\cot\theta$.

9. $\dfrac{(-1)^{n-1}(n-1)!}{a^n}\sin n\theta\sin^n\theta$, where $x=a\cot\theta$.

10. $\dfrac{(-1)^n(n-1)!}{a^2-b^2}\left[\dfrac{\sin(n+1)\theta\sin^{n+1}\theta}{b^{n+2}}-\dfrac{\sin(n+1)\phi\sin^{n+1}\phi}{a^{n+2}}\right]$,
where $x=b\cot\theta=a\cot\phi$.

11. $\dfrac{(-1)^n n!}{4a^3}\left[\dfrac{1}{(x-a)^{n+1}}-\dfrac{1}{(x+a)^{n+1}}-\dfrac{2}{a^{n+1}}\sin(n+1)\theta\sin^{n+1}\theta\right]$,
where $x=a\cot\theta$.

12. $\dfrac{(-1)^n}{2a}\left(\dfrac{2}{a\sqrt{3}}\right)^{n+2}n!\{\sin(n+1)\theta\sin^{n+1}\theta-\sin(n+1)\phi\sin^{n+1}\phi\}$,
where $x=\dfrac{a}{\sin\theta}\cos\left(\theta-\dfrac{\pi}{6}\right)=\dfrac{a}{\sin\phi}\cos\left(\phi+\dfrac{\pi}{6}\right)$.

PAGE 89.

1. $y_2=\dfrac{2(1-3x^4)}{(1+x^4)^2}$. 2. $y_3=\dfrac{2}{x}$.

3. $y_2=-\dfrac{(m+n)^2}{2}\sin(m+n)x-\dfrac{(m-n)^2}{2}\sin(m-n)x$. 4. $y_3=a^2e^{ax}(ax+3)$.

5. $\begin{cases}\text{If } r<n,\ y_r=n(n-1)\dots(n-r+1)x^{n-r}.\\ \text{If } r=n,\ y_r=n!.\\ \text{If } r>n,\ y_r=0.\end{cases}$

18. $y_n=a^{n-2}e^{ax}\left\{a^2x^2+2nax+n(n-1)\right\}$.

19. $y_n=a^{n-2}\left\{a^2x^2\sin\left(ax+\dfrac{n\pi}{2}\right)+2nax\sin\left(ax+\dfrac{\overline{n-1}\pi}{2}\right)\right.$
$$\left.+n(n-1)\sin\left(ax+\frac{\overline{n-2}\pi}{2}\right)\right\}.$$

20. $y_n=\dfrac{(-1)^n n!}{(a-b)}\left\{\dfrac{a^2}{(x-a)^{n+1}}-\dfrac{b^2}{(x-b)^{n+1}}\right\}$.

21. $y_n=(-1)^{n-1}n!\left\{\dfrac{(n+2)(n+1)}{2(x-1)^{n+3}}+\dfrac{n+1}{(x-1)^{n+2}}+\dfrac{1}{(x-1)^{n+1}}-\dfrac{1}{(x-2)^{n+1}}\right\}$.

22. $y_n=(-1)^n(n-2)!\sin^{n-1}\theta\left\{\sin\overline{n-1}\theta-(n-1)\cos n\theta\sin\theta\right\}$,
where $x=\cot\theta$.

23. $y_n=\dfrac{-1)^n(n-2)!}{x^{n-1}}+n!\left(\log x+1+\dfrac{1}{2}+\dfrac{1}{3}+\dots+\dfrac{1}{n}\right)$.

36. $y_n=a^{n+2}x^2e^{ax}$.

CHAPTER V.

PAGE 117.

21. $2\left(x+\dfrac{x^5}{5}+\dfrac{x^9}{9}+\dots\right)$.

22. Double the series n 21.

23. Treble the series in 21.

24. $\tan^{-1}\dfrac{p-qx}{q+px}=\tan^{-1}\dfrac{p}{q}-\tan^{-1}x=$etc.

25. $\tan^{-1}\dfrac{\sqrt{1+x^2}-1}{x}=\frac{1}{2}\tan^{-1}x=$ etc.

26. $\tan^{-1}\dfrac{x}{\sqrt{1-x^2}}=\sin^{-1}x=$ etc.

27. $\sec^{-1}\dfrac{1}{1-2x^2}=2\sin^{-1}x=$ etc.

28. $\sin^{-1}\dfrac{2x}{1+x^2}=2\tan^{-1}x=$ etc.

29. $\cos^{-1}\dfrac{x-x^{-1}}{x+x^{-1}}=2\cot^{-1}x=\pi-2\tan^{-1}x=$ etc.

30. $\sinh^{-1}(3x+4x^3)=3\sinh^{-1}x=$ etc. 32. $\dfrac{1}{24}x^5$.

34. Expansion of $e^{-kx}\cos bx$. (See question 4.)

36. $1 + nx + \frac{n^2x^2}{2!} + \frac{n(n^2-1^2)}{3!}x^3 + \frac{n^2(n^2-2^2)}{4!}x^4 + \frac{n(n^2-1^2)(n^2-3^2)}{5!}x^5 + \ldots$.

37. The relation between three consecutive coefficients is

$$2(n+1)a_{n+1} = 3a_n + (2n-1)a_{n-1}.$$

38. $y = \frac{\pi}{4} + \frac{a}{2}x + \frac{2b-a^2}{4}x^2 + \frac{a^3-6ab}{12}x^3 \ldots$.

63. $mx - \frac{m(m-1)(m-2)}{3!}x^3 + \frac{m(m-1)(m-2)(m-3)(m-4)}{5!}x^5 - \ldots$.

CHAPTER VI.

Page 149.

6. (1) $\frac{dy}{dx} = \frac{1+4xz}{1-2z}, \quad \frac{dz}{dx} = \frac{1+2x}{1-2z}.$

(2) $\frac{dx}{dy} = \frac{1-2z}{1+4xz}, \quad \frac{dz}{dy} = \frac{1+2x}{1+4xz}.$

(3) $\frac{dx}{dz} = \frac{1-2z}{1+2x}, \quad \frac{dy}{dz} = \frac{1+4xz}{1+2x}.$

7. $p = -\frac{a^3}{x^2y}, \quad q = -\frac{a^3}{xy^2}, \quad r = \frac{2a^3}{x^3y}, \quad s = \frac{a^3}{x^2y^2}, \quad t = \frac{2a^3}{xy^3}.$

Page 152.

2. (α) $-\frac{ax+hy}{hx+by}$.

(β) $-\frac{4x^3-5a^2y}{4y^3-5a^2x}$.

(γ) $\frac{y\tan x + \log\sin y}{\log\cos x - x\cot y}$.

(δ) $-\frac{y^x\log y + yx^{y-1} - (x+y)^{x+y}\log e(x+y)}{x^y\log x + xy^{x-1} - (x+y)^{x+y}\log e(x+y)}$.

(ε) $-\frac{x^{y-1}y^{x+1} + x^y \,.\, y^x\log y - \frac{\cos y}{x}x^{\cos y} - y^{\log x} \,.\, \frac{\log y}{x}}{x^{y+1}y^{x-1} + x^y \,.\, y^x\log x + x^{\cos y}\log x \,.\, \sin y - \frac{\log x}{y}y^{\log x}}$.

6. $\frac{\partial V}{\partial u} = \frac{\frac{\partial V}{\partial x}\cdot\frac{\partial v}{\partial y} - \frac{\partial V}{\partial y}\cdot\frac{\partial v}{\partial x}}{\frac{\partial u}{\partial x}\cdot\frac{\partial v}{\partial y} - \frac{\partial u}{\partial y}\cdot\frac{\partial v}{\partial x}}$.

16. $\frac{dy}{dz} = \frac{\sin z}{\cos y}\cdot\frac{c - b\cos z}{c - b\sin y}$.

18. $$\begin{cases} \frac{\partial z}{\partial x} = -\frac{c^n x^{n-1}}{a^n z^{n-1}}. \\ \frac{\partial^2 x}{\partial y \partial z} = -(n-1)\frac{a^{2n}}{b^n c^n}\frac{(yz)^{n-1}}{x^{2n-1}}. \\ \frac{dy}{dx} = \frac{b}{a}\,\frac{\left(\frac{x}{a}\right)^{n-1} + \left(\frac{z}{c}\right)^{n-1}}{\left(\frac{y}{b}\right)^{n-1} - \left(\frac{z}{c}\right)^{n-1}} \end{cases}$$

22. $\dfrac{a^2b^2}{c}\dfrac{x^2+y^2}{(a^2x^2+b^2y^2)^{\frac{3}{2}}}$.

CHAPTER VII.

Page 163.

Ex. 1. *Tangents.*

(1) $Xx+Yy=c^2$.

(2) $Yy=2a(X+x)$.

(3) $\dfrac{X}{x}+\dfrac{Y}{y}=2$.

(4) $Y-y=\sinh\dfrac{x}{c}(X-x)$.

(5) $X(2xy+y^2)+Y(x^2+2xy)=3a^3$.

(6) $Y-y=\cot x(X-x)$.

(7) $X(x^2-ay)+Y(y^2-ax)=axy$.

(8) $X\{2x(x^2+y^2)-a^2x\}+Y\{2y(x^2+y^2)+a^2y\}=a^2(x^2-y^2)$.

Normals.

(1) $\dfrac{X}{x}=\dfrac{Y}{y}$.

(2) $\dfrac{X-x}{2a}+\dfrac{Y-y}{y}=0$, etc.

2. Tangents are $Y=\pm\dfrac{3\sqrt{3}}{8}X-\dfrac{a}{8}$.
Normals are $Y=\mp\dfrac{8\sqrt{3}}{9}X+\dfrac{41}{36}a$.

4. For an ellipse, $r^2=a^2\cos^2\theta+b^2\sin^2\theta$.
For a rectangular hyperbola, $r^2=a^2\cos 2\theta$.

Page 192.

1. (α) Parallel at points of intersection with $ax+hy=0$.
Perpendicular at points of intersection with $hx+by=0$.

(β) Parallel where $\left(-\dfrac{a}{\sqrt[3]{2}}, \dfrac{3a\sqrt[3]{2}}{2}\right)$; perpendicular where $x=0$.

(γ) Parallel at $\left(\dfrac{4a}{3}, \dfrac{2\sqrt[3]{4}}{3}a\right)$.
Perpendicular at (0, 0), ($2a$, 0).

2. (α) $ax=\pm by$.
(β) $ax=\pm\sqrt{b^2-a^2}y$.
(γ) $x=0$ and $y=0$.

3. $\dfrac{c^4}{\sqrt{x^6+y^6}}$.

10. Area $=\frac{1}{2}\sqrt[3]{a^4xy}$.

11. $n=-2$; $n=1$.

21. (α) Tangent, $\dfrac{x\cos\theta}{a}+\dfrac{y\sin\theta}{b}=1$.
Normal, $ax\sec\theta-by\operatorname{cosec}\theta=a^2-b^2$.

(β) $\begin{cases} \text{Tangent, } x\sin\frac{\theta}{2} - y\cos\frac{\theta}{2} = a\theta\sin\frac{\theta}{2}. \\ \text{Normal, } x\cos\frac{\theta}{2} + y\sin\frac{\theta}{2} = a\theta\cos\frac{\theta}{2} + 2a\sin\frac{\theta}{2}. \end{cases}$

(γ) $\begin{cases} \text{Tangent, } x\sin\frac{A+B}{2B}\theta - Y\cos\frac{A+B}{2B}\theta = (A+B)\sin\frac{A-B}{2B}\theta. \\ \text{Normal, } x\cos\frac{A+B}{2B}\theta + Y\sin\frac{A+B}{2B}\theta = (A-B)\cos\frac{A-B}{2B}\theta. \end{cases}$

28. $lu = e\cos\theta$.

CHAPTER VIII.

PAGE 224.

1. $x + y = \frac{2a}{3}$.
2. $x + y = 0$.
3. $x + y = 0$.
4. $y = 0$.
5. $x = 0$.
6. $x = 2a$.
7. $x + y + a = 0$.
8. $x = 0, \quad y = 0, \quad x + y = 0$.
9. $y = 0$.
10. $x = \pm a$.
11. $x = a, \quad y = a, \quad x = y$.
12. $x = \pm a$.
13. $x = 0$.
14. $x = a$.
15. $x = \pm 1, \quad y = x$.
16. $x = 0, \ y = \pm\left(x + \frac{m}{2}\right)$.
17. $x + 2y = 0, \quad x + y = 1, \quad x - y = -1$.
18. $x = 0, \quad x - y = 0, \quad x - y + 1 = 0$.
19. $y = 0, \quad x = y, \quad x = y \pm 1$.
20. $x - 2y = 0, \quad x + 2y = \pm 2$.
21. $x + y = \pm 2\sqrt{2}, \quad x + 2y + 2 = 0$.
22. $y = 3x - 2a, \quad x + 3y = \pm a$.
23. $\theta = 0$.
24. $r\sin\theta = a$

25. $nr\sin\left(\theta - \frac{k\pi}{n}\right) = a\sec k\pi$, where k is any integer.

26. $r\sin\theta = a$. 27. $r\cos\theta = 2a$. 28. $\theta = \frac{\pi}{2}, \quad r\sin\theta = \frac{a}{2}$.

29. $r\sin\left(\theta - \frac{k\pi}{n}\right) = \frac{b}{n}$, where k is any integer.

30. $n\theta = k\pi$, where k is any integer.

34. $r = b$. 35. $x = \pm a, \quad y = x$. Above.

36. $y = x + a, \quad y = -x - a, \quad x = a$.
In the first quadrant above the first. In the fourth quadrant below the second.

40. $x^3 - 6x^2y + 11xy^2 - 6y^3 = x$. 42. $\frac{x^2}{a^2} + \frac{y^2}{b^2} = 1$.

44. $\left(\frac{a}{a'} + \frac{b}{b'} - 1\right)\left(\frac{x}{a} + \frac{y}{b} - 1\right)\frac{xy}{ab} = \frac{x}{a'} + \frac{y}{b'} - 1$.

45. $bxy(x^2 - y^2) = a(a^2 - b^2)(x^2 + y^2 - a^2)$.

46. $(x^2-y^2)^2=Ax$, or $r^3=a^3\dfrac{\cos\theta}{\cos^2 2\theta}$.

47. $2y^2(x^2-y^2)=3a^3x$.

CHAPTER IX.

Page 258.

1. (α) $y=0$.
   (β) $ax=by$.
   (γ) $y=\pm x$.
   (δ) $x=0,\ y=0$.

14. $x=a$ and $x=2a$.

17. $\theta=\pm\sin^{-1}\sqrt{\dfrac{2}{5}}$.

27. $x=7$ and $x=1$.

38. A straight line and a point.

40. A straight line and a conic.

48. $\begin{cases}\text{At } (0,\ a), & \tan\psi=\pm\dfrac{2}{\sqrt{3}}.\\ \text{At } (a,\ 0), & \tan\psi=\pm\sqrt{\dfrac{2}{3}}.\\ \text{At } (2a,\ a), & \tan\psi=\pm\dfrac{2}{\sqrt{3}}.\end{cases}$

50. Tangents at origin, $y=0$ and $y=\pm x$.
    Single cusp of first species at origin.
    $x+y$ is a factor.

CHAPTER X.

Page 279.

1. $\rho=a$; $\rho=a\cos\psi$; $\rho=a\sec^2\psi$.

2. $\rho=\dfrac{2(a+x)^{\frac{3}{2}}}{a^{\frac{1}{2}}}$; $\rho=\dfrac{y^2}{c}$.

4. $\rho=\dfrac{(a^2\sin^2\theta+b^2\cos^2\theta)^{\frac{3}{2}}}{ab}$.

5. $\rho=\dfrac{2r^{\frac{3}{2}}}{a^{\frac{1}{2}}}$, $\rho=\dfrac{a}{2}$, $\rho=\dfrac{a^m}{(m+1)r^{m-1}}$.

6. $\rho=a\dfrac{(\theta^2+1)^{\frac{3}{2}}}{\theta^4}$.

CHAPTER XI.

Page 322.

1. $256y^3+27x^4=0$.

2. $\dfrac{a^4}{x^2}+\dfrac{b^4}{y^2}=\dfrac{c^4}{a^2}$.

3. $y+\frac{1}{2}g\dfrac{x^2}{u^2}=\dfrac{u^2}{2g}$.

4. $\begin{cases}(1)\ 4x^3+27ay^2=0.\\ (2)\ y^2=4h(a+h-x).\end{cases}$

5. $y^2+4a(x-2a)=0$.

6. Two straight lines.

7. $\begin{cases}(1)\ \sqrt{x}+\sqrt{y}=\sqrt{k}.\\ (2)\ x^{\frac{n}{n+1}}+y^{\frac{n}{n+1}}=k^{\frac{n}{n+1}}.\\ (3)\ x^my^n=\dfrac{m^mn^n}{(m+n)^{m+n}}k^{m+n}.\end{cases}$

8. A parabola touching the axes.

9. A hyperbola.

10. $27ay^2 = 4(x-2a)^3$.

14. $\begin{cases} (1)\ x^{\frac{2}{3}} + y^{\frac{2}{3}} = k^{\frac{2}{3}}. \\ (2)\ x^{\frac{2}{5}} + y^{\frac{2}{5}} = k^{\frac{2}{5}}. \\ (3)\ x^{\frac{2m}{m+2}} + y^{\frac{2m}{m+2}} = k^{\frac{2m}{m+2}}. \\ (4)\ 2xy = k^2. \end{cases}$

15. $\begin{cases} (1)\ x^{\frac{1}{3}} + y^{\frac{1}{3}} = k^{\frac{1}{3}}. \\ (2)\ x^{\frac{m}{2m+1}} + y^{\frac{m}{2m+1}} = k^{\frac{m}{2m+1}}. \\ (3)\ 16xy = k^2. \end{cases}$

17. $x^{\frac{2}{3}} + y^{\frac{2}{3}} = a^{\frac{2}{3}}$.

20. $r^2 = a^2\cos^2\theta + b^2\sin^2\theta$.

46. $a^pb^q = \dfrac{(p+q)^{p+q}}{p^pq^q}k^{p+q}$.

47. A conic.

CHAPTER XIII.

PAGE 365.

1. $\log_b a$.
2. $\frac{3}{5}$.
3. $\frac{m}{n}$.
4. $\frac{1}{n}$.
5. 4.
6. 4.
7. 2.
8. 1.
9. $\frac{1}{2}$.
10. $\frac{3}{2}$.
11. $\frac{2}{3}$.
12. $\frac{1}{6}$.
13. 1.
14. 1.
15. $\frac{1}{15}$.
16. $-\frac{11}{6}$.
17. $\frac{13}{60}$.
18. $-\frac{2}{3}$.
19. $\frac{1}{2}$.
20. 1.
21. ∞.
22. 1.
23. $e^{-\frac{1}{6}}$.
24. 0.
25. e^{-1}.
26. $e^{\frac{1}{2}}$.

PAGE 376.

1. 2.
2. $\frac{1}{5}$.
3. 2.
4. $\frac{1}{2}$.
5. $-\frac{1}{a}$.
6. 4.
7. $m - \frac{4m^3}{3}$.
8. $\frac{1}{\sqrt{e}}$.
9. e.
10. 1.
11. $\begin{cases} \text{If } n > m, & \infty. \\ \quad n = m, & \frac{A}{a}. \\ \quad n < m, & 0. \end{cases}$
12. 1.
13. $e^{\frac{2}{a}}$.
14. e^2.
15. e^2.
16. $a_1a_2a_3 \ldots a_n$.
17. -1.
18. 0.
19. b, 0.
20. $\frac{1}{2}$.
21. $-a$.
22. $\frac{1}{3}$.
23. 0.
24. $\sqrt{a}$.
25. 1.
26. $\frac{1}{18}$.
27. $-\frac{e}{2}$.
28. $-\frac{7}{16}e$.
29. 0.
30. $-\frac{n(n-1)}{2!}a^{n-2}$.
31. 1.
33. $\frac{1}{2}(1 \pm \sqrt{-3})$.
34. 0 or ∞.
35. $\pm\frac{b}{\sqrt{a^2-b^2}}$.
36. 0 or ± 1.
39. $\frac{4}{15}$.
44. $e^{-\frac{a^2}{2b^2}}$.
45. 1, $b^{x-1}(b\cos bx - \sin bx)\cos^2 bx$.
46. $-\frac{3}{2}$.
47. $\frac{1}{\sqrt{2a}}$.
48. 1.

CHAPTER XIV.

PAGE 406.

1. Maximum Value $=34$, Minimum $=33$.

5. $x=-2,\ -1,\ 1,\ 2,$ give Maxima and Minima alternately.

6. At $x=1$, $y=$ Maximum.
 $x=3$, $y=$ Minimum.
 At $x=2$ and $x=4$ there are points of contrary flexure.

7. At $x=2$, $y=$ Minimum. At $x=\frac{4}{3}$, $y=$ Maximum.

8. $x=-\frac{7}{9}$ gives a Maximum, $x=1$ gives a Minimum.

9. It cannot lie between $\pm 2\sqrt{ab}$.

11. $x=2$ gives a Maximum, $x=5$ a Minimum.

12. Minimum Ordinate at $x=\frac{b}{\sqrt[3]{2}}$. A point of inflexion at $(-b, 0)$.

13. At $x=a$, $y=c$.
 At $x=\frac{a+6b}{7}$, $y=c\pm 6^3\left(\frac{a-b}{7}\right)^{\frac{7}{2}}$, a being supposed greater than b.

17. $(a+b)^2$.

23. Maximum Value $=\frac{1}{2}(a+b)^2$.
 Minimum Value $=-\frac{1}{2}(a-b)^2$.

29. A Maximum when the chords coincide with the transverse axis and lat. rectum.
 A Minimum when the chords are equally inclined to the transverse axis.

30. $\frac{ap}{p+q}$, $\frac{aq}{p+q}$.

32. n parts. Continued product $=e^n$.

33. $\frac{1}{\sqrt[p-1]{p}}$.

34. A Maximum when the segment is a semicircle.
 A Minimum when the radius is infinite.

35. The distances of the point from the extremities of the line are

$$\frac{2ar_1}{\sqrt{r_1^2+r_2^2}},\quad \frac{2ar_2}{\sqrt{r_1^2+r_2^2}}.$$

36. The point divides the line of centres in the ratio $r_1^{\frac{2}{3}} : r_2^{\frac{2}{3}}$, r_1 and r_2 being the radii.

37. $AO : AD = 1 : \sqrt{2}$.

38. If A be the smallest angle and b, c the adjacent sides, the distance of each end of the fence from $A = \sqrt{\frac{bc}{2}}$, and the length of the fence $= \sqrt{2bc}\sin\frac{A}{2}$.

42. $\frac{n+1}{n}a$ knots an hour.

47. Half the triangle formed by the chord and the tangents at its extremities, or three fourths of the area of the segment.

51. $a > b$, Maximum if $x = \frac{a+2b}{3}$.
$a < b$, Maximum if $x = a$.
$a = b$ gives a point of inflexion.

54. If $\cos\alpha$ be $> e$, Greatest $= \frac{l^2 \sin\alpha\cos\alpha}{(1 - e\cos\alpha)^2}$, Least $= \frac{l^2\sin\alpha\cos\alpha}{(1 + e\cos\alpha)^2}$.
If $\cos\alpha$ be $< e$, the above values are both Minima, and there are two Maxima each equal to $\frac{l^2\cot\alpha}{1 - e^2}$.

55. The tangent at P must be parallel to SQ.

56. If $h < 2a$, P is at the vertex.
If $h > 2a$, the abscissa of P is $h - 2a$, and the perpendicular is therefore the normal at P.

57. Maximum area $= 4r^2 \sin\alpha\cos^3\alpha$, where r is the radius of the circle and 2α the given angle.

60. $\sin AOQ = \frac{CA}{CO\sqrt{2}}$, C being the centre.

63 The height is three times the semiaxis to which the base is perpendicular.

GLASGOW: PRINTED AT THE UNIVERSITY PRESS BY ROBERT MACLEHOSE.

January, 1890.

A Catalogue

OF

Educational Books

PUBLISHED BY

Macmillan & Co.

BEDFORD STREET, STRAND, LONDON.

b

CONTENTS.

29 AND 30, BEDFORD STREET, COVENT GARDEN,
LONDON, W.C., *January*, 1890.

CLASSICS.

ELEMENTARY CLASSICS.

18mo, Eighteenpence each.

THIS SERIES FALLS INTO TWO CLASSES—

(1) First Reading Books for Beginners, provided not only with **Introductions and Notes**, but with **Vocabularies**, and in some cases with **Exercises** based upon the Text.

(2) Stepping-stones to the study of particular authors, intended for more advanced students who are beginning to read such authors as Terence, Plato, the Attic Dramatists, and the harder parts of Cicero, Horace, Virgil, and Thucydides.

These are provided with Introductions and Notes, **but no Vocabulary.** The Publishers have been led to provide the more strictly Elementary Books with Vocabularies by the representations of many teachers, who hold that beginners do not understand the use of a Dictionary, and of others who, in the case of middle-class schools where the cost of books is a serious consideration, advocate the Vocabulary system on grounds of economy. It is hoped that the two parts of the Series, fitting into one another, may together fulfil all the requirements of Elementary and Preparatory Schools, and the Lower Forms of Public Schools.

10.1.90.

The following Elementary Books, with Introductions, Notes, **and Vocabularies**, and in some cases with **Exercises**, are either ready or in preparation:—

Accidence, Latin, and Exercises Arranged for Be- GINNERS. By WILLIAM WELCH, M.A., and C. G. DUFFIELD, M.A., Assistant Masters at Surrey County School, Cranleigh.

Aeschylus.—PROMETHEUS VINCTUS. Edited by Rev. H. M. STEPHENSON, M.A.

Arrian.—SELECTIONS. Edited for the use of Schools, with Introduction, Notes, Vocabulary, and Exercises, by Rev. JOHN BOND, M.A., and A. S. WALPOLE, M.A.

Aulus Gellius, Stories from. Being Selections and Adaptations from the *Noctes Atticae*. Edited, with Exercises for the use of Lower Forms, by Rev. G. H. NALL, M.A., Assistant Master in Westminster School.

Cæsar.—THE HELVETIAN WAR. Being Selections from Book I. of the "De Bello Gallico." Adapted for the use of Beginners. With Notes, Exercises, and Vocabulary, by W. WELCH, M.A., and C. G. DUFFIELD, M.A.

THE INVASION OF BRITAIN. Being Selections from Books IV. and V. of the "De Bello Gallico." Adapted for the use of Beginners. With Notes, Vocabulary, and Exercises, by W. WELCH, M.A., and C. G. DUFFIELD, M.A.

THE GALLIC WAR. BOOK I. Edited by A. S. WALPOLE, M.A.

BOOKS II. AND III. Edited by the Rev. W. G. RUTHERFORD, M.A., LL.D., Head-Master of Westminster.

BOOK IV. Edited by CLEMENT BRYANS, M.A., Assistant-Master at Dulwich College.

BOOK V. By C. COLBECK, M.A., Assistant-Master at Harrow, formerly Fellow of Trinity College, Cambridge.

BOOK VI. By the same Editor.

SCENES FROM BOOKS V. AND VI. By the same Editor.

BOOK VII. Edited by Rev. JOHN BOND, M.A., and A. S. WALPOLE, M.A.

Cicero.—DE SENECTUTE. Edited by E. S. SHUCKBURGH, M.A., late Fellow of Emmanuel College, Cambridge.

DE AMICITIA. By the same Editor.

STORIES OF ROMAN HISTORY. Adapted for the Use of Beginners. With Notes, Vocabulary, and Exercises, by the Rev. G. E. JEANS, M.A., Fellow of Hertford College, Oxford, and A. V. JONES, M.A.; Assistant-Masters at Haileybury College.

Euripides.—HECUBA. Edited by Rev. JOHN BOND, M.A and A. S. WALPOLE, M.A.

Eutropius.—Adapted for the Use of Beginners. With Notes, Vocabulary, and Exercises, by WILLIAM WELCH, M.A., and C. G. DUFFIELD, M.A., Assistant-Masters at Surrey County School, Cranleigh.

Homer.—ILIAD. BOOK I. Edited by Rev. JOHN BOND, M.A., and A. S. WALPOLE, M.A.

ILIAD. BOOK XVIII. THE ARMS OF ACHILLES. Edited by S. R. JAMES, M.A., Assistant-Master at Eton College.

ODYSSEY. BOOK I. Edited by Rev. JOHN BOND, M.A., and A. S. WALPOLE, M.A.

Horace.—ODES. BOOKS I.—IV. Edited by T. E. PAGE, M.A., late Fellow of St. John's College, Cambridge; Assistant-Master at the Charterhouse. Each 1*s.* 6*d.*

Livy.—BOOK I. Edited by H. M. STEPHENSON, M.A., late Head Master of St. Peter's School, York.

BOOKS XXI. and XXII. (separately), with Notes adapted from Mr. Capes' Edition, for the use of junior students, and with Vocabularies by J. E. MELHUISH, M.A., Assistant-Master in St. Paul's School. [*Book XXI. in January.*

THE HANNIBALIAN WAR. Being part of the XXI. AND XXII. BOOKS OF LIVY, adapted for the use of beginners, by G. C. MACAULAY, M.A., late Fellow of Trinity College, Cambridge.

THE SIEGE OF SYRACUSE. Being part of the XXIV. AND XXV. BOOKS OF LIVY, adapted for the use of beginners. With Notes, Vocabulary, and Exercises, by GEORGE RICHARDS, M.A., and A. S. WALPOLE, M.A.

STORIES OF ANCIENT ROME, FROM LIVY. Adapted for the use of beginners. With Notes, Exercises, and Vocabulary, by HERBERT WILKINSON, M.A., formerly Postmaster of Merton College, Oxford. [*In the press.*

Lucian.—EXTRACTS FROM LUCIAN. Edited, with Notes, Exercises, and Vocabulary, by Rev. JOHN BOND, M.A., and A. S. WALPOLE, M.A.

Nepos.—SELECTIONS ILLUSTRATIVE OF GREEK AND ROMAN HISTORY. Edited for the use of beginners with Notes, Vocabulary and Exercises, by G. S. FARNELL, M.A.

Ovid.—SELECTIONS. Edited by E. S. SHUCKBURGH, M.A. late Fellow and Assistant-Tutor of Emmanuel College, Cambridge.

EASY SELECTIONS FROM OVID IN ELEGIAC VERSE. Arranged for the use of Beginners with Notes, Vocabulary, and Exercises, by HERBERT WILKINSON, M.A.

STORIES FROM THE METAMORPHOSES. Edited for the Use of Schools. With Notes, Exercises, and Vocabulary. By J. BOND, M.A., and A. S. WALPOLE, M.A.

Phædrus.—SELECT FABLES. Adapted for the Use of Beginners. With Notes, Exercises, and Vocabularies, by A. S. WALPOLE, M.A.

Thucydides.—THE RISE OF THE ATHENIAN EMPIRE. BOOK I. cc. LXXXIX. — CXVII. AND CXXVIII. — CXXXVIII. Edited with Notes, Vocabulary and Exercises, by F. H. COLSON, M.A., Head Master of Plymouth College; Fellow of St. John's College, Cambridge.

Virgil.—SELECTIONS. Edited by E. S. SHUCKBURGH, M.A.
GEORGICS. BOOK I. Edited by T. E. PAGE, M.A.
BOOK II. Edited by Rev. J. H. SKRINE, M.A., Warden of Trinity.
ÆNEID. BOOK I. Edited by A. S. WALPOLE, M.A.
BOOK II. Edited by T. E. PAGE, M.A., Assistant-Master at the Charterhouse.
BOOK III. Edited by T. E. PAGE, M.A. [*In preparation.*
BOOK IV. Edited by Rev. H. M. STEPHENSON, M.A.
BOOK V. Edited by Rev. A. CALVERT, M.A., late Fellow of St. John's College, Cambridge.
BOOK VI. Edited by T. E. PAGE, M.A.
BOOK VII. Edited by A. CALVERT, M.A.
BOOK VIII. Edited by A. CALVERT, M.A. [*In preparation.*
BOOK IX. Edited by Rev. H. M. STEPHENSON, M.A.
BOOK X. Edited by S. G. OWEN, M.A. [*In preparation.*

Xenophon.—ANABASIS. BOOK I. Edited by A. S. WALPOLE, M.A.
ANABASIS. BOOK I. Chaps. I.—VIII. for the use of Beginners, with Titles to the Sections, Notes, Vocabulary, and Exercises, by E. A. WELLS, M.A., Assistant Master in Durham School.
ANABASIS. BOOK II. Edited by A. S. WALPOLE, M.A.
ANABASIS. BOOK III. Edited by Rev. G. H. NALL. [*In preparation.*
ANABASIS, SELECTIONS FROM. BOOK IV. THE RETREAT OF THE TEN THOUSAND. Edited, with Notes, Vocabulary, and Exercises, by Rev. E. D. STONE, M.A., formerly Assistant-Master at Eton.
ANABASIS. BOOK IV. Edited by Rev. E. D. STONE. [*In preparation.*
SELECTIONS FROM THE CYROPÆDIA. Edited, with Notes, Vocabulary, and Exercises, by A. H. COOKE, M.A., Fellow and Lecturer of King's College, Cambridge.

The following more advanced Books, with Introductions and Notes, **but no Vocabulary,** are either ready, or in preparation:—

Cicero.—SELECT LETTERS. Edited by Rev. G. E. JEANS, M.A., Fellow of Hertford College, Oxford, and Assistant-Master at Haileybury College.

Herodotus.—SELECTIONS FROM BOOKS VII. AND VIII. THE EXPEDITION OF XERXES. Edited by A. H. COOKE, M.A., Fellow and Lecturer of King's College, Cambridge.

Horace. — SELECTIONS FROM THE SATIRES AND EPISTLES. Edited by Rev. W. J. V. BAKER, M.A., Fellow of St. John's College, Cambridge.
SELECT EPODES AND ARS POETICA. Edited by H. A. DALTON, M.A., formerly Senior Student of Christchurch; Assistant-Master in Winchester College.

Plato.—EUTHYPHRO AND MENEXENUS. Edited by C. E. GRAVES, M.A., Classical Lecturer and late Fellow of St. John's College, Cambridge.

Terence.—SCENES FROM THE ANDRIA. Edited by F. W. CORNISH, M.A., Assistant-Master at Eton College.

The Greek Elegiac Poets.— FROM CALLINUS TO CALLIMACHUS. Selected and Edited by Rev. HERBERT KYNASTON, D.D., Principal of Cheltenham College, and formerly Fellow of St. John's College, Cambridge.

Thucydides.—BOOK IV. CHS. I.—XLI. THE CAPTURE OF SPHACTERIA. Edited by C. E. GRAVES, M.A.

*** *Other Volumes to follow.*

CLASSICAL SERIES FOR COLLEGES AND SCHOOLS.

Fcap. 8vo.

Being select portions of Greek and Latin authors, edited with Introductions and Notes, for the use of Middle and Upper forms of Schools, or of candidates for Public Examinations at the Universities and elsewhere.

Æschines.— IN CTESIPHONTEM. Edited by Rev. T. GWATKIN, M.A., late Fellow of St. John's College, Cambridge, and E. S. Shuckburgh, M.A. [*In the press.*

Æschylus. — PERSÆ. Edited by A. O. PRICKARD, M.A. Fellow and Tutor of New College, Oxford. With Map. *3s. 6d.*
SEVEN AGAINST THEBES. SCHOOL EDITION. Edited by A. W. VERRALL, Litt.D., Fellow of Trinity College, Cambridge, and M. A. BAYFIELD, M.A., Head-Master's Assistant at Malvern College. *3s. 6d.*

Andocides.—DE MYSTERIIS. Edited by W. J. HICKIE, M.A., formerly Assistant-Master in Denstone College. *2s. 6d.*

Attic Orators.—Selections from ANTIPHON, ANDOCIDES· LYSIAS, ISOCRATES, AND ISAEUS. Edited, by R. C, JEBB, Litt.D., LL.D., Professor of Greek in the University of Cambridge. Second Edition. *6s.*

Cæsar.—THE GALLIC WAR. Edited, after Kraner, by Rev. JOHN BOND, M.A., and A. S. WALPOLE, M.A. With Maps. *6s.*

Catullus.—SELECT POEMS. Edited by F. P. SIMPSON, B.A., late Scholar of Balliol College, Oxford. New and Revised Edition. *5s.* The Text of this Edition is carefully adapted to School use.

Cicero.—THE CATILINE ORATIONS. From the German of KARL HALM. Edited, with Additions, by A. S. WILKINS, Litt.D., LL.D., Professor of Latin at the Owens College, Manchester, Examiner of Classics to the University of London. New Edition. 3*s.* 6*d.*

PRO LEGE MANILIA. Edited, after HALM, by Professor A. S. WILKINS, Litt.D., LL.D. 2*s.* 6*d.*

THE SECOND PHILIPPIC ORATION. From the German of KARL HALM. Edited, with Corrections and Additions, by JOHN E. B. MAYOR, Professor of Latin in the University of Cambridge, and Fellow of St. John's College. New Edition, revised. 5*s.*

PRO ROSCIO AMERINO. Edited, after HALM, by E. H. DONKIN, M.A., late Scholar of Lincoln College, Oxford, and Assistant-Master at Sherborne School. 4*s.* 6*d.*

PRO P. SESTIO. Edited by Rev. H. A. HOLDEN, M.A., LL.D., late Fellow of Trinity College, Cambridge; and late Classical Examiner to the University of London. 5*s.*

Demosthenes.—DE CORONA. Edited by B. DRAKE, M.A., late Fellow of King's College, Cambridge. Seventh and revised Edition, edited by E. S. SHUCKBURGH, M.A. 4*s.* 6*d.*

ADVERSUS LEPTINEM. Edited by Rev. J. R. KING, M.A., Fellow and Tutor of Oriel College, Oxford. 4*s.* 6*d.*

THE FIRST PHILIPPIC. Edited, after C. REHDANTZ, by Rev. T. GWATKIN, M.A., late Fellow of St. John's College, Cambridge. 2*s.* 6*d.*

IN MIDIAM. Edited by Prof. A. S. WILKINS, Litt.D., LL.D., and HERMAN HAGER, Ph.D., of the Owens College, Manchester. [*In preparation.*

Euripides.—HIPPOLYTUS. Edited by J. P. MAHAFFY, M.A., Fellow and Professor of Ancient History in Trinity College, Dublin, and J. B. BURY, Fellow of Trinity College, Dublin. 3*s.* 6*d.*

MEDEA. Edited by A. W. VERRALL, Litt.D., Fellow and Lecturer of Trinity College, Cambridge. 3*s.* 6*d.*

IPHIGENIA IN TAURIS. Edited by E. B. ENGLAND, M.A., Lecturer at the Owens College, Manchester. 4*s.* 6*d.*

ION. Edited by M. A. BAYFIELD, M.A., Headmaster's Assistant at Malvern College. 3*s.* 6*d.*

BACCHAE. Edited by R. Y. TYRRELL, M.A., Professor of Greek in the University of Dublin. [*In preparation.*

Herodotus.—BOOK III. Edited by G. C. MACAULAY, M.A., formerly Fellow of Trinity College, Cambridge. [*In preparation.*

BOOKS V. AND VI. Edited by J. STRACHAN, M.A., Professor of Greek in the Owens College, Manchester. [*In preparation.*

BOOKS VII. AND VIII. Edited by Mrs. MONTAGU BUTLER. [*In the press.*

Hesiod.—THE WORKS AND DAYS. Edited by W. T. LENDRUM, Assistant Master in Dulwich College. [*In preparation.*

Homer.—ILIAD. BOOKS I., IX., XI., XVI.—XXIV. THE STORY OF ACHILLES. Edited by the late J. H. PRATT, M.A., and WALTER LEAF, Litt.D., Fellows of Trinity College, Cambridge. 6*s.*

ODYSSEY. BOOK IX. Edited by Prof. JOHN E. B. MAYOR. 2*s.* 6*d.*

ODYSSEY. BOOKS XXI.—XXIV. THE TRIUMPH OF ODYSSEUS. Edited by S. G. HAMILTON, B.A., Fellow of Hertford College, Oxford. 3*s.* 6*d.*

Horace.—THE ODES. Edited by T. E. PAGE, M.A., formerly Fellow of St. John's College, Cambridge; Assistant-Master at the Charterhouse. 6*s.* (BOOKS I., II., III., and IV. separately, 2*s.* each.)

THE SATIRES. Edited by ARTHUR PALMER, M.A., Fellow of Trinity College, Dublin; Professor of Latin in the University of Dublin. 6*s.*

THE EPISTLES AND ARS POETICA. Edited by A. S. WILKINS, Litt.D., LL.D., Professor of Latin in Owens College, Manchester; Examiner in Classics to the University of London. 6*s.*

Isaeos.—THE ORATIONS. Edited by WILLIAM RIDGEWAY, M.A., Fellow of Caius College, Cambridge; and Professor of Greek in the University of Cork. [*In preparation.*

Juvenal. THIRTEEN SATIRES. Edited, for the Use of Schools, by E. G. HARDY, M.A., late Fellow of Jesus College, Oxford. 5*s.*

The Text of this Edition is carefully adapted to School use.

SELECT SATIRES. Edited by Professor JOHN E. B. MAYOR. X. AND XI. 3*s.* 6*d.* XII.—XVI. 4*s.* 6*d.*

Livy.—BOOKS II. AND III. Edited by Rev. H. M. STEPHENSON, M.A. 5*s.*

BOOKS XXI. AND XXII. Edited by the Rev. W. W. CAPES, M.A. With Maps. 5*s.*

BOOKS XXIII. AND XXIV. Edited by G. C. MACAULAY, M.A. With Maps. 5*s.*

THE LAST TWO KINGS OF MACEDON. EXTRACTS FROM THE FOURTH AND FIFTH DECADES OF LIVY. Selected and Edited, with Introduction and Notes, by F. H. RAWLINS, M.A., Fellow of King's College, Cambridge; and Assistant-Master at Eton. With Maps. 3*s.* 6*d.*

THE SUBJUGATION OF ITALY. SELECTIONS FROM THE FIRST DECADE. Edited by G. E. MARINDIN, M.A., formerly Assistant Master at Eton. [*In preparation.*

Lucretius. BOOKS I.—III. Edited by J. H. WARBURTON LEE, M.A., late Scholar of Corpus Christi College, Oxford, and Assistant-Master at Rossall. 4*s.* 6*d.*

Lysias.—SELECT ORATIONS. Edited by E. S. SHUCKBURGH, M.A., late Assistant-Master at Eton College, formerly Fellow and Assistant-Tutor of Emmanuel College, Cambridge. New Edition, revised. 6*s.*

Martial.—SELECT EPIGRAMS. Edited by Rev. H. M. STEPHENSON, M.A. New Edition, Revised and Enlarged. 6*s*. 6*d*.

Ovid.—FASTI. Edited by G. H. HALLAM, M.A., Fellow of St. John's College, Cambridge, and Assistant-Master at Harrow. With Maps. 5*s*.

HEROIDUM EPISTULÆ XIII. Edited by E. S. SHUCKBURGH, M.A. 4*s*. 6*d*.

TRISTIA. Edited by the same. [*In preparation.*

METAMORPHOSES. BOOKS I.—III. Edited by C. SIMMONS, M.A. [*In preparation.*

METAMORPHOSES. BOOKS XIII. AND XIV. Edited by C. SIMMONS, M.A. 4*s*. 6*d*.

Plato.—LACHES. Edited, with Introduction and Notes, by M. T. TATHAM, M.A., formerly Assistant Master at Westminster. 2*s*. 6*d*.

THE REPUBLIC. BOOKS I.—V. Edited by T. H. WARREN, M.A., President of St. Mary Magdalen College, Oxford. 6*s*.

Plautus.—MILES GLORIOSUS. Edited by R. Y. TYRRELL, M.A., Fellow of Trinity College, and Regius Professor of Greek in the University of Dublin. Second Edition Revised. 5*s*.

AMPHITRUO. Edited by ARTHUR PALMER, M.A., Fellow of Trinity College and Regius Professor of Latin in the University of Dublin. [*Nearly ready.*

Pliny.—LETTERS. BOOK III. Edited by Professor JOHN E. B. MAYOR. With Life of Pliny by G. H. RENDALL, M.A. 5*s*.

LETTERS. BOOKS I. and II. Edited by J. COWAN, M.A., Assistant-Master in the Manchester Grammar School. 5*s*.

Plutarch.—LIFE OF THEMISTOKLES. Edited by Rev. H. A. HOLDEN, M.A., LL.D. 5*s*.

LIVES OF GALBA AND OTHO. Edited by E. G. HARDY, M.A. [*In the press.*

Polybius.—THE HISTORY OF THE ACHÆAN LEAGUE AS CONTAINED IN THE REMAINS OF POLYBIUS. Edited by W. W. CAPES, M.A. 6*s*. 6*d*.

Propertius.—SELECT POEMS. Edited by Professor J. P. POSTGATE, M.A., Fellow of Trinity College, Cambridge. Second Edition, revised. 6*s*.

Sallust.—CATILINA AND JUGURTHA. Edited by C. MERIVALE, D.D., Dean of Ely. New Edition, carefully revised and enlarged, 4*s*. 6*d*. Or separately, 2*s*. 6*d*. each.

BELLUM CATULINAE. Edited by A. M. COOK, M.A., Assistant Master at St. Paul's School. 4*s*. 6*d*.

JUGURTHA. By the same Editor. [*In preparation.*

Tacitus.—AGRICOLA AND GERMANIA. Edited by A. J. CHURCH, M.A., and W. J. BRODRIBB, M.A., Translators of Tacitus. New Edition, 3*s*. 6*d*. Or separately, 2*s*. each.

THE ANNALS. BOOK VI. By the same Editors. 2*s*. 6*d*.

THE HISTORIES. BOOKS I. AND II. Edited by A. D. GODLEY, M.A., Fellow of Magdalen College, Oxford. 5*s*.

Tacitus.—THE HISTORIES. BOOKS III.—V. By the same Editor. 5*s*.

THE ANNALS. BOOKS I. AND II. Edited by J. S. REID, M.L., Litt.D. [*In preparation.*

Terence.—HAUTON TIMORUMENOS. Edited by E. S. SHUCKBURGH, M.A. 3*s*. With Translation, 4*s*. 6*d*.

PHORMIO. Edited by Rev. JOHN BOND, M.A., and A. S. WALPOLE, M.A. 4*s*. 6*d*.

Thucydides. BOOK IV. Edited by C. E. GRAVES, M.A., Classical Lecturer, and late Fellow of St. John's College, Cambridge. 5*s*.

BOOK V. By the same Editor. [*In the press.*

BOOKS I. AND III. Edited by C. BRYANS, M.A. [*In preparation.*

BOOK II. Edited by E. C. MARCHANT, M.A. [*In preparation.*

BOOKS VI. AND VII. THE SICILIAN EXPEDITION. Edited by the Rev. PERCIVAL FROST, M.A., late Fellow of St. John's College, Cambridge. New Edition, revised and enlarged, with Map. 5*s*.

BOOK VIII. Edited by Professor T. G. TUCKER. [*In preparation.*

Tibullus.—SELECT POEMS. Edited by Professor J. P. POSTGATE, M.A. [*In preparation.*

Virgil.—ÆNEID. BOOKS II. AND III. THE NARRATIVE OF ÆNEAS. Edited by E. W. HOWSON, M.A., Fellow of King's College, Cambridge, and Assistant-Master at Harrow. 3*s*.

Xenophon.—HELLENICA, BOOKS I. AND II. Edited by H. HAILSTONE, B.A., late Scholar of Peterhouse, Cambridge. With Map. 4*s*. 6*d*.

CYROPÆDIA. BOOKS VII. AND VIII. Edited by ALFRED GOODWIN, M.A., Professor of Greek in University College, London. 5*s*.

MEMORABILIA SOCRATIS. Edited by A. R. CLUER, B.A., Balliol College, Oxford. 6*s*.

THE ANABASIS. BOOKS I.—IV. Edited by Professors W. W. GOODWIN and J. W. WHITE. Adapted to Goodwin's Greek Grammar. With a Map. 5*s*.

HIERO. With Introduction, Summaries, Critical and Explanatory Notes and Indices, and Critical Appendix. Edited by Rev. H. A. HOLDEN, M.A., LL.D. Third Edition, revised. 3*s*. 6*d*.

OECONOMICUS. By the same Editor. With Introduction, Explanatory Notes, Critical Appendix, and Lexicon. Fourth Edition. 6*s*.

*** *Other Volumes will follow.*

CLASSICAL LIBRARY.

(1) **Texts**, Edited with **Introductions and Notes**, for the use of Advanced Students. (2) **Commentaries and Translations.**

Æschylus.—THE EUMENIDES. The Greek Text, with Introduction, English Notes, and Verse Translation. By BERNARD DRAKE, M.A., late Fellow of King's College, Cambridge. 8vo. 5*s.*

AGAMEMNON. Edited, with Introduction, Commentary, and Translation, by A. W. VERRALL, Litt.D. 8vo. 12*s.*

AGAMEMNON, CHOEPHORŒ, AND EUMENIDES. Edited, with Introduction and Notes, by A. O. PRICKARD, M.A., Fellow and Tutor of New College, Oxford. 8vo. [*In preparation.*

THE "SEVEN AGAINST THEBES." Edited, with Introduction, Commentary, and Translation, by A. W. VERRALL, Litt.D., Fellow of Trinity College, Cambridge. 8vo. 7*s.* 6*d.*

THE SUPPLICES. A Revised Text, with Introduction, Critical Notes, Commentary and Translation. By T. G. TUCKER, M.A., Fellow of St. John's College, Cambridge, and Professor of Classical Philology in the University of Melbourne. 8vo. 10*s.* 6*d.*

Antoninus, Marcus Aurelius.—BOOK IV. OF THE MEDITATIONS. The Text Revised, with Translation and Notes, by HASTINGS CROSSLEY, M.A., Professor of Greek in Queen's College, Belfast. 8vo. 6*s.*

Aristotle.—THE METAPHYSICS. BOOK I. Translated by a Cambridge Graduate. 8vo. 5*s.*

THE POLITICS. Edited, after SUSEMIHL, by R. D. HICKS, M.A., Fellow of Trinity College, Cambridge. 8vo. [*In the press.*

THE POLITICS. Translated, with Analysis and Critical Notes, by Rev. J. E. C. WELLDON, M.A., Fellow of King's College, Cambridge, and Head-Master of Harrow School. Second Edition, revised. Crown 8vo. 10*s.* 6*d.*

THE RHETORIC. Translated, with an Analysis and Critical Notes, by the same. Crown 8vo. 7*s.* 6*d.*

THE ETHICS. Translated, with an Analysis and Critical Notes, by the same. Crown 8vo. [*In preparation.*

AN INTRODUCTION TO ARISTOTLE'S RHETORIC. With Analysis, Notes, and Appendices. By E. M. COPE, Fellow and Tutor of Trinity College, Cambridge. 8vo. 14*s.*

THE SOPHISTICI ELENCHI. With Translation and Notes by E. POSTE, M.A., Fellow of Oriel College, Oxford. 8vo. 8*s.* 6*d.*

Aristophanes.—THE BIRDS. Translated into English Verse, with Introduction, Notes, and Appendices, by B. H. KENNEDY, D.D., Regius Professor of Greek in the University of Cambridge. Crown 8vo. 6*s.* Help Notes to the same, for the use of Students, 1*s.* 6*d.*

Attic Orators.—FROM ANTIPHON TO ISAEOS. By R. C. JEBB, Litt.D., LL.D., Professor of Greek in the University of Cambridge. 2 vols. 8vo. 25*s.*

Babrius.—Edited, with Introductory Dissertations, Critical Notes, Commentary and Lexicon. By Rev. W. GUNION RUTHERFORD, M.A., LL.D., Head-Master of Westminster. 8vo. 12*s*. 6*d*.

Cicero.—THE ACADEMICA. The Text revised and explained by J. S. REID, M.L., Litt.D., Fellow of Caius College, Cambridge. 8vo. 15*s*.

THE ACADEMICS. Translated by J. S. REID, M.L., Litt.D. 8vo. 5*s*. 6*d*.

SELECT LETTERS. After the Edition of ALBERT WATSON, M.A. Translated by G. E. JEANS, M.A., Fellow of Hertford College, Oxford, and late Assistant-Master at Haileybury. Second Edition. Revised. Crown 8vo. 10*s*. 6*d*.

(See also *Classical Series.*)

Euripides.—MEDEA. Edited, with Introduction and Notes, by A. W. VERRALL, Litt.D. 8vo. 7*s*. 6*d*.

IPHIGENIA IN AULIS. Edited, with Introduction and Notes, by E. B. ENGLAND, M.A. 8vo. [*In preparation.*

INTRODUCTION TO THE STUDY OF EURIPIDES. By Professor J. P. MAHAFFY. Fcap. 8vo. 1*s*. 6*d*. (*Classical Writers.*)

(See also *Classical Series.*)

Herodotus.—BOOKS I.—III. THE ANCIENT EMPIRES OF THE EAST. Edited, with Notes, Introductions, and Appendices, by A. H. SAYCE, Deputy-Professor of Comparative Philology, Oxford; Honorary LL.D., Dublin. Demy 8vo. 16*s*.

BOOKS IV.—IX. Edited by REGINALD W. MACAN, M.A., Lecturer in Ancient History at Brasenose College, Oxford. 8vo. [*In preparation.*

THE HISTORY. Translated into English with Notes and Indices. By G. C. MACAULAY, M.A., formerly Fellow of Trinity College, Cambridge. 2 vols. Crown 8vo. [*In the press.*

Homer.—THE ILIAD. Edited, with Introduction and English Notes, by WALTER LEAF, Litt.D., late Fellow of Trinity College, Cambridge. 8vo. In Two Volumes. Vol. I. Books I.—XII. 14*s*. Vol. II. Books XIII.—XIV. 14*s*.

THE ILIAD. Translated into English Prose. By ANDREW LANG, M.A., WALTER LEAF, Litt.D., and ERNEST MYERS, M.A. Crown 8vo. 12*s*. 6*d*.

THE ODYSSEY. Done into English by S. H. BUTCHER, M.A., Professor of Greek in the University of Edinburgh, and ANDREW LANG, M.A., late Fellow of Merton College, Oxford. Seventh and Cheaper Edition, revised and corrected. Crown 8vo. 6*s*.

INTRODUCTION TO THE STUDY OF HOMER. By the Right Hon. W. E. GLADSTONE, M.P. 18mo. 1*s*. (*Literature Primers.*)

HOMERIC DICTIONARY. For Use in Schools and Colleges. Translated from the German of Dr. G. AUTENRIETH, with Additions and Corrections, by R. P. KEEP, Ph.D. With numerous Illustrations. Crown 8vo. 6*s*.

(See also *Classical Series.*)

Horace.—THE WORKS OF HORACE RENDERED INTO ENGLISH PROSE. With Introductions, Running Analysis, Notes, &c. By J. LONSDALE, M.A., and S. LEE, M.A. (*Globe Edition.*) 3*s.* 6*d.*

STUDIES, LITERARY AND HISTORICAL, IN THE ODES OF HORACE. By A. W. VERRALL, Litt.D., Fellow of College, Cambridge. Demy 8vo. 8*s.* 6*d.*

(See also *Classical Series.*)

Juvenal.—THIRTEEN SATIRES OF JUVENAL. With a Commentary. By JOHN E. B. MAYOR, M.A., Professor of Latin in the University of Cambridge. Crown 8vo.

*** Vol. I. Fourth Edition, Revised and Enlarged. 10*s.* 6*d.* Vol. II. Second Edition. 10*s.* 6*d.*

*** The new matter consists of an Introduction (pp. 1—53), Additional Notes (pp. 333—466) and Index (pp. 467—526). It is also issued separately, as a Supplement to the previous edition, at 5*s.*

THIRTEEN SATIRES. Translated into English after the Text of J. E. B. MAYOR by ALEXANDER LEEPER, M.A., Warden of Trinity College, in the University of Melbourne. Crown 8vo. 3*s.* 6*d.*

(See also *Classical Series.*)

Ktesias.—THE FRAGMENTS OF THE PERSIKA OF KTESIAS. Edited with Introduction and Notes by JOHN GILMORE, M.A. 8vo. 8*s.* 6*d.*

Livy.—BOOKS I.—IV. Translated by Rev. H. M. STEPHENSON, M.A., late Head-Master of St. Peter's School, York. [*In preparation.*

BOOKS XXI.—XXV. Translated by ALFRED JOHN CHURCH, M.A., of Lincoln College, Oxford, Professor of Latin, University College, London, and WILLIAM JACKSON BRODRIBB, M.A., late Fellow of St. John's College, Cambridge. Cr. 8vo. 7*s.* 6*d.*

INTRODUCTION TO THE STUDY OF LIVY. By Rev. W. W. CAPES, Reader in Ancient History at Oxford. Fcap. 8vo. 1*s.* 6*d.* (*Classical Writers Series.*)

(See also *Classical Series.*)

Martial.—BOOKS I. AND II. OF THE EPIGRAMS. Edited, with Introduction and Notes, by Professor J. E. B. MAYOR, M.A. 8vo. [*In the press.*

(See also *Classical Series.*)

Pausanias.—DESCRIPTION OF GREECE. Translated by J. G. FRAZER, M.A., Fellow of Trinity College, Cambridge. [*In preparation.*

Phrynichus.—THE NEW PHRYNICHUS; being a Revised Text of the Ecloga of the Grammarian Phrynichus. With Introduction and Commentary by Rev. W. GUNION RUTHERFORD, M.A., LL.D., Head-Master of Westminster. 8vo. 18*s.*

Pindar.—THE EXTANT ODES OF PINDAR. Translated into English, with an Introduction and short Notes, by ERNEST MYERS, M.A., late Fellow of Wadham College, Oxford. Second Edition. Crown 8vo. 5*s.*

THE NEMEAN ODES. Edited, with Introduction and Notes, by JOHN B. BURY, M.A., Fellow of Trinity College, Dublin. 8vo. (*Classical Library.*) [*In the press.*

Plato.—PHÆDO. Edited, with Introduction, Notes, and Appendices, by R. D. ARCHER-HIND, M.A., Fellow of Trinity College. Cambridge. 8vo. 8*s.* 6*d.*

TIMAEUS.—Edited, with Introduction, Notes, and a Translation, by the same Editor. 8vo. 16*s.*

PHÆDO. Edited, with Introduction and Notes, by W. D. GEDDES, LL.D., Principal of the University of Aberdeen. Second Edition. Demy 8vo. 8*s.* 6*d.*

THE REPUBLIC OF PLATO. Translated into English, with an Analysis and Notes, by J. LL. DAVIES, M.A., and D. J. VAUGHAN, M.A. 18mo. 4*s.* 6*d.*

EUTHYPHRO, APOLOGY, CRITO, AND PHÆDO. Translated by F. J. CHURCH. 18mo. 4*s.* 6*d.*

PHÆDRUS, LYSIS, AND PROTAGORAS. A New and Literal Translation, mainly from the text of Bekker. By J. WRIGHT, M.A., Trinity College, Cambridge. 18mo. 4*s.* 6*d.*

(See also *Classical Series.*)

Plautus.—THE MOSTELLARIA OF PLAUTUS. With Notes, Prolegomena, and Excursus. By WILLIAM RAMSAY, M.A., formerly Professor of Humanity in the University of Glasgow. Edited by Professor GEORGE G. RAMSAY, M.A., of the University of Glasgow. 8vo. 14*s.*

(See also *Classical Series.*)

Pliny.—CORRESPONDENCE WITH TRAJAN. C. Plinii Caecilii Secundi Epistulae ad Traianum Imperatorem cum Eiusdem Responsis. Edited, with Notes and Introductory Essays, by E. G. HARDY, M.A., Fellow of Jesus College, Oxford, and formerly Head Master of Grantham School. 8vo. 10*s.* 6*d.*

(See also *Classical Series.*)

Polybius.—THE HISTORIES OF POLYBIUS. Translated from the Text of F. HULTSCH by E. S. SHUCKBURGH, M.A., late Fellow of Emmanuel College, Cambridge. 2 vols. Crown 8vo. 24*s.*

(See also *Classical Series.*)

Sallust.—CATILINE AND JUGURTHA. Translated, with Introductory Essays, by A. W. POLLARD, B.A. Crown 8vo. 6*s.* THE CATILINE (separately). Crown 8vo. 3*s.*

(See also *Classical Series.*)

Sophocles.—ŒDIPUS THE KING. Translated from the Greek of Sophocles into English Verse by E. D. A. MORSHEAD, M.A., late Fellow of New College, Oxford; Assistant Master at Winchester College. Fcap. 8vo. 3*s.* 6*d.*

Tacitus.—THE ANNALS. Edited, with Introductions and Notes, by G. O. HOLBROOKE, M.A., Professor of Latin in Trinity College, Hartford, U.S.A. With Maps. 8vo. 16*s.*

THE ANNALS. Translated by A. J. CHURCH, M.A., and W. J. BRODRIBB, M.A. With Notes and Maps. New Edition. Crown 8vo. 7*s.* 6*d.*

THE HISTORIES. Edited, with Introduction and Notes, by Rev. W. A. SPOONER, M.A., Fellow of New College, Oxford. 8vo. [*In the press.*

THE HISTORY. Translated by A. J. CHURCH, M.A., and W. J. BRODRIBB, M.A. With Notes and a Map. Crown 8vo. 6*s.*

THE AGRICOLA AND GERMANY, WITH THE DIALOGUE ON ORATORY. Translated by A. J. CHURCH, M.A., and W. J. BRODRIBB, M.A. With Notes and Maps. New and Revised Edition. Crown 8vo. 4*s.* 6*d.*

INTRODUCTION TO THE STUDY OF TACITUS. By A. J. CHURCH, M.A. and W. J. BRODRIBB, M.A. Fcap. 8vo. 1*s.* 6*d.* (*Classical Writers Series.*)

(See also *Classical Series.*)

Theocritus, Bion, and Moschus. Rendered into English Prose, with Introductory Essay, by A. LANG, M.A. 18mo. 4*s.* 6*d.*

*** Also an Edition on Large Paper. Crown 8vo. 9*s.*

Thucydides.—BOOK IV. A Revision of the Text, Illustrating the Principal Causes of Corruption in the Manuscripts of this Author. By WILLIAM GUNION RUTHERFORD, M.A., LL.D., Headmaster of Westminster. 8vo. 7*s.* 6*d.*

BOOK VIII. Edited with Introduction and Notes, by H. C. GOODHART, M.A., Fellow of Trinity College, Cambridge.

(See also *Classical Series.*) [*In preparation.*

Virgil.—THE WORKS OF VIRGIL RENDERED INTO ENGLISH PROSE, with Notes, Introductions, Running Analysis, and an Index, by JAMES LONSDALE, M.A., and SAMUEL LEE, M.A. New Edition. Globe 8vo. 3*s.* 6*d.*

THE ÆNEID. Translated by J. W. MACKAIL, M.A., Fellow of Balliol College, Oxford. Crown 8vo. 7*s.* 6*d.*

(See also *Classical Series.*)

Xenophon.—THE WORKS OF. Translated into English, with Introduction and Notes, by H. G. DAKYNS, M.A. In four vols. Vol. I., containing "The Anabasis;" and Books I. and II. of "The Hellenica." Crown 8vo. [*Just ready.*

(See also *Classical Series.*)

GRAMMAR, COMPOSITION, & PHILOLOGY.

Belcher.—SHORT EXERCISES IN LATIN PROSE COMPOSITION AND EXAMINATION PAPERS IN LATIN GRAMMAR, to which is prefixed a Chapter on Analysis of Sentences. By the Rev. H. BELCHER, M.A., Rector of the High School, Dunedin, N.Z. New Edition. 18mo. 1*s.* 6*d.*

KEY TO THE ABOVE (for Teachers only). 3*s.* 6*d.*

Belcher.—SHORT EXERCISES IN LATIN PROSE COMPOSITION. Part II., On the Syntax of Sentences, with an Appendix, including EXERCISES IN LATIN IDIOMS, &c. 18mo. 2*s.*

KEY TO THE ABOVE (for Teachers only). 3*s.*

Blackie.—GREEK AND ENGLISH DIALOGUES FOR USE IN SCHOOLS AND COLLEGES. By JOHN STUART BLACKIE, Emeritus Professor of Greek in the University of Edinburgh. New Edition. Fcap. 8vo. 2*s.* 6*d.*

Bryans.—LATIN PROSE EXERCISES BASED UPON CAESAR'S GALLIC WAR. With a Classification of Cæsar's Chief Phrases and Grammatical Notes on Cæsar's Usages. By CLEMENT BRYANS, M.A., Assistant-Master in Dulwich College. Second Edition, Revised and Enlarged. Extra fcap. 8vo. 2*s.* 6*d.*

KEY TO THE ABOVE (for Teachers only). New Edition. 4*s.* 6*d.*

GREEK PROSE EXERCISES based upon Thucydides. By the same Author. Extra fcap. 8vo. [*In preparation.*

Cookson.—A LATIN SYNTAX. By CHRISTOPHER COOKSON, M.A., Assistant-Master at St. Paul's School. 8vo. [*In preparation.*

Cornell University Studies in Classical Philology. Edited by ISAAC FLAGG, WILLIAM GARDNER HALE, and BENJAMIN IDE WHEELER. I. The *CUM*-Constructions: their History and Functions. Part 1, Critical. 1*s.* 8*d.* nett. Part 2, Constructive. By WILLIAM GARDNER HALE. 3*s.* 4*d.* nett. II. Analogy and the Scope of its Application in Language. By BENJAMIN IDE WHEELER. 1*s.* 3*d.* nett.

Eicke.—FIRST LESSONS IN LATIN. By K. M. EICKE, B.A., Assistant-Master in Oundle School. Globe 8vo. 2*s.*

England.—EXERCISES ON LATIN SYNTAX AND IDIOM. ARRANGED WITH REFERENCE TO ROBY'S SCHOOL LATIN GRAMMAR. By E. B. ENGLAND, M.A., Assistant Lecturer at the Owens College, Manchester. Crown 8vo. 2*s.* 6*d.* Key for Teachers only, 2*s.* 6*d.*

Giles.—A MANUAL OF GREEK AND LATIN PHILOLOGY. By P. GILES, M.A., Fellow of Gonville and Caius College, Cambridge. Crown 8vo. [*In preparation.*

Goodwin.—Works by W. W. GOODWIN, LL.D., Professor of Greek in Harvard University, U.S.A.

SYNTAX OF THE MOODS AND TENSES OF THE GREEK VERB. New Edition, Revised and Enlarged. 8vo. [*In the press.*

A GREEK GRAMMAR. New Edition, revised. Crown 8vo. 6*s.*

"It is the best Greek Grammar of its size in the English language."—ATHENÆUM.

A GREEK GRAMMAR FOR SCHOOLS. Crown 8vo. 3*s.* 6*d.*

Greenwood.—THE ELEMENTS OF GREEK GRAMMAR, including Accidence, Irregular Verbs, and Principles of Derivation and Composition; adapted to the System of Crude Forms. By J. G. GREENWOOD, sometime Principal of Owens College, Manchester. New Edition. Crown 8vo. 5*s.* 6*d.*

Hadley and Allen.—A GREEK GRAMMAR FOR SCHOOLS AND COLLEGES. By JAMES HADLEY, late Professor in Yale College. Revised and in part Rewritten by FREDERIC DE FOREST ALLEN, Professor in Harvard College. Crown 8vo. 6*s.*

Hodgson.—MYTHOLOGY FOR LATIN VERSIFICATION. A brief Sketch of the Fables of the Ancients, prepared to be rendered into Latin Verse for Schools. By F. HODGSON, B.D., late Provost of Eton. New Edition, revised by F. C. HODGSON, M.A. 18mo. 3*s.*

Jackson.—FIRST STEPS TO GREEK PROSE COMPOSITION. By BLOMFIELD JACKSON, M.A., Vicar of St. Bartholomew's, Moor Lane, E.C., formerly Senior Assistant Classical Master in King's College School, London. New Edition, revised and enlarged. 18mo. 1*s.* 6*d.*

KEY TO FIRST STEPS (for Teachers only). 18mo. 3*s.* 6*d.*

SECOND STEPS TO GREEK PROSE COMPOSITION, with Miscellaneous Idioms, Aids to Accentuation, and Examination Papers in Greek Scholarship. 18mo. 2*s.* 6*d.*

KEY TO SECOND STEPS (for Teachers only). 18mo. 3*s.* 6*d.*

Kynaston.—EXERCISES IN THE COMPOSITION OF GREEK IAMBIC VERSE by Translations from English Dramatists. By Rev. H. KYNASTON, D.D., Principal of Cheltenham College. With Introduction, Vocabulary, &c. New Edition, revised and enlarged. Extra fcap. 8vo. 5*s.*

KEY TO THE SAME (for Teachers only). Extra fcap. 8vo. 4*s.* 6*d.*

Lupton.—Works by J. H. LUPTON, M.A., Sur-Master of St. Paul's School, and formerly Fellow of St. John's College, Cambridge.

AN INTRODUCTION TO LATIN ELEGIAC VERSE COMPOSITION. Globe 8vo. 2*s.* 6*d.*

LATIN RENDERING OF THE EXERCISES IN PART II. (XXV.-C.). Globe 8vo. 3*s.* 6*d.*

AN INTRODUCTION TO LATIN LYRIC VERSE COMPOSITION. Globe 8vo. 3*s.*

KEY FOR THE USE OF TEACHERS ONLY. Globe 8vo. 4*s.* 6*d.*

Mackie.—PARALLEL PASSAGES FOR TRANSLATION INTO GREEK AND ENGLISH. Carefully graduated for the use of Colleges and Schools. With Indexes. By Rev. ELLIS C. MACKIE, Classical Master at Heversham Grammar School. Globe 8vo. 4*s.* 6*d.*

Macmillan.—FIRST LATIN GRAMMAR. By M. C. MACMILLAN, M.A., late Scholar of Christ's College, Cambridge; sometime Assistant-Master in St. Paul's School. New Edition, enlarged. Fcap. 8vo. 1*s.* 6*d.*

Macmillan's Greek Course.—Edited by Rev. W. GUNION RUTHERFORD, M.A., LL.D., Head Master of Westminster.

I.—FIRST GREEK GRAMMAR. By the Editor. New Edition, thoroughly revised. Globe 8vo. 2*s.*

Macmillan's Greek Course.—II.—EASY EXERCISES IN GREEK ACCIDENCE. By H. G. UNDERHILL, M.A., Assistant-Master St. Paul's Preparatory School. Globe 8vo. 2*s.*

III.—A SECOND GREEK EXERCISE BOOK. By the Rev. W. A. HEARD, M.A., Assistant-Master at Westminster School. Globe 8vo. [*In the press.*

IV.—MANUAL OF GREEK ACCIDENCE. By the Editor. [*In preparation.*

V.—MANUAL OF GREEK SYNTAX. By the Editor. [*In preparation.*

VI.—ELEMENTARY GREEK COMPOSITION. By the Editor. [*In preparation.*

Macmillan's Greek Reader.—STORIES AND LEGENDS. A First Greek Reader, with Notes, Vocabulary, and Exercises. By F. H. COLSON, M.A., late Fellow of St. John's College, Cambridge; Head Master of Plymouth College. Globe 8vo. 3*s.*

Macmillan's Latin Course. By A. M. COOK, M.A., Assistant-Master at St. Paul's School. I. First Year. New Edition, revised and enlarged. Globe 8vo. 3*s.* 6*d.* II. Second Year. 2*s.* 6*d.* [*Nearly ready.*

Macmillan's Latin Reader.—A LATIN READER FOR THE LOWER FORMS IN SCHOOLS. By H. J. HARDY, M.A., Assistant Master in Winchester College. Globe 8vo. 2*s.* 6*d.*

Macmillan's Shorter Latin Course. By A. M. COOK, M.A., Assistant-Master at St. Paul's School. Being an abridgment of "Macmillan's Latin Course," First Year. Globe 8vo. 1*s.* 6*d.*

Marshall.—A TABLE OF IRREGULAR GREEK VERBS. classified according to the arrangement of Curtius's Greek Grammar, By J. M. MARSHALL, M.A., Head Master of the Grammar School, Durham. New Edition. 8vo. 1*s.*

Mayor (John E. B.)—FIRST GREEK READER. Edited after KARL HALM, with Corrections and large Additions by Professor JOHN E. B. MAYOR, M.A., Fellow of St. John's College, Cambridge. New Edition, revised. Fcap. 8vo. 4*s.* 6*d.*

Mayor (Joseph B.)—GREEK FOR BEGINNERS. By the Rev. J. B. MAYOR, M.A., Professor of Classical Literature in King's College, London. Part I., with Vocabulary, 1*s.* 6*d.* Parts II. and III., with Vocabulary and Index, 3*s.* 6*d.* Complete in one Vol. fcap. 8vo. 4*s.* 6*d.*

Nixon.—PARALLEL EXTRACTS, Arranged for Translation into English and Latin, with Notes on Idioms. By J. E. NIXON, M.A., Fellow and Classical Lecturer, King's College, Cambridge. Part I.—Historical and Epistolary. New Edition, revised and enlarged. Crown 8vo. 3*s.* 6*d.*

PROSE EXTRACTS, Arranged for Translation into English and Latin, with General and Special Prefaces on Style and Idiom. I. Oratorical. II. Historical. III. Philosophical. IV. Anecdotes and Letters. By the same Author. 2nd Edition, enlarged to 280 pp. Crown 8vo. 4*s.* 6*d.*

Nixon.—SELECTIONS FROM PROSE EXTRACTS, including Easy Anecdotes and Letters with Notes and Hints. Pp. 120. 3*s*.

Translations of about 70 Extracts can be supplied to Schoolmasters (2*s*. 6*d*.), on application to the Author; and about 40 similarly of "Parallel Extracts." 1*s*. 6*d*. post free.

Pantin.—A FIRST LATIN VERSE BOOK. By W. E. P. PANTIN, M.A., Assistant-Master at St. Paul's School. Globe 8vo. [*In the press.*

Peile.—A PRIMER OF PHILOLOGY. By J. PEILE, Litt.D., Master of Christ's College, Cambridge. 18mo. 1*s*.

Postgate.—SERMO LATINUS. A Short Guide to Latin Prose Composition. Part I. Introduction. Part II. Selected Passages for Translation. By J. P. POSTGATE, M.A., Litt. D., Fellow and Classical Lecturer of Trinity College, Cambridge, Professor of Comparative Philology in University College, London. Globe 8vo. 2*s*. 6*d*.

KEY to "Selected Passages." 3*s*. 6*d*.

Postgate and Vince.—A DICTIONARY OF LATIN ETYMOLOGY. By J. P. POSTGATE, M.A., and C. A. VINCE, M.A. [*In preparation.*

Potts (A. W.)—Works by ALEXANDER W. POTTS, M.A., LL.D., late Fellow of St. John's College, Cambridge; Head Master of the Fettes College, Edinburgh.

HINTS TOWARDS LATIN PROSE COMPOSITION. New Edition. Extra fcap. 8vo. 3*s*.

PASSAGES FOR TRANSLATION INTO LATIN PROSE. Edited with Notes and References to the above. New Edition. Extra fcap. 8vo. 2*s*. 6*d*.

LATIN VERSIONS OF PASSAGES FOR TRANSLATION INTO LATIN PROSE (for Teachers only). 2*s*. 6*d*.

Preston.—EXERCISES IN LATIN VERSE OF VARIOUS KINDS. By Rev. G. PRESTON, M.A., late Fellow of Magdalene College, Cambridge, and Head Master of the King's School, Chester. Globe 8vo. 2*s*. 6*d*.

KEY, for the use of Teachers only. Globe 8vo. 5*s*.

Reid.—A GRAMMAR OF TACITUS. By J. S. REID, M.L., Fellow of Caius College, Cambridge. [*In preparation.*

A GRAMMAR OF VERGIL. By the same Author. [*In preparation.*

*** *Similar Grammars to other Classical Authors will probably follow.*

Roby.—A GRAMMAR OF THE LATIN LANGUAGE, from Plautus to Suetonius. By H. J. ROBY, M.A., late Fellow of St. John's College, Cambridge. In Two Parts. Part I. Fifth Edition, containing:—Book I. Sounds. Book II. Inflexions. Book III. Word-formation. Appendices. Crown 8vo. 9*s*.

Part II. Syntax, Prepositions, &c. Crown 8vo. 10*s*. 6*d*.

"Marked by the clear and practised insight of a master in his art. A book that would do honour to any country."—ATHENÆUM.

SCHOOL LATIN GRAMMAR. By the same Author. Cr. 8vo. 5*s*.

Rush.—SYNTHETIC LATIN DELECTUS. A First Latin Construing Book arranged on the Principles of Grammatical Analysis. With Notes and Vocabulary. By E. RUSH, B.A. With Preface by the Rev. W. F. MOULTON, M.A., D.D. New and Enlarged Edition. Extra fcap. 8vo. *2s. 6d.*

Rust.—FIRST STEPS TO LATIN PROSE COMPOSITION. By the Rev. G. RUST, M.A., of Pembroke Coll., Oxford, Master of the Lower School, King's Coll., London. New Ed. 18mo. *1s. 6d.*

KEY TO THE ABOVE. By W. M. YATES, Assistant-Master in the High School, Sale. 18mo. *3s. 6d.*

Rutherford.—Works by the Rev. W. GUNION RUTHERFORD, M.A., LL.D., Head-Master of Westminster.

REX LEX. A Short Digest of the principal Relations between Latin, Greek, and Anglo-Saxon Sounds. 8vo. [*In preparation.*

THE NEW PHRYNICHUS; being a Revised Text of the Ecloga of the Grammarian Phrynichus. With Introduction and Commentary. 8vo. *18s.* (See also Macmillan's Greek Course.)

Shuckburgh.—PASSAGES FROM LATIN AUTHORS FOR TRANSLATION INTO ENGLISH. Selected with a view to the needs of Candidates for the Cambridge Previous, Local, and Schools' Examinations. By E. S. SHUCKBURGH, M.A., Emmanuel College, Cambridge. Crown 8vo. *2s.*

Simpson.—LATIN PROSE AFTER THE BEST AUTHORS. By F. P. SIMPSON, B.A., late Scholar of Balliol College, Oxford. Part I. CÆSARIAN PROSE. Extra fcap. 8vo. *2s. 6d.*

KEY TO THE ABOVE, for Teachers only. Extra fcap. 8vo. *5s.*

Strachan and Wilkins.— ANALECTA. Passages for Translation. Selected by JOHN S. STRACHAN, M.A., Pembroke College, Cambridge, and A. S. WILKINS, Litt.D., LL.D., Fellow of St. John's College, Cambridge, Professors in Owens College, Manchester. Crown 8vo. *5s.*

Thring.—Works by the Rev. E. THRING, M.A., late Head-Master of Uppingham School.

A LATIN GRADUAL. A First Latin Construing Book for Beginners. New Edition, enlarged, with Coloured Sentence Maps. Fcap. 8vo. *2s. 6d.*

A MANUAL OF MOOD CONSTRUCTIONS. Fcap. 8vo. *1s. 6d.*

Welch and Duffield.—LATIN ACCIDENCE AND EXERCISES ARRANGED FOR BEGINNERS. By WILLIAM WELCH, M.A., and C. G. DUFFIELD, M.A., Assistant Masters at Cranleigh School. 18mo. *1s. 6d.*

This book is intended as an introduction to Macmillan's *Elementary Classics*, and is the development of a plan which has been in use for some time and has been worked satisfactorily.

White.—FIRST LESSONS IN GREEK. Adapted to GOODWIN'S GREEK GRAMMAR, and designed as an introduction to the ANABASIS OF XENOPHON. By JOHN WILLIAMS WHITE, Ph.D., Assistant-Professor of Greek in Harvard University. Crown 8vo. *4s. 6d.*

Wright.—Works by J. WRIGHT, M.A., late Head Master of Sutton Coldfield School.

A HELP TO LATIN GRAMMAR; or, The Form and Use of Words in Latin, with Progressive Exercises. Crown 8vo. 4*s*. 6*d*.

THE SEVEN KINGS OF ROME. An Easy Narrative, abridged from the First Book of Livy by the omission of Difficult Passages; being a First Latin Reading Book, with Grammatical Notes and Vocabulary. New and revised Edition. Fcap. 8vo. 3*s*. 6*d*.

FIRST LATIN STEPS; OR, AN INTRODUCTION BY A SERIES OF EXAMPLES TO THE STUDY OF THE LATIN LANGUAGE. Crown 8vo. 3*s*.

ATTIC PRIMER. Arranged for the Use of Beginners. Extra fcap. 8vo. 2*s*. 6*d*.

A COMPLETE LATIN COURSE, comprising Rules with Examples, Exercises, both Latin and English, on each Rule, and Vocabularies. Crown 8vo. 2*s*. 6*d*.

ANTIQUITIES, ANCIENT HISTORY, AND PHILOSOPHY.

Arnold.—Works by W. T. ARNOLD, M.A.

A HANDBOOK OF LATIN EPIGRAPHY. [*In preparation.*

THE ROMAN SYSTEM OF PROVINCIAL ADMINISTRATION TO THE ACCESSION OF CONSTANTINE THE GREAT. Crown 8vo. 6*s*.

Arnold (T.)—THE SECOND PUNIC WAR. Being Chapters on THE HISTORY OF ROME. By the late THOMAS ARNOLD, D.D., formerly Head Master of Rugby School, and Regius Professor of Modern History in the University of Oxford. Edited, with Notes, by W. T. ARNOLD, M.A. With 8 Maps. Crown 8vo. 8*s*. 6*d*.

Beesly.—STORIES FROM THE HISTORY OF ROME. By Mrs. BEESLY. Fcap. 8vo. 2*s*. 6*d*.

Burn.—ROMAN LITERATURE IN RELATION TO ROMAN ART. By Rev. ROBERT BURN, M.A., Fellow of Trinity College, Cambridge. With numerous Illustrations. Extra Crown 8vo. 14*s*.

Bury.—A HISTORY OF THE LATER ROMAN EMPIRE FROM ARCADIUS TO IRENE, A.D. 395—800. By JOHN B. BURY, M.A., Fellow of Trinity College, Dublin. 2 vols. 8vo. 32*s*.

Classical Writers.—Edited by JOHN RICHARD GREEN, M.A., LL.D. Fcap. 8vo. 1*s*. 6*d*. each.

EURIPIDES. By Professor MAHAFFY.
MILTON. By the Rev. STOPFORD A. BROOKE, M.A.
LIVY. By the Rev. W. W. CAPES, M.A.
VIRGIL. By Professor NETTLESHIP, M.A.
SOPHOCLES. By Professor L. CAMPBELL, M.A.
DEMOSTHENES. By Professor S. H. BUTCHER, M.A.
TACITUS. By Professor A. J. CHURCH, M.A., and W. J. BRODRIBB, M.A.

Freeman.—Works by EDWARD A. FREEMAN, D.C.L., LL.D., Hon. Fellow of Trinity College, Oxford, Regius Professor of Modern History in the University of Oxford.

HISTORY OF ROME. (*Historical Course for Schools.*) 18mo. [*In preparation.*

A SCHOOL HISTORY OF ROME. Crown 8vo. [*In preparation.*

HISTORICAL ESSAYS. Second Series. [Greek and Roman History.] 8vo. 10*s.* 6*d.*

Fyffe.—A SCHOOL HISTORY OF GREECE. By C. A. FYFFE, M.A. Crown 8vo. [*In preparation.*

Gardner.—SAMOS AND SAMIAN COINS. An Essay. By PERCY GARDNER, M.A., Litt.D., Professor of Archæology in the University of Oxford. With Illustrations. 8vo. 7*s.* 6*d.*

Geddes. — THE PROBLEM OF THE HOMERIC POEMS. By W. D. GEDDES, Principal of the University of Aberdeen. 8vo. 14*s.*

Gladstone.—Works by the Rt. Hon. W. E. GLADSTONE, M.P.

THE TIME AND PLACE OF HOMER. Crown 8vo. 6*s.* 6*d.*

A PRIMER OF HOMER. 18mo. 1*s.*

Gow.—A COMPANION TO SCHOOL CLASSICS. By JAMES GOW, Litt.D., Master of the High School, Nottingham; late Fellow of Trinity College, Cambridge. With Illustrations. Second Edition. Revised. Crown 8vo. 6*s.*

Harrison and Verrall.—CULTS AND MONUMENTS OF ANCIENT ATHENS. By Miss J. E. HARRISON and Mrs. A. W. VERRALL. Illustrated. Crown 8vo. [*In the press.*

Jebb.—Works by R. C. JEBB, Litt.D., LL.D., Professor of Greek in the University of Cambridge.

THE ATTIC ORATORS FROM ANTIPHON TO ISAEOS. 2 vols. 8vo. 25*s.*

A PRIMER OF GREEK LITERATURE. 18mo. 1*s.*

(See also *Classical Series.*)

Kiepert.—MANUAL OF ANCIENT GEOGRAPHY, Translated from the German of Dr. HEINRICH KIEPERT. Crown 8vo. 5*s.*

Lanciani.—ANCIENT ROME IN THE LIGHT OF RECENT DISCOVERIES. By RODOLFO LANCIANI, LL.D. (Harv.), Professor of Archæology in the University of Rome, Director of Excavations for the National Government and the Municipality of Rome, &c. With 100 Illustrations. Small 4to. 24*s.*

Mahaffy.—Works by J. P. MAHAFFY, M.A., D.D., Fellow and Professor of Ancient History in Trinity College, Dublin, and Hon. Fellow of Queen's College, Oxford.

SOCIAL LIFE IN GREECE; from Homer to Menander. Fifth Edition, revised and enlarged. Crown 8vo. 9*s.*

GREEK LIFE AND THOUGHT; from the Age of Alexander to the Roman Conquest. Crown 8vo. 12*s.* 6*d.*

Mahaffy.—RAMBLES AND STUDIES IN GREECE. With Illustrations. Third Edition, Revised and Enlarged. With Map. Crown 8vo. 10*s.* 6*d.*

A HISTORY OF CLASSICAL GREEK LITERATURE. In Two Vols. Crown 8vo. 9*s.* each. Vol. I. The Poets, with an Appendix on Homer by Prof. SAYCE. Vol. II. The Prose Writers.

A PRIMER OF GREEK ANTIQUITIES. With Illustrations. 18mo. 1*s.*

EURIPIDES. 18mo. 1*s.* 6*d.* (*Classical Writers Series.*)

Mayor (J. E. B.)—BIBLIOGRAPHICAL CLUE TO LATIN LITERATURE. Edited after HÜBNER, with large Additions, by Professor JOHN E. B. MAYOR. Crown 8vo. 10*s.* 6*d.*

Newton.—ESSAYS IN ART AND ARCHÆOLOGY. By Sir CHARLES NEWTON, K.C.B., D.C.L., formerly Professor of Archæology in University College, London, and Keeper of Greek and Roman Antiquities at the British Museum. 8vo. 12*s.* 6*d.*

Sayce.—THE ANCIENT EMPIRES OF THE EAST. By A. H. SAYCE, Deputy-Professor of Comparative Philosophy, Oxford, Hon. LL.D. Dublin. Crown 8vo. 6*s.*

Stewart.—THE TALE OF TROY. Done into English by AUBREY STEWART, M.A., late Fellow of Trinity Colleges Cambridge. Globe 8vo. 3*s.* 6*d.*

Waldstein.—CATALOGUE OF CASTS IN THE MUSEUM OF CLASSICAL ARCHÆOLOGY. By CHARLES WALDSTEIN, Litt.D., Ph.D., LL.D., University Reader in Classical Archæology. Crown 8vo. 1*s.* 6*d.*

*** Also an Edition on Large Paper, small 4to. 5*s.*

Wilkins.—A PRIMER OF ROMAN ANTIQUITIES. By Professor WILKINS, Litt. D., LL.D. Illustrated. 18mo. 1*s.*

A PRIMER OF ROMAN LITERATURE. By the same Author. [*Just ready.*

MATHEMATICS.

(1) Arithmetic and Mensuration, (2) Algebra, (3) Euclid and Elementary Geometry, (4) Trigonometry, (5) Higher Mathematics.

ARITHMETIC AND MENSURATION.

Aldis.—THE GREAT GIANT ARITHMOS. A most Elementary Arithmetic for Children. By MARY STEADMAN ALDIS. With Illustrations. Globe 8vo. 2*s.* 6*d.*

Army Preliminary Examination, Specimens of Papers Set at the, 1882—88.—With Answers to the Mathematical Questions. Subjects: Arithmetic, Algebra, Euclid, Geometrical Drawing, Geography, French, English Dictation. Crown 8vo. 3*s*. 6*d*.

This selection of papers at recent Preliminary Examinations for the admission to the Army are intended for the use of Candidates preparing for these Examinations.

Bradshaw.—A COURSE OF EASY ARITHMETICAL EXAMPLES FOR BEGINNERS. By J. G. BRADSHAW, B.A., formerly Scholar of Jesus College, Cambridge, Assistant Master at Clifton College. Globe 8vo. 2*s*. With Answers, 2*s*. 6*d*.

Brooksmith (J.).—ARITHMETIC IN THEORY AND PRACTICE. By J. BROOKSMITH, M.A., LL.B., St. John's College, Cambridge; Barrister-at-Law; one of the Masters of Cheltenham College. New Edition, revised. Crown 8vo. 4*s*. 6*d*.

Brooksmith. — ARITHMETIC FOR BEGINNERS. By J. BROOKSMITH, M.A., LL.B., St. John's College, Cambridge; Barrister-at-Law; and sometime one of the Masters of Cheltenham College; and E. J. BROOKSMITH, B.A., LL.M., St. John's College, Cambridge; Instructor of Mathematics at the Royal Military Academy, Woolwich. Globe 8vo. 1*s*. 6*d*.

Candler.—HELP TO ARITHMETIC. Designed for the use of Schools. By H. CANDLER, M.A., Mathematical Master of Uppingham School. Second Edition. Extra fcap. 8vo. 2*s*. 6*d*.

Dalton.—RULES AND EXAMPLES IN ARITHMETIC. By the Rev. T. DALTON, M.A., Assistant-Master in Eton College. New Edition, with Answers. 18mo. 2*s*. 6*d*.

Goyen.—HIGHER ARITHMETIC AND ELEMENTARY MENSURATION. By P. GOYEN, Inspector of Schools, Dunedin, New Zealand. Crown 8vo. 5*s*.

Hall and Knight.—ARITHMETICAL EXERCISES AND EXAMINATION PAPERS. With an Appendix containing Questions in LOGARITHMS and MENSURATION. By H. S. HALL, M.A., formerly Scholar of Christ's College, Cambridge, Master of the Military and Engineering Side, Clifton College; and S. R. KNIGHT, B.A., formerly Scholar of Trinity College, Cambridge, late Assistant Master at Marlborough College, Authors of "Elementary Algebra," "Algebraical Exercises and Examination Papers," and "Higher Algebra," &c. Globe 8vo. 2*s*. 6*d*.

Lock.—Works by Rev. J. B. LOCK, M.A., Senior Fellow, Assistant Tutor, and Lecturer in Gonville and Caius College, Cambridge, formerly Master at Eton.

ARITHMETIC FOR SCHOOLS. With Answers and 1000 additional Examples for Exercise. Third Edition, revised. Stereotyped. Globe 8vo. 4*s*. 6*d*. Or in Two Parts:—Part I. Up to and including Practice. Globe 8vo. 2*s*. Part II. With 1000 additional Examples for Exercise. Globe 8vo. 3*s*.

*** *The complete book and the Parts can also be obtained* without answers *at the same price. But the editions* with answers *will always be supplied unless the others are specially asked for.*

Lock.—A KEY TO "ARITHMETIC FOR SCHOOLS." By the Rev. R. G. WATSON, M.A. Crown 8vo. 10*s*. 6*d*.

ARITHMETIC FOR BEGINNERS. A School Class Book of Commercial Arithmetic. Globe 8vo. 2*s*. 6*d*.

A KEY TO "ARITHMETIC FOR BEGINNERS." By Rev. R. G. WATSON, M.A. Crown 8vo. 8*s*. 6*d*.

A SHILLING CLASS-BOOK OF ARITHMETIC ADAPTED FOR USE IN ELEMENTARY SCHOOLS. 18mo. [*In the press.*

Pedley.—EXERCISES IN ARITHMETIC for the Use of Schools. Containing more than 7,000 original Examples. By S. PEDLEY, late of Tamworth Grammar School. Crown 8vo. 5*s*.

Also in Two Parts 2*s*. 6*d*. each.

Smith.—Works by the Rev. BARNARD SMITH, M.A., late Rector of Glaston, Rutland, and Fellow and Senior Bursar of S. Peter's College, Cambridge.

ARITHMETIC AND ALGEBRA, in their Principles and Application; with numerous systematically arranged Examples taken from the Cambridge Examination Papers, with especial reference to the Ordinary Examination for the B.A. Degree. New Edition, carefully Revised. Crown 8vo. 10*s*. 6*d*.

ARITHMETIC FOR SCHOOLS. New Edition. Crown 8vo. 4*s*. 6*d*.

A KEY TO THE ARITHMETIC FOR SCHOOLS. New Edition. Crown 8vo. 8*s*. 6*d*.

EXERCISES IN ARITHMETIC. Crown 8vo, limp cloth, 2*s*. With Answers, 2*s*. 6*d*. Answers separately, 6*d*.

SCHOOL CLASS-BOOK OF ARITHMETIC. 18mo, cloth. 3*s*. Or sold separately, in Three Parts, 1*s*. each.

KEYS TO SCHOOL CLASS-BOOK OF ARITHMETIC. Parts I., II., and III., 2*s*. 6*d*. each.

SHILLING BOOK OF ARITHMETIC FOR NATIONAL AND ELEMENTARY SCHOOLS. 18mo, cloth. Or separately, Part I. 2*d*.; Part II. 3*d*.; Part III. 7*d*. Answers, 6*d*.

THE SAME, with Answers complete. 18mo, cloth. 1*s*. 6*d*.

KEY TO SHILLING BOOK OF ARITHMETIC. 18mo. 4*s*. 6*d*.

EXAMINATION PAPERS IN ARITHMETIC. 18mo. 1*s*. 6*d*. The same, with Answers, 18mo, 2*s*. Answers, 6*d*.

KEY TO EXAMINATION PAPERS IN ARITHMETIC. 18mo. 4*s*. 6*d*.

THE METRIC SYSTEM OF ARITHMETIC, ITS PRINCIPLES AND APPLICATIONS, with numerous Examples, written expressly for Standard V. in National Schools. New Edition. 18mo, cloth, sewed. 3*d*.

A CHART OF THE METRIC SYSTEM, on a Sheet, size 42 in. by 34 in. on Roller, mounted and varnished. New Edition. Price 3*s*. 6*d*.

Also a Small Chart on a Card, price 1*d*.

Smith.—EASY LESSONS IN ARITHMETIC, combining Exercises in Reading, Writing, Spelling, and Dictation. Part I. for Standard I. in National Schools. Crown 8vo. 9*d.*

EXAMINATION CARDS IN ARITHMETIC. (Dedicated to Lord Sandon.) With Answers and Hints.

Standards I. and II. in box, 1*s.* Standards III., IV., and V., in boxes, 1*s.* each. Standard VI. in Two Parts, in boxes, 1*s.* each.

A and B papers, of nearly the same difficulty, are given so as to prevent copying, and the colours of the A and B papers differ in each Standard, and from those of every other Standard, so that a master or mistress can see at a glance whether the children have the proper papers.

Stevens.—ELEMENTARY MENSURATION. With exercises in the mensuration of Plane and Solid Figures. By F. H. STEVENS, M.A., formerly Scholar of Queens' College, Cambridge, Master of the Military and Engineering side, Clifton College. Globe 8vo. [*In the press.*

Thornton.—FIRST LESSONS IN BOOK-KEEPING. By J. THORNTON. New Edition. Crown 8vo. 2*s.* 6*d.*

A KEY TO THE ABOVE. Oblong 4to. 10*s.* 6*d.*

Todhunter.—MENSURATION FOR BEGINNERS. By I TODHUNTER, M.A., F.R.S., D.Sc., late of St. John's College, Cambridge. With Examples. New Edition. 18mo. 2*s.* 6*d.*

KEY TO MENSURATION FOR BEGINNERS. By the Rev. FR. LAWRENCE MCCARTHY, Professor of Mathematics in St Peter's College, Agra. Crown 8vo. 7*s.* 6*d.*

ALGEBRA.

Dalton.—RULES AND EXAMPLES IN ALGEBRA. By the Rev. T. DALTON, M.A., Assistant-Master of Eton College. Part I. New Edition. 18mo. 2*s.* Part II. 18mo. 2*s.* 6*d.*

*** *A Key to Part I. for Teachers only*, 7*s.* 6*d.*

Hall and Knight.—Works by H. S. HALL, M.A., formerly Scholar of Christ's College, Cambridge, Master of the Military and Engineering Side, Clifton College; and S. R. KNIGHT, B.A., formerly Scholar of Trinity College, Cambridge, late Assistant-Master at Marlborough College.

ELEMENTARY ALGEBRA FOR SCHOOLS. Fifth Edition, Revised and Corrected. Globe 8vo, bound in maroon coloured cloth, 3*s.* 6*d.*; with Answers, bound in green coloured cloth, 4*s.* 6*d.*

KEY. [*In the Press.*

ALGEBRAICAL EXERCISES AND EXAMINATION PAPERS. To accompany ELEMENTARY ALGEBRA. Second Edition, revised. Globe 8vo. 2*s.* 6*d.*

HIGHER ALGEBRA. A Sequel to "ELEMENTARY ALGEBRA FOR SCHOOLS." Second Edition. Crown 8vo. 7*s.* 6*d.*

SOLUTIONS OF THE EXAMPLES IN "HIGHER ALGEBRA." Crown 8vo. 10*s.* 6*d.*

Jones and Cheyne.—ALGEBRAICAL EXERCISES. Progressively Arranged. By the Rev. C. A. JONES, M.A., and C. H. CHEYNE, M.A., F.R.A.S., Mathematical Masters of Westminster School. New Edition. 18mo. 2*s.* 6*d.*

SOLUTIONS AND HINTS FOR THE SOLUTION OF SOME OF THE EXAMPLES IN THE ALGEBRAICAL EXERCISES OF MESSRS. JONES AND CHEYNE. By Rev. W. FAILES, M.A., Mathematical Master at Westminster School, late Scholar of Trinity College, Cambridge. Crown 8vo. 7*s.* 6*d.*

Smith (Barnard).—ARITHMETIC AND ALGEBRA, in their Principles and Application; with numerous systematically arranged Examples taken from the Cambridge Examination Papers, with especial reference to the Ordinary Examination for the B.A. Degree. By the Rev. BARNARD SMITH, M.A., late Rector of Glaston, Rutland, and Fellow and Senior Bursar of St. Peter's College, Cambridge. New Edition, carefully Revised. Crown 8vo. 10*s.* 6*d.*

Smith (Charles).—Works by CHARLES SMITH, M.A., Fellow and Tutor of Sidney Sussex College, Cambridge.

ELEMENTARY ALGEBRA. Globe 8vo. 4*s.* 6*d.*

In this work the author has endeavoured to explain the principles of Algebra in as simple a manner as possible for the benefit of beginners, bestowing great care upon the explanations and proofs of the fundamental operations and rules.

A TREATISE ON ALGEBRA. Crown 8vo. 7*s.* 6*d.*

SOLUTIONS OF THE EXAMPLES IN "A TREATISE ON ALGEBRA." Crown 8vo. 10*s.* 6*d.*

Todhunter.—Works by I. TODHUNTER, M.A., F.R.S., D.Sc. late of St. John's College, Cambridge.

"Mr. Todhunter is chiefly known to Students of Mathematics as the author of a eries of admirable mathematical text-books, which possess the rare qualities of being clear in style and absolutely free from mistakes, typographical or other."—SATURDAY REVIEW.

ALGEBRA FOR BEGINNERS. With numerous Examples. New Edition. 18mo. 2*s.* 6*d.*

KEY TO ALGEBRA FOR BEGINNERS. Crown 8vo. 6*s.* 6*d.*

ALGEBRA. For the Use of Colleges and Schools. New Edition. Crown 8vo. 7*s.* 6*d.*

KEY TO ALGEBRA FOR THE USE OF COLLEGES AND SCHOOLS. Crown 8vo. 10*s.* 6*d.*

EUCLID, & ELEMENTARY GEOMETRY.

Constable.—GEOMETRICAL EXERCISES FOR BEGINNERS. By SAMUEL CONSTABLE. Crown 8vo. 3*s.* 6*d.*

Cuthbertson.—EUCLIDIAN GEOMETRY. By FRANCIS CUTHBERTSON, M.A., LL.D., Head Mathematical Master of the City of London School. Extra fcap. 8vo. 4*s.* 6*d.*

Dodgson.—Works by CHARLES L. DODGSON, M.A., Student and late Mathematical Lecturer of Christ Church, Oxford.

EUCLID. BOOKS I. AND II. Sixth Edition, with words substituted for the Algebraical Symbols used in the First Edition. Crown 8vo. 2*s*.

*** The text of this Edition has been ascertained, by counting the words, to be *less than five-sevenths* of that contained in the ordinary editions.

EUCLID AND HIS MODERN RIVALS. Second Edition. Crown 8vo. 6*s*.

CURIOSA MATHEMATICA. Part I. A New Theory of Parallels. Second Edition. Crown 8vo. 2*s*.

Dupuis.—ELEMENTARY SYNTHETIC GEOMETRY OF THE POINT, LINE AND CIRCLE IN THE PLANE. By N. F. DUPUIS, M.A., F.R.S.C., Professor of Pure Mathematics in the University of Queen's College, Kingston, Canada. Globe 8vo. 4*s*. 6*d*.

Eagles.—CONSTRUCTIVE GEOMETRY OF PLANE CURVES. By T. H. EAGLES, M.A., Instructor in Geometrical Drawing, and Lecturer in Architecture at the Royal Indian Engineering College, Cooper's Hill. With numerous Examples. Crown 8vo. 12*s*.

Hall and Stevens.—A TEXT BOOK OF EUCLID'S ELEMENTS. Including Alternative Proofs, together with additional Theorems and Exercises, classified and arranged. By H. S. HALL, M.A., and F. H. STEVENS, M.A., Masters of the Military and Engineering Side, Clifton College. Globe 8vo. Book I., 1*s*.; Books I. and II., 1*s*. 6*d*.; Books I.—IV., 3*s*.; Books III.—VI., 3*s*.; Books I.—VI. and XI., 4*s*. 6*d*.; Book XI. 1*s*.

Halsted.—THE ELEMENTS OF GEOMETRY. By GEORGE BRUCE HALSTED, Professor of Pure and Applied Mathematics in the University of Texas. 8vo. 12*s*. 6*d*.

Kitchener.—A GEOMETRICAL NOTE-BOOK, containing Easy Problems in Geometrical Drawing preparatory to the Study of Geometry. For the Use of Schools. By F. E. KITCHENER, M.A., Head-Master of the Newcastle-under-Lyme High School. New Edition. 4to. 2*s*.

Lock.—EUCLID FOR BEGINNERS. Being an Introduction to existing Text-books. By Rev. J. B. LOCK, M.A., author of "Arithmetic for Schools," "Trigonometry," "Dynamics," "Elementary Statics," &c. [*In preparation.*

Mault.—NATURAL GEOMETRY: an Introduction to the Logical Study of Mathematics. For Schools and Technical Classes. With Explanatory Models, based upon the Tachymetrical works of Ed. Lagout. By A. MAULT. 18mo. 1*s*.

Models to Illustrate the above, in Box, 12*s*. 6*d*.

Millar.—ELEMENTS OF DESCRIPTIVE GEOMETRY. By J. B. MILLAR, M.E., Civil Engineer, Lecturer on Engineering in the Victoria University, Manchester. Second Edition. Cr. 8vo. 6*s*.

Stevens.—Works by F. H. STEVENS, M.A., formerly Scholar of Queen's College, Cambridge, Master of the Military and Engineering side, Clifton College.

EUCLID. BOOK XI. Propositions 1—21 with alternative proofs, Exercises, and Additional Theorems and Examples. Globe 8vo. 1*s.*

ELEMENTARY MENSURATION, with exercises in the mensuration of Plane and Solid Figures. Globe 8vo. [*In the press.*

Syllabus of Plane Geometry (corresponding to Euclid, Books I.—VI.). Prepared by the Association for the Improvement of Geometrical Teaching. New Edition. Crown 8vo. 1*s.* [*In the press.*

Syllabus of Modern Plane Geometry.—Association for the Improvement of Geometrical Teaching. Crown 8vo. Sewed. 1*s.*

Todhunter.—THE ELEMENTS OF EUCLID. For the Use of Colleges and Schools. By I. TODHUNTER, M.A., F.R.S., D.Sc., of St. John's College, Cambridge. New Edition. 18mo. 3*s.* 6*d.*

KEY TO EXERCISES IN EUCLID. Crown 8vo. 6*s.* 6*d.*

Wilson (J. M.).—ELEMENTARY GEOMETRY. BOOKS I.—V. Containing the Subjects of Euclid's first Six Books. Following the Syllabus of the Geometrical Association. By the Rev. J. M. WILSON, M.A., Head Master of Clifton College. New Edition. Extra fcap. 8vo. 4*s.* 6*d.*

TRIGONOMETRY.

Beasley.—AN ELEMENTARY TREATISE ON PLANE TRIGONOMETRY. With Examples. By R. D. BEASLEY, M.A. Ninth Edition, revised and enlarged. Crown 8vo. 3*s.* 6*d.*

Johnson.—A TREATISE ON TRIGONOMETRY. By W. E. JOHNSON, M.A., late Scholar and Assistant Lecturer in Mathematics of King's College, Cambridge. Crown 8vo. 8*s.* 6*d.*

Lock.—Works by Rev. J. B. LOCK, M.A., Senior Fellow, Assistant Tutor and Lecturer in Gonville and Caius College, Cambridge; formerly Master at Eton.

TRIGONOMETRY FOR BEGINNERS, as far as the Solution of Triangles. Third Edition. Globe 8vo. 2*s.* 6*d.*

A KEY TO "TRIGONOMETRY FOR BEGINNERS." Adapted from the key to "Elementary Trigonometry." Crown 8vo. 6*s.* 6*d.*

ELEMENTARY TRIGONOMETRY. Sixth Edition (in this edition the chapter on logarithms has been carefully revised.) Globe 8vo. 4*s.* 6*d.*

Mr. E. J. ROUTH, D.Sc., F.R.S., writes:—"It is an able treatise. It takes the difficulties of the subject one at a time, and so leads the young student easily along."

Lock.—A KEY TO "ELEMENTARY TRIGONOMETRY." By HENRY CARR, B.A., of the Grammar School, Lagos, West Africa. Crown 8vo. 8*s.* 6*d.*

HIGHER TRIGONOMETRY. Fifth Edition. Globe 8vo. 4*s.* 6*d.* Both Parts complete in One Volume. Globe 8vo. 7*s.* 6*d.*

(See also under *Arithmetic*, *Higher Mathematics*, and *Euclid*.)

M'Clelland and Preston.—A TREATISE ON SPHERICAL TRIGONOMETRY. With applications to Spherical Geometry and numerous Examples. By WILLIAM J. M'CLELLAND, M.A., Principal of the Incorporated Society's School, Santry, Dublin, and THOMAS PRESTON, M.A. In Two Parts. Crown 8vo. Part I. To the End of Solution of Triangles, 4*s.* 6*d.* Part II., 5*s.* Both Parts in one Volume. 8*s.* 6*d.*

Palmer.—TEXT-BOOK OF PRACTICAL LOGARITHMS AND TRIGONOMETRY. By J. H. PALMER, Head Schoolmaster R.N., H.M.S. *Cambridge*, Devonport. Globe 8vo. 4*s.* 6*d.*

Snowball.—THE ELEMENTS OF PLANE AND SPHERICAL TRIGONOMETRY. By J. C. SNOWBALL, M.A. Fourteenth Edition. Crown 8vo. 7*s.* 6*d.*

Todhunter.—Works by I. TODHUNTER, M.A., F.R.S., D.Sc., late of St. John's College, Cambridge.

TRIGONOMETRY FOR BEGINNERS. With numerous Examples. New Edition. 18mo. 2*s.* 6*d.*

KEY TO TRIGONOMETRY FOR BEGINNERS. Cr. 8vo. 8*s.* 6*d.*

PLANE TRIGONOMETRY. For Schools and Colleges. New Edition. Crown 8vo. 5*s.*

KEY TO PLANE TRIGONOMETRY. Crown 8vo. 10*s.* 6*d.*

A TREATISE ON SPHERICAL TRIGONOMETRY. New Edition, enlarged. Crown 8vo. 4*s.* 6*d.*

(See also under *Arithmetic and Mensuration*, *Algebra*, and *Higher Mathematics*.)

HIGHER MATHEMATICS.

Airy.—Works by Sir G. B. AIRY, K.C.B., formerly Astronomer-Royal.

ELEMENTARY TREATISE ON PARTIAL DIFFERENTIAL EQUATIONS. Designed for the Use of Students in the Universities. With Diagrams. Second Edition. Crown 8vo. 5*s.* 6*d.*

ON THE ALGEBRAICAL AND NUMERICAL THEORY OF ERRORS OF OBSERVATIONS AND THE COMBINATION OF OBSERVATIONS. Second Edition, revised. Crown 8vo. 6*s.* 6*d.*

Alexander (T.).—ELEMENTARY APPLIED MECHANICS. Being the simpler and more practical Cases of Stress and Strain wrought out individually from first principles by means of Elementary Mathematics. By T. ALEXANDER, C.E., Professor of Civil Engineering, Trinity College, Dublin. Part I. Crown 8vo. 4*s.* 6*d.*

Alexander and Thomson.—ELEMENTARY APPLIED MECHANICS. By THOMAS ALEXANDER, C.E., Professor of Civil Engineering, Trinity College, Dublin, and ARTHUR WATSON THOMSON, C.E., B.Sc., Professor of Engineering at the Technical College, Glasgow. Part II. TRANSVERSE STRESS. Crown 8vo. 10*s.* 6*d.*

Army Preliminary Examination, 1882-1888, Specimens of Papers set at the. With answers to the Mathematical Questions. Subjects: Arithmetic, Algebra, Euclid, Geometrical Drawing, Geography, French, English Dictation. Cr. 8vo. 3*s.* 6*d.*

Ball (W. W. R.).—A SHORT ACCOUNT OF THE HISTORY OF MATHEMATICS. By W. W. ROUSE BALL, Fellow and Assistant Tutor of Trinity College, Cambridge, and of the Inner Temple, Barrister-at-Law. Crown 8vo. 10*s.* 6*d.*

Ball (Sir R. S.).—EXPERIMENTAL MECHANICS. A Course of Lectures delivered at the Royal College of Science for Ireland. By SIR ROBERT STAWELL BALL, LL.D., F.R.S., Astronomer-Royal of Ireland. New Edition. With Illustrations. Crown 8vo. 6*s.*

Boole.—THE CALCULUS OF FINITE DIFFERENCES. By G. BOOLE, D.C.L., F.R.S., late Professor of Mathematics in the Queen's University, Ireland. Third Edition, revised by J. F. MOULTON, Q.C. Crown 8vo. 10*s.* 6*d.*

Cambridge Senate-House Problems and Riders, with Solutions:—

1875—PROBLEMS AND RIDERS. By A. G. GREENHILL, M.A. Crown 8vo. 8*s.* 6*d.*

1878—SOLUTIONS OF SENATE-HOUSE PROBLEMS. By the Mathematical Moderators and Examiners. Edited by J. W. L. GLAISHER, M.A., Fellow of Trinity College, Cambridge. 12*s.*

Carll.—A TREATISE ON THE CALCULUS OF VARIATIONS. Arranged with the purpose of Introducing, as well as Illustrating, its Principles to the Reader by means of Problems, and Designed to present in all Important Particulars a Complete View of the Present State of the Science. By LEWIS BUFFETT CARLL, A.M. Demy 8vo. 21*s.*

Cheyne.—AN ELEMENTARY TREATISE ON THE PLANETARY THEORY. By C. H. H. CHEYNE, M.A., F.R.A.S. With a Collection of Problems. Third Edition. Edited by Rev. A. FREEMAN, M.A., F.R.A.S. Crown 8vo. 7*s.* 6*d.*

Christie.—A COLLECTION OF ELEMENTARY TEST-QUESTIONS IN PURE AND MIXED MATHEMATICS; with Answers and Appendices on Synthetic Division, and on the Solution of Numerical Equations by Horner's Method. By JAMES R. CHRISTIE, F.R.S., Royal Military Academy, Woolwich. Crown 8vo. 8*s.* 6*d.*

Clausius.—MECHANICAL THEORY OF HEAT. By R. CLAUSIUS. Translated by WALTER R. BROWNE, M.A., late Fellow of Trinity College, Cambridge. Crown 8vo. 10*s*. 6*d*.

Clifford.—THE ELEMENTS OF DYNAMIC. An Introduction to the Study of Motion and Rest in Solid and Fluid Bodies. By W. K. CLIFFORD, F.R.S., late Professor of Applied Mathematics and Mechanics at University College, London. Part I.—KINEMATIC. Crown 8vo. Books I—III. 7*s*. 6*d*.; Book IV. and Appendix 6*s*.

Cockshott and Walters.—A TREATISE ON GEOMETRICAL CONICS. In accordance with the Syllabus of the Association for the Improvement of Geometrical Teaching By A. COCKSHOTT, M.A., formerly Fellow and Assistant-Tutor of Trinity College, Cambridge; Assistant-Master at Eton; and Rev. F. B. WALTERS, M.A., Fellow of Queens' College, Cambridge, and Principal of King William's College, Isle of Man. With Diagrams. Crown 8vo. 5*s*.

Cotterill.—APPLIED MECHANICS: an Elementary General Introduction to the Theory of Structures and Machines. By JAMES H. COTTERILL, F.R.S., Associate Member of the Council of the Institution of Naval Architects, Associate Member of the Institution of Civil Engineers, Professor of Applied Mechanics in the Royal Naval College, Greenwich. Medium 8vo. 18*s*.

Cotterill and Slade.—ELEMENTARY MANUAL OF APPLIED MECHANICS. By JAMES H. COTTERILL, F.R.S. and J. H. SLADE. Crown 8vo. [*In the press.*

Day (R. E.)—ELECTRIC LIGHT ARITHMETIC. By R. E. DAY, M.A., Evening Lecturer in Experimental Physics at King's College, London. Pott 8vo. 2*s*.

Day.—PROPERTIES OF CONIC SECTIONS PROVED GEOMETRICALLY. Part I. The Ellipse with an ample collection of Problems. By the Rev. H. G. DAY, M.A., formerly Fellow of St. John's College, Cambridge. Crown 8vo. 3*s*. 6*d*.

Dodgson.—CURIOSA MATHEMATICA. Part I. A New Theory of Parallels. By CHARLES L. DODGSON, M.A., Student and late Mathematical Lecturer of Christ Church, Oxford, Author of "Euclid and his Modern Rivals," "Euclid, Books I. and II.," &c. Second Edition. Crown 8vo. 2*s*.

Drew.—GEOMETRICAL TREATISE ON CONIC SECTIONS. By W. H. DREW, M.A., St. John's College, Cambridge. New Edition, enlarged. Crown 8vo. 5*s*.

Dyer.—EXERCISES IN ANALYTICAL GEOMETRY. Compiled and arranged by J. M. DYER, M.A. With Illustrations. Crown 8vo. 4*s*. 6*d*.

Eagles.—CONSTRUCTIVE GEOMETRY OF PLANE CURVES. By T. H. EAGLES, M.A., Instructor in Geometrical Drawing, and Lecturer in Architecture at the Royal Indian Engineering College, Cooper's Hill. With numerous Examples. Crown 8vo. 12*s*.

Edgar (J. H.) and Pritchard (G. S.).—NOTE-BOOK ON PRACTICAL SOLID OR DESCRIPTIVE GEOMETRY. Containing Problems with help for Solutions. By J. H. EDGAR, M.A., Lecturer on Mechanical Drawing at the Royal School of Mines, and G. S. PRITCHARD. Fourth Edition, revised by ARTHUR MEEZE. Globe 8vo. 4*s.* 6*d.*

Edwards.—THE DIFFERENTIAL CALCULUS With Applications and numerous Examples. An Elementary Treatise by JOSEPH EDWARDS, M.A., formerly Fellow of Sidney Sussex College, Cambridge. Crown 8vo. 10*s.* 6*d.*

Ferrers.—Works by the Rev. N. M. FERRERS, M.A., Master of Gonville and Caius College, Cambridge.

AN ELEMENTARY TREATISE ON TRILINEAR CO-ORDINATES, the Method of Reciprocal Polars, and the Theory of Projectors. New Edition, revised. Crown 8vo. 6*s.* 6*d.*

AN ELEMENTARY TREATISE ON SPHERICAL HARMONICS, AND SUBJECTS CONNECTED WITH THEM. Crown 8vo. 7*s.* 6*d.*

Forsyth.—A TREATISE ON DIFFERENTIAL EQUATIONS. By ANDREW RUSSELL FORSYTH, M.A., F.R.S., Fellow and Assistant Tutor of Trinity College, Cambridge. Second Edition. 8vo. 14*s.*

Frost.—Works by PERCIVAL FROST, M.A., D.Sc., formerly Fellow of St. John's College, Cambridge; Mathematical Lecturer at King's College.

AN ELEMENTARY TREATISE ON CURVE TRACING. 8vo. 12*s.*

SOLID GEOMETRY. Third Edition. Demy 8vo. 16*s.*

HINTS FOR THE SOLUTION OF PROBLEMS in the Third Edition of SOLID GEOMETRY. 8vo. 8*s.* 6*d.*

Greaves.—A TREATISE ON ELEMENTARY STATICS. By JOHN GREAVES, M.A., Fellow and Mathematical Lecturer of Christ's College, Cambridge. Second Edition. Crown 8vo. 6*s.* 6*d.*

STATICS FOR BEGINNERS. By the Same Author. Globe 8vo. 3*s.* 6*d.*

Greenhill.—DIFFERENTIAL AND INTEGRAL CALCULUS. With Applications. By A. G. GREENHILL, M.A., Professor of Mathematics to the Senior Class of Artillery Officers, Woolwich, and Examiner in Mathematics to the University of London. Crown 8vo. 7*s.* 6*d.*

Hayward.—THE ALGEBRA OF CO-PLANAR VECTORS AND TRIGONOMETRY. By R. B. HAYWARD, F.R.S., Assistant Master at Harrow. [*In preparation.*

Hicks.—DYNAMICS OF PARTICLES AND SOLIDS. By W. M. HICKS, Principal, and Professor of Mathematics and Physics, Firth College, Sheffield, late Fellow of St. John's College, Cambridge. Crown 8vo. [*In the press.*

Ibbetson.—THE MATHEMATICAL THEORY OF PERFECTLY ELASTIC SOLIDS, with a short account of Viscous Fluids. An Elementary Treatise. By WILLIAM JOHN IBBETSON, M.A., Fellow of the Royal Astronomical Society, and of the Cambridge Philosophical Society, Member of the London Mathematical Society, late Senior Scholar of Clare College, Cambridge. 8vo. 21*s.*

Jellett (John H.).—A TREATISE ON THE THEORY OF FRICTION. By JOHN H. JELLETT, B.D., late Provost of Trinity College, Dublin; President of the Royal Irish Academy. 8vo. 8*s.* 6*d.*

Johnson.—Works by WILLIAM WOOLSEY JOHNSON, Professor of Mathematics at the U.S. Naval Academy, Annapolis, Maryland.

INTEGRAL CALCULUS, an Elementary Treatise on the; Founded on the Method of Rates or Fluxions. Demy 8vo. 9*s.*

CURVE TRACING IN CARTESIAN CO-ORDINATES. Crown 8vo. 4*s.* 6*d.*

A TREATISE ON ORDINARY AND DIFFERENTIAL EQUATIONS. Extra crown 8vo. 15*s.*

Jones.—EXAMPLES IN PHYSICS. By D. E. JONES, B.Sc., Professor of Physics in University College of Wales, Aberystwyth. Fcap. 8vo. 3*s.* 6*d.*

Kelland and Tait.—INTRODUCTION TO QUATERNIONS, with numerous examples. By P. KELLAND, M.A., F.R.S., and P. G. TAIT, M.A., Professors in the Department of Mathematics in the University of Edinburgh. Second Edition. Crown 8vo. 7*s.* 6*d.*

Kempe.—HOW TO DRAW A STRAIGHT LINE: a Lecture on Linkages. By A. B. KEMPE. With Illustrations. Crown 8vo. 1*s.* 6*d.* (*Nature Series.*)

Kennedy.—THE MECHANICS OF MACHINERY. By A. B. W. KENNEDY, F.R.S., M.Inst.C.E., late Professor of Engineering and Mechanical Technology in University College, London. With Illustrations. Crown 8vo. 12*s.* 6*d.*

Knox.—DIFFERENTIAL CALCULUS FOR BEGINNERS. By ALEXANDER KNOX. Fcap. 8vo. 3*s.* 6*d.*

Lock.—Works by the Rev. J. B. LOCK, M.A., Author of "Trigonometry," "Arithmetic for Schools," &c.

HIGHER TRIGONOMETRY. Fifth Edition. Globe 8vo. 4*s.* 6*d.*

DYNAMICS FOR BEGINNERS. Second Edition. (Stereotyped.) Globe 8vo. 3*s.* 6*d.*

ELEMENTARY STATICS. Globe 8vo. 4*s.* 6*d.*

(See also under *Arithmetic*, *Euclid*, and *Trigonometry*.)

Lupton.—CHEMICAL ARITHMETIC. With 1,200 Examples. By SYDNEY LUPTON, M.A., F.C.S., F.I.C., formerly Assistant-Master in Harrow School. Second Edition. Fcap. 8vo. 4*s.* 6*d.*

Macfarlane.—PHYSICAL ARITHMETIC. By ALEXANDER MACFARLANE, M.A., D.Sc., F.R.S.E., Examiner in Mathematics to the University of Edinburgh. Crown 8vo. 7*s*. 6*d*.

MacGregor.—KINEMATICS AND DYNAMICS. An Elementary Treatise. By JAMES GORDON MACGREGOR, M.A., D.Sc., Fellow of the Royal Societies of Edinburgh and of Canada, Munro Professor of Physics in Dalhousie College, Halifax, Nova Scotia. With Illustrations. Crown 8vo. 10*s*. 6*d*.

Merriman.—A TEXT BOOK OF THE METHOD OF LEAST SQUARES. By MANSFIELD MERRIMAN, Professor of Civil Engineering at Lehigh University, Member of the American Philosophical Society, American Association for the Advancement of Science, &c. Demy 8vo. 8*s*. 6*d*.

Millar.—ELEMENTS OF DESCRIPTIVE GEOMETRY By J.B. MILLAR, C.E., Assistant Lecturer in Engineering in Owens College, Manchester. Second Edition. Crown 8vo. 6*s*.

Milne.—Works by the Rev. JOHN J. MILNE, M.A., Private Tutor, late Scholar, of St. John's College, Cambridge, &c., &c., formerly Second Master of Heversham Grammar School.

WEEKLY PROBLEM PAPERS. With Notes intended for the use of students preparing for Mathematical Scholarships, and for the Junior Members of the Universities who are reading for Mathematical Honours. Pott 8vo. 4*s*. 6*d*.

SOLUTIONS TO WEEKLY PROBLEM PAPERS. Crown 8vo. 10*s*. 6*d*.

COMPANION TO "WEEKLY PROBLEM PAPERS." Crown 8vo. 10*s*. 6*d*.

Muir.—A TREATISE ON THE THEORY OF DETERMINANTS. With graduated sets of Examples. For use in Colleges and Schools. By THOS. MUIR, M.A., F.R.S.E., Mathematical Master in the High School of Glasgow. Crown 8vo. 7*s*. 6*d*.

Palmer.—TEXT BOOK OF PRACTICAL LOGARITHMS AND TRIGONOMETRY. By J. H. PALMER, Head Schoolmaster R.N., H.M.S. *Cambridge*, Devonport. Globe 8vo. 4*s*. 6*d*.

Parkinson.—AN ELEMENTARY TREATISE ON MECHANICS. For the Use of the Junior Classes at the University and the Higher Classes in Schools. By S. PARKINSON, D.D., F.R.S., Tutor and Prælector of St. John's College, Cambridge. With a Collection of Examples. Sixth Edition, revised. Crown 8vo. 9*s*. 6*d*.

Pirie.—LESSONS ON RIGID DYNAMICS. By the Rev. G. PIRIE, M.A., late Fellow and Tutor of Queen's College, Cambridge; Professor of Mathematics in the University of Aberdeen. Crown 8vo. 6*s*.

Puckle.—AN ELEMENTARY TREATISE ON CONIC SECTIONS AND ALGEBRAIC GEOMETRY. With Numerous Examples and Hints for their Solution; especially designed for the Use of Beginners. By G. H. PUCKLE, M.A. Fifth Edition, revised and enlarged. Crown 8vo. 7*s*. 6*d*.

Reuleaux.—THE KINEMATICS OF MACHINERY. Outlines of a Theory of Machines. By Professor F. REULEAUX Translated and Edited by Professor A. B. W. KENNEDY, F.R.S. C.E. With 450 Illustrations. Medium 8vo. *21s.*

Rice and Johnson.—DIFFERENTIAL CALCULUS, an Elementary Treatise on the; Founded on the Method of Rates or Fluxions. By JOHN MINOT RICE, Professor of Mathematics in the United States Navy, and WILLIAM WOOLSEY JOHNSON, Professor of Mathematics at the United States Naval Academy. Third Edition, Revised and Corrected. Demy 8vo. 18*s.* Abridged Edition, 9*s.*

Robinson.—TREATISE ON MARINE SURVEYING. Prepared for the use of younger Naval Officers. With Questions for Examinations and Exercises principally from the Papers of the Royal Naval College. With the results. By Rev. JOHN L. ROBINSON, Chaplain and Instructor in the Royal Naval College, Greenwich. With Illustrations. Crown 8vo. 7*s.* 6*d.*

CONTENTS.—Symbols used in Charts and Surveying—The Construction and Use of Scales—Laying off Angles—Fixing Positions by Angles — Charts and Chart-Drawing—Instruments and Observing — Base Lines—Triangulation—Levelling—Tides and Tidal Observations—Soundings—Chronometers—Meridian Distances—Method of Plotting a Survey—Miscellaneous Exercises—Index.

Routh.—Works by EDWARD JOHN ROUTH, D.Sc., LL.D., F.R.S., Fellow of the University of London, Hon. Fellow of St. Peter's College, Cambridge.

A TREATISE ON THE DYNAMICS OF THE SYSTEM OF RIGID BODIES. With numerous Examples. Fourth and enlarged Edition. Two Vols. 8vo. Vol. I.—Elementary Parts. 14*s.* Vol. II.—The Advanced Parts. 14*s.*

STABILITY OF A GIVEN STATE OF MOTION, PARTICULARLY STEADY MOTION. Adams' Prize Essay for 1877. 8vo. 8*s.* 6*d.*

Sanderson.—HYDROSTATICS FOR BEGINNERS. By F. W. SANDERSON, M.A., late Fellow of the University of Durham, and Scholar of Christ's College, Cambridge; Assistant Master in Dulwich College. Globe 8vo. 4*s.* 6*d.*

Smith (C.).—Works by CHARLES SMITH, M.A., Fellow and Tutor of Sidney Sussex College, Cambridge.

CONIC SECTIONS. Seventh Edition. Crown 8vo. 7*s.* 6*d.*

SOLUTIONS TO CONIC SECTIONS. Crown 8vo. 10*s.* 6*d.*

AN ELEMENTARY TREATISE ON SOLID GEOMETRY. Second Edition. Crown 8vo. 9*s.* 6*d.* (See also under *Algebra.*)

Syllabus of Elementary Dynamics. Drawn up by the Association for the Improvement of Geometrical Teaching. 4to. Part I. [*In preparation.*

Tait and Steele.—A TREATISE ON DYNAMICS OF A PARTICLE. With numerous Examples. By Professor TAIT, M.A., and W. J. STEELE, B.A. Sixth Edition, revised. Crown 8vo. 12*s.*

Thomson.—Works by J. J. THOMSON, Fellow of Trinity College, Cambridge, and Professor of Experimental Physics in the University.

A TREATISE ON THE MOTION OF VORTEX RINGS. An Essay to which the Adams Prize was adjudged in 1882 in the University of Cambridge. With Diagrams. 8vo. 6*s*.

APPLICATIONS OF DYNAMICS TO PHYSICS AND CHEMISTRY. Crown 8vo. 7*s*. 6*d*.

Todhunter.—Works by I. TODHUNTER, M.A., F.R.S., D.Sc., late of St. John's College, Cambridge.

"Mr. Todhunter is chiefly known to students of Mathematics as the author of a series of admirable mathematical text-books, which possess the rare qualities of being clear in style and absolutely free from mistakes, typographical and other."—SATURDAY REVIEW.

MECHANICS FOR BEGINNERS. With numerous Examples. New Edition. 18mo. 4*s*. 6*d*.

KEY TO MECHANICS FOR BEGINNERS. Crown 8vo. 6*s*. 6*d*.

AN ELEMENTARY TREATISE ON THE THEORY OF EQUATIONS. New Edition, revised. Crown 8vo. 7*s*. 6*d*.

PLANE CO-ORDINATE GEOMETRY, as applied to the Straight Line and the Conic Sections. With numerous Examples. New Edition, revised and enlarged. Crown 8vo. 7*s*. 6*d*.

KEY TO PLANE CO-ORDINATE GEOMETRY. By C. W. BOURNE, M.A. Head Master of the College, Inverness. Crown 8vo. 10*s*. 6*d*.

A TREATISE ON THE DIFFERENTIAL CALCULUS. With numerous Examples. New Edition. Crown 8vo. 10*s*. 6*d*.

A KEY TO DIFFERENTIAL CALCULUS. By H. ST. J. HUNTER, M.A. Crown 8vo. 10*s*. 6*d*.

A TREATISE ON THE INTEGRAL CALCULUS AND ITS APPLICATIONS. With numerous Examples. New Edition, revised and enlarged. Crown 8vo. 10*s*. 6*d*.

A KEY TO INTEGRAL CALCULUS. By H. ST. JOHN HUNTER, M.A. Crown 8vo. 10*s*. 6*d*.

EXAMPLES OF ANALYTICAL GEOMETRY OF THREE DIMENSIONS. New Edition, revised. Crown 8vo. 4*s*.

A TREATISE ON ANALYTICAL STATICS. With numerous Examples. Fifth Edition. Edited by Professor J. D. EVERETT, F.R.S. Crown 8vo. 10*s*. 6*d*.

A HISTORY OF THE MATHEMATICAL THEORY OF PROBABILITY, from the time of Pascal to that of Laplace. 8vo. 18*s*.

A HISTORY OF THE MATHEMATICAL THEORIES OF ATTRACTION, AND THE FIGURE OF THE EARTH, from the time of Newton to that of Laplace. 2 vols. 8vo. 24*s*.

AN ELEMENTARY TREATISE ON LAPLACE'S, LAME'S, AND BESSEL'S FUNCTIONS. Crown 8vo. 10*s*. 6*d*.

See also under *Arithmetic and Mensuration*, *Algebra*, and *Trigonometry*.)

Wilson (J. M.).—SOLID GEOMETRY AND CONIC SECTIONS. With Appendices on Transversals and Harmonic Division. For the Use of Schools. By Rev. J. M. WILSON, M.A. Head Master of Clifton College. New Edition. Extra fcap. 8vo. 3*s*. 6*d*.

Woolwich Mathematical Papers, for Admission into the Royal Military Academy, Woolwich, 1880—1888 inclusive. Edited by E. J. BROOKSMITH, B.A., LL.M., St. John's College, Cambridge; Instructor in Mathematics at the Royal Military Academy, Woolwich. Crown 8vo. 6*s*.

Wolstenholme.—MATHEMATICAL PROBLEMS, on Subjects included in the First and Second Divisions of the Schedule of subjects for the Cambridge Mathematical Tripos Examination. Devised and arranged by JOSEPH WOLSTENHOLME, D.Sc., late Fellow of Christ's College, sometime Fellow of St. John's College, and Professor of Mathematics in the Royal Indian Engineering College. New Edition, greatly enlarged. 8vo. 18*s*.

EXAMPLES FOR PRACTICE IN THE USE OF SEVEN-FIGURE LOGARITHMS. For the Use of Colleges and Schools. By the same Author. 8vo. 5*s*.

SCIENCE.

(1) Natural Philosophy, (2) Astronomy, (3) Chemistry, (4) Biology, (5) Medicine, (6) Military and Naval Text Books, (7) Anthropology, (8) Physical Geography and Geology, (9) Agriculture.

NATURAL PHILOSOPHY.

Airy.—Works by Sir G. B. AIRY, K.C.B., formerly Astronomer Royal.

ON SOUND AND ATMOSPHERIC VIBRATIONS. With the Mathematical Elements of Music. Designed for the Use of Students in the University. Second Edition, revised and enlarged. Crown 8vo. 9*s*.

A TREATISE ON MAGNETISM. Designed for the Use of Students in the University. Crown 8vo. 9*s*. 6*d*.

GRAVITATION: an Elementary Explanation of the Principal Perturbations in the Solar System. 2nd Ed. Cr. 8vo. 7*s*. 6*d*.

Alexander (T.).—ELEMENTARY APPLIED MECHANICS. Being the simpler and more practical Cases of Stress and Strain wrought out individually from first principles by means of Elementary Mathematics. By T. ALEXANDER, C.E., Professor of Civil Engineering, Trinity College, Dublin. Crown 8vo. Part I. 4*s*. 6*d*.

Alexander — Thomson. — ELEMENTARY APPLIED MECHANICS. By Professor THOMAS ALEXANDER, C.E., and Professor ARTHUR WATSON THOMSON, C.E., B.Sc. Part II. TRANSVERSE STRESS; upwards of 150 Diagrams, and 200 Examples carefully worked out. Crown 8vo. 10*s*. 6*d*.

Ball (R. S.).—EXPERIMENTAL MECHANICS. A Course of Lectures delivered at the Royal College of Science for Ireland. By Sir R. S. BALL, LL.D., F.R.S., Astronomer Royal of Ireland. Second and Cheaper Edition. With Illustrations. Crown 8vo. 6*s*.

Bottomley.—FOUR-FIGURE MATHEMATICAL TABLES. Comprising Logarithmic and Trigonometrical Tables, and Tables of Squares, Square Roots, and Reciprocals. By J. T. BOTTOMLEY, M.A., F.R.S.E., F.C.S., Lecturer in Natural Philosophy in the University of Glasgow. 8vo. 2*s*. 6*d*.

Chisholm. — THE SCIENCE OF WEIGHING AND MEASURING, AND THE STANDARDS OF MEASURE AND WEIGHT. By H. W. CHISHOLM, Warden of the Standards. With numerous Illustrations. Crown 8vo. 4*s*. 6*d*. (*Nature Series.*)

Clarke.—A TABLE OF SPECIFIC GRAVITY FOR SOLIDS AND LIQUIDS. (Constants of Nature: Part I.) New Edition. Revised and Enlarged. By FRANK WIGGLESWORTH CLARKE, Chief Chemist, U.S. Geological Survey. 8vo, 12*s*. 6*d*. (Published for the Smithsonian Institution, Washington, U.S. of America.)

Clausius.—MECHANICAL THEORY OF HEAT. By R. CLAUSIUS. Translated by WALTER R. BROWNE, M.A., late Fellow of Trinity College, Cambridge. Crown 8vo 10*s*. 6*d*.

Cotterill.—APPLIED MECHANICS: an Elementary General Introduction to the Theory of Structures and Machines. By JAMES H. COTTERILL, F.R.S., Associate Member of the Council of the Institution of Naval Architects, Associate Member of the Institution of Civil Engineers, Professor of Applied Mechanics in the Royal Naval College, Greenwich. Medium 8vo. 18*s*.

Cotterill and Slade.—ELEMENTARY MANUAL OF APPLIED MECHANICS. By JAMES H. COTTERILL, F.R.S and J. H. SLADE. Crown 8vo. [*In the press.*

Cumming.—AN INTRODUCTION TO THE THEORY OF ELECTRICITY. By LINNÆUS CUMMING, M.A., one of the Masters of Rugby School. With Illustrations. Crown 8vo. 8*s*. 6*d*.

Daniell.—A TEXT-BOOK OF THE PRINCIPLES OF PHYSICS. By ALFRED DANIELL, M.A., LL.B., D.Sc., F.R.S.E., late Lecturer on Physics in the School of Medicine, Edinburgh. With Illustrations. Second Edition. Revised and Enlarged. Medium 8vo. 21*s*.

Day.—ELECTRIC LIGHT ARITHMETIC. By R. E. DAY, M.A., Evening Lecturer in Experimental Physics at King's College, London. Pott 8vo. 2*s*.

Everett.—UNITS AND PHYSICAL CONSTANTS. By J. D. EVERETT, M.A., D.C.L., F.R.S., F.R.S.E., Professor of Natural Philosophy, Queen's College, Belfast. Second Edition. Extra fcap. 8vo. 5*s*.

Ganguillet and Kutter.—A GENERAL FORMULA FOR THE UNIFORM FLOW OF WATER IN RIVERS AND OTHER CHANNELS. By E. GANGUILLET and W. R. KUTTER, Engineers, in Berne, Switzerland. Translated from the German, with numerous Additions, including Tables and Diagrams, and the Elements of over 1,200 Gaugings of Rivers, Small Channels and Pipes in English Measure, by RUDOLPH HERING, Assoc. Am. Soc. C.E., M.Inst. C.E., and JOHN C. TRAUTWINE, Jun., Assoc. Am. Soc. C.E., Assoc. Inst. C.E. 8vo. 17*s*.

Gray.—THE THEORY AND PRACTICE OF ABSOLUTE MEASUREMENTS IN ELECTRICITY AND MAGNETISM. By ANDREW GRAY, M.A., F.R S.E., Professor of Physics in the University College of North Wales. Two Vols. Crown 8vo. Vol. I. 12*s*. 6*d*. [Vol. II. *In the press.*

ABSOLUTE MEASUREMENTS IN ELECTRICITY AND MAGNETISM. By the same Author. Second Edition. Revised and greatly enlarged. Fcap. 8vo. 5*s*. 6*d*.

Greaves.—ELEMENTARY STATICS, A TREATISE ON. By JOHN GREAVES, M.A., Fellow and Mathematical Lecturer of Christ's College, Cambridge. Second Edition, revised. Crown 8vo. 6*s*. 6*d*.

STATICS FOR BEGINNERS. By the same. Crown 8vo. 3*s*. 6*d*.

Grove.—A DICTIONARY OF MUSIC AND MUSICIANS. (A.D. 1450—1889.) By Eminent Writers, English and Foreign. Edited by Sir GEORGE GROVE, D.C.L., Director of the Royal College of Music, &c. In four volumes. Demy 8vo.

Vols. I., II., III., and IV. Price 21*s*. each.

Parts I. to XIV., Parts XIX—XXII., price 3*s*. 6*d*. each. Parts XV., XVI., price 7*s*. Parts XVII., XVIII., price 7*s*. Parts XXIII.—XXV. (Appendix), 9*s*.

*** The Appendix just published completes the Dictionary. A general Index will be published as a separate volume.

"Dr. Grove's Dictionary will be a boon to every intelligent lover of music."—SATURDAY REVIEW.

Huxley.—INTRODUCTORY PRIMER OF SCIENCE. By T. H. HUXLEY, F.R.S., &c. 18mo. 1*s*.

Ibbetson.—THE MATHEMATICAL THEORY OF PERFECTLY ELASTIC SOLIDS, with a Short Account of Viscous Fluids. An Elementary Treatise. By WILLIAM JOHN IBBETSON, B.A., F.R.A.S., Senior Scholar of Clare College, Cambridge. 8vo. 21*s*.

Jones.—Works by D. E. JONES, B.Sc., Professor of Physics in University College of Wales, Aberystwyth.

EXAMPLES IN PHYSICS. Fcap. 8vo. 3*s*. 6*d*.

SOUND, LIGHT, AND HEAT. An Elementary Text-book. With Illustrations. Fcap. 8vo. [*In the press.*

Kempe.—HOW TO DRAW A STRAIGHT LINE; a Lecture on Linkages. By A. B. KEMPE. With Illustrations. Crown 8vo. 1*s.* 6*d.* (*Nature Series.*)

Kennedy.—THE MECHANICS OF MACHINERY. By A. B. W. KENNEDY, F.R.S., M.Inst.C.E., late Professor of Engineering and Mechanical Technology in University College, London. With numerous Illustrations. Crown 8vo. 12*s.* 6*d.*

Lang.—EXPERIMENTAL PHYSICS. By P. R. SCOTT LANG, M.A., Professor of Mathematics in the University of St. Andrews. With Illustrations. Crown 8vo. [*In the press.*

Lock.—Works by Rev. J. B. LOCK, M.A., Senior Fellow, Assistant Tutor and Lecturer in Gonville and Caius College, Cambridge, formerly Master at Eton, &c.

DYNAMICS FOR BEGINNERS. Second Edition. Globe 8vo. 3*s.* 6*d.*

ELEMENTARY STATICS. Globe 8vo. 4*s.* 6*d.*

Lodge.—MODERN VIEWS OF ELECTRICITY. By OLIVER J. LODGE, LL.D., D.Sc., F.R.S., Professor of Experimental Physics in University College, Liverpool. Illustrated. Crown 8vo. 6*s.* 6*d.* (*Nature Series.*)

Loewy.—Works by B. LOEWY, F.R.A.S., Science Master at the London International College, and Examiner in Experimental Physics to the College of Preceptors.

QUESTIONS AND EXAMPLES ON EXPERIMENTAL PHYSICS: Sound, Light, Heat, Electricity, and Magnetism. Fcap. 8vo. 2*s.*

A GRADUATED COURSE OF NATURAL SCIENCE FOR ELEMENTARY AND TECHNICAL SCHOOLS AND COLLEGES. In Three Parts. Part I. FIRST YEAR'S COURSE. Globe 8vo. 2*s.*

Lupton.—NUMERICAL TABLES AND CONSTANTS IN ELEMENTARY SCIENCE. By SYDNEY LUPTON, M.A., F.C.S., F.I.C., Assistant Master at Harrow School. Extra fcap. 8vo. 2*s.* 6*d.*

Macfarlane.—PHYSICAL ARITHMETIC. By ALEXANDER MACFARLANE, D.Sc., Examiner in Mathematics in the University of Edinburgh. Crown 8vo. 7*s.* 6*d.*

Macgregor.—KINEMATICS AND DYNAMICS. An Elementary Treatise. By JAMES GORDON MACGREGOR, M.A., D. Sc., Fellow of the Royal Societies of Edinburgh and of Canada, Munro Professor of Physics in Dalhousie College, Halifax, Nova Scotia. With Illustrations. Crown 8vo. 10*s.* 6*d.*

Mayer.—SOUND: a Series of Simple, Entertaining, and Inexpensive Experiments in the Phenomena of Sound, for the Use of Students of every age. By A. M. MAYER, Professor of Physics in the Stevens Institute of Technology, &c. With numerous Illustrations. Crown 8vo. 3*s.* 6*d.* (*Nature Series.*)

Mayer and Barnard.—LIGHT: a Series of Simple, Entertaining, and Inexpensive Experiments in the Phenomena of Light, for the Use of Students of every age. By A. M. MAYER and C. BARNARD. With numerous Illustrations. Crown 8vo. 2*s*. 6*d*. (*Nature Series.*)

Newton.—PRINCIPIA. Edited by Professor Sir W. THOMSON and Professor BLACKBURNE. 4to. 31*s*. 6*d*.

THE FIRST THREE SECTIONS OF NEWTON'S PRINCIPIA. With Notes and Illustrations. Also a Collection of Problems, principally intended as Examples of Newton's Methods. By PERCIVAL FROST, M.A. Third Edition. 8vo. 12*s*.

Parkinson.—A TREATISE ON OPTICS. By S. PARKINSON, D.D., F.R.S., Tutor and Prælector of St. John's College, Cambridge. Fourth Edition, revised and enlarged. Crown 8vo. 10*s*. 6*d*.

Peabody.—THERMODYNAMICS OF THE STEAM ENGINE AND OTHER HEAT-ENGINES. By CECIL H. PEABODY, Associate Professor of Steam Engineering, Massachusetts Institute of Technology. 8vo. 21*s*.

Perry.—STEAM. AN ELEMENTARY TREATISE. By JOHN PERRY, C.E., Whitworth Scholar, Fellow of the Chemical Society, Professor of Mechanical Engineering and Applied Mechanics at the Technical College, Finsbury. With numerous Woodcuts and Numerical Examples and Exercises. 18mo. 4*s*. 6*d*.

Rayleigh.—THE THEORY OF SOUND. By LORD RAYLEIGH, M.A., F.R.S., formerly Fellow of Trinity College, Cambridge. 8vo. Vol. I. 12*s*. 6*d*. Vol. II. 12*s*. 6*d*. [*Vol. III. in the press.*

Reuleaux.—THE KINEMATICS OF MACHINERY. Outlines of a Theory of Machines. By Professor F. REULEAUX. Translated and Edited by Professor A. B. W. KENNEDY, F.R.S., C.E. With 450 Illustrations. Medium 8vo. 21*s*.

Roscoe and Schuster.—SPECTRUM ANALYSIS. Lectures delivered in 1868 before the Society of Apothecaries of London. By Sir HENRY E. ROSCOE, LL.D., F.R.S., formerly Professor of Chemistry in the Owens College, Victoria University, Manchester. Fourth Edition, revised and considerably enlarged by the Author and by ARTHUR SCHUSTER, F.R.S., Ph.D., Professor of Applied Mathematics in the Owens College, Victoria University. With Appendices, numerous Illustrations, and Plates. Medium 8vo. 21*s*.

Sanderson.—HYDROSTATICS FOR BEGINNERS. By F. W. SANDERSON, M.A., late Fellow of the University of Durham, and Scholar of Christ's College, Cambridge; Assistant Master in Dulwich College. Globe 8vo. 4*s*. 6*d*.

Shann.—AN ELEMENTARY TREATISE ON HEAT, IN RELATION TO STEAM AND THE STEAM-ENGINE. By G. SHANN, M.A. With Illustrations. Crown 8vo. 4*s*. 6*d*.

Spottiswoode.—POLARISATION OF LIGHT. By the late W. SPOTTISWOODE, F.R.S. With Illustrations. New Edition. Crown 8vo. 3*s*. 6*d*. (*Nature Series.*)

Stewart (Balfour).—Works by BALFOUR STEWART, F.R.S., late Langworthy Professor of Physics in the Owens College, Victoria University, Manchester.

PRIMER OF PHYSICS. With numerous Illustrations. New Edition, with Questions. 18mo. 1*s.* (*Science Primers.*)

LESSONS IN ELEMENTARY PHYSICS. With numerous Illustrations and Chromolitho of the Spectra of the Sun, Stars, and Nebulæ. New and Enlarged Edition. Fcap. 8vo. 4*s.* 6*d.*

QUESTIONS ON BALFOUR STEWART'S ELEMENTARY LESSONS IN PHYSICS. By Prof. THOMAS H. CORE, Owens College, Manchester. New Edition. Fcap. 8vo. 2*s.*

Stewart and Gee.—ELEMENTARY PRACTICAL PHYSICS, LESSONS IN. By BALFOUR STEWART, F.R.S., and W. W. HALDANE GEE, B.Sc. Crown 8vo.

Vol. I.—GENERAL PHYSICAL PROCESSES. 6*s.*
Vol. II.—ELECTRICITY AND MAGNETISM. 7*s.* 6*d.*
Vol. III.—OPTICS, HEAT, AND SOUND. [*In the press.*

PRACTICAL PHYSICS FOR SCHOOLS AND THE JUNIOR STUDENTS OF COLLEGES. By the same Authors. Globe 8vo.

Vol. I.—ELECTRICITY AND MAGNETISM. 2*s.* 6*d.*
Vol. II.—HEAT, LIGHT, AND SOUND. [*In the press.*

Stokes.—ON LIGHT. Being the Burnett Lectures, delivered in Aberdeen in 1883, 1884, 1885. By Sir GEORGE GABRIEL STOKES, M.A., P.R.S., &c., Fellow of Pembroke College, and Lucasian Professor of Mathematics in the University of Cambridge. First Course: ON THE NATURE OF LIGHT.—Second Course: ON LIGHT AS A MEANS OF INVESTIGATION.—Third Course: ON THE BENEFICIAL EFFECTS OF LIGHT. Complete in one volume. Crown 8vo. 7*s.* 6*d.* (*Nature Series.*)

*** The 2nd and 3rd Courses may be had separately. 8vo. 2*s.* 6*d.* each.

Stone.—AN ELEMENTARY TREATISE ON SOUND. By W. H. STONE, M.D. With Illustrations. 18mo. 3*s.* 6*d.*

Tait.—HEAT. By P. G. TAIT, M.A., Sec. R.S.E., formerly Fellow of St. Peter's College, Cambridge, Professor of Natural Philosophy in the University of Edinburgh. Crown 8vo. 6*s.*

Thompson.—ELEMENTARY LESSONS IN ELECTRICITY AND MAGNETISM. By SILVANUS P. THOMPSON, Principal and Professor of Physics in the Technical College, Finsbury. With Illustrations. New Edition, Revised. Forty-Third Thousand. Fcap. 8vo. 4*s.* 6*d.*

Thomson, Sir W.—Works by Sir WILLIAM THOMSON, D.C.L., LL.D., F.R.S., F.R.S.E., Fellow of St. Peter's College, Cambridge, and Professor of Natural Philosophy in the University of Glasgow.

ELECTROSTATICS AND MAGNETISM, REPRINTS OF PAPERS ON. Second Edition. Medium 8vo. 18*s.*

Thomson.—POPULAR LECTURES AND ADDRESSES. 3 vols. With Illustrations. Crown 8vo. Vol. I. Constitution of Matter. 6*s.* (*Nature Series.*)

Thomson, J. J.—Works by J. J. THOMSON, Fellow of Trinity College, Cavendish Professor of Experimental Physics in the University.

THE MOTION OF VORTEX RINGS, A TREATISE ON. An Essay to which the Adams Prize was adjudged in 1882 in the University of Cambridge. With Diagrams. 8vo. 6*s.*

APPLICATIONS OF DYNAMICS TO PHYSICS AND CHEMISTRY. Crown 8vo. 7*s.* 6*d.*

Todhunter.—NATURAL PHILOSOPHY FOR BEGINNERS. By I. TODHUNTER, M.A., F.R.S., D.Sc.

Part I. The Properties of Solid and Fluid Bodies. 18mo. 3*s.* 6*d.*
Part II. Sound, Light, and Heat. 18mo. 3*s.* 6*d.*

Turner.—HEAT AND ELECTRICITY, A COLLECTION OF EXAMPLES ON. By H. H. TURNER, B.A., Fellow of Trinity College, Cambridge. Crown 8vo. 2*s.* 6*d.*

Witham.—STEAM ENGINE DESIGN. For the Use of Mechanical Engineers, Students, and Draughtsmen. By J. M. WITHAM, Member of the Society of American Engineers; Professor of Engineering, Arkansas Industrial University; late Assistant Engineer, U.S. Navy. With 210 Illustrations. 8vo. 25*s.*

Wright (Lewis).—LIGHT; A COURSE OF EXPERIMENTAL OPTICS, CHIEFLY WITH THE LANTERN. By LEWIS WRIGHT. With nearly 200 Engravings and Coloured Plates. Crown 8vo. 7*s.* 6*d.*

ASTRONOMY.

Airy.—POPULAR ASTRONOMY. With Illustrations by Sir G. B. AIRY, K.C.B., formerly Astronomer-Royal. New Edition. 18mo. 4*s.* 6*d.*

Forbes.—TRANSIT OF VENUS. By G. FORBES, M.A., Professor of Natural Philosophy in the Andersonian University, Glasgow. Illustrated. Crown 8vo. 3*s.* 6*d.* (*Nature Series.*)

Godfray.—Works by HUGH GODFRAY, M.A., Mathematical Lecturer at Pembroke College, Cambridge.

A TREATISE ON ASTRONOMY, for the Use of Colleges and Schools. Fourth Edition. 8vo. 12*s.* 6*d.*

AN ELEMENTARY TREATISE ON THE LUNAR THEORY, with a Brief Sketch of the Problem up to the time of Newton. Second Edition, revised. Crown 8vo. 5*s.* 6*d.*

Lockyer.—Works by J. NORMAN LOCKYER, F.R.S.

PRIMER OF ASTRONOMY. With numerous Illustrations. New Edition. 18mo. 1*s.* (*Science Primers.*)

Lockyer.—ELEMENTARY LESSONS IN ASTRONOMY. With Coloured Diagram of the Spectra of the Sun, Stars, and Nebulæ, and numerous Illustrations. New Edition (36th thousand), revised throughout. Fcap. 8vo. 5*s*. 6*d*.

QUESTIONS ON LOCKYER'S ELEMENTARY LESSONS IN ASTRONOMY. For the Use of Schools. By JOHN FORBES-ROBERTSON. 18mo, cloth limp. 1*s*. 6*d*.

THE CHEMISTRY OF THE SUN. With Illustrations. 8vo. 14*s*.

THE METEORIC HYPOTHESIS. With Illustrations. 8vo. [*In the press.*

THE EVOLUTION OF THE HEAVENS AND THE EARTH. Crown 8vo. With Illustrations. [*In the press.*

Newcomb.—POPULAR ASTRONOMY. By S. NEWCOMB, LL.D., Professor U.S. Naval Observatory. With 112 Illustrations and 5 Maps of the Stars. Second Edition, revised. 8vo. 18*s*.

"It is unlike anything else of its kind, and will be of more use in circulating a knowledge of Astronomy than nine-tenths of the books which have appeared on the subject of late years."—SATURDAY REVIEW.

CHEMISTRY.

Armstrong.—A MANUAL OF INORGANIC CHEMISTRY. By HENRY ARMSTRONG, Ph.D., F.R.S., Professor of Chemistry in the City and Guilds of London Technical Institute. Crown 8vo. [*In preparation.*

Cohen.—THE OWENS COLLEGE COURSE OF PRACTICAL ORGANIC CHEMISTRY. By JULIUS B. COHEN, Ph.D., F.C.S., Assistant Lecturer on Chemistry in the Owens College, Manchester. With a Preface by Sir HENRY ROSCOE, F.R.S., and C. SCHORLEMMER, F.R.S. Fcap. 8vo. 2*s*. 6*d*.

Cooke.—ELEMENTS OF CHEMICAL PHYSICS. By JOSIAH P. COOKE, Junr., Erving Professor of Chemistry and Mineralogy in Harvard University. Fourth Edition. Royal 8vo. 21*s*.

Fleischer.—A SYSTEM OF VOLUMETRIC ANALYSIS. By EMIL FLEISCHER. Translated, with Notes and Additions, from the Second German Edition by M. M. PATTISON MUIR, F.R.S.E. With Illustrations. Crown 8vo. 7*s*. 6*d*.

Frankland.—AGRICULTURAL CHEMICAL ANALYSIS. A Handbook of. By PERCY FARADAY FRANKLAND, Ph.D., B.Sc., F.C.S., Associate of the Royal School of Mines, and Demonstrator of Practical and Agricultural Chemistry in the Normal School of Science and Royal School of Mines, South Kensington Museum. Founded upon *Leitfaden für die Agriculture Chemiche Analyse*, von Dr. F. KROCKER. Crown 8vo. 7*s*. 6*d*.

Hartley.—A COURSE OF QUANTITATIVE ANALYSIS FOR STUDENTS. By W. NOEL HARTLEY, F.R.S., Professor of Chemistry, and of Applied Chemistry, Science and Art Department, Royal College of Science, Dublin. Globe 8vo. 5*s*.

Hiorns.—Works by ARTHUR H. HIORNS, Principal of the School of Metallurgy, Birmingham and Midland Institute.

PRACTICAL METALLURGY AND ASSAYING. A Text-Book for the use of Teachers, Students, and Assayers. With Illustrations. Globe 8vo. 6*s.*

A TEXT-BOOK OF ELEMENTARY METALLURGY FOR THE USE OF STUDENTS. To which is added an Appendix of Examination Questions, embracing the whole of the Questions set in the three stages of the subject by the Science and Art Department for the past twenty years. Globe 8vo. 4*s.*

IRON AND STEEL MANUFACTURE. A Text-Book for Beginners. Illustrated. Globe 8vo. 3*s.* 6*d.*

Jones.—Works by FRANCIS JONES, F.R.S.E., F.C.S., Chemical Master in the Grammar School, Manchester.

THE OWENS COLLEGE JUNIOR COURSE OF PRACTICAL CHEMISTRY. With Preface by Sir HENRY ROSCOE, F.R.S., and Illustrations. New Edition. 18mo. 2*s.* 6*d.*

QUESTIONS ON CHEMISTRY. A Series of Problems and Exercises in Inorganic and Organic Chemistry. Fcap. 8vo. 3*s.*

Landauer.—BLOWPIPE ANALYSIS. By J. LANDAUER. Authorised English Edition by J. TAYLOR and W. E. KAY, of Owens College, Manchester. [*New Edition in preparation.*

Lupton.—CHEMICAL ARITHMETIC. With 1,200 Problems. By SYDNEY LUPTON, M.A., F.C.S., F.I.C., formerly Assistant-Master at Harrow. Second Edition, Revised and Abridged. Fcap. 8vo. 4*s.* 6*d.*

Meldola.—THE CHEMISTRY OF PHOTOGRAPHY. By RAPHAEL MELDOLA, F.R.S., Professor of Chemistry in the Technical College, Finsbury, City and Guilds of London Institute for the Advancement of Technical Education. Crown 8vo. 6*s.* (*Nature Series.*)

Meyer.—HISTORY OF CHEMISTRY. By ERNST VON MEYER. Translated by GEORGE MCGOWAN. 8vo. [*In prep.*

Mixter.—AN ELEMENTARY TEXT-BOOK OF CHEMISTRY. By WILLIAM G. MIXTER, Professor of Chemistry in the Sheffield Scientific School of Yale College. Second and Revised Edition. Crown 8vo. 7*s.* 6*d.*

Muir.—PRACTICAL CHEMISTRY FOR MEDICAL STUDENTS. Specially arranged for the first M.B. Course. By M. M. PATTISON MUIR, F.R.S.E. Fcap. 8vo. 1*s.* 6*d.*

Muir and Wilson.—THE ELEMENTS OF THERMAL CHEMISTRY. By M. M. PATTISON MUIR, M.A., F.R.S.E., Fellow and Prælector of Chemistry in Gonville and Caius College. Cambridge; Assisted by DAVID MUIR WILSON. 8vo. 12*s.* 6*d.*

Ramsay.—EXPERIMENTAL PROOFS OF CHEMICAL THEORY FOR BEGINNERS. By WILLIAM RAMSAY, Ph.D., Professor of Chemistry in University Coll., London. Pott 8vo. 2*s.* 6*d.*

Remsen.—Works by IRA REMSEN, Professor of Chemistry in the Johns Hopkins University.

COMPOUNDS OF CARBON; or, Organic Chemistry, an Introduction to the Study of. Crown 8vo. *6s. 6d.*

AN INTRODUCTION TO THE STUDY OF CHEMISTRY (INORGANIC CHEMISTRY). Crown 8vo. *6s. 6d.*

THE ELEMENTS OF CHEMISTRY. A Text Book for Beginners. New Edition. Fcap. 8vo. *2s. 6d.*

A TEXT-BOOK OF INORGANIC CHEMISTRY. Medium 8vo. *16s.*

Roscoe.—Works by Sir HENRY E. ROSCOE, F.R.S., formerly Professor of Chemistry in the Victoria University the Owens College, Manchester.

PRIMER OF CHEMISTRY. With numerous Illustrations. New Edition. With Questions. 18mo. *1s.* (*Science Primers.*)

LESSONS IN ELEMENTARY CHEMISTRY, INORGANIC AND ORGANIC. With numerous Illustrations and Chromolitho of the Solar Spectrum, and of the Alkalies and Alkaline Earths. New Edition. Fcap. 8vo. *4s. 6d.* (*See under* THORPE.)

Roscoe and Schorlemmer.—INORGANIC AND ORGANIC CHEMISTRY. A Complete Treatise on Inorganic and Organic Chemistry. By Sir HENRY E. ROSCOE, F.R.S., and Prof. C. SCHORLEMMER, F.R.S. With Illustrations. Medium 8vo.

Vols. I. and II.—INORGANIC CHEMISTRY.

Vol. I.—The Non-Metallic Elements. Second Edition, revised. *21s.* Vol. II. Part I.—Metals. *18s.* Vol. II. Part II.—Metals. New and thoroughly Revised Edition. *18s.*

Vol. III.—ORGANIC CHEMISTRY.
THE CHEMISTRY OF THE HYDROCARBONS and their Derivatives, or ORGANIC CHEMISTRY. With numerous Illustrations. Five Parts. Parts I., II., and IV. *21s.* each. Parts III. and V. *18s.* each.

Thorpe.—A SERIES OF CHEMICAL PROBLEMS, prepared with Special Reference to Sir H. E. Roscoe's Lessons in Elementary Chemistry, by T. E. THORPE, Ph.D., F.R.S., Professor of Chemistry in the Normal School of Science, South Kensington, adapted for the Preparation of Students for the Government, Science, and Society of Arts Examinations. With a Preface by Sir HENRY E. ROSCOE, F.R.S. New Edition, with Key. 18mo. *2s.*

Thorpe and Rücker.—A TREATISE ON CHEMICAL PHYSICS. By T. E. THORPE, Ph.D., F.R.S., Professor of Chemistry in the Normal School of Science, and Professor A. W. RÜCKER. Illustrated. 8vo. [*In preparation.*

Wright.—METALS AND THEIR CHIEF INDUSTRIAL APPLICATIONS. BY C. ALDER WRIGHT, D.Sc., &c., Lecturer on Chemistry in St. Mary's Hospital Medical School. Extra fcap. 8vo. *3s. 6d.*

BIOLOGY.

Allen.—ON THE COLOUR OF FLOWERS, as Illustrated in the British Flora. By GRANT ALLEN. With Illustrations. Crown 8vo. 3*s*. 6*d*. (*Nature Series.*)

Balfour.—A TREATISE ON COMPARATIVE EMBRYOLOGY. By F. M. BALFOUR, M.A., F.R.S., Fellow and Lecturer of Trinity College, Cambridge. With Illustrations. Second Edition, reprinted without alteration from the First Edition. In 2 vols. 8vo. Vol. I. 18*s*. Vol. II. 21*s*.

Balfour and Ward.—A GENERAL TEXT BOOK OF BOTANY. By ISAAC BAYLEY BALFOUR, F.R.S., Professor of Botany in the University of Edinburgh, and H. MARSHALL WARD, F.R.S., Fellow of Christ's College, Cambridge, and Professor of Botany in the Royal Indian Engineering College, Cooper's Hill. 8vo. [*In preparation.*

Bettany.—FIRST LESSONS IN PRACTICAL BOTANY By G. T. BETTANY, M.A., F.L.S., formerly Lecturer in Botany at Guy's Hospital Medical School. 18mo. 1*s*.

Bower.—A COURSE OF PRACTICAL INSTRUCTION IN BOTANY. By F. O. BOWER, D.Sc., F.L.S., Regius Professor of Botany in the University of Glasgow. Crown 8vo. 10*s*. 6*d*.

Darwin (Charles).—MEMORIAL NOTICES OF CHARLES DARWIN, F.R.S., &c. By THOMAS HENRY HUXLEY, F.R.S., G. J. ROMANES, F.R.S., ARCHIBALD GEIKIE, F.R.S., and W. T. THISELTON DYER, F.R.S. Reprinted from *Nature.* With a Portrait, engraved by C. H. JEENS. Crown 8vo. 2*s*. 6*d*. (*Nature Series.*)

Eimer.—ORGANIC EVOLUTION AS THE RESULT OF THE INHERITANCE OF ACQUIRED CHARACTER ACCORDING TO THE LAWS OF ORGANIC GROWTH. By Dr. G. H. THEODOR EIMER, Professor of Zoology and Comparative Anatomy in Tubingen. Translated by J. T. CUNNINGHAM, M.A., F.R.S.E., late Fellow of University College, Oxford. 8vo. Part I., with 6 Figures in the Text. [*In the press.*

Fearnley.—A MANUAL OF ELEMENTARY PRACTICAL HISTOLOGY. By WILLIAM FEARNLEY. With Illustrations. Crown 8vo. 7*s*. 6*d*.

Flower and Gadow.—AN INTRODUCTION TO THE OSTEOLOGY OF THE MAMMALIA. By WILLIAM HENRY FLOWER, LL.D., F.R.S., Director of the Natural History Departments of the British Museum, late Hunterian Professor of Comparative Anatomy and Physiology in the Royal College of Surgeons of England. With numerous Illustrations. Third Edition. Revised with the assistance of HANS GADOW, Ph.D., M.A., Lecturer on the Advanced Morphology of Vertebrates and Strickland Curator in the University of Cambridge. Crown 8vo. 10*s*. 6*d*.

Foster.—Works by MICHAEL FOSTER, M.D., Sec. R.S., Professor of Physiology in the University of Cambridge.

PRIMER OF PHYSIOLOGY. With numerous Illustrations. New Edition. 18mo. 1*s.*

A TEXT-BOOK OF PHYSIOLOGY. With Illustrations. Fifth Edition, largely revised. In Three Parts. Part I., comprising Book I. Blood—The Tissues of Movement, The Vascular Mechanism. 10*s.* 6*d.* Part II., comprising Book II. The Tissues of Chemical Action, with their Respective Mechanisms—Nutrition. 10*s.* 6*d.*

Part III. is in the press preparing for early publication.

Foster and Balfour.—THE ELEMENTS OF EMBRYOLOGY. By MICHAEL FOSTER, M.A., M.D., LL.D., Sec. R.S., Professor of Physiology in the University of Cambridge, Fellow of Trinity College, Cambridge, and the late FRANCIS M. BALFOUR, M.A., LL.D., F.R.S., Fellow of Trinity College, Cambridge, and Professor of Animal Morphology in the University. Second Edition, revised. Edited by ADAM SEDGWICK, M.A., Fellow and Assistant Lecturer of Trinity College, Cambridge, and WALTER HEAPE, Demonstrator in the Morphological Laboratory of the University of Cambridge. With Illustrations. Crown 8vo. 10*s.* 6*d.*

Foster and Langley.—A COURSE OF ELEMENTARY PRACTICAL PHYSIOLOGY AND HISTOLOGY. By Prof. MICHAEL FOSTER, M.D., F.R.S., &c., and J. N. LANGLEY, M.A., F.R.S., Fellow of Trinity College, Cambridge. Sixth Edition. Crown 8vo. 7*s.* 6*d.*

Gamgee.—A TEXT-BOOK OF THE PHYSIOLOGICAL CHEMISTRY OF THE ANIMAL BODY. Including an Account of the Chemical Changes occurring in Disease. By A. GAMGEE, M.D., F.R.S., formerly Professor of Physiology in the Victoria University the Owens College, Manchester. 2 Vols. 8vo. With Illustrations. Vol. I. 18*s.* [*Vol. II. in the press.*

Gray.—STRUCTURAL BOTANY, OR ORGANOGRAPHY ON THE BASIS OF MORPHOLOGY. To which are added the principles of Taxonomy and Phytography, and a Glossary of Botanical Terms. By Professor ASA GRAY, LL.D. 8vo. 10*s.* 6*d.*

THE SCIENTIFIC PAPERS OF ASA GRAY. Selected by CHARLES SPRAGUE SARGENT. 2 vols. Vol. I. Reviews of Works on Botany and Related Subjects, 1834–1887. Vol. II. Essays; Biographical Sketches, 1841–1886. Demy 8vo. 21*s.*

Hamilton.—A SYSTEMATIC AND PRACTICAL TEXT-BOOK OF PATHOLOGY. By D. J. HAMILTON, F.R.C.S.E., F.R.S.E., Professor of Pathological Anatomy, University of Aberdeen. Copiously Illustrated. 8vo. Volume I. 25*s.*

Hooker.—Works by Sir J. D. HOOKER, K.C.S.I., C.B., M.D., F.R.S., D.C.L.

PRIMER OF BOTANY. With numerous Illustrations. New Edition. 18mo. 1*s.* (*Science Primers.*)

THE STUDENT'S FLORA OF THE BRITISH ISLANDS. Third Edition, revised. Globe 8vo. 10*s.* 6*d.*

Howes.—AN ATLAS OF PRACTICAL ELEMENTARY BIOLOGY. By G. B. HOWES, Assistant Professor of Zoology, Normal School of Science and Royal School of Mines. With a Preface by THOMAS HENRY HUXLEY, F.R.S. Royal 4to. 14*s.*

Huxley.—Works by THOMAS HENRY HUXLEY, F.R.S.

INTRODUCTORY PRIMER OF SCIENCE. 18mo. 1*s.* (*Science Primers.*)

LESSONS IN ELEMENTARY PHYSIOLOGY. With numerous Illustrations. New Edition Revised. Fcap. 8vo. 4*s.* 6*d.*

QUESTIONS ON HUXLEY'S PHYSIOLOGY FOR SCHOOLS. By T. ALCOCK, M.D. New Edition. 18mo. 1*s.* 6*d.*

Huxley and Martin.—A COURSE OF PRACTICAL INSTRUCTION IN ELEMENTARY BIOLOGY. By T. H. HUXLEY, F.R.S., LL.D., assisted by H. N. MARTIN, M.A., M.B., D.Sc., F.R.S., Fellow of Christ's College, Cambridge. New Edition, revised and extended by G. B. HOWES, Assistant Professor of Zoology, Normal School of Science, and Royal School of Mines, and D. H. SCOTT, M.A., PH.D., Assistant Professor of Botany, Normal School of Science, and Royal School of Mines. New Edition, thoroughly revised. With a Preface by T. H. HUXLEY, F.R.S. Crown 8vo. 10*s.* 6*d.*

Klein.—Works by E. KLEIN, M.D., F.R.S., Lecturer on General Anatomy and Physiology in the Medical School of St. Bartholomew's Hospital, Professor of Bacteriology at the College of State Medicine, London.

MICRO-ORGANISMS AND DISEASE. An Introduction into the Study of Specific Micro-Organisms. With 121 Illustrations. Third Edition, Revised. Crown 8vo. 6*s.*

THE BACTERIA IN ASIATIC CHOLERA. Crown 8vo. 5*s.*

Lankester.—Works by Professor E. RAY LANKESTER, F.R.S.

A TEXT BOOK OF ZOOLOGY. 8vo. [*In preparation.*

DEGENERATION: A CHAPTER IN DARWINISM. Illustrated. Crown 8vo. 2*s.* 6*d.* (*Nature Series.*)

Lubbock.—Works by Sir JOHN LUBBOCK, M.P., F.R.S., D.C.L.

THE ORIGIN AND METAMORPHOSES OF INSECTS. With numerous Illustrations. New Edition. Crown 8vo. 3*s.* 6*d.* (*Nature Series.*)

ON BRITISH WILD FLOWERS CONSIDERED IN RELATION TO INSECTS. With numerous Illustrations. New Edition. Crown 8vo. 4*s.* 6*d.* (*Nature Series.*)

Lubbock.—FLOWERS, FRUITS, AND LEAVES. With Illustrations. Second Edition. Crown 8vo. 4*s.* 6*d.* (*Nature Series.*)

Martin and Moale.—ON THE DISSECTION OF VERTEBRATE ANIMALS. By Professor H. N. MARTIN and W. A. MOALE. Crown 8vo. [*In preparation.*

Mivart.—Works by ST. GEORGE MIVART, F.R.S., Lecturer on Comparative Anatomy at St. Mary's Hospital.

LESSONS IN ELEMENTARY ANATOMY. With upwards of 400 Illustrations. Fcap. 8vo. 6*s.* 6*d.*

THE COMMON FROG. Illustrated. Cr. 8vo. 3*s.* 6*d.* (*Nature Series.*)

Müller.—THE FERTILISATION OF FLOWERS. By Professor HERMANN MÜLLER. Translated and Edited by D'ARCY W. THOMPSON, B.A., Professor of Biology in University College, Dundee. With a Preface by CHARLES DARWIN, F.R.S. With numerous Illustrations. Medium 8vo. 21*s.*

Oliver.—Works by DANIEL OLIVER, F.R.S., &c., Professor of Botany in University College, London, &c.

FIRST BOOK OF INDIAN BOTANY. With numerous Illustrations. Extra fcap. 8vo. 6*s.* 6*d.*

LESSONS IN ELEMENTARY BOTANY. With nearly 200 Illustrations. New Edition. Fcap. 8vo. 4*s.* 6*d.*

Parker.—A COURSE OF INSTRUCTION IN ZOOTOMY (VERTEBRATA). By T. JEFFREY PARKER, B.Sc. London, Professor of Biology in the University of Otago, New Zealand. With Illustrations. Crown 8vo. 8*s.* 6*d.*

LESSONS IN ELEMENTARY BIOLOGY. By the same Author. With Illustrations. 8vo. [*In the press.*

Parker and Bettany.—THE MORPHOLOGY OF THE SKULL. By Professor W. K. PARKER, F.R.S., and G. T. BETTANY, M.A., F.L.S. Illustrated. Crown 8vo. 10*s.* 6*d.*

Romanes.—THE SCIENTIFIC EVIDENCES OF ORGANIC EVOLUTION. By GEORGE J. ROMANES, M.A., LL.D., F.R.S., Zoological Secretary of the Linnean Society. Crown 8vo. 2*s.* 6*d.* (*Nature Series.*)

Sedgwick.—A SUPPLEMENT TO F. M. BALFOUR'S TREATISE ON EMBRYOLOGY. By ADAM SEDGWICK, M.A., F.R.S., Fellow and Lecturer of Trinity College, Cambridge. Illustrated. 8vo. [*In preparation.*

Smith (W. G.).—DISEASES OF FIELD AND GARDEN CROPS, CHIEFLY SUCH AS ARE CAUSED BY FUNGI. By WORTHINGTON G. SMITH, F.L.S., M.A.I., Member of the Scientific Committee R.H.S. With 143 New Illustrations drawn and engraved from Nature by the Author. Fcap. 8vo. 4*s.* 6*d.*

Stewart—Corry.—A FLORA OF THE NORTH-EAST OF IRELAND. Including the Phanerogamia, the Cryptogamia Vascularia, and the Muscineæ. By SAMUEL ALEXANDER STEWART, Fellow of the Botanical Society of Edinburgh, Curator of the Collections in the Belfast Museum, and Honorary Associate of the Belfast Natural History and Philosophical Society; and the late THOMAS HUGHES CORRY, M.A., F.L.S., F.Z.S., M.R.I.A., F.B.S. Edin., Lecturer on Botany in the University Medical and Science Schools, Cambridge, Assistant Curator of the University Herbarium, &c., &c. Crown 8vo. 5*s.* 6*d.*

Wallace.—DARWINISM. An Exposition of the Theory of Natural Selection with some of its applications. By ALFRED RUSSEL WALLACE, LL.D., F.L.S., Author of "The Malay Archipelago," "The Geographical Distribution of Animals," "Island Life," &c. Third Edition. Crown 8vo. 9*s.*

Ward.—TIMBER AND SOME OF ITS DISEASES. By H. MARSHALL WARD, M.A., F.R.S., F.L.S., late Fellow of Christ's College, Cambridge; Professor of Botany in the Royal Indian Engineering College, Cooper's Hill. With Illustrations. Crown 8vo. 6*s.* (*Nature Series.*)

Wiedersheim (Prof.).—ELEMENTS OF THE COMPARATIVE ANATOMY OF VERTEBRATES. Adapted from the German of ROBERT WIEDERSHEIM, Professor of Anatomy, and Director of the Institute of Human and Comparative Anatomy in the University of Freiburg-in-Baden, by W. NEWTON PARKER, Professor of Biology in the University College of South Wales and Monmouthshire. With Additions by the Author and Translator. With Two Hundred and Seventy Woodcuts. Medium 8vo. 12*s.* 6*d.*

MEDICINE.

Brunton.—Works by T. LAUDER BRUNTON, M.D., D.Sc., F.R.C.P., F.R.S., Assistant Physician and Lecturer on Materia Medica at St. Bartholomew's Hospital; Examiner in Materia Medica in the University of London, in the Victoria University, and in the Royal College of Physicians, London; late Examiner in the University of Edinburgh.

A TEXT-BOOK OF PHARMACOLOGY, THERAPEUTICS, AND MATERIA MEDICA. Adapted to the United States Pharmacopœia, by FRANCIS H. WILLIAMS, M.D., Boston, Mass. Third Edition. Adapted to the New British Pharmacopœia, 1885. Medium 8vo. 21*s.*

TABLES OF MATERIA MEDICA: A Companion to the Materia Medica Museum. With Illustrations. New Edition Enlarged. Cheaper Issue. 8vo. 5*s.*

Griffiths.—LESSONS ON PRESCRIPTIONS AND THE ART OF PRESCRIBING. By W. HANDSEL GRIFFITHS, Ph.D., L.R.C.P.E. New Edition. Adapted to the Pharmacopœia, 1885. 18mo. 3*s*. 6*d*.

Hamilton.—A TEXT-BOOK OF PATHOLOGY, SYSTEMATIC AND PRACTICAL. By D. J. HAMILTON, M.B., F.R.C.S.E., F.R.S.E., Professor of Pathological Anatomy, University of Aberdeen. Copiously Illustrated. Vol I. 8vo. 25*s*.

Klein.—Works by E. KLEIN, M.D., F.R.S., Lecturer on General Anatomy and Physiology in the Medical School of St. Bartholomew's Hospital, London.

MICRO-ORGANISMS AND DISEASE. An Introduction into the Study of Specific Micro-Organisms. With 121 Illustrations. Third Edition, Revised. Crown 8vo. 6*s*.

THE BACTERIA IN ASIATIC CHOLERA. By the Same Author. Crown 8vo. 5*s*.

White.—A TEXT-BOOK OF GENERAL THERAPEUTICS. By W. HALE WHITE, M.D., F.R.C.P., Senior Assistant Physician to and Lecturer in Materia Medica at Guy's Hospital. With Illustrations. Crown 8vo. 8*s*. 6*d*.

Ziegler-Macalister.—TEXT-BOOK OF PATHOLOGICAL ANATOMY AND PATHOGENESIS. By Professor ERNST ZIEGLER of Tübingen. Translated and Edited for English Students by DONALD MACALISTER, M.A., M.D., B.Sc., F.R.C.P., Fellow and Medical Lecturer of St. John's College, Cambridge, Physician to Addenbrooke's Hospital, and Teacher of Medicine in the University. With numerous Illustrations. Medium 8vo.

Part I.—GENERAL PATHOLOGICAL ANATOMY. Second Edition. 12*s*. 6*d*.

Part II.—SPECIAL PATHOLOGICAL ANATOMY. Sections I.—VIII. Second Edition. 12*s*. 6*d*. Sections IX.—XII. 12*s*. 6*d*.

MILITARY AND NAVAL SCIENCE.

Aitken (Sir W.).—THE GROWTH OF THE RECRUIT AND YOUNG SOLDIER. With a view to the selection of "Growing Lads" for the Army, and a Regulated System of Training for Recruits. By Sir WILLIAM AITKEN, Knt., M.D., F.R.S., Professor of Pathology in the Army Medical School, Examiner in Medicine for the Military Services of the Queen; Fellow of the Sanitary Institute of Great Britain; formerly Demonstrator of Anatomy in the University of Glasgow; and Pathologist attached to the Military Hospitals of Scutari during the Russian War. Crown 8vo. 8*s*. 6*d*.

Army Preliminary Examination, 1882–1888, Specimens of Papers set at the. With answers to the Mathematical Questions. Subjects: Arithmetic, Algebra, Euclid, Geometrical Drawing, Geography, French, English Dictation. Crown 8vo. 3*s*. 6*d*.

Mercur.—ELEMENTS OF THE ART OF WAR. Prepared for the use of Cadets of the United States Military Academy. By JAMES MERCUR, Professor of Civil Engineering at the United States Academy, West Point, New York. Second Edition, Revised and Corrected. 8vo. 17*s*.

Palmer.—TEXT BOOK OF PRACTICAL LOGARITHMS AND TRIGONOMETRY. By J. H. PALMER, Head Schoolmaster R.N., H.M.S. *Cambridge*, Devonport. Globe 8vo. 4*s*. 6*d*.

Robinson.—TREATISE ON MARINE SURVEYING. Prepared for the use of younger Naval Officers. With Questions for Examinations and Exercises principally from the Papers of the Royal Naval College. With the results. By Rev. JOHN L. ROBINSON, Chaplain and Instructor in the Royal Naval College, Greenwich. With Illustrations. Crown 8vo. 7*s*. 6*d*.

CONTENTS.—Symbols used in Charts and Surveying—The Construction and Use of Scales—Laying off Angles—Fixing Positions by Angles—Charts and Chart-Drawing—Instruments and Observing—Base Lines—Triangulation—Levelling—Tides and Tidal Observations—Soundings—Chronometers—Meridian Distances—Method of Plotting a Survey—Miscellaneous Exercises—Index.

Shortland.—NAUTICAL SURVEYING. By the late Vice-Admiral SHORTLAND, LL.D., late Fellow of Pembroke College, Cambridge. 8vo. [*In the press.*

Wolseley.—Works by General Viscount WOLSELEY, G.C.M.G., G.C.B.

THE SOLDIER'S POCKET-BOOK FOR FIELD SERVICE. Fifth Edition, Revised and Enlarged. 16mo, roan. 5*s*.

FIELD POCKET-BOOK FOR THE AUXILIARY FORCES. 16mo, cloth. 1*s*. 6*d*.

Woolwich Mathematical Papers, for Admission into the Royal Military Academy, Woolwich, 1880–1888 inclusive. Edited by E. J. BROOKSMITH, B.A., LL.M., St. John's College, Cambridge; Instructor in Mathematics at the Royal Military Academy, Woolwich. Crown 8vo. 6*s*.

ANTHROPOLOGY.

Flower.—FASHION IN DEFORMITY, as Illustrated in the Customs of Barbarous and Civilised Races. By Professor FLOWER, F.R.S., F.R.C.S. With Illustrations. Crown 8vo. 2*s*. 6*d*. (*Nature Series.*)

Tylor.—ANTHROPOLOGY. An Introduction to the Study of Man and Civilisation. By E. B. TYLOR, D.C.L., F.R.S. With numerous Illustrations. Crown 8vo. 7*s*. 6*d*.

PHYSICAL GEOGRAPHY & GEOLOGY.

Blanford.—THE RUDIMENTS OF PHYSICAL GEOGRAPHY FOR THE USE OF INDIAN SCHOOLS; with a Glossary of Technical Terms employed. By H. F. BLANFORD, F.G.S. New Edition, with Illustrations. Crown 8vo. 2*s*. 6*d*.

Ferrel.—A POPULAR TREATISE ON THE WINDS. Comprising the General Motions of the Atmosphere, Monsoons, Cyclones, Tornadoes, Waterspouts, Hailstorms, &c. By WILLIAM FERREL, M.A., Ph.D., late Professor and Assistant in the Signal Service; Member of the American National Academy of Sciences, and of other Home and Foreign Scientific Societies. Demy 8vo. 18*s*.

Fisher.—PHYSICS OF THE EARTH'S CRUST. By the Rev. OSMOND FISHER, M.A., F.G.S., Rector of Harlton, Hon. Fellow of King's College, London, and late Fellow and Tutor of Jesus College, Cambridge. Second Edition, altered and enlarged. 8vo. 12*s*.

Geikie.—Works by ARCHIBALD GEIKIE, LL.D., F.R.S., Director-General of the Geological Survey of Great Britain and Ireland, and Director of the Museum of Practical Geology, London, formerly Murchison Professor of Geology and Mineralogy in the University of Edinburgh, &c.

PRIMER OF PHYSICAL GEOGRAPHY. With numerous Illustrations. New Edition. With Questions. 18mo. 1*s*. (*Science Primers.*)

ELEMENTARY LESSONS IN PHYSICAL GEOGRAPHY. With numerous Illustrations. New Edition. Fcap. 8vo. 4*s*. 6*d*. QUESTIONS ON THE SAME. 1*s*. 6*d*.

PRIMER OF GEOLOGY. With numerous Illustrations. New Edition. 18mo. 1*s*. (*Science Primers.*)

CLASS BOOK OF GEOLOGY. With upwards of 200 New Illustrations. Crown 8vo. 10*s*. 6*d*.

TEXT-BOOK OF GEOLOGY. With numerous Illustrations. Second Edition, Seventh Thousand, Revised and Enlarged. 8vo. 28*s*.

OUTLINES OF FIELD GEOLOGY. With Illustrations. New Edition. Extra fcap. 8vo. 3*s*. 6*d*.

THE SCENERY AND GEOLOGY OF SCOTLAND, VIEWED IN CONNEXION WITH ITS PHYSICAL GEOLOGY. With numerous Illustrations. Crown 8vo. 12*s*. 6*d*.

(See also under *Geography*.)

Huxley.—PHYSIOGRAPHY. An Introduction to the Study of Nature. By THOMAS HENRY HUXLEY, F.R.S. With numerous Illustrations, and Coloured Plates. New and Cheaper Edition. Crown 8vo. 6*s*.

Lockyer.—OUTLINES OF PHYSIOGRAPHY—THE MOVEMENTS OF THE EARTH. By J. NORMAN LOCKYER, F.R.S., Correspondent of the Institute of France, Foreign Member of the Academy of the Lyncei of Rome, &c., &c.; Professor of Astronomical Physics in the Normal School of Science, and Examiner in Physiography for the Science and Art Department. With Illustrations. Crown 8vo. Sewed, 1*s.* 6*d.*

Phillips.—A TREATISE ON ORE DEPOSITS. By J. ARTHUR PHILLIPS, F.R.S., V.P.G.S., F.C.S., M.Inst.C.E., Ancien Élève de l'École des Mines, Paris; Author of "A Manual of Metallurgy," "The Mining and Metallurgy of Gold and Silver," &c. With numerous Illustrations. 8vo. 25*s.*

Rosenbusch — Iddings. — MICROSCOPICAL PHYSIOGRAPHY OF THE ROCK-MAKING MINERALS: AN AID TO THE MICROSCOPICAL STUDY OF ROCKS. By H. ROSENBUSCH. Translated and Abridged for Use in Schools and Colleges. By JOSEPH P. IDDINGS. Illustrated by 121 Woodcuts and 26 Photomicrographs. 8vo, 24*s.*

AGRICULTURE.

Frankland.—AGRICULTURAL CHEMICAL ANALYSIS, A Handbook of. By PERCY FARADAY FRANKLAND, Ph.D., B.Sc., F.C.S., Associate of the Royal School of Mines, and Demonstrator of Practical and Agricultural Chemistry in the Normal School of Science and Royal School of Mines, South Kensington Museum. Founded upon *Leitfaden für die Agriculture Chemiche Analyse*, von Dr. F. KROCKER. Crown 8vo. 7*s.* 6*d.*

Smith (Worthington G.).—DISEASES OF FIELD AND GARDEN CROPS, CHIEFLY SUCH AS ARE CAUSED BY FUNGI. By WORTHINGTON G. SMITH, F.L.S., M.A.I., Member of the Scientific Committee of the R.H.S. With 143 Illustrations, drawn and engraved from Nature by the Author. Fcap. 8vo. 4*s.* 6*d.*

Tanner.—Works by HENRY TANNER, F.C.S., M.R.A.C., Examiner in the Principles of Agriculture under the Government Department of Science; Director of Education in the Institute of Agriculture, South Kensington, London; sometime Professor of Agricultural Science, University College, Aberystwith.

ELEMENTARY LESSONS IN THE SCIENCE OF AGRICULTURAL PRACTICE. Fcap. 8vo. 3*s.* 6*d.*

FIRST PRINCIPLES OF AGRICULTURE. 18mo. 1*s.*

THE PRINCIPLES OF AGRICULTURE. A Series of Reading Books for use in Elementary Schools. Prepared by HENRY TANNER, F.C.S., M.R.A.C. Extra fcap. 8vo.

I. The Alphabet of the Principles of Agriculture. 6*d.*

II. Further Steps in the Principles of Agriculture. 1*s.*

III. Elementary School Readings on the Principles of Agriculture for the third stage. 1*s.*

Ward.—TIMBER AND SOME OF ITS DISEASES. By H. MARSHALL WARD, M.A., F.L.S., F.R.S., Fellow of Christ's College, Cambridge, Professor of Botany at the Royal Indian Engineering College, Cooper's Hill. With Illustrations. Crown 8vo. 6*s*. (*Nature Series.*)

POLITICAL ECONOMY.

Böhm-Bauerk.—CAPITAL AND INTEREST. Translated by WILLIAM SMART. 8vo. [*In the press.*

Cairnes.—THE CHARACTER AND LOGICAL METHOD OF POLITICAL ECONOMY. By J. E. CAIRNES, LL.D., Emeritus Professor of Political Economy in University College, London. New Edition. Crown 8vo. 6*s*.

Cossa.—GUIDE TO THE STUDY OF POLITICAL ECONOMY. By Dr. LUIGI COSSA, Professor in the University of Pavia. Translated from the Second Italian Edition. With a Preface by W. STANLEY JEVONS, F.R.S. Crown 8vo. 4*s*. 6*d*.

Donisthorpe.—INDIVIDUALISM: A System of Politics. By WORDSWORTH DONISTHORPE. 8vo. 14*s*.

Fawcett (Mrs.).—Works by Mrs. HENRY FAWCETT.—POLITICAL ECONOMY FOR BEGINNERS, WITH QUESTIONS. Fourth Edition. 18mo. 2*s*. 6*d*.

TALES IN POLITICAL ECONOMY. Crown 8vo. 3*s*.

Fawcett.—A MANUAL OF POLITICAL ECONOMY. By Right Hon. HENRY FAWCETT, F.R.S. Seventh Edition, revised, with a chapter on "State Socialism and the Nationalisation of the Land," and an Index. Crown 8vo. 12*s*.

AN EXPLANATORY DIGEST of the above. By CYRIL A. WATERS, B.A. Crown 8vo. 2*s*. 6*d*.

Gunton.—WEALTH AND PROGRESS: A CRITICAL EXAMINATION OF THE WAGES QUESTION AND ITS ECONOMIC RELATION TO SOCIAL REFORM. By GEORGE GUNTON. Crown 8vo. 6*s*.

Jevons.—Works by W. STANLEY JEVONS, LL.D. (Edinb.), M.A. (Lond.), F.R.S., late Professor of Political Economy in University College, London, Examiner in Mental and Moral Science in the University of London.

PRIMER OF POLITICAL ECONOMY. New Edition. 18mo. 1*s*. (*Science Primers.*)

THE THEORY OF POLITICAL ECONOMY. Third Edition. Revised. Demy 8vo. 10*s*. 6*d*.

Keynes.—THE SCOPE AND METHOD OF POLITICAL ECONOMY. By J. N. KEYNES, M.A. [*In preparation.*

Marshall.—THE ECONOMICS OF INDUSTRY. By A. MARSHALL, M.A., Professor of Political Economy in the University of Cambridge, and MARY P. MARSHALL, late Lecturer at Newnham Hall, Cambridge. Extra fcap. 8vo. 2*s*. 6*d*.

Marshall.—ECONOMICS. By ALFRED MARSHALL, M.A., Professor of Political Economy in the University of Cambridge. 2 vols. 8vo. [*In the press.*

Palgrave.—A DICTIONARY OF POLITICAL ECONOMY. By various Writers. Edited by R. H. INGLIS PALGRAVE. [*In the press.*

Sidgwick.—THE PRINCIPLES OF POLITICAL ECONOMY. By HENRY SIDGWICK, M.A., LL.D., Knightbridge Professor of Moral Philosophy in the University of Cambridge, &c., Author of "The Methods of Ethics." Second Edition, revised. 8vo. 16*s.*

THE ELEMENTS OF POLITICS. By the same Author. 8vo. [*In the press.*

Walker.—Works by FRANCIS A. WALKER, M.A., Ph.D., Author of "Money," "Money in its Relation to Trade," &c.

POLITICAL ECONOMY. Second Edition, revised and enlarged. 8vo. 12*s.* 6*d.*

A BRIEF TEXT-BOOK OF POLITICAL ECONOMY. Crown 8vo. 6*s.* 6*d.*

THE WAGES QUESTION. 8vo. 14*s.*

Wicksteed.—ALPHABET OF ECONOMIC SCIENCE. By PHILIP H. WICKSTEED, M.A. Part I. Elements of the Theory of Value or Worth. Globe 8vo. 2*s.* 6*d*

MENTAL AND MORAL PHILOSOPHY.

Boole.—THE MATHEMATICAL ANALYSIS OF LOGIC. Being an Essay towards a Calculus of Deductive Reasoning. By GEORGE BOOLE. 8vo. Sewed. 5*s.*

Calderwood.—HANDBOOK OF MORAL PHILOSOPHY. By the Rev. HENRY CALDERWOOD, LL.D., Professor of Moral Philosophy, University of Edinburgh. Fourteenth Edition, largely rewritten. Cr. 8vo. 6*s.*

Clifford.—SEEING AND THINKING. By the late Professor W. K. CLIFFORD, F.R.S. With Diagrams. Crown 8vo. 3*s.* 6*d.* (*Nature Series.*)

Jardine.—THE ELEMENTS OF THE PSYCHOLOGY OF COGNITION. By the Rev. ROBERT JARDINE, B.D., D.Sc. (Edin.), Ex-Principal of the General Assembly's College, Calcutta. Third Edition, revised and improved. Crown 8vo. 6*s.* 6*d.*

Jevons.—Works by the late W. STANLEY JEVONS, LL.D., M.A., F.R.S.

PRIMER OF LOGIC. New Edition. 18mo. 1*s.* (*Science Primers.*)

ELEMENTARY LESSONS IN LOGIC; Deductive and Inductive, with copious Questions and Examples, and a Vocabulary of Logical Terms. New Edition. Fcap. 8vo. 3*s.* 6*d.*

THE PRINCIPLES OF SCIENCE. A Treatise on Logic and Scientific Method. New and Revised Edition. Crown 8vo. 12*s.* 6*d.*

Jevons.—STUDIES IN DEDUCTIVE LOGIC. Second Edition. Crown 8vo. 6*s.*

LOGICAL PAPERS. 8vo. [*In preparation.*

Kant—Max Müller.—CRITIQUE OF PURE REASON. By IMMANUEL KANT. In commemoration of the Centenary of its first Publication. Translated into English by F. MAX MÜLLER. With an Historical Introduction by LUDWIG NOIRÉ. 2 vols. 8vo. 16*s.* each.

Volume I. HISTORICAL INTRODUCTION, by LUDWIG NOIRÉ; &c., &c.

Volume II. CRITIQUE OF PURE REASON, translated by F. MAX MÜLLER.

For the convenience of students these volumes are now sold separately.

Kant—Mahaffy and Bernard.—KANT'S CRITICAL PHILOSOPHY FOR ENGLISH READERS. By J. P. MAHAFFY, D.D., Fellow and Tutor of Trinity College, Dublin, Professor of Ancient History in the University of Dublin, and JOHN H. BERNARD, B.D., Fellow of Trinity College, Dublin, Archbishop King's Lecturer in Divinity in the University of Dublin. A new and completed Edition in 2 vols. Crown 8vo.

Vol. I. THE KRITIK OF PURE REASON EXPLAINED AND DEFENDED. 7*s.* 6*d.*

Vol. II. THE "PROLEGOMENA." Translated with Notes and Appendices. 6*s.*

Keynes.—FORMAL LOGIC, Studies and Exercises in. Including a Generalisation of Logical Processes in their application to Complex Inferences. By JOHN NEVILLE KEYNES, M.A., late Fellow of Pembroke College, Cambridge. Second Edition, Revised and Enlarged. Crown 8vo. 10*s.* 6*d.*

McCosh.—Works by JAMES McCOSH, D.D., LL.D., Litt.D., President of Princeton College, Author of "Intuitions of the Mind," "Laws of Discursive Thought," &c.

PSYCHOLOGY. Crown 8vo.

I. THE COGNITIVE POWERS. 6*s.* 6*d.*

II. THE MOTIVE POWERS. Crown 8vo. 6*s.* 6*d.*

FIRST AND FUNDAMENTAL TRUTHS: being a Treatise on Metaphysics. Extra crown 8vo. 9*s.*

Ray.—A TEXT-BOOK OF DEDUCTIVE LOGIC FOR THE USE OF STUDENTS. By P. K. RAY, D.Sc. (Lon. and Edin.), Professor of Logic and Philosophy, Presidency College Calcutta. Fourth Edition. Globe 8vo. 4*s.* 6*d.*

The SCHOOLMASTER says:—"This work . . . is deservedly taking a place among the recognised text-books on Logic."

Sidgwick.—Works by HENRY SIDGWICK, M.A., LL.D., Knightbridge Professor of Moral Philosophy in the University of Cambridge.

THE METHODS OF ETHICS. Third Edition. 8vo. 14*s.* A Supplement to the Second Edition, containing all the important Additions and Alterations in the Third Edition. Demy 8vo. 6*s.*

Sidgwick.—OUTLINES OF THE HISTORY OF ETHICS, for English Readers. Second Edition, revised. Crown 8vo. 3*s*. 6*d*.

ELEMENTS OF POLITICS. Demy 8vo. [*In the press.*

Venn.—Works by JOHN VENN, Sc.D., F.R.S., M.A., Fellow and Lecturer in Moral Sciences in Gonville and Caius College, Cambridge, Examiner in Moral Philosophy in the University of London.

THE LOGIC OF CHANCE. An Essay on the Foundations and Province of the Theory of Probability, with special Reference to its Logical Bearings and its Application to Moral and Social Science. Third Edition, rewritten and greatly enlarged. Crown 8vo. 10*s*. 6*d*.

SYMBOLIC LOGIC. Crown 8vo. 10*s*. 6*d*.

THE PRINCIPLES OF EMPIRICAL OR INDUCTIVE LOGIC. Medium 8vo. 18*s*.

GEOGRAPHY.

Bartholomew.—Works by JOHN BARTHOLOMEW, F.R.G.S.

THE ELEMENTARY SCHOOL ATLAS. 4to. 1*s*.

This Elementary Atlas is designed to illustrate the principal text-books on Elementary Geography.

PHYSICAL AND POLITICAL SCHOOL ATLAS, Consisting of 80 Maps and complete Index. Prepared for the use of Senior Pupils. Royal 4to. [*In the press.*

THE LIBRARY REFERENCE ATLAS OF THE WORLD. A complete Series of 84 Modern Maps. With Geographical Index to 100,000 places. Half-morocco. Gilt edges. Folio. £2 12*s*. 6*d*. net.

*** This work has been designed with the object of supplying the public with a thoroughly complete and accurate atlas of Modern Geography, in a convenient reference form, and at a moderate price.

Clarke.—CLASS-BOOK OF GEOGRAPHY. By C. B. CLARKE. F.R.S. New Edition, revised 1889, with Eighteen Coloured Maps, Fcap. 8vo. Paper covers, 3*s*. ; cloth, 3*s*. 6*d*.

Elderton.—MAP DRAWING AND MAP MAKING. By WILLIAM A. ELDERTON. Globe 8vo. [*In the press.*

Geikie.—Works by ARCHIBALD GEIKIE, F.R.S., Director-General of the Geological Survey of the United Kingdom, and Director of the Museum of Practical Geology, Jermyn Street, London; formerly Murchison Professor of Geology and Mineralogy in the University of Edinburgh.

Geikie.—THE TEACHING OF GEOGRAPHY. A Practical Handbook for the use of Teachers. Crown 8vo. 2*s*. Being Volume I. of a New Geographical Series Edited by ARCHIBALD GEIKIE, F.R.S.

*** The aim of this volume is to advocate the claims of geography as an educational discipline of a high order, and to show how these claims may be practically recognised by teachers.

GEOGRAPHY OF THE BRITISH ISLES. 18mo. 1*s*.

Green. — A SHORT GEOGRAPHY OF THE BRITISH ISLANDS. By JOHN RICHARD GREEN and ALICE STOPFORD GREEN. With Maps. Fcap. 8vo. 3*s*. 6*d*.

Grove.—A PRIMER OF GEOGRAPHY. By Sir GEORGE GROVE, D.C.L. With Illustrations. 18mo. 1*s*. (*Science Primers.*)

Kiepert.—A MANUAL OF ANCIENT GEOGRAPHY. From the German of Dr. H. KIEPERT. Crown 8vo. 5*s*.

Macmillan's Geographical Series. Edited by ARCHIBALD GEIKIE, F.R.S., Director-General of the Geological Survey of the United Kingdom.

The following List of Volumes is contemplated :—

THE TEACHING OF GEOGRAPHY. A Practical Handbook for the use of Teachers. By ARCHIBALD GEIKIE, F.R.S., Director-General of the Geological Survey of the United Kingdom, and Director of the Museum of Practical Geology, Jermyn Street, London ; formerly Murchison Professor of Geology and Mineralogy in the University of Edinburgh. Crown 8vo. 2*s*.

*** The aim of this volume is to advocate the claims of geography as an educational discipline of a high order, and to show how these claims may be practically recognized by teachers.

GEOGRAPHY OF THE BRITISH ISLES. By ARCHIBALD GEIKIE, F.R.S. 18mo. 1*s*.

THE ELEMENTARY SCHOOL ATLAS. With 24 Maps in Colours, specially designed to illustrate all Elementary Text-books of Geography. By JOHN BARTHOLOMEW. F.R.G.S. 4to. 1*s*.

AN ELEMENTARY CLASS-BOOK OF GENERAL GEOGRAPHY. By HUGH ROBERT MILL, D.Sc. Edin. With Illustrations. Crown 8vo. 3*s*. 6*d*.

GEOGRAPHY OF THE BRITISH COLONIES. By GEORGE M. DAWSON and ALEXANDER SUTHERLAND. [*In preparation.*

GEOGRAPHY OF EUROPE. By JAMES SIME, M.A. With Illustrations. Globe 8vo. [*In the press.*

GEOGRAPHY OF NORTH AMERICA. By Professor N. S. SHALER. [*In preparation.*

MAPS AND MAP-MAKING. By WILLIAM A. ELDERTON. [*In the press.*

GEOGRAPHY OF ASIA.

GEOGRAPHY OF AFRICA.

GEOGRAPHY OF THE OCEANS AND OCEANIC ISLANDS.

Macmillan's Geographical Series—*continued.*

ADVANCED CLASS-BOOK OF THE GEOGRAPHY OF BRITAIN.

GEOGRAPHY OF AUSTRALIA AND NEW ZEALAND.

GEOGRAPHY OF AMERICA.

GEOGRAPHY OF INDIA. By H. F. BLANFORD, F.G.S.

ADVANCED CLASS-BOOK OF THE GEOGRAPHY OF EUROPE.

Mill.—AN ELEMENTARY CLASS-BOOK OF GENERAL GEOGRAPHY. By HUGH ROBERT MILL, D.Sc., F.R.S.E., Lecturer on Physiography and on Commercial Geography in the Heriot-Watt College, Edinburgh. With Illustrations. Crown 8vo. 3*s*. 6*d*.

Sime.—A GEOGRAPHY OF EUROPE. By JAMES SIME, M.A. With Illustrations. Globe 8vo. [*In the press.*

Strachey.—LECTURES ON GEOGRAPHY. By General R. STRACHEY, R.E., C.S.I., President of the Royal Geographical Society. Crown 8vo. 4*s*. 6*d*.

HISTORY.

Arnold (T.).—THE SECOND PUNIC WAR. Being Chapters from THE HISTORY OF ROME. By THOMAS ARNOLD, D.D. Edited, with Notes, by W. T. ARNOLD, M.A. With 8 Maps. Crown 8vo. 8*s*. 6*d*.

Arnold (W. T.).—THE ROMAN SYSTEM OF PROVINCIAL ADMINISTRATION TO THE ACCESSION OF CONSTANTINE THE GREAT. By W. T. ARNOLD, M.A. Crown 8vo. 6*s*.

"Ought to prove a valuable handbook to the student of Roman history."—GUARDIAN.

Beesly.—STORIES FROM THE HISTORY OF ROME. By Mrs. BEESLY. Fcap. 8vo. 2*s*. 6*d*.

Bryce.—Works by JAMES BRYCE, D.C.L., Fellow of Oriel College, and Regius Professor of Civil Law in the University of Oxford.

THE HOLY ROMAN EMPIRE. Ninth Edition. Crown 8vo. 7*s*. 6*d*.

*** Also a *Library Edition.* Demy 8vo. 14*s*.

THE AMERICAN COMMONWEALTH. New and Cheaper Edition. 2 vols. Extra crown 8vo. 25*s*. Part I.—The National Government. Part II.—The State Governments. Part III.—The Party System. Part IV.—Public Opinion. Part V.—Illustrations and Reflections. Part VI.—Social Institutions.

Buckley.—A HISTORY OF ENGLAND FOR BEGINNERS. By ARABELLA B. BUCKLEY. Author of "A Short History of Natural Science," &c. With Coloured Maps, Chronological and Genealogical Tables. Globe 8vo. 3*s*.

Bury.—A HISTORY OF THE LATER ROMAN EMPIRE FROM ARCADIUS TO IRENE, A.D. 395-800. By JOHN B. BURY, Fellow of Trinity Coll., Dublin. 2 vols. 8vo. 32*s.*

Du Pré.—OUTLINES OF ENGLISH HISTORY. By ARTHUR M. D. DU PRÉ, Assistant-Master in the Hull and East Riding College. Globe 8vo. [*In the press.*

Eggleston.—THE HOUSEHOLD HISTORY OF THE UNITED STATES AND ITS PEOPLE. By EDWARD EGGLESTON. With Illustrations. 4to. 12*s.*

English Statesmen, Twelve.—A Series of Short Biographies, not designed to be a complete roll of Famous Statesmen, but to present in historic order the lives and work of those leading actors in our affairs who by their direct influence have left an abiding mark on the policy, the institutions, and the position of Great Britain among States.

The following list of subjects is the result of careful selection. The great movements of national history are made to follow one another in a connected course, and the series is intended to form a continuous narrative of English freedom, order, and power. The volumes as follow, Crown 8vo, 2*s.* 6*d.* each, are ready or in preparation:—

WILLIAM THE CONQUEROR. By EDWARD A. FREEMAN, D.C.L., LL.D. [*Ready.*
HENRY II. By Mrs. J. R. GREEN. [*Ready.*
EDWARD I. By F. YORK POWELL.
HENRY VII. By JAMES GAIRDNER. [*Ready.*
CARDINAL WOLSEY. By Professor M. CREIGHTON. [*Ready.*
ELIZABETH. By E. S. BEESLEY.
OLIVER CROMWELL. By FREDERIC HARRISON. [*Ready.*
WILLIAM III. By H. D. TRAILL. [*Ready.*
WALPOLE. By JOHN MORLEY. [*Ready.*
CHATHAM. By JOHN MORLEY.
PITT. By JOHN MORLEY.
PEEL. By J. R. THURSFIELD. [*In the press.*

Fiske.—Works by JOHN FISKE, formerly Lecturer on Philosophy at Harvard University, author of "Outlines of Cosmic Philosophy, based on the Doctrine of Evolution, with Criticisms on the Positive Philosophy," "Darwinism, and other Essays," "American Political Ideas viewed from the Standpoint of Universal History."

THE CRITICAL PERIOD IN AMERICAN HISTORY, 1783—1789. Extra crown 8vo. 10*s.* 6*d.*

THE BEGINNINGS OF NEW ENGLAND; or, The Puritan Theocracy in its Relations to Civil and Religious Liberty. Crown 8vo. 7*s.* 6*d.*

Freeman.—Works by EDWARD A. FREEMAN, D.C.L., LL.D., Regius Professor of Modern History in the University of Oxford, &c.

OLD ENGLISH HISTORY. With Five Coloured Maps. New Edition. Extra fcap. 8vo. 6*s.*

Freeman.—A SCHOOL HISTORY OF ROME. Crown 8vo. [*In preparation.*

METHODS OF HISTORICAL STUDY. A Course of Lectures. 8vo. 10*s*. 6*d*.

THE CHIEF PERIODS OF EUROPEAN HISTORY. Six Lectures read in the University of Oxford in Trinity Term, 1885. With an Essay on Greek Cities under Roman Rule. 8vo. 10*s*. 6*d*.

HISTORICAL ESSAYS. First Series. Fourth Edition. 8vo. 10*s*. 6*d*.

Contents:—The Mythical and Romantic Elements in Early English History—The Continuity of English History—The Relations between the Crown of England and Scotland—St. Thomas of Canterbury and his Biographers, &c.

HISTORICAL ESSAYS. Second Series. Third Edition, with additional Essays. 8vo. 10*s*. 6*d*.

Contents:—Ancient Greece and Mediæval Italy—Mr. Gladstone's Homer and the Homeric Ages—The Historians of Athens—The Athenian Democracy—Alexander the Great—Greece during the Macedonian Period—Mommsen's History of Rome—Lucius Cornelius Sulla—The Flavian Cæsars, &c., &c.

HISTORICAL ESSAYS. Third Series. 8vo. 12*s*.

Contents:—First Impressions of Rome—The Illyrian Emperors and their Land—Augusta Treverorum—The Goths at Ravenna—Race and Language—The Byzantine Empire—First Impressions of Athens—Mediæval and Modern Greece—The Southern Slaves—Sicilian Cycles—The Normans at Palermo.

THE GROWTH OF THE ENGLISH CONSTITUTION FROM THE EARLIEST TIMES. Fourth Edition. Crown 8vo. 5*s*.

GENERAL SKETCH OF EUROPEAN HISTORY. New Edition. Enlarged, with Maps, &c. 18mo. 3*s*. 6*d*. (Vol. I. of Historical Course for Schools.)

EUROPE. 18mo. 1*s*. (*History Primers.*)

Fyffe.—A SCHOOL HISTORY OF GREECE. By C. A. FYFFE, M.A. Crown 8vo. [*In preparation.*

Green.—Works by JOHN RICHARD GREEN, M.A., LL.D., late Honorary Fellow of Jesus College, Oxford.

A SHORT HISTORY OF THE ENGLISH PEOPLE. New and Thoroughly Revised Edition. With Coloured Maps, Genealogical Tables, and Chronological Annals. Crown 8vo. 8*s*. 6*d*. 148th Thousand.

Also the same in Four Parts. Parts I., II., and III. ready; Part IV. shortly. With the corresponding portion of Mr. Tait's "Analysis." Crown 8vo. 3*s*. each. Part I. 607—1265. Part II. 1204—1553. Part III. 1540—1689. Part IV. 1660—1873.

HISTORY OF THE ENGLISH PEOPLE. In four vols. 8vo.

Vol. I.—EARLY ENGLAND, 449—1071—Foreign Kings 1071-1214—The Charter, 1214-1201—The Parliament, 1307-1461. With eight Coloured Maps. 8vo. 16*s*.

Vol. II.—THE MONARCHY, 1461-1540—The Reformation, 1540-1603. 8vo. 16*s*.

Vol. III.—PURITAN ENGLAND, 1603-1660—The Revolution, 1660-1688. With four Maps. 8vo. 16*s*.

Vol. IV.—The Revolution, 1688-1760—Modern England, 1760-1815. With Maps and Index. 8vo. 16*s*.

Green.—THE MAKING OF ENGLAND. With Maps. 8vo. 16*s*.

THE CONQUEST OF ENGLAND. With Maps and Portrait. 8vo. 18*s*.

ANALYSIS OF ENGLISH HISTORY, based on Green's "Short History of the English People." By C. W. A. TAIT, M.A., Assistant-Master, Clifton College. Crown 8vo. 3*s*. 6*d*.

READINGS FROM ENGLISH HISTORY. Selected and Edited by JOHN RICHARD GREEN. Three Parts. Globe 8vo. 1*s*. 6*d*. each. I. Hengist to Cressy. II. Cressy to Cromwell. III. Cromwell to Balaklava.

Guest.—LECTURES ON THE HISTORY OF ENGLAND. By M. J. GUEST. With Maps. Crown 8vo. 6*s*.

Historical Course for Schools—Edited by EDWARD A. FREEMAN, D.C.L., LL.D., late Fellow of Trinity College, Oxford, Regius Professor of Modern History in the University of Oxford.

I.—GENERAL SKETCH OF EUROPEAN HISTORY. By EDWARD A. FREEMAN, D.C.L. New Edition, revised and enlarged, with Chronological Table, Maps, and Index. 18mo. 3*s*. 6*d*.

II.—HISTORY OF ENGLAND. By EDITH THOMPSON. New Ed., revised and enlarged, with Coloured Maps. 18mo. 2*s*. 6*d*.

III.—HISTORY OF SCOTLAND. By MARGARET MACARTHUR. New Edition. 18mo. 2*s*.

IV.—HISTORY OF ITALY. By the Rev. W. HUNT, M.A. New Edition, with Coloured Maps. 18mo. 3*s*. 6*d*.

V.—HISTORY OF GERMANY. By J. SIME, M.A. New Edition Revised. 18mo. 3*s*.

VI.—HISTORY OF AMERICA. By JOHN A. DOYLE. With Maps. 18mo. 4*s*. 6*d*.

VII.—EUROPEAN COLONIES. By E. J. PAYNE, M.A. With Maps. 18mo. 4*s*. 6*d*.

VIII.—FRANCE. By CHARLOTTE M. YONGE. With Maps. 18mo. 3*s*. 6*d*.

GREECE. By EDWARD A. FREEMAN, D.C.L. [*In preparation.*

ROME. By EDWARD A. FREEMAN, D.C.L. [*In preparation.*

History Primers—Edited by JOHN RICHARD GREEN, M.A., LL.D., Author of "A Short History of the English People."

ROME. By the Rev. M. CREIGHTON, M.A., Dixie Professor of Ecclesiastical History in the University of Cambridge. With Eleven Maps. 18mo. 1*s*.

GREECE. By C. A. FYFFE, M.A., Fellow and late Tutor of University College, Oxford. With Five Maps. 18mo. 1*s*.

EUROPEAN HISTORY. By E. A. FREEMAN, D.C.L., LL.D. With Maps. 18mo. 1*s*.

GREEK ANTIQUITIES. By the Rev. J. P. MAHAFFY, M.A. Illustrated. 18mo. 1*s*.

CLASSICAL GEOGRAPHY. By H. F. TOZER, M.A. 18mo. 1*s*.

GEOGRAPHY. By Sir G. GROVE, D.C.L. Maps. 18mo. 1*s*.

ROMAN ANTIQUITIES. By Professor WILKINS, Litt.D., LL.D. Illustrated. 18mo. 1*s*.

FRANCE. By CHARLOTTE M. YONGE. 18mo. 1*s*.

Hole.—A GENEALOGICAL STEMMA OF THE KINGS OF ENGLAND AND FRANCE. By the Rev. C. HOLE. On Sheet. 1*s.*

Jennings—CHRONOLOGICAL TABLES. A synchronistic arrangement of the events of Ancient History (with an Index). By the Rev. ARTHUR C. JENNINGS, Rector of King's Stanley, Gloucestershire, Author of "A Commentary on the Psalms," "Ecclesia Anglicana," "Manual of Church History," &c. 8vo. 5*s.*

Labberton.—NEW HISTORICAL ATLAS AND GENERAL HISTORY. By R. H. LABBERTON, Litt.Hum.D. 4to. New Edition Revised and Enlarged. 15*s.*

Lethbridge.—A SHORT MANUAL OF THE HISTORY OF INDIA. With an Account of INDIA AS IT IS. The Soil, Climate, and Productions; the People, their Races, Religions, Public Works, and Industries; the Civil Services, and System of Administration. By Sir ROPER LETHBRIDGE, M.A., C.I.E., late Scholar of Exeter College, Oxford, formerly Principal of Kishnaghur College, Bengal, Fellow and sometime Examiner of the Calcutta University. With Maps. Crown 8vo. 5*s.*

Mahaffy.—GREEK LIFE AND THOUGHT FROM THE AGE OF ALEXANDER TO THE ROMAN CONQUEST. By the Rev. J. P. MAHAFFY, M.A., D.D., Fellow of Trinity College, Dublin, Author of "Social Life in Greece from Homer to Menander," "Rambles and Studies in Greece," &c. Crown 8vo. 12*s.* 6*d.*

Marriott.—THE MAKERS OF MODERN ITALY: MAZZINI, CAVOUR, GARIBALDI. Three Lectures delivered at Oxford. By J. A. R. MARRIOTT, M.A., New College and Worcester College, Oxford, Lecturer in Modern History and Political Economy. Crown 8vo. 1*s.* 6*d.*

Michelet.—A SUMMARY OF MODERN HISTORY. Translated from the French of M. MICHELET, and continued to the Present Time, by M. C. M. SIMPSON Globe 8vo. 4*s.* 6*d.*

Norgate.—ENGLAND UNDER THE ANGEVIN KINGS. By KATE NORGATE. With Maps and Plans. 2 vols. 8vo. 32*s.*

Otté.—SCANDINAVIAN HISTORY. By E. C. OTTÉ. With Maps. Globe 8vo. 6*s.*

Seeley.—Works by J. R. SEELEY, M.A., Regius Professor of Modern History in the University of Cambridge.

THE EXPANSION OF ENGLAND. Crown 8vo. 4*s.* 6*d.*

OUR COLONIAL EXPANSION. Extracts from the above. Crown 8vo Sewed. 1*s.*

Tait.—ANALYSIS OF ENGLISH HISTORY, based on Green's "Short History of the English People." By C. W. A. TAIT, M.A., Assistant-Master, Clifton College. Crown 8vo. 3*s.* 6*d.*

Wheeler.—Works by J. TALBOYS WHEELER.

A SHORT HISTORY OF INDIA AND OF THE FRONTIER STATES OF AFGHANISTAN, NEPAUL, AND BURMA. With Maps. Crown 8vo. 12*s.*

COLLEGE HISTORY OF INDIA, ASIATIC AND EUROPEAN. With Maps. Crown 8vo. 3*s.* 6*d.*

Yonge (Charlotte M.). — CAMEOS FROM ENGLISH HISTORY. By CHARLOTTE M. YONGE, Author of "The Heir of Redclyffe." Extra fcap. 8vo. New Edition. 5*s.* each. (1) FROM ROLLO TO EDWARD II. (2) THE WARS IN FRANCE. (3) THE WARS OF THE ROSES. (4) REFORMATION TIMES. (5) ENGLAND AND SPAIN. (6) FORTY YEARS OF STUART RULE (1603—1643). (7) THE REBELLION AND RESTORATION (1642—1678.) [*Shortly.*

EUROPEAN HISTORY. Narrated in a Series of Historical Selections from the Best Authorities. Edited and arranged by E. M. SEWELL and C. M. YONGE. First Series, 1003—1154. New Edition. Crown 8vo. 6*s.* Second Series, 1088—1228. New Edition. Crown 8vo. 6*s.*

THE VICTORIAN HALF CENTURY—A JUBILEE BOOK. With a New Portrait of the Queen. Crown 8vo., paper covers, 1*s.* Cloth, 1*s.* 6*d.*

LAW.

Anglo-Saxon Law, Essays on.—Contents: Anglo-Saxon Law Courts, Land and Family Law, and Legal Procedure. Medium 8vo. 18*s.*

Ball.—THE STUDENT'S GUIDE TO THE BAR. By WALTER W. R. BALL, M.A., of the Inner Temple, Barrister-at-Law; Fellow and Assistant Tutor of Trinity College, Cambridge, and Fellow of University College, London. Fourth Edition Revised. Crown 8vo. 2*s.* 6*d.*

Bigelow. — HISTORY OF PROCEDURE IN ENGLAND FROM THE NORMAN CONQUEST. The Norman Period, 1066-1204. By MELVILLE MADISON BIGELOW, Ph.D., Harvard University. Demy 8vo. 16*s.*

Bryce.—THE AMERICAN COMMONWEALTH. By JAMES BRYCE, M.P., D.C.L., Regius Professor of Civil Law in the University of Oxford, Author of "The Holy Roman Empire." Two Volumes. Extra Crown 8vo. 25*s.* Part I. THE NATIONAL GOVERNMENT. Part II. THE STATE GOVERNMENTS. Part III. THE PARTY SYSTEM. Part IV. PUBLIC OPINION. Part V. ILLUSTRATIONS AND REFLECTIONS. Part VI. SOCIAL INSTITUTIONS.

Buckland.—OUR NATIONAL INSTITUTIONS. A Short Sketch for Schools. By ANNA BUCKLAND. New Edition. With Glossary. 18mo. 1*s.*

Dicey.—INTRODUCTION TO THE STUDY OF THE LAW OF THE CONSTITUTION. By A. V. DICEY, B.C.L., of the Inner Temple; Barrister-at-Law; Vinerian Professor of English Law in the University of Oxford; Fellow of All Souls' College; Hon. LL.D. Glasgow. Third Edition. Demy 8vo. 12*s.* 6*d.*

English Citizen, The.—A Series of Short Books on his Rights and Responsibilities. Edited by HENRY CRAIK, C.B. M.A. (Oxon.), LL.D. (Glasgow). Crown 8vo. 3*s.* 6*d.* each.

CENTRAL GOVERNMENT. By H. D. TRAILL, D.C.L., late Fellow of St. John's College, Oxford.

THE ELECTORATE AND THE LEGISLATURE. By SPENCER WALPOLE, Author of "The History of England from 1815."

THE POOR LAW. By the Rev. T. W. FOWLE, M.A.

THE NATIONAL BUDGET; THE NATIONAL DEBT; TAXES AND RATES. By A. J. WILSON.

THE STATE IN RELATION TO LABOUR. By W. STANLEY JEVONS, LL.D., F.R.S.

THE STATE AND THE CHURCH. By the Hon. ARTHUR ELLIOT, M.P.

FOREIGN RELATIONS. By SPENCER WALPOLE.

THE STATE IN ITS RELATION TO TRADE. By Sir T. H. FARRER, Bart., Permanent Secretary to the Board of Trade.

LOCAL GOVERNMENT. By M. D. CHALMERS, M.A.

THE STATE IN ITS RELATION TO EDUCATION. By HENRY CRAIK, C.B., M.A., LL.D.

THE LAND LAWS. By Sir F. POLLOCK, Bart. late Fellow of Trinity College, Cambridge, Professor of Jurisprudence in the University of Oxford, &c. Second Edition.

COLONIES AND DEPENDENCIES. Part I. INDIA. By J. S. COTTON, M.A. II. THE COLONIES. By E. J. PAYNE, M.A.

JUSTICE AND POLICE. By F. W. MAITLAND.

THE PUNISHMENT AND PREVENTION OF CRIME. By Colonel Sir EDMUND DU CANE, K.C.B., R.E., Chairman of Commissioners of Prisons, Chairman of Directors of Prisons, Inspector-General of Military Prisons, Surveyor-General of Prisons.

Holmes.—THE COMMON LAW. By O. W. HOLMES, Jun. Demy 8vo. 12*s.*

Maitland.—PLEAS OF THE CROWN FOR THE COUNTY OF GLOUCESTER BEFORE THE ABBOT OF READING AND HIS FELLOW JUSTICES ITINERANT, IN THE FIFTH YEAR OF THE REIGN OF KING HENRY THE THIRD, AND THE YEAR OF GRACE 1221. Edited by F. W. MAITLAND. 8vo. 7*s.* 6*d.*

(See also under *English Citizen Series.*)

Paterson.—WORKS BY JAMES PATERSON, Barrister-at-Law.

COMMENTARIES ON THE LIBERTY OF THE SUBJECT, AND THE LAWS OF ENGLAND RELATING TO THE SECURITY OF THE PERSON. Cheaper Issue. Two Vols. Crown 8vo. 21*s.*

THE LIBERTY OF THE PRESS, SPEECH, AND PUBLIC WORSHIP. Being Commentaries on the Liberty of the Subject and the Laws of England. Crown 8vo. 12*s.*

Phillimore.—PRIVATE LAW AMONG THE ROMANS. From the Pandects. By JOHN GEORGE PHILLIMORE, Q.C. 8vo. 16*s.*

Pollock.—ESSAYS IN JURISPRUDENCE AND ETHICS. By Sir FREDERICK POLLOCK, Bart., Barrister-at-Law, M.A., Hon. LL.D. Edin.; Corpus Christi Professor of Jurisprudence in the University of Oxford; late Fellow of Trinity College, Cambridge. 8vo. 10*s.* 6*d.*

(See also under *English Citizen Series.*)

Richey.—THE IRISH LAND LAWS. By ALEXANDER G. RICHEY, Q.C., LL.D., Deputy Regius Professor of Feudal English Law in the University of Dublin. Crown 8vo. 3*s.* 6*d.*

Stephen.—Works by Sir J. FITZJAMES STEPHEN, Q.C., K.C.S.I., a Judge of the High Court of Justice, Queen's Bench Division.

A DIGEST OF THE LAW OF EVIDENCE. Fifth Edition, revised and enlarged. Crown 8vo. 6*s.*

A DIGEST OF THE CRIMINAL LAW: CRIMES AND PUNISHMENTS. Fourth Edition, revised. 8vo. 16*s.*

A DIGEST OF THE LAW OF CRIMINAL PROCEDURE IN INDICTABLE OFFENCES. By Sir JAMES F. STEPHEN, K.C.S.I., &c., and HERBERT STEPHEN, LL.M., of the Inner Temple, Barrister-at-Law. 8vo. 12*s.* 6*d.*

A HISTORY OF THE CRIMINAL LAW OF ENGLAND. New and Revised Edition. Three Vols. 8vo. 48*s.*

GENERAL VIEW OF THE CRIMINAL LAW OF ENGLAND. Second Edition. 8vo. The first edition of this work was published in 1863. The new edition will be substantially a new work, intended as a text-book on the Criminal Law for University and other Students, adapted to the present day. [*In the press.*

MODERN LANGUAGES AND LITERATURE.

(1) English, (2) French, (3) German, (4) Modern Greek, (5) Italian, (6) Spanish.

ENGLISH.

Abbott.—A SHAKESPEARIAN GRAMMAR. An attempt to illustrate some of the Differences between Elizabethan and Modern English. By the Rev. E. A. ABBOTT, D.D., formerly Head Master of the City of London School. New Edition. Extra fcap. 8vo. 6*s.*

Bacon.—ESSAYS. Edited by F. G. SELBY, M.A., Professor of Logic and Moral Philosophy, Deccan College, Poona. Globe 8vo. 3*s.* 6*d.*

Burke.—REFLECTIONS ON THE FRENCH REVOLUTION. Edited by F. G. SELBY, M.A. Globe 8vo. [*In the press.*

Brooke.—PRIMER OF ENGLISH LITERATURE. By the Rev. STOPFORD A. BROOKE, M.A. 18mo. 1*s.* (*Literature Primers.*)

Butler.—HUDIBRAS. Edited, with Introduction and Notes, by ALFRED MILNES, M.A. Lon., late Student of Lincoln College, Oxford. Extra fcap 8vo. Part I. 3*s.* 6*d.* Parts II. and III. 4*s.* 6*d.*

Campbell.—SELECTIONS. Edited by CECIL M. BARROW M.A., Principal and Professor of English and Classics, Doveton College, Madras. [*In preparation.*

Cowper's TASK: AN EPISTLE TO JOSEPH HILL, ESQ.; TIROCINIUM, or a Review of the Schools; and THE HISTORY OF JOHN GILPIN. Edited, with Notes, by WILLIAM BENHAM, B.D. Globe 8vo. 1*s.* (*Globe Readings from Standard Authors.*)

THE TASK. Edited by F. J. ROWE, M.A., and W. T. WEBB, M.A., Professors of English Literature, Presidency College, Calcutta. [*In preparation.*

Dowden.—SHAKESPEARE. By Professor DOWDEN. 18mo. 1*s.* (*Literature Primers.*)

Dryden.—SELECT PROSE WORKS. Edited, with Introduction and Notes, by Professor C. D. YONGE. Fcap. 8vo. 2*s.* 6*d.*

Gladstone.—SPELLING REFORM FROM AN EDUCATIONAL POINT OF VIEW. By J. H. GLADSTONE, Ph.D., F.R.S., Member of the School Board for London. New Edition. Crown 8vo. 1*s.* 6*d.*

Globe Readers. For Standards I.—VI. Edited by A. F. MURISON, sometime English Master at the Aberdeen Grammar School. With Illustrations. Globe 8vo.

Primer I. (48 pp.) 3*d.*	Book III. (232 pp.) 1*s.* 3*d.*
Primer II. (48 pp.) 3*d.*	Book IV. (328 pp.) 1*s.* 9*d.*
Book I. (96 pp.) 6*d.*	Book V. (416 pp.) 2*s.*
Book II. (136 pp.) 9*d.*	Book VI. (448 pp.) 2*s.* 6*d.*

"Among the numerous sets of readers before the public the present series is honourably distinguished by the marked superiority of its materials and the careful ability with which they have been adapted to the growing capacity of the pupils. The plan of the two primers is excellent for facilitating the child's first attempts to read. In the first three following books there is abundance of entertaining reading. . . . . Better food for young minds could hardly be found."—THE ATHENÆUM.

***The Shorter Globe Readers.**—With Illustrations. Globe 8vo.

Primer I. (48 pp.) 3*d.*	Standard III. (178 pp.) 1*s.*
Primer II. (48 pp.) 3*d.*	Standard IV. (182 pp.) 1*s.*
Standard I. (92 pp.) 6*d.*	Standard V. (216 pp.) 1*s.* 3*d.*
Standard II. (124 pp.) 9*d.*	Standard VI. (228 pp.) 1*s.* 6*d.*

* This Series has been abridged from "The Globe Readers" to meet the demand for smaller reading books.

Goldsmith.—THE TRAVELLER, or a Prospect of Society; and THE DESERTED VILLAGE. By OLIVER GOLDSMITH. With Notes, Philological and Explanatory, by J. W. HALES, M.A. Crown 8vo. 6*d.*

THE VICAR OF WAKEFIELD. With a Memoir of Goldsmith by Professor MASSON. Globe 8vo. 1*s.* (*Globe Readings from Standard Authors.*)

SELECT ESSAYS. Edited, with Introduction and Notes, by Professor C. D. YONGE. Fcap. 8vo. 2*s.* 6*d.*

THE TRAVELLER AND THE DESERTED VILLAGE. Edited by ARTHUR BARRETT, B.A., Professor of English Literature, Elphinstone College, Bombay. Globe 8vo. 1*s.* 6*d.*

THE VICAR OF WAKEFIELD. Edited by HAROLD LITTLEDALE, B.A., Professor of History and English Literature, Baroda College. Globe 8vo. [*In preparation.*

Gosse.—A HISTORY OF EIGHTEENTH CENTURY LITERATURE (1660-1780). By EDMUND GOSSE, M.A. Crown 8vo. 7*s.* 6*d.*

Gray.—POEMS. By JOHN BRADSHAW, LL.D. [*In preparation.*

Hales.—LONGER ENGLISH POEMS, with Notes, Philological and Explanatory, and an Introduction on the Teaching of English, Chiefly for Use in Schools. Edited by J. W. HALES, M.A., Professor of English Literature at King's College, London. New Edition. Extra fcap. 8vo. 4*s.* 6*d.*

Helps.—ESSAYS WRITTEN IN THE INTERVALS OF BUSINESS. Edited by F. J. ROWE, M.A., and W. T. WEBB, M.A., Professors of English Literature, Presidency College, Calcutta. Globe 8vo. 2*s.* 6*d.*

Johnson's LIVES OF THE POETS. The Six Chief Lives (Milton, Dryden, Swift, Addison, Pope, Gray), with Macaulay's "Life of Johnson." Edited with Preface and Notes by MATTHEW ARNOLD. New and cheaper edition. Crown 8vo. 4*s.* 6*d.*

Lamb (Charles).—TALES FROM SHAKESPEARE. Edited, with Preface, by the Rev. CANON AINGER, M.A. Globe 8vo. 2*s.* (*Globe Readings from Standard Authors.*)

Literature Primers—Edited by JOHN RICHARD GREEN, M.A., LL.D., Author of "A Short History of the English People."

ENGLISH COMPOSITION. By Professor NICHOL. 18mo. 1*s.*

EXERCISES ON ENGLISH COMPOSITION. By the same. [*In the press.*

ENGLISH GRAMMAR. By the Rev. R. MORRIS, LL.D., sometime President of the Philological Society. 18mo. 1*s.*

ENGLISH GRAMMAR EXERCISES. By R. MORRIS, LL.D., and H. C. BOWEN, M.A. 18mo. 1*s.*

EXERCISES ON MORRIS'S PRIMER OF ENGLISH GRAMMAR. By JOHN WETHERELL, of the Middle School, Liverpool College. 18mo. 1*s.*

ENGLISH LITERATURE. By STOPFORD BROOKE, M.A. New Edition. 18mo. 1*s.*

SHAKSPERE. By Professor DOWDEN. 18mo. 1*s.*

THE CHILDREN'S TREASURY OF LYRICAL POETRY. Selected and arranged with Notes by FRANCIS TURNER PALGRAVE. In Two Parts. 18mo. 1*s.* each.

PHILOLOGY. By J. PEILE, M.A. 18mo. 1*s.*

ROMAN LITERATURE. By Professor A. S. WILKINS, Litt.D., LL.D. [*Just ready.*

A History of English Literature in Four Volumes. Crown 8vo.

EARLY ENGLISH LITERATURE. By STOPFORD BROOKE, M.A. [*In preparation.*

ELIZABETHAN LITERATURE. 1560—1665. By GEORGE SAINTSBURY. 7*s.* 6*d.*

EIGHTEENTH CENTURY LITERATURE. 1660—1780. By EDMUND GOSSE. M.A. 7*s.* 6*d.*

THE MODERN PERIOD. By PROFESSOR E. DOWDEN. [*In prep.*

Macmillan's Reading Books.—Adapted to the English and Scotch Codes. Bound in Cloth.

PRIMER. 18mo. (48 pp.) 2*d.*

BOOK I. for Standard I. 18mo. (96 pp.) 4*d.*

BOOK II. for Standard II. 18mo. (144 pp.) 5*d.*

BOOK III. for Standard III. 18mo. (160 pp.) 6*d.*

BOOK IV. for Standard IV. 18mo. (176 pp.) 8*d.*

BOOK V. for Standard V. 18mo. (380 pp.) 1*s.*

BOOK VI. for Standard VI. Cr. 8vo. (430 pp.) 2*s.*

Book VI. is fitted for higher Classes, and as an Introduction to English Literature.

Macmillan's Copy-Books—

Published in two sizes, viz. :—

1. Large Post 4to. Price 4*d.* each.
2. Post Oblong. Price 2*d.* each.

1. INITIATORY EXERCISES AND SHORT LETTERS.
2. WORDS CONSISTING OF SHORT LETTERS
*3. LONG LETTERS. With Words containing Long Letters—Figures.
*4. WORDS CONTAINING LONG LETTERS.
4a. PRACTISING AND REVISING COPY-BOOK. For Nos. 1 to 4.
*5. CAPITALS AND SHORT HALF-TEXT. Words beginning with a Capital.
*6. HALF-TEXT WORDS beginning with Capitals—Figures.
*7. SMALL-HAND AND HALF-TEXT. With Capitals and Figures.
*8. SMALL-HAND AND HALF-TEXT. With Capitals and Figures.
8a. PRACTISING AND REVISING COPY-BOOK. For Nos. 5 to 8.
*9. SMALL-HAND SINGLE HEADLINES—Figures.
10. SMALL-HAND SINGLE HEADLINES - Figures.
11. SMALL-HAND DOUBLE HEADLINES—Figures.
12. COMMERCIAL AND ARITHMETICAL EXAMPLES, &c.
12a. PRACTISING AND REVISING COPY-BOOK. For Nos. 8 to 12.

* *These numbers may be had with Goodman's Patent Sliding Copies.* Large Post 4to. Price 6*d.* each.

Martin.—THE POET'S HOUR: Poetry selected and arranged for Children. By FRANCES MARTIN. New Edition. 18mo. 2*s.* 6*d.*

SPRING-TIME WITH THE POETS: Poetry selected by FRANCES MARTIN. New Edition. 18mo. 3*s* 6*d.*

Milton.—By STOPFORD BROOKE, M.A. Fcap. 8vo. 1*s.* 6*d.* (*Classical Writers Series.*)

Milton.—PARADISE LOST. Books I. and II. Edited, with Introduction and Notes, by MICHAEL MACMILLAN, B.A. Oxon, Professor of Logic and Moral Philosophy, Elphinstone College, Bombay. Globe 8vo. 2*s.* 6*d.*, or separately 1*s.* 6*d.* each.

L'ALLEGRO, IL PENSEROSO, LYCIDAS, ARCADES, SONNETS, &c. Edited by WILLIAM BELL, M.A., Prof. of Philosophy and Logic, Government College, Lahore. Globe 8vo. 2*s.*

COMUS. By the same Editor. 1*s.* 6*d.*

SAMSON AGONISTES. By H. M. PERCIVAL, M.A., Professor of English Literature, Presidency College, Calcutta. [*Just ready.*

Morley.—ON THE STUDY OF LITERATURE. The Annual Address to the Students of the London Society for the Extension of University Teaching. Delivered at the Mansion House, February 26, 1887. By JOHN MORLEY. Globe 8vo. Cloth. 1*s.* 6*d.*

Also a Popular Edition in Pamphlet form for Distribution, price 2*d.*

APHORISMS. By the same. Being an Address delivered before the Philosophical Society of Edinburgh, November 11, 1887. Globe 8vo. 1*s.* 6*d.*

Morris.—Works by the Rev. R. MORRIS, LL.D.

HISTORICAL OUTLINES OF ENGLISH ACCIDENCE, comprising Chapters on the History and Development of the Language, and on Word-formation. New Edition. Extra fcap. 8vo. 6*s.*

Morris.—ELEMENTARY LESSONS IN HISTORICAL ENGLISH GRAMMAR, containing Accidence and Word-formation. New Edition. 18mo. 2*s*. 6*d*.

PRIMER OF ENGLISH GRAMMAR. 18mo. 1*s*. (See also *Literature Primers*.)

Morris and Kellner.—HISTORICAL OUTLINES OF ENGLISH SYNTAX. By Rev. R. MORRIS and Dr. L. KELLNER. [*In preparation.*

Oliphant.—THE OLD AND MIDDLE ENGLISH. A New Edition of "THE SOURCES OF STANDARD ENGLISH," revised and greatly enlarged. By T. L. KINGTON OLIPHANT. Extra fcap. 8vo. 9*s*.

THE NEW ENGLISH. By the same Author. 2 vols. Cr. 8vo. 21*s*.

Palgrave.—THE CHILDREN'S TREASURY OF LYRICAL POETRY. Selected and arranged, with Notes, by FRANCIS TURNER PALGRAVE. 18mo. 2*s*. 6*d*. Also in Two Parts. 1*s*. each.

Patmore.—THE CHILDREN'S GARLAND FROM THE BEST POETS. Selected and arranged by COVENTRY PATMORE. Globe 8vo. 2*s*. (*Globe Readings from Standard Authors.*)

Plutarch.—Being a Selection from the Lives which Illustrate Shakespeare. North's Translation. Edited, with Introductions, Notes, Index of Names, and Glossarial Index, by the Rev. W. W. SKEAT, M.A. Crown 8vo. 6*s*.

Ryland.—CHRONOLOGICAL OUTLINES OF ENGLISH LITERATURE, by F. RYLAND, M.A. Cr. 8vo. [*In the press.*

Saintsbury.—A HISTORY OF ELIZABETHAN LITERATURE. 1560-1665. By GEORGE SAINTSBURY. Cr. 8vo. 7*s*. 6*d*.

Scott.—LAY OF THE LAST MINSTREL, and THE LADY OF THE LAKE. Edited, with Introduction and Notes, by FRANCIS TURNER PALGRAVE. Globe 8vo. 1*s*. (*Globe Readings from Standard Authors.*)

MARMION; and THE LORD OF THE ISLES. By the same Editor. Globe 8vo. 1*s*. (*Globe Readings from Standard Authors.*)

MARMION. Edited, with Introduction and Notes, by MICHAEL MACMILLAN, B.A. Globe 8vo. 3*s*. 6*d*.

THE LADY OF THE LAKE. Edited by G. H. STUART, M.A., Professor of English Literature, Presidency College, Madras. Globe 8vo. [*In the press.*

THE LAY OF THE LAST MINSTREL. By the same Editor. Globe 8vo. Cantos I. to III. 1*s*. 6*d*.

ROKEBY. By MICHAEL MACMILLAN, B.A. Globe 8vo. 3*s*. 6*d*.

Shakespeare.—A SHAKESPEARIAN GRAMMAR. By Rev. E. A. ABBOTT, D.D., Head Master of the City of London School. Globe 8vo. 6*s*.

A SHAKESPEARE MANUAL. By F. G. FLEAY, M.A., late Head Master of Skipton Grammar School. Second Edition. Extra fcap. 8vo. 4*s*. 6*d*.

PRIMER OF SHAKESPEARE. By Professor DOWDEN. 18mo. 1*s*. (*Literature Primers.*)

Shakespeare—*continued.*

THE TEMPEST. Edited by K. DEIGHTON, late Principal of Agra College. Globe 8vo. 1*s.* 6*d.*

MUCH ADO ABOUT NOTHING. By the same Editor. Globe 8vo. 2*s.*

TWELFTH NIGHT. By the same Editor. Globe 8vo. 1*s.* 6*d.*

THE WINTER'S TALE. By the same Editor. Globe 8vo. 2*s.* 6*d.*

KING JOHN. By the same Editor. [*In preparation.*

HENRY V. By the same Editor. Globe 8vo. 2*s.*

RICHARD III. Edited by C. H. TAWNEY, M.A., Principal and Professor of English Literature, Elphinstone College, Calcutta. Globe 8vo. 2*s.* 6*d.*

JULIUS CÆSAR. By the same Editor. [*In preparation.*

MACBETH. By K. DEIGHTON. [*In preparation.*

OTHELLO. By the same Editor. Globe 8vo. 2*s.* 6*d.*

CYMBELINE. By the same Editor. Globe 8vo. 2*s.* 6*d.*

Sonnenschein and Meiklejohn. — THE ENGLISH METHOD OF TEACHING TO READ. By A. SONNENSCHEIN and J. M. D. MEIKLEJOHN, M.A. Fcap. 8vo.

COMPRISING :

THE NURSERY BOOK, containing all the Two-Letter Words in the Language. 1*d.* (Also in Large Type on Sheets for School Walls. 5*s.*)

THE FIRST COURSE, consisting of Short Vowels with Single Consonants. 6*d.*

THE SECOND COURSE, with Combinations and Bridges, consisting of Short Vowels with Double Consonants. 6*d.*

THE THIRD AND FOURTH COURSES, consisting of Long Vowels, and all the Double Vowels in the Language. 6*d.*

"These are admirable books, because they are constructed on a principle, and that the simplest principle on which it is possible to learn to read English."—SPECTATOR.

Southey.—LIFE OF NELSON. Edited by MICHAEL MACMILLAN, B.A. [*In the press.*

Taylor.—WORDS AND PLACES; or, Etymological Illustrations of History, Ethnology, and Geography. By the Rev. ISAAC TAYLOR, M.A., Litt. D., Hon. LL.D., Canon of York. Third and Cheaper Edition, revised and compressed. With Maps. Globe 8vo. 6*s.*

Tennyson.—The COLLECTED WORKS of LORD TENNYSON, Poet Laureate. An Edition for Schools. In Four Parts. Crown 8vo. 2*s.* 6*d.* each.

SELECTIONS FROM LORD TENNYSON'S POEMS. Edited with Notes for the Use of Schools. By the Rev. ALFRED AINGER, M.A., LL.D., Canon of Bristol. [*In preparation.*

SELECT POEMS OF LORD TENNYSON. With Introduction and Notes. By F. J. ROWE, M.A., and W. T. WEBB, M.A., Professors of English Literature, Presidency College, Calcutta. Globe 8vo. 3*s.* 6*d.*

This selection contains:—"Recollections of the Arabian Nights," "The Lady of Shalott," "Oenone," "The Lotos Eaters," "Ulysses," "Tithonus," "Morte d'Arthur," "Sir Galahad," "Dora," "Ode on the Death of the Duke of Wellington," and "The Revenge."

Thring.—THE ELEMENTS OF GRAMMAR TAUGHT IN ENGLISH. By EDWARD THRING, M.A., late Head Master of Uppingham. With Questions. Fourth Edition. 18mo. 2*s*.

Vaughan (C.M.).—WORDS FROM THE POETS. By C. M. VAUGHAN. New Edition. 18mo, cloth. 1*s*.

Ward.—THE ENGLISH POETS. Selections, with Critical Introductions by various Writers and a General Introduction by MATTHEW ARNOLD. Edited by T. H. WARD, M.A. 4 Vols. Vol. I. CHAUCER TO DONNE.—Vol. II. BEN JONSON TO DRYDEN.—Vol. III. ADDISON TO BLAKE.—Vol. IV. WORDSWORTH TO ROSSETTI. Crown 8vo. Each 7*s*. 6*d*.

Wetherell.—EXERCISES ON MORRIS'S PRIMER OF ENGLISH GRAMMAR. By JOHN WETHERELL, M.A. 18mo. 1*s*. (*Literature Primers.*)

Woods.—A FIRST POETRY BOOK. Compiled by M. A. WOODS, Head Mistress of the Clifton High School for Girls. Fcap. 8vo. 2*s*. 6*d*.

A SECOND POETRY BOOK. Compiled by the same. Fcap. 8vo. In two Parts, 2*s*. 6*d*. each.

A THIRD POETRY BOOK. Compiled by the same. Fcap. 8vo. 4*s*. 6*d*.

Wordsworth.—SELECTIONS. Edited by WILLIAM WORDSWORTH, B.A., Principal and Professor of History and Political Economy, Elphinstone College, Bombay. [*In preparation.*

Yonge (Charlotte M.).—THE ABRIDGED BOOK OF GOLDEN DEEDS. A Reading Book for Schools and general readers. By the Author of "The Heir of Redclyffe." 18mo, cloth. 1*s*. *Globe Readings Edition.* Globe 8vo. 2*s*.

FRENCH.

Beaumarchais.—LE BARBIER DE SEVILLE. Edited, with Introduction and Notes, by L. P. BLOUET, Assistant Master in St. Paul's School. Fcap. 8vo. 3*s*. 6*d*.

Bowen.—FIRST LESSONS IN FRENCH. By H. COURTHOPE BOWEN, M.A., Principal of the Finsbury Training College for Higher and Middle Schools. Extra fcap. 8vo. 1*s*.

Breymann.—Works by HERMANN BREYMANN, Ph.D., Professor of Philology in the University of Munich.

A FRENCH GRAMMAR BASED ON PHILOLOGICAL PRINCIPLES. Second Edition. Extra fcap. 8vo. 4*s*. 6*d*.

FIRST FRENCH EXERCISE BOOK. Extra fcap. 8vo. 4*s*. 6*d*.

SECOND FRENCH EXERCISE BOOK. Extra fcap. 8vo. 2*s*. 6*d*.

Fasnacht.—Works by G. EUGÈNE FASNACHT, Author of "Macmillan's Progressive French Course," Editor of "Macmillan's Foreign School Classics," &c.

THE ORGANIC METHOD OF STUDYING LANGUAGES. Extra fcap. 8vo. I. French. 3*s.* 6*d.*

A SYNTHETIC FRENCH GRAMMAR FOR SCHOOLS. Crown 8vo. 3*s.* 6*d.*

GRAMMAR AND GLOSSARY OF THE FRENCH LANGUAGE OF THE SEVENTEENTH CENTURY. Crown 8vo. [*In preparation.*

Macmillan's Primary Series of French Reading Books.—Edited by G. EUGÈNE FASNACHT, formerly Assistant-Master in Westminster School. With Illustrations. Globe 8vo.

CORNAZ—NOS ENFANTS ET LEURS AMIS. PAR SUZANNE CORNAZ. Edited by EDITH HARVEY. With Notes, Vocabulary, and Exercises. Globe 8vo. 1*s.* 6*d.*

DE MAISTRE—LA JEUNE SIBÉRIENNE ET LE LÉPREUX DE LA CITÉ D'AOSTE. Edited, with Introduction, Notes, and Vocabulary. By STEPHANE BARLET, B.Sc. Univ. Gall. and London; Assistant-Master at the Mercers' School, Examiner to the College of Preceptors, the Royal Naval College, &c. 1*s.* 6*d.*

FLORIAN—FABLES. Selected and Edited, with Notes, Vocabulary, Dialogues, and Exercises, by the Rev. CHARLES YELD, M.A., Head Master of University School, Nottingham. Illustrated. 1*s.* 6*d.*

LA FONTAINE—A SELECTION OF FABLES. Edited, with Introduction, Notes, and Vocabulary, by L. M. MORIARTY, B.A., Assistant Master at Harrow. 2*s.* 6*d.*

MOLESWORTH.—FRENCH LIFE IN LETTERS. By Mrs. MOLESWORTH. With Notes on Idioms, &c. 1*s.* 6*d.*

PERRAULT—CONTES DE FÉES. Edited, with Introduction, Notes, and Vocabulary, by G. E. FASNACHT. New Edition with Exercises. 1*s.* 6*d.*

(See also *German Authors*, page 81.)

Macmillan's Progressive French Course.—By G. EUGÈNE FASNACHT, formerly Assistant-Master in Westminster School.

I.—FIRST YEAR, containing Easy Lessons on the Regular Accidence. New and thoroughly revised Edition. Extra fcap. 8vo. 1*s.*

II.—SECOND YEAR, containing an Elementary Grammar with copious Exercises, Notes, and Vocabularies. A new Edition, enlarged and thoroughly revised. Extra fcap. 8vo. 2*s.*

III.—THIRD YEAR, containing a Systematic Syntax, and Lessons in Composition. Extra fcap. 8vo. 2*s.* 6*d.*

THE TEACHER'S COMPANION TO MACMILLAN'S PROGRESSIVE FRENCH COURSE. With Copious Notes, Hints for Different Renderings, Synonyms, Philological Remarks, &c. By G. E. FASNACHT. Globe 8vo. *First Year* 4*s.* 6*d.*, *Second Year* 4*s.* 6*d.*, *Third Year* 4*s.* 6*d.*

Macmillan's French Composition.—By G. E. FASNACHT. Part I. Elementary. Extra Fcap. 8vo. 2*s*. 6*d*. [*Ready*. Part II. Advanced. [*Part II. in the press*.

THE TEACHER'S COMPANION TO MACMILLAN'S COURSE OF FRENCH COMPOSITION. By G. EUGENE FASNACHT. First Course. Extra fcap. 8vo. 4*s*. 6*d*.

Macmillan's Progressive French Readers. By G. EUGÈNE FASNACHT.

I.—FIRST YEAR, containing Tales, Historical Extracts, Letters, Dialogues, Ballads, Nursery Songs, &c., with Two Vocabularies: (1) in the order of subjects; (2) in alphabetical order. A new and thoroughly revised Edition, with Imitative Exercises. Extra fcap. 8vo. 2*s*. 6*d*.

II.—SECOND YEAR, containing Fiction in Prose and Verse, Historical and Descriptive Extracts, Essays, Letters, Dialogues, &c. New Edition, with Imitative Exercises. Extra fcap. 8vo. 2*s*. 6*d*.

Macmillan's Foreign School Classics. Edited by G. EUGÈNE FASNACHT. 18mo.

FRENCH.

CORNEILLE—LE CID. Edited by G. E. FASNACHT. 1*s*.

DUMAS—LES DEMOISELLES DE ST. CYR. Edited by VICTOR OGER, Lecturer in University College, Liverpool. 1*s*. 6*d*.

LA FONTAINE'S FABLES. Books I.—VI. Edited by L. M. MORIARTY, B.A., Assistant Master at Harrow. [*In preparation*.

MOLIÈRE—L'AVARE. By the same Editor. 1*s*.

MOLIÈRE—LE BOURGEOIS GENTILHOMME. By the same Editor. 1*s*. 6*d*.

MOLIÈRE—LES FEMMES SAVANTES. By G. E. FASNACHT. 1*s*.

MOLIÈRE—LE MISANTHROPE. By the same Editor. 1*s*.

MOLIÈRE—LE MÉDECIN MALGRE LUI. By the same Editor. 1*s*.

RACINE—BRITANNICUS. Edited by EUGÈNE PELLISSIER, Assistant-Master in Clifton College, and Lecturer in University College, Bristol. 2*s*.

FRENCH READINGS FROM ROMAN HISTORY. Selected from Various Authors and Edited by C. COLBECK, M.A., late Fellow of Trinity College, Cambridge; Assistant-Master at Harrow. 4*s*. 6*d*.

SAND, GEORGE—LA MARE AU DIABLE. Edited by W. E. RUSSELL, M.A., Assistant-Master in Haileybury College. 1*s*.

SANDEAU, JULES—MADEMOISELLE DE LA SEIGLIERE. Edited by H. C. STEEL, Assistant-Master in Winchester College. 1*s*. 6*d*.

THIERS'S HISTORY OF THE EGYPTIAN EXPEDITION. Edited by Rev. H. A. BULL, M.A. Assistant-Master in Wellington College. [*In preparation*.

Macmillan's Foreign School Classics (*continued*)—

VOLTAIRE—CHARLES XII. Edited by G. E. FASNACHT. 3*s.* 6*d.*

*** *Other volumes to follow.*

(See also *German Authors*, page 81.)

Masson (Gustave).—A COMPENDIOUS DICTIONARY OF THE FRENCH LANGUAGE (French-English and English-French). Adapted from the Dictionaries of Professor ALFRED ELWALL. Followed by a List of the Principal Diverging Derivations, and preceded by Chronological and Historical Tables. By GUSTAVE MASSON, Assistant-Master and Librarian, Harrow School. New Edition. Crown 8vo. 6*s.*

Molière.—LE MALADE IMAGINAIRE. Edited, with Introduction and Notes, by FRANCIS TARVER, M.A., Assistant-Master at Eton. Fcap. 8vo. 2*s.* 6*d.*

(See also *Macmillan's Foreign School Classics.*)

Pellissier.—FRENCH ROOTS AND THEIR FAMILIES. A Synthetic Vocabulary, based upon Derivations, for Schools and Candidates for Public Examinations. By EUGÈNE PELLISSIER, M.A., B.Sc., LL.B., Assistant-Master at Clifton College, Lecturer at University College, Bristol. Globe 8vo. 6*s.*

GERMAN.

Huss.—A SYSTEM OF ORAL INSTRUCTION IN GERMAN, by means of Progressive Illustrations and Applications of the leading Rules of Grammar. By HERMANN C. O. HUSS, Ph.D. Crown 8vo. 5*s.*

Macmillan's Progressive German Course. By G. EUGÈNE FASNACHT.

PART I.—FIRST YEAR. Easy Lessons and Rules on the Regular Accidence. Extra fcap. 8vo. 1*s.* 6*d.*

Part II.—SECOND YEAR. Conversational Lessons in Systematic Accidence and Elementary Syntax. With Philological Illustrations and Etymological Vocabulary. New Edition, enlarged and thoroughly recast. Extra fcap. 8vo. 3*s.* 6*d.*

Part III.—THIRD YEAR. [*In the press.*

TEACHER'S COMPANION TO MACMILLAN'S PROGRESSIVE GERMAN COURSE. With copious Notes, Hints for Different Renderings. Synonyms, Philological Remarks, &c. By G. E. FASNACHT. Extra Fcap. 8vo. FIRST YEAR. 4*s.* 6*d.* SECOND YEAR. 4*s.* 6*d.*

Macmillan's Progressive German Readers. By G. E. FASNACHT.

I.—FIRST YEAR, containing an Introduction to the German order of Words, with Copious Examples, extracts from German Authors in Prose and Poetry; Notes, and Vocabularies. Extra Fcap. 8vo., 2*s.* 6*d.*

Macmillan's Primary German Reading Books.

(See also *French Authors*, page 78.)

GRIMM—KINDER UND HAUSMÄRCHEN. Selected and Edited, with Notes, and Vocabulary, by G. E. FASNACHT. New Edition, with Exercises. 2*s*. 6*d*.

HAUFF.—DIE KARAVANE. Edited, with Notes and Vocabulary, by HERMAN HAGER, Ph.D. Lecturer in the Owens College, Manchester. New Edition, with Exercises, arranged by G. E. FASNACHT. 3*s*.

SCHMID, CHR. VON.—H. VON EICHENFELS. Edited, with Notes, Vocabulary, and Exercises, by G. E. Fasnacht. 2*s*. 6*d*.

Macmillan's Foreign School Classics. Edited by G. EUGÈNE FASNACHT, 18mo.

GERMAN.

FREYTAG (G.).—DOKTOR LUTHER. Edited by FRANCIS STORR, M.A., Head Master of the Modern Side, Merchant Taylors' School. [*In preparation.*

GOETHE—GÖTZ VON BERLICHINGEN. Edited by H. A. BULL, M.A., Assistant Master at Wellington College. 2*s*.

GOETHE—FAUST. PART I., followed by an Appendix on PART II. Edited by JANE LEE, Lecturer in German Literature at Newnham College, Cambridge. 4*s*. 6*d*.

HEINE—SELECTIONS FROM THE REISEBILDER AND OTHER PROSE WORKS. Edited by C. COLBECK, M.A., Assistant-Master at Harrow, late Fellow of Trinity College, Cambridge. 2*s*. 6*d*.

LESSING.—MINNA VON BARNHELM. Edited by JAMES SIME, M.A. [*In preparation.*

SCHILLER—SELECTIONS FROM SCHILLER'S LYRICAL POEMS. Edited, with Notes and a Memoir of Schiller, by E. J. TURNER, B.A., and E. D. A. MORSHEAD, M.A. Assistant-Masters in Winchester College. 2*s*. 6*d*.

SCHILLER—DIE JUNGFRAU VON ORLEANS. Edited by JOSEPH GOSTWICK. 2*s*. 6*d*.

SCHILLER—MARIA STUART. Edited by C. SHELDON, M.A., D.Lit., of the Royal Academical Institution, Belfast. 2*s*. 6*d*.

SCHILLER—WILHELM TELL. Edited by G. E. FASNACHT. 2*s*. 6*d*.

SCHILLER.—WALLENSTEIN. Part I. DAS LAGER. Edited by H. B. COTTERILL, M.A. 2*s*.

UHLAND—SELECT BALLADS. Adapted as a First Easy Reading Book for Beginners. With Vocabulary. Edited by G. E. FASNACHT. 1*s*.

*** *Other Volumes to follow.*

(See also *French Authors*, page 79.)

Pylodet.—NEW GUIDE TO GERMAN CONVERSATION; containing an Alphabetical List of nearly 800 Familiar Words; followed by Exercises; Vocabulary of Words in frequent use; Familiar Phrases and Dialogues; a Sketch of German Literature, Idiomatic Expressions, &c. By L. PYLODET. 18mo, cloth limp. 2*s*. 6*d*.

Whitney.—Works by W. D. WHITNEY, Professor of Sanskrit and Instructor in Modern Languages in Yale College.

A COMPENDIOUS GERMAN GRAMMAR. Crown 8vo. 4*s*. 6*d*.

A GERMAN READER IN PROSE AND VERSE. With Notes and Vocabulary. Crown 8vo. 5*s*.

Whitney and Edgren.—A COMPENDIOUS GERMAN AND ENGLISH DICTIONARY, with Notation of Correspondences and Brief Etymologies. By Professor W. D. WHITNEY, assisted by A. H. EDGREN. Crown 8vo. 7*s*. 6*d*.

THE GERMAN-ENGLISH PART, separately, 5*s*.

MODERN GREEK.

Vincent and Dickson. — HANDBOOK TO MODERN GREEK. By Sir EDGAR VINCENT, K.C.M.G. and T. G. DICKSON, M.A. Second Edition, revised and enlarged, with Appendix on the relation of Modern and Classical Greek by Professor JEBB. Crown 8vo. 6*s*.

ITALIAN.

Dante. — THE PURGATORY OF DANTE. Edited, with Translation and Notes, by A. J. BUTLER, M.A., late Fellow of Trinity College, Cambridge. Crown 8vo. 12*s*. 6*d*.

THE PARADISO OF DANTE. Edited, with Translation and Notes, by the same Author. Crown 8vo. 12*s*. 6*d*.

READINGS ON THE PURGATORIO OF DANTE. Chiefly based on the Commentary of Benvenuto Da Imola. By the Hon. WILLIAM WARREN VERNON, M.A. With an Introduction by the Very Rev. the DEAN OF ST. PAUL'S. 2 vols. Crown 8vo. 24*s*.

SPANISH.

Calderon.—FOUR PLAYS OF CALDERON. Edited, with Introduction and Notes, by NORMAN MACCOLL, M.A., late Fellow of Downing College, Cambridge. Crown 8vo. 14*s*.

The four plays here given are *El Principe Constante*, *La Vida es Sueno*, *El Alcalde de Zalamea*, and *El Escondido y La Tapada*.

DOMESTIC ECONOMY.

Barker.—FIRST LESSONS IN THE PRINCIPLES OF COOKING. By LADY BARKER. New Edition. 18mo. 1*s*.

Berners.—FIRST LESSONS ON HEALTH. By J. BERNERS. New Edition. 18mo. 1*s*.

Cookery Book.—THE MIDDLE-CLASS COOKERY BOOK. Edited by the Manchester School of Domestic Cookery. Post 8vo. 1*s*. 6*d*.

Craven.—A GUIDE TO DISTRICT NURSES. By Mrs. DACRE CRAVEN (*née* Florence Sarah Lees), Hon. Associate of the Order of St. John of Jerusalem, &c. Crown 8vo. 2*s*. 6*d*.

Fawcett.—TALES IN POLITICAL ECONOMY. By Mrs. HENRY FAWCETT. Globe 8vo. 3*s*.

Frederick.—HINTS TO HOUSEWIVES ON SEVERAL POINTS, PARTICULARLY ON THE PREPARATION OF ECONOMICAL AND TASTEFUL DISHES. By Mrs. FREDERICK. Crown 8vo. 1*s*.

"This unpretending and useful little volume distinctly supplies a desideratum . . . . The author steadily keeps in view the simple aim of 'making every-day meals at home, particularly the dinner, attractive,' without adding to the ordinary household expenses."—SATURDAY REVIEW.

Grand'homme.—CUTTING-OUT AND DRESSMAKING. From the French of Mdlle. E. GRAND'HOMME. With Diagrams. 18mo. 1*s*.

Jex-Blake.—THE CARE OF INFANTS. A Manual for Mothers and Nurses. By SOPHIA JEX-BLAKE, M.D., Member of the Irish College of Physicians; Lecturer on Hygiene at the London School of Medicine for Women. 18mo. 1*s*.

Tegetmeier.—HOUSEHOLD MANAGEMENT AND COOKERY. With an Appendix of Recipes used by the Teachers of the National School of Cookery. By W. B. TEGETMEIER. Compiled at the request of the School Board for London. 18mo. 1*s*.

Thornton.—FIRST LESSONS IN BOOK-KEEPING. By J. THORNTON. New Edition. Crown 8vo. 2*s*. 6*d*.

The object of this volume is to make the theory of Book-keeping sufficiently plain for even children to understand it.

A KEY TO THE ABOVE FOR THE USE OF TEACHERS AND PRIVATE STUDENTS. Containing all the Exercises worked out, with brief Notes. By J. THORNTON. Oblong 4to. 10*s*. 6*d*.

Wright.—THE SCHOOL COOKERY-BOOK. Compiled and Edited by C. E. GUTHRIE WRIGHT, Hon Sec. to the Edinburgh School of Cookery. 18mo. 1*s*.

ART AND KINDRED SUBJECTS.

Anderson.—LINEAR PERSPECTIVE, AND MODEL DRAWING. A School and Art Class Manual, with Questions and Exercises for Examination, and Examples of Examination Papers. By LAURENCE ANDERSON. With Illustrations. Royal 8vo. 2*s*.

Collier.—A PRIMER OF ART. With Illustrations. By JOHN COLLIER. 18mo. 1*s.*

Cook.—THE NATIONAL GALLERY: A POPULAR HANDBOOK TO. By EDWARD T. COOK, with a Preface by JOHN RUSKIN, LL.D., and Selections from his Writings. Second Edition. Crown 8vo. Half Morocco, 14*s.*

*** Also an Edition on large paper, limited to 250 copies. 2 vols. 8vo.

Delamotte.—A BEGINNER'S DRAWING BOOK. By P. H. DELAMOTTE, F.S.A. Progressively arranged. New Edition improved. Crown 8vo. 3*s.* 6*d.*

Ellis.—SKETCHING FROM NATURE. A Handbook for Students and Amateurs. By TRISTRAM J. ELLIS. With a Frontispiece and Ten Illustrations, by H. STACY MARKS, R.A., and Thirty Sketches by the Author. New Edition, revised and enlarged. Crown 8vo. 3*s.* 6*d.*

Hunt.—TALKS ABOUT ART. By WILLIAM HUNT. With a Letter from Sir J. E. MILLAIS, Bart., R.A. Crown 8vo. 3*s.* 6*d.*

Meldola.—THE CHEMISTRY OF PHOTOGRAPHY. By RAPHAEL MELDOLA, F.R.S., Professor of Chemistry in the Technical College, Finsbury, City and Guilds of London Institute for the Advancement of Technical Education. Crown 8vo. 6*s.* (*Nature Series.*)

Taylor.—A PRIMER OF PIANOFORTE PLAYING. By FRANKLIN TAYLOR. Edited by Sir GEORGE GROVE. 18mo. 1*s.*

Taylor.—A MANUAL OF SIGHT SINGING. By SEDLEY TAYLOR, M.A., Author of "Sound and Music." [*In the press.*

Tyrwhitt.—OUR SKETCHING CLUB. Letters and Studies on Landscape Art. By the Rev. R. ST. JOHN TYRWHITT, M.A. With an authorised Reproduction of the Lessons and Woodcuts in Prof. Ruskin's "Elements of Drawing." Fourth Edition. Crown 8vo. 7*s.* 6*d.*

WORKS ON TEACHING.

Arnold.—REPORTS ON ELEMENTARY SCHOOLS. 1852–1882. By MATTHEW ARNOLD, D.C.L., LL.D. Edited by the Right Hon. Sir FRANCIS SANDFORD, K.C.B. Cheaper Issue. Crown 8vo. 3*s.* 6*d.*

Ball.—THE STUDENT'S GUIDE TO THE BAR. By WALTER W. R. BALL, M.A., of the Inner Temple, Barrister-at-Law; Fellow and Assistant Tutor of Trinity College, Cambridge, and Fellow of University College, London. Fourth Edition Revised. Crown 8vo. 2*s.* 6*d.*

Blakiston—THE TEACHER. Hints on School Management. A Handbook for Managers, Teachers' Assistants, and Pupil Teachers. By J. R. BLAKISTON, M.A. Crown 8vo. 2*s*. 6*d*. (Recommended by the London, Birmingham, and Leicester School Boards.)

"Into a comparatively small book he has crowded a great deal of exceedingly useful and sound advice. It is a plain, common-sense book, full of hints to the teacher on the management of his school and his children."—SCHOOL BOARD CHRONICLE.

Calderwood.—ON TEACHING. By Professor HENRY CALDERWOOD. New Edition. Extra fcap. 8vo. 2*s*. 6*d*.

Geikie.—THE TEACHING OF GEOGRAPHY. A Practical Handbook for the use of Teachers. By ARCHIBALD GEIKIE, F.R.S., Director-General of the Geological Survey of the United Kingdom, &c. (Being the Introductory Volume to Macmillan's Geographical Series.) Crown 8vo. 2*s*.

Gladstone.—OBJECT TEACHING. A Lecture delivered at the Pupil-Teacher Centre, William Street Board School, Hammersmith. By J. H. GLADSTONE, Ph.D., F.R.S., Member of the London School Board. With an Appendix. Crown 8vo. 3*d*.

"It is a short but interesting and instructive publication, and our younger teachers will do well to read it carefully and thoroughly. There is much in these few pages which they can learn and profit by."—THE SCHOOL GUARDIAN.

Hertel.—OVERPRESSURE IN HIGH SCHOOLS IN DENMARK. By Dr. HERTEL, Municipal Medical Officer, Copenhagen. Translated from the Danish by C. GODFREY SÖRENSEN. With Introduction by Sir J. CRICHTON-BROWNE, M.D., LL.D., F.R.S. Crown 8vo. 3*s*. 6*d*.

DIVINITY.

*** For other Works by these Authors, see THEOLOGICAL CATALOGUE.

Abbott (Rev. E. A.)—BIBLE LESSONS. By the Rev. E. A. ABBOTT, D.D., formerly Head Master of the City of London School. New Edition. Crown 8vo. 4*s*. 6*d*.

"Wise, suggestive, and really profound initiation into religious thought."—GUARDIAN.

Abbott—Rushbrooke.—THE COMMON TRADITION OF THE SYNOPTIC GOSPELS, in the Text of the Revised Version. By EDWIN A. ABBOTT, D.D., formerly Fellow of St. John's College, Cambridge, and W. G. RUSHBROOKE, M.L., formerly Fellow of St. John's College, Cambridge. Cr. 8vo. 3*s*. 6*d*.

The Acts of the Apostles.—Being the Greek Text as revised by Professors WESTCOTT and HORT. With Explanatory Notes for the Use of Schools, by T. E. PAGE, M.A., late Fellow of St. John's College, Cambridge; Assistant Master at the Charterhouse. Fcap. 8vo. 4*s*. 6*d*.

Arnold.—Works by MATTHEW ARNOLD, D.C.L., formerly Professor of Poetry in the University of Oxford, and Fellow of Oriel.

A BIBLE-READING FOR SCHOOLS.—THE GREAT PROPHECY OF ISRAEL'S RESTORATION (Isaiah, Chapters xl.—lxvi.). Arranged and Edited for Young Learners. New Edition. 18mo, cloth. 1*s.*

ISAIAH XL.—LXVI. With the Shorter Prophecies allied to it. Arranged and Edited, with Notes. Crown 8vo. 5*s.*

ISAIAH OF JERUSALEM, IN THE AUTHORISED ENGLISH VERSION. With Introduction, Corrections, and Notes. Crown 8vo. 4*s.* 6*d.*

Benham.—A COMPANION TO THE LECTIONARY. Being a Commentary on the Proper Lessons for Sundays and Holy Days. By Rev. W. BENHAM, B.D., Rector of S. Edmund with S. Nicholas Acons, &c. New Edition. Crown 8vo. 4*s.* 6*d.*

Cassel.—MANUAL OF JEWISH HISTORY AND LITERATURE; preceded by a BRIEF SUMMARY OF BIBLE HISTORY. By DR. D. CASSEL. Translated by Mrs. HENRY LUCAS. Fcap. 8vo. 2*s.* 6*d.*

Cheetham.—A CHURCH HISTORY OF THE FIRST SIX CENTURIES. By the Ven. ARCHDEACON CHEETHAM, Crown 8vo. [*In the press.*

Cross.—BIBLE READINGS SELECTED FROM THE PENTATEUCH AND THE BOOK OF JOSHUA. By the Rev. JOHN A. CROSS. Second Edition enlarged, with Notes. Globe 8vo. 2*s.* 6*d.*

Curteis.—MANUAL OF THE THIRTY-NINE ARTICLES. By G. H. CURTEIS, M.A., Principal of the Lichfield Theological College. [*In preparation.*

Davies.—THE EPISTLES OF ST. PAUL TO THE EPHESIANS, THE COLOSSIANS, AND PHILEMON; with Introductions and Notes, and an Essay on the Traces of Foreign Elements in the Theology of these Epistles. By the Rev. J. LLEWELYN DAVIES, M.A., Rector of Christ Church, St. Marylebone; late Fellow of Trinity College, Cambridge. Second Edition. Demy 8vo. 7*s.* 6*d.*

Drummond.—THE STUDY OF THEOLOGY, INTRODUCTION TO. By JAMES DRUMMOND, LL.D., Professor of Theology in Manchester New College, London. Crown 8vo. 5*s.*

Gaskoin.—THE CHILDREN'S TREASURY OF BIBLE STORIES. By Mrs. HERMAN GASKOIN. Edited with Preface by Rev. G. F. MACLEAR, D.D. PART I.—OLD TESTAMENT HISTORY. 18mo. 1*s.* PART II.—NEW TESTAMENT. 18mo. 1*s.* PART III.—THE APOSTLES: ST. JAMES THE GREAT, ST. PAUL, AND ST. JOHN THE DIVINE. 18mo. 1*s.*

Golden Treasury Psalter.—Students' Edition. Being an Edition of "The Psalms Chronologically arranged, by Four Friends," with briefer Notes. 18mo. 3*s*. 6*d*.

Greek Testament.—Edited, with Introduction and Appendices, by CANON WESTCOTT and Dr. F. J. A. HORT. Two Vols. Crown 8vo. 10*s*. 6*d*. each.

Vol. I. The Text.

Vol. II. Introduction and Appendix.

Greek Testament.—Edited by Canon WESTCOTT and Dr. HORT. School Edition of Text. 12mo. cloth. 4*s*. 6*d*. 18mo. roan, red edges. 5*s*. 6*d*.

GREEK TESTAMENT, SCHOOL READINGS IN THE. Being the outline of the life of our Lord, as given by St. Mark, with additions from the Text of the other Evangelists. Arranged and Edited, with Notes and Vocabulary, by the Rev. A. CALVERT, M.A., late Fellow of St. John's College, Cambridge. Fcap. 8vo. 4*s*. 6*d*.

THE ACTS OF THE APOSTLES. Being the Greek Text as revised by Drs. WESTCOTT and HORT. With Explanatory Notes by T. E. PAGE, M.A., Assistant Master at the Charterhouse. Fcap. 8vo. 4*s*. 6*d*.

THE GOSPEL ACCORDING TO ST. MATTHEW. Being the Greek Text as revised by Drs. WESTCOTT and HORT. With Explanatory Notes by Rev. A. SLOMAN, M.A., Head Master of Birkenhead School. [*Just ready*.

THE GOSPEL ACCORDING TO ST. MARK. Being the Greek Text as revised by Drs. WESTCOTT and HORT. With Explanatory Notes by Rev. J. O. F. MURRAY, M.A., Lecturer in Emmanuel College, Cambridge. Fcap. 8vo. [*In preparation*.

THE GOSPEL ACCORDING TO ST. LUKE. Being the Greek Text as revised by Drs. WESTCOTT and HORT. With Explanatory Notes by Rev. JOHN BOND, M.A. [*In preparation*.

Hardwick.—Works by Archdeacon HARDWICK :—

A HISTORY OF THE CHRISTIAN CHURCH. Middle Age. From Gregory the Great to the Excommunication of Luther. Edited by WILLIAM STUBBS, M.A., Regius Professor of Modern History in the University of Oxford. With Four Maps. New Edition. Crown 8vo. 10*s*. 6*d*.

A HISTORY OF THE CHRISTIAN CHURCH DURING THE REFORMATION. Ninth Edition. Edited by Professor STUBBS. Crown 8vo. 10*s*. 6*d*.

Hoole.—THE CLASSICAL ELEMENT IN THE NEW TESTAMENT. Considered as a Proof of its Genuineness, with an Appendix on the Oldest Authorities used in the Formation of the Canon. By CHARLES H. HOOLE, M.A., Student of Christ Church, Oxford. 8vo. 10*s*. 6*d*.

Jennings and Lowe.—THE PSALMS, WITH INTRODUCTIONS AND CRITICAL NOTES. By A. C. JENNINGS, M.A.; assisted in parts by W. H. LOWE, M.A. In 2 vols. Second Edition Revised. Crown 8vo. 10*s.* 6*d.* each.

Kay.—ST. PAUL'S TWO EPISTLES TO THE CORINTHIANS, A COMMENTARY ON. By the late Rev. W. KAY, D.D., Rector of Great Leghs, Essex, and Hon. Canon of St. Albans; formerly Principal of Bishop's College, Calcutta; and Fellow and Tutor of Lincoln College. Demy 8vo. 9*s.*

Kuenen.—PENTATEUCH AND BOOK OF JOSHUA: an Historico-Critical Inquiry into the Origin and Composition of the Hexateuch. By A. KUENEN, Professor of Theology at Leiden. Translated from the Dutch, with the assistance of the Author, by PHILLIP H. WICKSTEED, M.A. 8vo. 14*s.*

The OXFORD MAGAZINE says:—"The work is absolutely indispensable to all special students of the Old Testament."

Lightfoot.—Works by the Right Rev. J. B. LIGHTFOOT, D.D., D.C.L., LL.D., late Bishop of Durham.

ST. PAUL'S EPISTLE TO THE GALATIANS. A Revised Text, with Introduction, Notes, and Dissertations. Ninth Edition, revised. 8vo. 12*s.*

ST. PAUL'S EPISTLE TO THE PHILIPPIANS. A Revised Text, with Introduction, Notes, and Dissertations. Ninth Edition, revised. 8vo. 12*s.*

ST. CLEMENT OF ROME—THE TWO EPISTLES TO THE CORINTHIANS. A Revised Text, with Introduction and Notes. 8vo. 8*s.* 6*d.*

ST. PAUL'S EPISTLES TO THE COLOSSIANS AND TO PHILEMON. A Revised Text, with Introductions, Notes, and Dissertations. Eighth Edition, revised. 8vo. 12*s.*

THE APOSTOLIC FATHERS. Part II. S. IGNATIUS—S. POLYCARP. Revised Texts, with Introductions, Notes, Dissertations, and Translations. Second Edition. 3 vols. 8vo. 48*s.*

APOSTOLIC FATHERS. Abridged Edition. With short Introductions, Greek Text, and English Translation. By the same Author. 8vo. [*In the press.*

ESSAYS ON THE WORK ENTITLED "SUPERNATURAL RELIGION." (Reprinted from the *Contemporary Review.*) 8vo. 10*s.* 6*d.*

Maclear.—Works by the Rev. G. F. MACLEAR, D.D., Canon of Canterbury, Warden of St. Augustine's College, Canterbury, and late Head-Master of King's College School, London:—

ELEMENTARY THEOLOGICAL CLASS-BOOKS.

A CLASS-BOOK OF OLD TESTAMENT HISTORY. New Edition, with Four Maps. 18mo. 4*s.* 6*d.*

A CLASS-BOOK OF NEW TESTAMENT HISTORY, including the Connection of the Old and New Testaments. With Four Maps. New Edition. 18mo. 5*s.* 6*d.*

AN INTRODUCTION TO THE THIRTY-NINE ARTICLES. [*In the press.*

CLASS BOOK OF THE CREEDS. 18mo. 2*s.* 6*d.*

Maclear.—A SHILLING BOOK OF OLD TESTAMENT HISTORY, for National and Elementary Schools. With Map. 18mo, cloth. New Edition.

A SHILLING BOOK OF NEW TESTAMENT HISTORY, for National and Elementary Schools. With Map. 18mo, cloth. New Edition.

These works have been carefully abridged from the Author's large manuals.

CLASS-BOOK OF THE CATECHISM OF THE CHURCH OF ENGLAND. New Edition. 18mo. 1*s*. 6*d*.

A FIRST CLASS-BOOK OF THE CATECHISM OF THE CHURCH OF ENGLAND. With Scripture Proofs, for Junior Classes and Schools. New Edition. 18mo. 6*d*.

A MANUAL OF INSTRUCTION FOR CONFIRMATION AND FIRST COMMUNION. WITH PRAYERS AND DEVOTIONS. 32mo, cloth extra, red edges. 2*s*.

Maurice.—THE LORD'S PRAYER, THE CREED, AND THE COMMANDMENTS. A Manual for Parents and Schoolmasters. To which is added the Order of the Scriptures. By the Rev. F. DENISON MAURICE, M.A. 18mo, cloth, limp. 1*s*.

Pentateuch and Book of Joshua: an Historico-Critical Inquiry into the Origin and Composition of the Hexateuch. By A. KUENEN, Professor of Theology at Leiden. Translated from the Dutch, with the assistance of the Author, by PHILIP H. WICKSTEED, M.A. 8vo. 14*s*.

Procter.—A HISTORY OF THE BOOK OF COMMON PRAYER, with a Rationale of its Offices. By Rev. F. PROCTER. M.A. 18th Edition, revised and enlarged. Crown 8vo. 10*s*. 6*d*.

Procter and Maclear.—AN ELEMENTARY INTRODUCTION TO THE BOOK OF COMMON PRAYER. Re-arranged and supplemented by an Explanation of the Morning and Evening Prayer and the Litany. By the Rev. F. PROCTER and the Rev. Dr. MACLEAR. New and Enlarged Edition, containing the Communion Service and the Confirmation and Baptismal Offices. 18mo. 2*s*. 6*d*.

Psalms, The, with Introductions and Critical Notes.—By A. C. JENNINGS, M.A., Jesus College, Cambridge, Tyrwhitt Scholar, Crosse Scholar, Hebrew University Prizeman, and Fry Scholar of St. John's College, Carus and Scholefield Prizeman, Vicar of Whittlesford, Cambs.; assisted in Parts by W. H. LOWE, M.A., Hebrew Lecturer and late Scholar of Christ's College, Cambridge, and Tyrwhitt Scholar. In 2 vols. Second Edition Revised. Crown 8vo. 10*s*. 6*d*. each.

Ramsay.—THE CATECHISER'S MANUAL; or, the Church Catechism Illustrated and Explained, for the Use of Clergymen, Schoolmasters, and Teachers. By the Rev. ARTHUR RAMSAY, M.A. New Edition. 18mo. 1*s*. 6*d*.

Rendall.—THE EPISTLE TO THE HEBREWS. English Text with Commentary. By the Rev. F. RENDALL, M.A., formerly Fellow of Trinity College, Cambridge, and Master of Harrow School. Crown 8vo. 7*s*. 6*d*.

Ryle.—AN INTRODUCTION TO THE CANON OF THE OLD TESTAMENT. By Rev. H. E. RYLE, M.A., Fellow of King's College, and Hulsean Professor of Divinity in the University of Cambridge. Crown 8vo. [*In preparation.*

Simpson.—AN EPITOME OF THE HISTORY OF THE CHRISTIAN CHURCH DURING THE FIRST THREE CENTURIES, AND OF THE REFORMATION IN ENGLAND. Compiled for the use of Students in Schools and Universities by the Rev. WILLIAM SIMPSON, M.A., Queen's College, Cambridge. Seventh Edition. Fcap. 8vo. 3*s*. 6*d*.

St. James' Epistle.—The Greek Text with Introduction and Notes. By Rev. JOSEPH MAYOR, M.A., Professor of Moral Philosophy in King's College, London. 8vo. [*In preparation.*

St. John's Epistles.—The Greek Text with Notes and Essays, by BROOKE FOSS WESTCOTT, D.D., Regius Professor of Divinty, and Fellow of King's College, Cambridge, Canon of Westminstier &c. Second Edition Revised. 8vo. 12*s*. 6*d*.

St. Paul's Epistles.—Greek Text, with Introduction and Notes.

THE EPISTLE TO THE GALATIANS. Edited by the Right Rev. J. B. LIGHTFOOT, D.D., Bishop of Durham. Ninth Edition. 8vo. 12*s*.

THE EPISTLE TO THE PHILIPPIANS. By the same Editor. Ninth Edition. 8vo. 12*s*.

THE EPISTLE TO THE COLOSSIANS AND TO PHILEMON. By the same Editor. Eighth Edition. 8vo. 12*s*.

THE EPISTLE TO THE ROMANS. Edited by the Very Rev. C. J. VAUGHAN, D.D., Dean of Llandaff, and Master of the Temple. Fifth Edition. Crown 8vo. 7*s*. 6*d*.

THE EPISTLE TO THE PHILIPPIANS, with Translation, Paraphrase, and Notes for English Readers. By the same Editor. Crown 8vo. 5*s*.

THE EPISTLE TO THE THESSALONIANS, COMMENTARY ON THE GREEK TEXT. By JOHN EADIE, D.D., LL.D. Edited by the Rev. W. YOUNG, M.A., with Preface by Professor CAIRNS. 8vo. 12*s*.

St. Paul's Epistles—*continued.*

THE EPISTLES TO THE EPHESIANS, THE COLOSSIANS, AND PHILEMON; with Introductions and Notes, and an Essay on the Traces of Foreign Elements in the Theology of these Epistles. By the Rev. J. LLEWELYN DAVIES, M.A., Rector of Christ Church, St. Marylebone; late Fellow of Trinity College, Cambridge. Second Edition, revised. Demy 8vo. 7*s.* 6*d.*

THE TWO EPISTLES TO THE CORINTHIANS, A COMMENTARY ON. By the late Rev. W. KAY, D.D., Rector of Great Leghs, Essex, and Hon. Canon of St. Albans; formerly Principal of Bishop's College, Calcutta; and Fellow and Tutor of Lincoln College. Demy 8vo. 9*s.*

The Epistle to the Hebrews. In Greek and English. With Critical and Explanatory Notes. Edited by Rev. FREDERIC RENDALL, M.A., formerly Fellow of Trinity College, Cambridge, and Assistant-Master at Harrow School. Crown 8vo. 6*s.*

THE ENGLISH TEXT, WITH COMMENTARY. By the same Editor. Crown 8vo. 7*s.* 6*d.*

The Epistle to the Hebrews. With Notes by C. J. VAUGHAN, D.D., Dean of Llandaff, Master of the Temple. Crown 8vo. [*In the press.*

*** A Companion Volume to that on the Epistle to the Romans, of which the Sixth Edition was published in 1885.

The Epistle to the Hebrews. The Greek Text with Notes and Essays by Canon WESTCOTT, D.D. 8vo. 14*s.*

Westcott.—Works by BROOKE FOSS WESTCOTT, D.D., Canon of Westminster, Regius Professor of Divinity, and Fellow of King's College, Cambridge.

A GENERAL SURVEY OF THE HISTORY OF THE CANON OF THE NEW TESTAMENT DURING THE FIRST FOUR CENTURIES. Sixth Edition. With Preface on "Supernatural Religion." Crown 8vo. 10*s.* 6*d.*

INTRODUCTION TO THE STUDY OF THE FOUR GOSPELS. Seventh Edition. Crown 8vo. 10*s.* 6*d.*

THE BIBLE IN THE CHURCH. A Popular Account of the Collection and Reception of the Holy Scriptures in the Christian Churches. New Edition. 18mo, cloth. 4*s.* 6*d.*

THE EPISTLES OF ST. JOHN. The Greek Text, with Notes and Essays. Second Edition Revised. 8vo. 12*s.* 6*d.*

THE EPISTLE TO THE HEBREWS. The Greek Text, with Notes and Essays. 8vo. 14*s.*

SOME THOUGHTS FROM THE ORDINAL. Cr. 8vo. 1*s.* 6*d.*

Westcott and Hort. — THE NEW TESTAMENT IN THE ORIGINAL GREEK. The Text Revised by B. F. WESTCOTT, D.D., Regius Professor of Divinity, Canon of Westminster, and F. J. A. HORT, D.D., Lady Margaret Professor of Divinity Fellow of Emmanuel College, Cambridge: late Fellows of Trinity College, Cambridge. 2 vols. Crown 8vo. 10*s*. 6*d*. each.

Vol. I. Text.

Vol. II. Introduction and Appendix.

Westcott and Hort.—THE NEW TESTAMENT IN THE ORIGINAL GREEK, FOR SCHOOLS. The Text Revised by BROOKE FOSS WESTCOTT, D.D., and FENTON JOHN ANTHONY HORT, D.D. 12mo. cloth. 4*s*. 6*d*. 18mo. roan, red edges. 5*s*. 6*d*.

Wilson. — THE BIBLE STUDENT'S GUIDE to the more Correct Understanding of the English Translation of the Old Testament, by reference to the original Hebrew. By WILLIAM WILSON, D.D., Canon of Winchester, late Fellow of Queen's College, Oxford. Second Edition, carefully revised. 4to. cloth. 25*s*.

Wright.—THE BIBLE WORD-BOOK: A Glossary of Archaic Words and Phrases in the Authorised Version of the Bible and the Book of Common Prayer. By W. ALDIS WRIGHT, M.A., Vice-Master of Trinity College, Cambridge. Second Edition, Revised and Enlarged. Crown 8vo. 7*s*. 6*d*.

Yonge (Charlotte M.).—SCRIPTURE READINGS FOR SCHOOLS AND FAMILIES. By CHARLOTTE M. YONGE. Author of "The Heir of Redclyffe." In Five Vols.

FIRST SERIES. GENESIS TO DEUTERONOMY. Extra fcap. 8vo. 1*s*. 6*d*. With Comments, 3*s*. 6*d*.

SECOND SERIES. From JOSHUA to SOLOMON. Extra fcap. 8vo. 1*s*. 6*d*. With Comments, 3*s*. 6*d*.

THIRD SERIES. The KINGS and the PROPHETS. Extra fcap. 8vo. 1*s*. 6*d*. With Comments, 3*s*. 6*d*.

FOURTH SERIES. The GOSPEL TIMES. 1*s*. 6*d*. With Comments. Extra fcap. 8vo, 3*s*. 6*d*.

FIFTH SERIES. APOSTOLIC TIMES. Extra fcap. 8vo. 1*s*. 6*d*. With Comments, 3*s*. 6*d*.

Zechariah—Lowe.—THE HEBREW STUDENT'S COMMENTARY ON ZECHARIAH, HEBREW AND LXX. With Excursus on Syllable-dividing, Metheg, Initial Dagesh, and Siman Rapheh. By W. H. LOWE, M.A., Hebrew Lecturer at Christ's College, Cambridge. Demy 8vo. 10*s*. 6*d*.

RICHARD CLAY AND SONS, LIMITED, LONDON AND BUNGAY.

www.ingramcontent.com/pod-product-compliance
Lightning Source LLC
LaVergne TN
LVHW011253110826
845149LV00001B/118

* 9 7 8 1 4 1 8 1 8 5 3 2 9 *